THE COMPLETE
Blackpowder Handbook
3RD EDITION

By Sam Fadala

DBI BOOKS
a division of Krause Publications, Inc.

Staff

SENIOR STAFF EDITORS
Harold A. Murtz
Ray Ordorica

PRODUCTION MANAGER
John L. Duoba

ELECTRONIC PUBLISHING DIRECTOR
Sheldon L. Factor

ELECTRONIC PUBLISHING MANAGER
Nancy J. Mellem

ELECTRONIC PUBLISHING ASSOCIATE
Laura M. Mielzynski

GRAPHIC DESIGN
John Duoba
Bill Brodt
Bill Limbaugh
Jim Billy

PHOTOGRAPHY
Nancy Fadala

COVER PHOTOGRAPHY
John Hanusin

MANAGING EDITOR
Pamela J. Johnson

PUBLISHER
Charles T. Hartigan

Arms and Armour Press, London, G.B., exclusive licensees and distributor in Britain and Europe, India and Pakistan; Book Services International, Sandton, Transvaal, exclusive distributor in South Africa and Zimbabwe. Forrester Books N.Z. Limited, Auckland, exclusive distributor in New Zealand.

ISBN 0-87349-175-0 **Library of Congress Catalog #79-54268**

Table of Contents

The Way Things Were and Are

Chapter 1: The Roots of All Shooting 7
Chapter 2: How the West Was Won—Really 11

Blackpowder Basics

Chapter 3: Get Ready for Muzzle-Loading 18
Chapter 4: The Blackpowder Battery 24
Chapter 5: Blackpowder Hunting Rules State by State 30
Chapter 6: Outfit for the Blackpowder Shooter 39
Chapter 7: Blackpowder Accoutrements 47
Chapter 8: Safe and Sane Blackpowder Shooting 53

What Kind of Muzzleloader Is It?

Chapter 9: Replica Blackpowder Guns 61
Chapter 10: Non-Replica Frontloaders 67
Chapter 11: The Modern Muzzleloader 72
Chapter 12: The Custom Muzzleloader 79
Chapter 13: Collecting Blackpowder Guns 86

Blackpowder Components and Shooting Facts

Chapter 14: Blackpowder—The Propellant 94
Chapter 15: Pyrodex—The Replica Blackpowder 103
    Special Report: Testing the Pyrodex Pellet 109
Chapter 16: The Surprising Round Ball 110
Chapter 17: The Blackpowder Conical 118
Chapter 18: Patches for Blackpowder Guns 125
Chapter 19: Blackpowder Chemicals Then and Now 132
Chapter 20: Surefire Blackpowder Loading Methods 139
Chapter 21: The Optimum Load 147
Chapter 22: Blackpowder Ballistics 155
Chapter 23: Blackpowder Accuracy 163
Chapter 24: Casting Blackpowder Bullets 171

Muzzleloader Mechanics

Chapter 25: How the Muzzle-Loading Rifle Works 180
Chapter 26: Basic Types of Blackpowder Locks 187
Chapter 27: Understanding Rifling Twist 193
Chapter 28: Ignition 199
Chapter 29: Sights for Blackpowder Guns 205
Chapter 30: Sighting-In the Blackpowder Firearm 212

The Blackpowder Battery

Chapter 31: The Great Little Muzzle-Loading Smallbore 220
Chapter 32: The Blackpowder Big Bore 225
Chapter 33: Rifled Muskets—Rugged Hunting/Target Guns 233
Chapter 34: The Single Shot Breechloader 240

Table of Contents

The Blackpowder Battery (cont.)

Chapter 35: The Blackpowder Shotgun 249
Chapter 36: The Blackpowder Sidearm 261
Chapter 37: Blackpowder Smoothbores 270

Maintenance and Cleaning

Chapter 38: Muzzleloader Maintenance 278
    Special Report: Blackpowder Corrosion Tests 286
Chapter 39: Troubleshooting the Frontloader 289
Chapter 40: Muzzleloader Modifications 296
Chapter 41: Modern Old-Time Gunsmiths 304

Hunting with Muzzleloaders

Chapter 42: Big Game with Blackpowder 309
Chapter 43: Small Game and Varmints with Blackpowder 318
Chapter 44: Wild Turkeys and Other Birds with Blackpowder 325

Great Blackpowder Guns, Cartridges and Men

Chapter 45: The Amazing 45-70 Government Cartridge 332
Chapter 46: Blackpowder Ivory Hunters 339
Chapter 47: American Sharpshooters of the Past 344
Chapter 48: The Hawken Rifle 348

For Fun and Knowing

Chapter 49: Rendezvous Shooting Games 355
Chapter 50: Blackpowder Cartridge Games 361

Reference Section I
Catalog of Blackpowder Firearms

Blackpowder Handguns 368
Blackpowder Muskets & Rifles 374
Blackpowder Shotguns 389

Reference Section II
Appendicies

Appendix A: Manufacturers' Directory of Blackpowder Guns and Accessories 392
Appendix B: Frontloader's Library 394
Appendix C: Blackpowder Periodical Publications 397
Appendix D: Blackpowder Arms Associations 397
Appendix E: Glossary of Blackpowder Terms 398

Introduction

IF YOU ALREADY shoot muzzleloaders or blackpowder cartridge guns, you understand how interesting and enjoyable the adventure is. But if you haven't fired the "old-timers," you don't know what you're missing, and I doubt that any writer could put the experience into words. It's something you have to find out for yourself. Around five million shooters in America alone—some say more—burn blackpowder at least part-time, with a large number of dedicated fans using nothing but yesterday's guns. I never viewed sootburning as a takeover. I always thought of it as a perfect addition to, and completion of, the general shooting game. That's why this Third Edition is dedicated to serve all who carry the old-style guns today, as well as those who haven't even picked one up yet.

If I could, I'd like to be at every gun range in the world tomorrow morning with a battery of blackpowder firearms ready to go—replicas, non-replicas, modern muzzleloaders, old-time cartridge guns—the works. All I'd ask shooters to do is try one shot. Just one. If they didn't enjoy it, I'd pick up my guns and go home. But I wouldn't worry about them not liking muzzleloaders or blackpowder cartridge arms. I've yet to see a dedicated, interested marksman turn his nose up at a blackpowder gun after trying one. On the contrary, I've watched faces light up with smiles that said, "Hey, that was OK! Can I do it again?" Most do just that. They shoot again, and again, and again.

Then they start asking questions. It was those questions that brought about the original COMPLETE BLACKPOWDER HANDBOOK. Here are a few examples of the usual queries:

Why would I want to shoot blackpowder? Are those old myths about muzzleloaders true? Can you really hunt, I mean really take big game with a blackpowder gun—cleanly and quickly? How powerful are those old-time guns anyway? Is there any such thing as true accuracy from a muzzleloader? How good were those old Sharps and Remington single shot rifles the buffalo hunters used? I'm a handgun hunter—are there any muzzle-loading handguns strong enough to take big game? Are blackpowder shotguns efficient on upland birds, on wild turkeys? Who were the mountain men, and what did they shoot? Does it take a long time, and a lot of trouble, to clean a blackpowder gun after shooting? Do I have to spend a lot of money to get started in muzzle-loading?

These, and many other questions, are answered in the course of the Third Edition. However, let's touch on a few basic queries right now, just to get started.

Why Shoot Blackpowder?

There are many good reasons to join the ranks of the blackpowder enthusiast. The major reason was already recited above: to complete the game of shooting. Muzzleloaders are now, and have always been, a part of the shooting world. The idea that frontloaders became entirely defunct when cartridge guns came out is false. There have always been blackpowder guns around, not only in America, but in Canada as well as many other countries. Anyone doubting the fact that blackpowder guns continued to interest shooters in our land can take a look at old issues of *The American Rifleman* magazine. Blackpowder articles date back to the beginning of the NRA, and they continue to grace the pages of the NRA's major publication today.

Two more reasons for shooting muzzleloaders: They are interesting and teach us a lot. After all, to know more about muzzle-loading is to know more about shooting in general. Remember that it is the modern firearm that is simple. Ammo is pre-made in the form of a cartridge, and once the cartridge is loaded into a rifle or handgun, the firearm is ready to function. Likewise, the modern shotgun with its readymade fodder. Of course, the same is true of the blackpowder cartridge—in a way—you prepare the round and then simply load it into the gun. However, this old-timer demands special attention and a surprising amount of knowledge for mastery. For example: Do you use hot or cool primers for blackpowder cartridges? What alloy is best for bullets? Which powder? How do you make a lubricating disk? Dedicated blackpowder cartridge fans know all about these and many other topics.

But the muzzleloader is truly "handloaded" for every shot, from the front end, with powder and projectile, as well as lubes and, for the shotgun, a wad system. And if you want to get the most from the frontloader, you simply must know what you're doing. After you walk the trails in this book, you will. That's the goal and promise of this work, because blackpowder shooting without accuracy and power is just making smoke.

OK, so muzzleloaders are interesting, and they require considerable knowledge to master. But are they practical? In one regard, no. If practical means shoot, go home and put the gun away, frontloaders are not. But there is a practical side to the sootburner—special hunts. Lots of them. All over the country. For example, Colorado has a wonderful blackpowder-only elk hunt that takes place during the rut. It's such an important hunt that guides, such as Steve Pike of Gunnison's Tenderfoot Outfitter and Guide Service, specialize in that season, leading hunters to fine bulls in September. Just about every state has blackpowder-only hunts, often several per year.

Think about it. In many areas, due mainly to human concentration, long-range, high-power smokeless-powder cartridge guns are out of the question, while muzzleloaders are acceptable. That's because blackpowder guns have plenty of power for big game, but they do not propel their projectiles nearly as far as modern guns do, making them more ideal than shell-shuckers in developed regions. While there are many good reasons for taking up a frontloader, chances are that the majority of today's shooters buy one to take advantage of a blackpowder-only season—often in areas closed to modern arms. Some of these hunts are held during special times of the year, too, either before or after regular seasons.

There are many other good reasons to shoot blackpowder, including plinking fun, basic practice, informal target shooting, serious competition, benchrest (with special heavy-barrel rifles), silhouette (with single shot blackpowder cartridge rifles), small game and big game during regular seasons, as well as historical interest (emulating the past), and pure challenge. A number of shooters enjoy the idea of using firearms that fit niches of past history, such as the Civil War, Daniel Boone's days on the Eastern Seaboard, the fur trade era of the mountain men, and so forth. Other shooters relish the challenge of learning how to load a smokepole to perform at its best. They like getting close to their game by stalking. They prefer having to make that one shot count with a slow-loading firearm that probably won't afford another opportunity before the "target" moves on. There's also a terrific challenge in shooting far with the smokepole at inanimate targets, which is what the blackpowder silhouette game is all about.

Over the last decade, a whole new world of blackpowder shooting has opened up for the person who just didn't want to get into the true old-timey way of doing things. He wasn't interested in historical replicas, or mountain man history, or buckskinning, or competing at the silhouette match or benchrest shoot. This person simply wanted to fire blackpowder in the most practical way possible. While the newfangled old-time guns give dyed-in-the-buckskin boys a fever just to look at them, they tickle the fancy of those who care only about shooting blackpowder with as little hassle as possible. The modern muzzleloader caught on in such a big way that most major blackpowder companies now offer one or more in their lineup. These look-like-modern blackpowder guns require the same, or at least similar, hands-on shooting that is demanded of older-style sootburners, but there is no way to confuse them with the guns of Kit Carson's era. Some of them even have bolts.

While the range of the blackpowder gun is reduced, due to modest velocity and projectiles with relatively low ballistic coefficients, the fun of shooting is extended. Blackpowder guns are downright enjoyable! Is it the healthy roar? The smoke? The old-time basic nature of the firearm? Of course it is these and other factors that make the old-time shooting iron a lot of fun to fire.

Why would I *not* want to shoot blackpowder? Cleaning the guns afterward ranks number one among negatives concerning muzzleloading. But there's good news about that, so keep on reading.

What Do I Have to Buy?

There are a number of new reasonably priced blackpowder guns on the market today, let alone the many used bargains waiting at gunshops everywhere. The used guns usually end up on the for-sale rack because their owners moved on to upgraded shooting irons, from slightly more sophisticated models all the way to handmade customs. In short, it doesn't take a fat wallet to get going in blackpowder. One smallbore rifle, or one big bore, or one shotgun, or even one sidearm will do it, although the well-outfitted marksman won't be happy without an example of each type. If economy is the goal, the 32-caliber smallbore squirrel rifle—which is great for plinking, informal target work, just pure fun, and putting lots of game from cottontails and squirrels to wild turkeys in the freezer—can actually be fired for fewer pennies per shot than a 22 rifle. That's because of the do-it-yourself nature of black-

powder shooting: Make your own cast bullets; even construct your own percussion caps.

Myths, Misconceptions and Downright Lies

Throughout the Third Edition, an attempt is made to pour water on old wives' tales, as well as often-repeated nonsenses about blackpowder. Many of these outdated ideas took life in the days of prescience. They hibernated during the long era when smokeless-powder guns took over as our main firearms, then got lively again with the rebirth of blackpowder shooting. It's easier to spread interesting myths than it is to to fill the world with the truth. That's why so many foolish ideas about smokepoles survive to this day. Where appropriate, these little tales are cut down to size.

The few shooting laws and rules laid out in the pages of this edition are based on data, not hearsay, inuendo, or campfire palaver. Information in the Third Edition is based upon the best researched data available, plus demonstrations, tests and repeated experiences.

Safety and Maintenance

The only good shooting is safe shooting. That's why this edition is dedicated to rules that make blackpowder shooting as accident-free as any shooting activity can be. Tests conducted with special non-sporting firearms are not to be repeated with sporting guns, of course. This notice is clearly sounded throughout the text.

After-shooting cleaning methods described in the Third Edition are based on an extended test, where five barrels were subjected to six months of shooting and cleaning in a program designed to find out what truly happens to guns that burn muzzleloader propellants. There is good news: Due to high-tech lubes and solvents, blackpowder guns are easier to clean than ever, using the old water-only method, the newer solvent-only way, or a combination of the two.

Where Is Blackpowder Shooting Going?

Up, up and away, that's where. More special blackpowder hunts and hunting areas are in the making. More shooters are learning how much fun it is to shoot smokepoles. Muzzleloader choices are wider than ever. Expect to see more blackpowder guns from companies famous for modern smokeless-powder cartridge arms, such as Remington. The modern muzzleloader has brought in a whole new army of shooters who are interested in old-time shooting, but not with old-time guns. Meanwhile, custom gunmakers remain busy crafting beauty and romance with steel and wood to emulate the Golden Era of Shooting. And it's hardly an American/Canadian game anymore. European shooters are realizing that there's a place for old-time guns in their world, while Australians and New Zealanders also embrace the frontloader more than ever.

The Third Edition promises to be not only bigger, but better. There are over twenty new chapters, from silhouette shooting to collecting, and a major overhaul in illustrations. But THE COMPLETE BLACKPOWDER HANDBOOK has not forgotten its original mission: to make muzzleloading more succesful and enjoyable for everyone.

If you already shoot muzzleloaders, keep it up. If you don't, give it a try. And do consider blackpowder cartridge guns. They've really come into their own recently with the exciting games of silhouette and cowboy action shooting.

Sam Fadala, Wyoming

The Roots Of All Shooting

NOBODY KNOWS WHERE or when the first smoke cloud from a charge of blackpowder lifted against the sky. China? Maybe. Greece? Could be. But we can be fairly certain that the first real "engines of war" to use the propellant, which came to be known simply as gunpowder, came from Europe a very long time ago. That those first "hand-guns," as they were called, were truly effective is questionable. They were simply tubes held by hand, as opposed to canons or other stationary weapons. They seem to have burned relatively minuscule powder charges. In a way, they were mortars, although the true mortar is not normally held aloft by the shooter. We don't really know all there is to know about these early "guns."

That's because the entire history of shooting is an exasperating study at best. Even relatively recent events are about as clear as a mud-hole. A couple examples are in order, just to prove this important point. Did you know that there is argument concerning the exact birthday of the world famous 22 Long Rifle cartridge? We have a pretty good handle on what company developed the amazing little rimfire cartridge, and roughly when this work took place, but we really don't know precisely when the round reached gunshops for sale. The also-famous 30-30, originally called the 30 WCF (30-caliber Winchester Center Fire or Center Fire) isn't totally documented, either. Many believe it was originally a blackpowder round.

I do not think it was ever a blackpowder cartridge. It carried the nomenclature of old-time rounds, and that's why it was popularly

Guns were hardly the first tools to throw missiles. Bows long preceded firearms and, in fact, the earliest guns were inferior to good bows, such as the English-style longbow depicted here.

called the 30-30, being 30-caliber with 30 grains of *smokeless* powder. That's what I think, anyway, but I've read "facts" to the contrary. Also, it did not come out, really, in 1894, although it was no doubt ready in that year. The Model 1894 Winchester rifle that fired the round came along in that year, but the 30-30 wasn't on the street until 1895 in a nickel barrel version of the '94. And by the way, don't call the 30-30 the first smokeless powder cartridge. It wasn't. The 8mm Lebel beat it, and so did the 30-40 Krag. Truly, it wasn't even the first *sporting* smokeless cartridge, since its little sister, the 25-35 Winchester, came out simultaneously.

I present these two historical gun notes to prove a point: gun history is spotty at best. We simply do what we can with the documents we have. Since gunpowder preceded guns, let's talk about that chemical mixture for just a moment. India and China no doubt had something that went boom! a heck of a long time ago. In the 11th century there probably was some form of gunpowder, perhaps weak, and from what I gather, definitely not a propellant that fired bullets, arrows, or anything else, through the air. Smoke? Sure. Fire? Yes. Thunderous noise?

Gun history is often convoluted and unclear. The famous 30-30, which began life as the 30 W.C.F. (Winchester Center Fire) in 1895, is often credited as the first smokeless powder cartridge in the world. It was not.

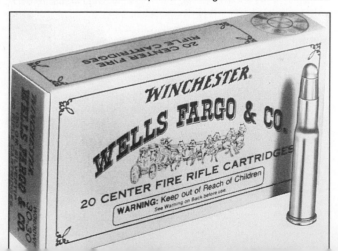

Gunmaking in early America was a one-at-a-time affair in simple shop. The gunmaker was often a tradesman of many talents, capable of repairing as well as building firearms.

Men like Eliphalet Remington built fine muzzleloaders in the early 19th century. Their efforts evolved into more modern arms. The great Remington company, known for its fine up-to-date cartridge firearms, now offers a modern muzzleloader.

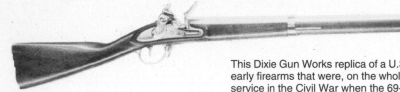

This Dixie Gun Works replica of a U.S. 1816 flintlock smoothbore musket gives a good idea of early firearms that were, on the whole, quite effective. Made until 1844, this musket even saw service in the Civil War when the 69-caliber longarm was converted to percussion ignition.

You bet. But no references to smoke, fire, and noise in the very early times leads us to believe that anything was propelled. In short, gunpowder of some sort probably did exist as early as the 11th century, or earlier, but it wasn't used in guns, because there were no guns.

There were few to no "real" guns in the 13th or 14th centuries, either, but incendiaries were quite popular for warfare. And I think these burnables caused some historical mayhem. I wouldn't be surprised if long-ago notes pertaining to fire, smoke, and noise on the battlefield had to do entirely with incendiaries. References to fire are clear enough. A ball of combustible stuff ignited would certainly provide the fire that historians spoke of, but these flaming objects could be tossed by a catapult. They would make not only fire, but smoke, and as for the boom! it could be from an exploding missile. I wonder how long it took the old-time warrior to figure out that since gunpowder exploded, you could build some sort of shell, put gunpowder in it, light it ablaze, and launch it against the enemy? When the thing landed, it either flared up viciously, or perhaps blew up. In short, references to fire, smoke, and thunder do not impress me with the idea that they refer to guns.

Most certainly, gunpowder was around long before guns. Roger Bacon, the English friar, was a student at Oxford, and a lecturer there. The scholar wrote about explosives in various papers. We're talking about the middle 1200s here, which makes Bacon's observations very interesting. We don't know, unfortunately, if his original works were tampered with or not. Sometimes chapters were added to certain writings long after the original author had passed on. However, we do have an interesting *documented* note from Bacon. He definitely knew about gunpowder, and what's more, he knew about it in the 1200s. He understood that the chemical mixture "blew up." Here is what he said in a treatise that Blackmore (mentioned below) pins down as being written between 1266 and 1268:

There is a child's toy of sound and fire made in various parts of the world with powders of saltpeters, sulphur, and charcoal of hazelwood. The powder is enclosed in an instrument of parchment the size of a finger, and since this can make such a noise that it seriously distresses the ears of men if one is taken unawares, and the terrible flash is also very alarming, if an instrument of large size were used no one could stand the terror of the noise and the flash. If the instrument were made of solid materials the violence of the explosion would be much greater.

Obviously, Bacon was right. The "firecracker" he described, if larger and in a stronger container, could be used to improve the explosive quality of the powder. The end result could be a hand grenade of sorts. But you don't hear Bacon actually talking about a gun. He's making note of gunpowder, obviously used in a rather innocuous manner to make noise, a little smoke, and a flash of fire. Interestingly, the three major ingredients that comprise blackpowder today, and for centuries earlier, were already noted as saltpeter, sulfur, and charcoal.

We end up with assumptions in many cases when referring to gun history, because we are forced to assume when pieces of a puzzle are missing. I've always been fascinated with how things got started—who invented what, and when was it invented? We assume, probably with good reason, that ancient man tossed rocks, and then found out that sharpened rocks tied to the ends of sticks could be used to beat upon animals and fellow humans, and later thrust as a weapon, and eventually thrown. Throwing spears was OK, but if only they could be tossed farther than the arm allowed. And so someone came up with a plan to launch a spear-like missile with a device today called an atlatl. Atlatls are still in use in a sporing way. In fact, I attended the national atlatl championship in Casper, Wyoming not long ago. The atlatl was a launching pad of sorts. How a bent stick and string got into the picture is anybody's guess. Did the original bow throw darts that were first made for launching from atlatls? We don't know, and probably never will.

But you get the picture. We don't know about atlatl or bow origins, and we will also never know who made the first gun, either, or how it was built. So, again, we're forced to make assumptions based upon as much fact as we can gather, and as much logic as we can muster. Records show there were bamboo tubes that were used to hold gunpowder. Later on, these tubes were apparently made of heavy paper. Perhaps some were of paper, others of bamboo. That we are not sure

(Above) Blackpowder shooting did not die out in America. Men like Elmer Keith wrote about muzzleloaders and blackpowder cartridge guns early in this century.

Although blackpowder shooting games continue to grow in popularity, hunting with a frontloader remains the number one reason for getting into the sport of shooting old-style guns.

of, but it's pretty clear that they held gunpowder. Did they shoot projectiles? Hard to say. Probably not to begin with. They may have been more like hand grenades with objects inside that acted like shrapnel. But further references suggest that these tubes, which had one open end and one blocked end, did propel arrows.

Here again, we rely on good guesses. Just as we like to think of the atlatl as a device that followed the spear, and the bow as a tool that logically came after the atlatl, it's nice to think of these early tubes as throwing a missile that was common to man—an arrow. And they probably did. Records seem to indicate just that. Of course, we can never rely entirely on those records, as they were sometimes hazy, sometimes clouded over with superstitious nonsense, and sometimes just plain lies.

However, it does seem that some of these bamboo or paper weapons had a name, one being the "lotus." Other similar tools were around, but their names were lost. Howard L. Blackmore refers to the "lotus," and another unnamed weapon, in his book, *Guns And Rifles of the World*. He says, "The 'Lotus' was a reinforced cylinder container on the end of a wooden stick. It was loaded with all sorts of powder and ejected fire, smoke, poison and iron arrows a foot long." Notice that the arrows Blackmore mentions were made of iron, rather than wood or another lighter material. It seems that the makers of the lotus and other "powder tubes" found that gunpowder had force enough to launch fairly heavy objects.

I said superstition clouded the picture, and that point requires a little bit of documentation. Blackmore refers to the other weapon as follows: "Another gun, without a name, consisted of a copper tube three feet long with a straight wooden handle which shot one arrow at a time a distance of 200-300 paces." Blackmore also points out that this "gun" used gunpowder with the three basic ingredients we still make blackpowder out of today: saltpeter, sulfur, and charcoal. But the Chinese added superfluous materials to the mixture, based, one can readily assume, on some sort of absolutely non-scientific reasoning. In Blackmore's words: "The ingredients of this powder are worthy of note. They were the chemicals of gunpowder—saltpeter, sulphur, charcoal—plus

white arsenic, stone coal, various bitter substances, four kinds of ginger and *human sperm*." (Italics his.) It's easy to see that shooting was as much alchemy as science to the ancient Chinese. They had the elements of gunpowder all right, but their additives were ludicrous.

The early tubes, or hand-guns, plugged on one end, open on the other, certainly sound like gun barrels to me. After all, isn't the muzzleloader barrel a tube with one open and one closed end? So we call these devices guns, and yet they probably didn't shoot bullets for quite some time, firing arrows and other missiles. There's something else we have to realize: these and other early guns were not as all-round effective as the bow and arrow. That's an important fact, because it explains why guns did not take over the military field for a long time. It took many years and a great deal of development before the gun whipped the bow. Certainly the guns carried into Japan in the late 1500s and early 1600s were no match against the Japanese bow and archer. Ship's canons were certainly formidable, but hand-held arms carried by English, Dutch, and Spanish sailors were not as accurate as Japanese bows, and of course, could not fire nearly as rapidly as the bow.

While it is obvious that the gun eventually outran the bow by leagues in terms of power (force), firepower (number of shots fired rapidly), and range (distance reached by the projectile), for many years the bow reigned supreme. The English longbow is legendary for its sway over guns, and there are many accounts of this simple weapon outdoing early firearms. English longbows of heavy draw weight, shooting broadhead arrows, penetrated armor. Arrows could be fired rapidly, and from quite a distance. They often came down upon the enemy right out of the sky. Of course, crossbows also bested guns at first. "A bolt from the blue" does not refer to lightning, as we often think toady. The bolt from the blue was a crossbow bolt coming down out of the blue heavens above—suddenly and without warning. These deadly bolts literally rained down upon the enemy.

Make no mistake that even in more modern times the bow held its own against the gun. This is an important aspect of gun history, because we often think of the firearm coming along and abruptly shoving every other weapon into immediate decline. It just wasn't so.

Today, the blackpowder spirit lives on in many different ways, from cowboy shooting with blackpowder cartridge repeaters to customs like this long rifle.

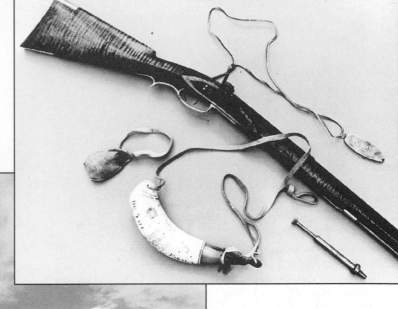

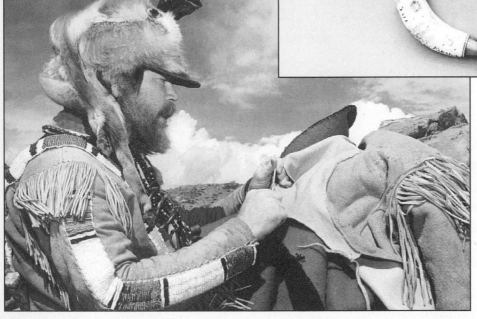

The mountain man movement in the 19th-century Far West spurred 20th-century interest that will without doubt continue into the 21st century. Jim "Bear Claw" O'Meara poses here with his fur trapper's outfit, which he made himself.

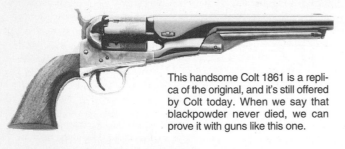

This handsome Colt 1861 is a replica of the original, and it's still offered by Colt today. When we say that blackpowder never died, we can prove it with guns like this one.

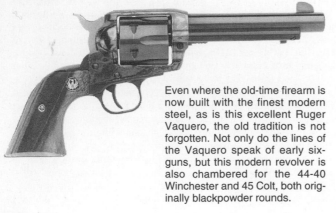

Even where the old-time firearm is now built with the finest modern steel, as is this excellent Ruger Vaquero, the old tradition is not forgotten. Not only do the lines of the Vaquero speak of early six-guns, but this modern revolver is also chambered for the 44-40 Winchester and 45 Colt, both originally blackpowder rounds.

Our own native Americans were exceedingly deadly with their bows. Tennessee sharpshooters that literally sliced the British to pieces in the open field with their accurate muzzle-loading long rifles, were themselves turned into human porcupines when they went up against native American archers in the forest. The latter could put an arrow on target quickly and from just about any posture, even leaning around a tree. And it didn't take long to "reload" those bows!

Finally, however, the world did have a "real gun." I believe that, as with the ball patch, Europe deserves credit for producing the first true shooting machine. As to who really did invent this gun, or precisely what it was, we can never be certain. After all, along with mistakes and bad scholarship, plus misplaced and lost documents, as well as other pitfalls of gun history, we also had the pretenders. I refer to at least two figures who are often credited with early gunpowder and shooting, Mark the Greek (also known as Marcus Graecus) and Black Berthold (also known as Berthold Schwarz). Chances are, these "experts" never existed at all! They were made up by somebody writing "history." Such tomfoolery really muddied the waters of firearm research.

Once the first true guns came into use, they were continually developed and improved. Many wonderful firearms were born in the era of blackpowder, before smokeless came along, and some of them are treated in chapters of this book. Matchlocks, wheellocks, flintlocks, and percussion systems are discussed in Chapter 26, for example. There were many interesting guns carried by early American soldiers, explorers, pioneers, and hunters. These firearms were a far cry from the little bamboo and paper sticks of the distant past. They were so good that many never died. We are still shooting them today, with great joy and interest. ●

chapter 2

How the West Was Won—Really

THE QUEST FOR the west started back east. It had to, because that's where the European immigrants began life in America. They were from England to begin with, but soon enough other old country citizens got the urge to start all over again in the new land, and so they sailed the ocean, bringing their guns with them. The German *Jaeger* was one of those guns. It had much to do with the development of the Kentucky rifle, which was really born in Pennsylvania. Our look at American muzzleloader rifle development could concentrate in the east, for most certainly the fantastic Kentucky/Pennsylvania long rifle was the epitome of the golden era in American arms. But we look west because it was a rifle of less beauty, although as much romance and historical interest, that gave sway to the blackpowder guns we shoot today.

The fur trade era of the early to middle 1800s, its rendezvous, and the "plains rifle" championed by the "doings" in the Far West, had so much impact on America and American shooting that when blackpowder boomed again, beginning in the late 1950s, Hawken-type rifles came to the fore. Of course, it was really replica handguns that got the ball rolling, but riflewise, the comparatively short, stout, big bore hunting shoulder arm was king. But before talking about the fur trade, its rendezvous and rifle, let's look for a moment into the window of the distant past. Hopefully, the fact that gun history is incomplete was established in Chapter 1, so it will be no surprise to learn that we don't know, for sure, who invented the European rifle that forecast the great Pennsylvania rifle that turned into the plains rifle.

Roots

One thing is certain: it all started with rifling. After all, that's where the rifle got its name, from the spiral grooves within the bore. Where did they come from? Rifling (never spelled *riflings*) was another European invention, perhaps of German origin, possibly of Austrian birth. Some sources say Gaspard Kollner of Vienna, late in the 15th century, got the idea. Others credit Augustus Kotter of Nuremburg for inventing rifling in 1520. Yet another researcher mixes the two names, coming up with a "Gaspard Koller" of Nuremburg. No matter, really, but it is interesting that early rifling may have had nothing to do with spinning a projectile. Originators of the no-spin theory can produce old-time guns with straight, not spiral, rifling. The idea? To aid cleaning, they say. I find smoothbores easier to clean than rifled arms, no matter how the grooves are cut, and I'll bet shooters of the past figured that out, too. However, straight-groove rifling did exist.

Regardless of theories, the European rifle was the forerunner of the long-barreled beauty of early America. Leatherstocking lads, such as

Daniel Boone, fought with the long gun, and also hunted with it. Trailblazers of the Eastern Seaboard cut many new paths into the frontier with their accurate rifles in hand. Although various sources list different calibers for the Pennsylvania long rifle, I contend that the 45 was prevalent. That is based on a couple decades of looking at original guns in museums, at gun shows, and in collections, and 45 came up more frequently than others.

A super marksman could take the little 45—and 45 is little when shooting a round ball—and bring home the bacon with it, usually in the form of eastern whitetailed deer and wild turkeys. Of course, 45-caliber was also ample size for warfare. But when the initial shove west became a big push in the early 1800s, a new breed of adventurer answered the call. He relied on the longarm of the hour—at first—as well as a multitude of lesser firearms. But later on, he opted for a different type of rifle. These men were the original buckskinners. While we think of Daniel Boone softly treading the forest floors of the East, we see Jim Bridger riding his horse over the rugged mountain passes of the West. Boone was often afoot, and could carry only so much gear. Bridger was ahorse, often with pack animals to tote his supplies. The lean, long, and rather lightweight, Pennsylvania rifle was not entirely correct for the West. And so the plains rifle was born: shorter for horseback, larger-caliber for bigger-than-whitetail animals, capable of burning heavy doses of blackpowder to gain reasonable range and power. Shots were longer out West, and at bigger animals. The mother of invention—necessity—struck again. The trapper had his rifle.

Those who took trails west in search of beaver were called mountain men, and in spirit, if not replication, it is their rifle we most often carry today. The name "Hawken," which in reality fits only those rifles built by the Hawken brothers, became generic in modern times. And why not? In general design, it is a prime blackpowder rifle for hunting. But we can't leave the men who carried the rifle in the 19th century out of this picture. In my opinion they were the most interesting explorers of American history. This is not an east/west prejudice, for the fur hunters of the Rocky Mountains were Eastern boys to begin with. It's just that they were indeed a wild and crazy bunch of characters willing to risk all for a big time and a free life. They had a fantastic time in a vast unexplored territory filled with truly large game such as elk, grizzlies, and the heaviest four-footed animal on the continent, the American bison or buffalo. They "staked claim" to the American west for Americans, winning respect with their plains rifle. A great many never returned to their Eastern homes, falling to grizzly bear attacks, bitter winters, rushing streams of icy water, and run-ins with the people who already inhabited the land.

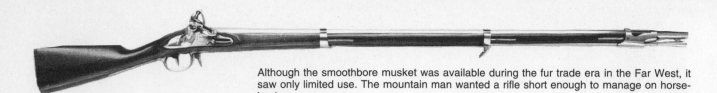

Although the smoothbore musket was available during the fur trade era in the Far West, it saw only limited use. The mountain man wanted a rifle short enough to manage on horseback.

Dennis Mulford stands with a custom replica of a Jaeger rifle. The Jaeger was short and stout, often had a large bore, but was not normally loaded with a very heavy powder charge. It was one of the forerunners of the American muzzleloader.

Two important historical figures, and a third man whose contribution is seldom taught in grade schools, played significant roles in westward expansion and the mountain man movement. The first two were Thomas Jefferson and John Jacob Astor. As explained in *Powder River*, a 1938 book written by Struthers Burt, Jefferson was a far-sighted man. He was the Seward of his time, for like Seward, who saw Alaska as more than a huge ice cube, Jefferson knew that the "other half" of America, the West, was worth having. Seward's Folly, the purchase of Alaska, haunted the man to his grave. Jefferson was more fortunate. While no one in his time realized the potential of Western America, at least the president was in a position to have his way about its exploration. Astor, on the other hand, was a businessman. He wanted to ensure that the rich

fur trade of the West went to American advantage, not French or Canadian companies. The third person to play a role in winning the west was the historically underrated John Colter, a one-man exploration party.

Jefferson got the ball rolling by dispatching the great Lewis and Clark Expedition. He saw to it that the men were well-supplied to carry out their duties. The surface goal was to go west until reaching the Pacific Ocean. The underlying goal was to study the flora, fauna and geography of the Far West. And they were to make their presence known to French and English trappers who had an eye on the place for themselves and their countries. Lewis and Clark were the head of a spear aimed at the unknown. That spear penetrated the entire breadth of Western America. The expedition's spirit of adventure may have been on the wane by August of 1806, but not John Colter's, who made the acquaintance of two trappers, Forest Hancock and Joseph Dickson, on the Yellowstone River. These men had gone farther West than any American at that time, except for Lewis and Clark. Colter asked Lewis and Clark if they would honorably discharge him from their employ so he could join Hancock and Dickson in a trapping venture. He got what now seems a rather remarkable response from his commanders.

The Lewis and Clark journals contain the following quote under August 14 and 15 entries: "The offer [to Colter by Hancock and Dickson] was a very advantageous one, and, as he [Colter] had always performed his duty, and his services might be dispensed with, we agreed that he might go, provided none of the rest would ask or expect a similar indulgence. To this they cheerfully answered that they wished Colter every success and would not apply for liberty to separate before we reached St. Louis." Lewis and Clark supplied Colter, as well as his new partners, with gunpowder, lead, and a "variety of articles which might be useful to him, and he left us the next day." August 16, 1806, was the beginning of Colter's adventure. He spent seven years of his life, from 1803 to 1810, exploring the Far West. He discovered in 1872 what we call Western Wyoming, including the land that was to become Yellowstone Park. When he later told about the Yellowstone area, with its geysers and scalding gushing waters, he was branded a liar. People called his "imagined" land Colter's Hell.

John Colter's story is much too long to tell here, but it is fascinating. Briefly, when asked by Manuel Lisa to drum up some fur trade business, John walked an estimated 500 miles in wintertime visiting various Indian camps in search of traders. He traveled on "webs," which we call snowshoes, and carried his provisions in a pack on his back. With rifle in hand—he always had his rifle—Colter covered an immense territory, which is now part of Montana and Wyoming, making it back to Manuel Lisa's trading post by spring. Obviously, the natives of the region would have killed Colter had he not been a special man. Colter explored, drew maps, and made perhaps the most exciting "run for your life" in American history. At the time of the run, John had a partner named John Potts. The two encountered a party of Blackfeet Indians. Potts resisted and was immediately killed.

Having a sense of humor, as Burt puts it, the Indians stripped Colter of his clothing, gave him a 100-yard head start, and told him to run for it. Cactus is not confined to the desert, and the Indians saw to it that Colter's path of escape took him through plenty of the spiny plants. His feet filled with needles, John continued to run. He was fast, and he stayed ahead for a long while. However, one of his pursuers was faster and caught up to the white man. Colter turned on a dime. His chaser stumbled and broke his lance. John picked it up and dispatched the man with his own weapon. Colter survived by reaching the Madison River, diving in, and coming up in a beaver dam where he found both air to breathe and a hidden sanctuary.

The mountain man found a different world "out west." The black

Some long-barreled Kentucky rifles, like this flintlock, did make it on the trail West. But the long tom was not right for horseback travel, and it gave way to a new breed of muzzleloader.

(Below) The plains rifle made sense for action in the Far West. It was not a lightweight by any means, but it was shorter than the Pennsylvania/Kentucky longarm, and therefore easier to put into action.

bear of the East was represented, but so was his big and deadly cousin, the grizzly. You didn't drop a grizzly with a "peashooter." You needed a stout rifle to do the job—a bigger gun than normally carried. Some camps had such a rifle, a special grizzly bear gun, I am told. I saw one in a museum. It was huge of bore, massive of barrel, and it could fire a tremendous hunk of lead at good muzzle velocity. There were deer, too, the familiar whitetailed variety often referred to as the common deer, and also mule deer, which were called blacktails. There was an even larger deer out West—the wapiti or elk, as well as western moose. While there may have been a few pockets of bison east of the Mississippi River, the mountain man found vast herds of the animals in the West. The bison, largest quadruped in North America, was bigger than a grizzly or moose. A one-ton bull was common. Herd bulls up to 3000 pounds have been documented by scale.

The real terror, though, was the grizzly, "Old Ephraim." Here was an animal that could shoot back. He was sometimes known as the "white bear," and he won battles with many men in spite of their firesticks. Rifles that served well in the East were often less than perfect west of the Mississippi. That includes the wonderful Pennsylvania long rifle. Although handsome, light in the hand, swift to the shoulder, and accurate, this work of art was no match for the bear "that walks like a man." A 45-caliber round ball in the 133-grain class was a mere flyweight against the flak jacket of hide and muscle worn by the grizzly. Something much stouter was needed. The plains rifle answered the needs of the mountain man.

The Plains Rifle is Born

It is impossible to neatly categorize the rifles that went West with those early mountain men. There were naturally many smallbores in the hands of those early explorers, because they carried what they owned. There were also military muskets, and all manner of guns, some of them noted as over 50-caliber in the *Journals of Lewis and Clark*. There were smoothbores, muskets, shotguns, and different types of sidearms, the latter mostly single shot pistols. Most of the guns were flintlocks, but there were some caplocks. The guns that went West were built by French, English, Americans, and other gunmakers. Many of these guns were not that accurate, or even that reliable. So the plains rifle was indeed a new breed created to serve the needs of the mountain man.

First consideration was caliber, generally 50 and larger. The bigger bore handled a bigger round ball, which in turn created a heavier barrel, much wider across the flats than the usual Pennsylvania longarm. Greater barrel wall thickness was desired for strength, and the octagonal barrel of a plains rifle could be more than an inch across the flats, perhaps 1 1/8 inches. The big barrel demanded a stock to match, with a

wrist that was thick enough to withstand the recoil of heavy powder charges. The new rifle was accurate enough, albeit probably not as accurate, across the board, as the Eastern rifle. The sturdy, tough, rugged gun was not all that good looking, with its shorter, heavier, barrel, but it managed well across the saddle, and that counted for more than beauty when lives were at stake.

While barrels in the 44-inch and even longer realm were common on Pennsylvania/Kentucky rifles, the plains rifle wore a tube of about 34 to 36 inches. Always bear in mind that these rifles were made one at a time. Therefore, variations were the rule, not standardizations as with mass production. For example, I saw an original plains rifle with a 33-inch barrel, another with a 36-inch barrel. Obviously, there was no one standard length. Along with the larger barrel and bigger stock came, naturally, more weight. The plains rifle weighed in the vicinity of 10 to 11 pounds, but I've seen a couple of original Hawken plains rifles that tipped the balance at 15 pounds. The style was plain, although a few were embellished with carving and inlays. The half-stock design, iron furniture, and no patch box was about as flashy as a potato sack. Straight-grip stocks prevailed, but a few late Hawkens had pistol grips. There were flintlock plains rifles, but most were percussion.

So here is your plains rifle: it is relatively short of barrel and overall length, stout, of half-stock design, percussion ignition, rugged, plenty accurate for big game hunting, and very reliable. That's a loose stereotype with many exceptions, such as the full-stock J&S Hawken rifle pictured on page 27 of John Baird's *Hawken Rifles* book, not to mention flintlock models. Caliber 53 is assumed as average, which is all right in general, but other calibers were not rare. I saw a 55-caliber Hawken and have been told of a 60-caliber gun. Furthermore, rifles could be freshed out, which simply means rebored to a larger caliber due to a worn or pitted bore. Therefore, a plains rifle which began as a 50 or 53 could end up 55-caliber or larger.

The plains rifle was touted as "flat-shooting" to 150 yards, which it was not. No rifle, including the 220 Swift, shoots flat to 150 yards. However, the term was applied to the plains rifle because with heavy powder charges and a patched round ball, it could be sighted in for about a hundred yards, and a practiced rifleman using a bit of "Arkansas elevation" could hit a target at 200 yards.

The Hawken Rifle

Sam and Jake Hawken made superb plains rifles. They did so well that their name has become intertwined with the very style and period of this firearm. You can mean plains rifle and say Hawken and get away with it, with few faulting your terminology. The famous name now marks an entire genre of firearms, including muzzleloaders that look as much like a Hawken as a Weatherby resembles an AK-47. Today's "Hawken" rifle may have a barrel under 20 inches long, or it might be 36 inches long. It could be mass-produced or handmade. This is not to suggest that original Hawkens are always easy to distinguish from other plains rifles. For one thing, Sam and Jake did not make every part by hand. A lock might come from here, a piece of furniture from there. I am assured by collectors, however, that every Hawken was rifled with a 1:48-inch rate of twist, not because a turn in 4 feet is magic, as we shall see in Chapter 27, but because the brothers only had one rifling machine.

The Mountain Men

Now about the men who carried the plains rifle. They were trappers. It's impossible to extricate them from the Lewis and Clark expedition, because it was the information imparted by these two men that prompted beaver trapping in the first place, so you must link Jefferson to the story, because he sent Lewis and Clark on their mission of discovery. Beaver trapping in the Far West was, as hinted to earlier, for more than pelts. It created an American presence in a land not entirely claimed. Sure, Jefferson wanted beaver pelts, but the President also wanted a group of rough and ready American boys roaming around where there were no street signs. Lewis and Clark passed through the country. The mountain men set up housekeeping there with permanent camps in the Rockies. Why did they go West? Lured by the prospect of riches? Maybe, but I think they bought into the adventure of it. As admitted earlier, they were the most outrageous, courageous,

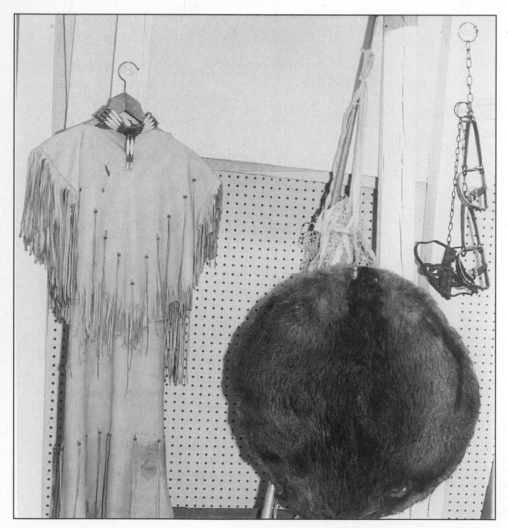

A beaver "blanket," to the right of the buckskin dress, was the object of the mountain man's adventure into the Far West of America.

The fur trade was just that—a trade of goods for beaver plews. The wagons came in with tobacco, galena, knives, and especially gewgaws for the Indian ladies. They went back with "cash" in the form of pelts.

(Below) There remains some argument concerning the actual dress of the mountain man. Some say he was clean-shaven because shaving kits were found among these trapper's outfits. Plus, the Indian brave was not bearded. Did the mountain man shave to compete for the Indian maiden? Today, the image presented is extremely colorful, and no doubt more than a little accurate. Here, Jim "Bear Claw" O'Meara poses in his modern day mountain man getup.

devil-may-care group to set foot in North America. They won the West, not with a repeater, but with a plains rifle.

Not to say that the mountain men weren't after beaver pelts. They certainly were, because the well-dressed man-about-town topped off his dress with a beaver hat crown—a fashion dating back to the 1400s. Hair from the beaver could be worked into an excellent felt, smooth and shiny, a material that retained its shape nicely. Now there was a new supply of beaver, in America! Out West, the streams were swollen with them. Lewis and Clark said so. Fair is fair, however, and it's only right to admit that it was the French trapper who preceded all others in working the waters of the West for beaver pelts. These men usually trapped in concert with natives of the region. They did not penetrate as far as Lewis and Clark, or make John Colter's discoveries, but they certainly were there.

So how did beaver pelts make it from the West to the East? Ask who turns the wheels of America, and you have your answer. Businessmen, of course, and so it was that beaver trapping became a business. A few years after the return of Lewis and Clark, two men decided to take a chance. But they had no intention of trapping anything, except perhaps men to work for them. These gentlemen were William H. Ashley and Andrew Henry. Ashley was the more enterprising, perhaps, while Henry was probably a bit more adventuresome. How would they find their trappers? What do you do when you're looking for something in America? Advertise! And so the St. Louis papers of 1822 carried this ad:

To Enterprising Young Men

The subscriber wishes to engage ONE HUNDRED MEN, to ascend the river Missouri to its source, there to be employed for one, two or three years—For particulars, enquire of Major Andrew Henry, near the Lead Mines, in the County of Washington, (who will ascend with, and command the party) or to the subscriber at St. Louis.

Wm. H. Ashley

There were free trappers, or Ashley Men, as they were called. Ashley did not set fixed wages for his trappers. Rather, he agreed to let them keep half of all furs in exchange for supplies and transportation to the mountains. The true mountain man was, as they put it in those days, "on his own hook." He worked for himself without wages, although there was a company in the background to supply trade goods in a summer meeting that came to be known as the rendezvous. If not fearless, the free trapper certainly acted and reacted with reckless abandon. Life was balanced like a coin on its edge. Certain mistakes could be made only once. There was no second chance. Reports filtered back to

civilization concerning these wayward souls. A greenhorn at one summer rendezvous saw a card game played on the back of a dead man. The dearly departed was set up "on all fours" by his comrades, his stiff body serving as a playing surface. Life was a thread dangling among razor blades.

During a skirmish one mountain man was struck. "I'm hit," he cried out to his friends. They pretended not to hear him correctly, returning answers that offered no comfort whatsoever to the injured party.

The mountain man followed the Plains Indian in many ways, for the Native American knew how to survive in the Far West. One example of this is the tipi, a shelter against wind and storm.

The rendezvous was a gathering of Native Americans and mountain men, and high times were the order of the day. Celebration was in order. Hard work was left behind for a little while.

"He says he hit one!"

"No, I said I'm hit!"

"Where'd you hit 'im?"

"No, it's me who's hit!"

Another time a rabid wolf sprang into camp. The men knew the animal was sick from hydrophobia, and they were aware of the danger of a bite from the crazed canine. Instead of putting a round ball through the wolf, or climbing a tree to get out of harm's way, sport ensued. "He's over here," one mountain man shouted, burying himself beneath his sleeping robes to ward off the bite of death. Shouts of "Over here," and "No, over here" continued. Two men were bitten. They wandered away from camp and were never seen again. It was a tough life, but just right for the fur trappers. Many of them returned to civilization only to find their old world dull. They came back to the mountains to face all the dangers.

They were bold and foolhardy and had a pocket full of time, and no master to tell them how to spend it. And they made marks in the land that live to this day, scraping out the trails that one day became highways from the East to the West. Lewis and Clark, yes, but it was individual trappers that truly mapped the Far West. Their names still grace the land: Bridger National Forest, Henry's Fork of the Snake River,

Jackson, Wyoming, Mount Fitzpatrick, Jedediah Smith Wilderness Area. A mountain man museum/memorial resides in Pinedale, Wyoming, a tribute to the taming of the West. It is housed in a large and beautiful building displaying treasures of the era. The superb Museum of the Fur Trade in Chardon, Nebraska also praises the free trapper of the West.

This is the romance of the mountain man. It is no wonder that some modern blackpowder shooters emulate him. There were no greater adventurers in the early days of this country. Today, his dress, firearms, camping style, and his summer gathering time, the rendezvous, are copied by people who call themselves buckskinners. Some of these men and women live parts of their daily lives, if not in every way, at least in spirit, as the mountain man did. I know several who wear the beards and part of the mountain man dress all year long. The buckskinner is also a student of the fur trade, although he does not adhere to every nuance of the mountain man code. The fact is, mountain men were the same as any other group—there was much individual variation. Frederick Remington did more, perhaps, than any other painter to give us an image of the fur hunter; however, he was late. He painted only a reflection on the waters of life, for the true picture had passed with the ripples of time. Bearded? Maybe, but shaving razors were

Coleman coolers in a tipi? You bet, when its a modern rendezvous. It's a mix of the old and the new.

found in mountain man kits. An Indian maiden was not used to much facial hair. Long hair? Probably. Indian clothes? I'm sure the mountain man adopted part of the Indian costume, but I'm equally sure he didn't take to all of it.

The Original Rendezvous

Before talking about the modern rendezvous, more needs to be said of the original "shining time" the mountain men had at their summer trade gathering. Remember Ashley and Henry. Their St. Louis newspaper advertisement worked. William Ashley would stay behind working out the business details, while Major Andrew Henry led a party of free trappers into the West. The trappers would gather at a preestablished point in summer for their rendezvous, where goods were wagon-drawn from back East. The first rendezvous took place in 1825. General Ashley met his free trappers on what is now known as Henry's Fork of the Green River not far from Daniel, Wyoming, which is near Pinedale. The men had trapped until freeze-up, then they took refuge until spring.

Now spring had come and gone, and here they were with their plews of beaver and big doings. They would trade their hard-earned furs for worldly goods. They were also anxious to hear what was hap-

pening back home. The rendezvous fur trade was on! Pelts were traded for dollar value, that is, inflated dollar value. The trappers needed galena (lead) for bullets, powder, butcher knives, clothing, maybe even a new rifle. His precious smoking tobacco, sugar, and flour needed replenishment. Gewgaws—trinkets that the Indian ladies admired— were important. And all trading was not from trapper to eastern businessman, either. One mountain man, the story goes, gave up $2000 worth of pelts for the hand of a chief's daughter. She must have been a knockout.

How much could a free trapper earn, plunging his hands into the frigid boreal waters of the Far West? One thousand, perhaps two in a season for rheumatism at best, a Blackfoot arrow at worst. Actually, the pay itself was not so bad, since a worker of the period made about a buck and a half a day, but the trapper really didn't get money. He traded his catch to the company store for goods, and the goods were swapped at grossly inflated prices. Whisky worth 30 cents a gallon in St. Louis was cut with water and sold for $3 a pint at rendezvous. Tobacco, coffee, and sugar at 10 cents a pound back East went for $2 a pound. Gunpowder ran the same for English Diamond Grade, while American du Pont, considered better, traded for $12 a pound. Gewgaws carried a 2000 percent markup. A cheap Indian trade rifle worth $10 back home fetched several times that price at rendezvous, while a fine Hawken rifle worth $40 in St. Louis commanded $80.

But money was not the real issue with the mountain men. Lifestyle was. The entire season's effort evaporated in as little as one day, but the men had their goods, and a fine time gambling at cards, talking with friends they had not seen for months, learning of the goings on back East, arguing, competing in shooting games, 'hawk tossin', knife throwing, bragging, fist fighting, and impressing the Indian women. The deal wasn't so bad, they supposed. After all, there was great risk hauling loaded wagons across hostile country. The investors deserved their returns. After all, a lost wagon brought nothing.

The Modern Rendezvous

Finally, the beaver hat slipped from fashion and the mountain men were no more. Their work was done. They had been a presence out West, and they had turned furs into dollars. When an era is gone, it's gone. But it can be emulated, and that's what happened in modern times. The rendezvous was born again. I credit the NMLRA (National Muzzle Loading Rifle Association) with its birth. A handful of blackpowder enthusiasts revived the old-time fur fair. Once again, people met on common ground to trade goods, enter shooting, knife throwing, and fire-starting contests, and to have a big time the old way. Rendezvous sprang up both East and West, and continue to this hour. The NMLRA (see Reference Section) continues to inform about these modern shining times, and blackpowder clubs virtually all over the world hold rendezvous of their own. A booshway, or leader, is chosen to run the affair, and rules are set down.

For the purposes of this work, it is the shooting contests that mean the most. Some are off-hand at the common black bullseye, others are more animated. Want to have fun? Try splitting a fired round ball on the blade of an axe imbedded in a vertically-oriented stump. On either side of the blade, balloons are set. If you split the ball well enough, the two pieces will break the two balloons. Or cut a wooden stake in the ground in half with bullets from your frontloader. Or ring a gong, sometimes way out at several hundred yards.

There in the valley a few dozen tipis stand as if they were a painting. Smoke rises from campfires all over, hovering in thin wisps on the air, catching and reflecting the natural light of day, and the man-made light of the fire when the sun goes down. At dark, tiny sparks from flint 'n' steel striking together play at ground level, with curls of smoke rising like miniature smoke signals. Standing on a hill, you see evening campfires reborn with long tongues of flame licking at dry wood. Listen, and you hear talking and laughing, as well as music on the air from dulcimers and flutes. Stew is bubbling. It's time to join your friends around the campfire before resting back in your tipi on your buffalo robe. ●

17

chapter 3

Get Ready for Muzzle-Loading

YOU GET STARTED in blackpowder shooting just like any other sport by deciding how you want to play the game, then gathering the gear that lets you do it right. Blackpowder shooting is more complicated then firing modern arms, but in this case, complication breeds fascination. I recall my own beginnings. The road was bumpy, with many detours. I got my hands on some poorly designed guns of questionable manufacture. They didn't "go off" all the time, and a couple of them weren't entirely safe to shoot. I considered dropping out before becoming a dedicated player. Then I got lucky. University of Alaska professor, and Dean of the College of Arts and Letters, Charles J. Keim, invited me to his home to see an original muzzleloader, as well as a custom replica flintlock long rifle.

Professor Keim handed me the rifles one at a time. "Oh," I thought to myself. "This is what it's really all about." I recall sitting by his fireplace examining the workmanship of the guns. "This rifle," he said, "belonged to my wife's grandfather." He held in his hand a sleek half-stock plains rifle, a handsome original. The flickering rays of firelight played over the metalwork and stock with heliographic messages. A charm exuded from those fine muzzleloaders that I never saw in the blackpowder guns I'd been shooting. I was entranced.

"You can shoot them," Professor Keim offered. I didn't want to risk the rifles at the range, but I did spend the next couple of hours talking about the old days and guns. That was the real beginning for me in blackpowder shooting. The path ahead was not totally blazed, but at least I knew where I wanted to go. Unfortunately, a few traps lay enroute.

An early trap for me was reading what modern gun writers were saying about the old-style guns. Unfortunately, not all of it was accurate. Much advice was clothed in old wives' tales and near voodoo. Then I met another man named John Baird, publisher of a little magazine called *The Buckskin Report*. John took exception to some of my unkindly remarks on the inefficiency of the patched round ball. Logic, plus what I'd been reading, suggested that the sphere of lead had to be worthless. Baird set me straight, and I was wise enough to take his comments to heart. A period of listening, learning and discovering began—especially discovery. I began testing, for myself, and reading older texts on muzzle-loading recommended by Baird. When I doubted something that I heard or read, I tested the concept for myself. I was beginning to really learn about blackpowder shooting.

Starting out in muzzle-loading is best accomplished, as many new ventures in life are begun, with the help of a knowledgeable veteran— emphasis on *knowledgeable*. Just because some guy has been busting caps for twenty years doesn't mean he knows all about muzzle-loading. But real experts are as near as the local blackpowder club. These dyed-in-the-buckskin downwind shooters learned through experience, and they're willing to pass their knowledge on. Take advantage of that know-how, but always with an open mind. Nobody knows everything there is to know about muzzle-loading. You may find that dedicated frontloader expert at a local blackpowder club. Join up. Listen, but also watch how the veterans handle their firearms, and for that matter, which guns and gear they prefer. You may save a lot of wasted steps, not to mention dollars, by avoiding the mistakes others have made. Also watch for all safety measures long-time blackpowder shooters take. Keep an eye out for the "familiarity breeds contempt" problem. Doing something for a long time can lead to relaxing the rules, and that can be bad.

So join a blackpowder club and go to the shoots. Ask a lot of questions. Listen and discover, and get help from advisors who really know what they are doing. But don't forget to read and study. And if you're an old-time blackpowder shooter, take it upon yourself to pass on what you know.

How young can you get started? Well, perhaps not as young as the author's daughter Nicole, who seems pretty thrilled with her father's trophy blackpowder antelope, but smokepole shooting is a good way to spend some time at just about any age.

Shooters who enjoy blackpowder guns also like to talk about them and help newcomers into the sport. Anyone interested in truly getting into blackpowder, muzzleloader or breechloader, should think about attending a shoot like this one.

Realistic Expectations

There are three major reasons for failure in the muzzle-loading sport. First, accuracy can seem downright lacking compared with modern arms. Initially, the newcomer is content with four-to six-inch groups at 100 yards because he is fascinated by the smoke and boom of the old frontstoker. Later, this shotgun group becomes discouraging and the shooter gives up the sport. Take heart. The well-constructed muzzleloader, properly fed, can achieve perfectly acceptable accuracy. In fact, certain frontloaders (heavy benchrest slug guns, especially) produce amazingly close clusters. Second, the blackpowder level of performance, not only ballistically, but in terms of the mechanical function of the firearm, may seem lacking, which is true. The newcomer is often disappointed with both power and function, and he gives up.

Ballistic performance and slick-action repeated fire are out of the question here. Muzzleloaders don't shoot as far, or as flat, or as pow-

Can you really get "blackpowder close?" Of course you can. It's a matter of hunting wisely and knowing your game.

erfully as modern cartridge guns do. However, blackpowder ballistics can be extremely gratifying if a person knows how to make the most of them. As a matter of fact, within range limitations, charcoal power can be astonishing. A blackpowder attitude must be adopted with a full acceptance of the sport. Third, cleanup after shooting may appear burdensome. The sportsman envisions tubs of hot water, rotten egg smell, rubbing, scrubbing—what a mess! He decides not to bother. However, he should take a look at Chapter 38 before deciding against blackpowder. He'll be happy to know that this edition describes three cleaning methods that *do not* demand a great deal of labor: water/solvent; water only; solvent only. He can make up his own mind, after trying each, which method he prefers.

Learning the Blackpowder Pace

Ours is an exciting, interesting world. It is also a fast-paced time, where speed seems to be more important than just about anything else. Promises of the twenty-hour work week, echoed in the 1950s, have not materialized. The forty-hour week is now considered short for many. Just driving to work can be an Indy 500 experience. But we don't have to shoot blackpowder that way. Muzzle-loading enjoys a slower pace. The newcomer to the sport must understand that in order to have a good time he simply must slow down and savor the moment. It's one shot at a time, friends. Even at rendezvous, where multiple shots are often necessary in competition, everyone understands that the blackpowder gun is not a semi-auto. It takes a little time to pour the powder charge, seat a bullet or shot charge, and cap or prime before the shot can be delivered. Learning to load with deliberation and care, making every missile count downrange, is the right attitude, and making that one shot count is the enjoyable and safe way to go. Of course, with experience and practice comes proficiency, and fairly fast second shots, even at game, do become a reality.

Dealing with Range, Sights, and Firepower

For the most part, the old-time blackpowder firearm is simply not on par with the modern shooting instrument when it comes to range, sights and firepower. It is shorter-shooting. Sights are often open irons, although nowadays peeps and scopes are common, and you cannot expect to fire a great many shots in a small span of time. While these facts seem to downgrade the blackpowder experience, they don't. In a way, they enhance the game. The shooter actually has more fun getting closer to his target, choosing the sights he prefers to work with, and making that one shot really count.

Range

Effective range is treated throughout this book, but we do need to touch on the subject a bit more here, because getting started in muzzleloading requires a knowledge of blackpowder trajectory. How flat and how far will the old guns shoot? Of course, this is a relative question, the answer predicated upon the rifle in question, the load it fires, and the marksman. A practiced blackpowder rifleman can shoot surprisingly far on inanimate targets. Some shooters consistently ring gongs out to 500 yards, while the metallic silhouette contest, allowing offhand shooting only, is played by blackpowder *cartridge* fans who knock the ram over at 500 meters—with iron sighted rifles. But the truth is, a shot on big game beyond about 125 yards is probably a shot too far for most of us.

Blackpowder shotgunners can also enjoy great success. The grand old man of the sootburning scattergun, V.M. Starr, actually won contests against modern shotgun shooters. Having said all of this, I'm forced to conclude, however, that there is no way the old-time smokepole, shotgun or otherwise, is going to outdistance the modern firearm. It simply isn't in the cards. No frontloader shoots as flat as a 7mm Remington Magnum, for example, and only a few deliver nearly as much downrange impact, although certain huge-bore blackpowder rifles can pack the mail out to 300 yards and farther.

More About Sights

When I first got into blackpowder shooting, iron sights were the only types available, generally speaking. Irons are still most popular, and they are "the law" on many blackpowder-only hunts where scopes are legally forbidden. However, the advent and surging popularity of the modern muzzleloader brought with it a desire to mount scopes on these guns. Scopes are now very popular on muzzleloaders, and are legal during *some* primitive hunts. Check local laws before going scoped, of course. I have not yet tried a scope on a hunting frontloader, but I did learn some time ago that in testing muzzleloader accuracy a scope was vital, and I have mounted many on test rifles. A favorite scope for this work is a Bausch & Lomb 6-24x, set on 24x for best bullet grouping. High magnification is the name of the bullet-grouping game, and that's why you never see low-power scopes at a benchrest match. Sights and success go hand in hand. Getting started right demands sights that a shooter can live with. For more on the topic, see Chapter 29.

Firepower

Rapid shooting is out of the question, except for getting two quick shots off with a double rifle or shotgun, or a multiple-load pistol or revolver. But fast reloading is possible, especially with "readyloads," which are discussed in Chapter 7. It's a matter of practice. One well-placed round ball from the proper distance, or one good conical, will drop a big game animal in its tracks. However, the rule when hunting is to reload immediately after you shoot at anything, even a rabbit. That rule payed off for me on an exciting hunt. I had no interest in shooting a "contained" bison, but I did want the blackpowder experience of dropping a shaggy with a muzzleloader. And I searched around until I found such a hunt.

By the way, those who say the American bison is a barnyard milk cow don't know what they are talking about. This fact is grimly proved almost every year. I live not far from Yellowstone Park, and it's a rare season when someone isn't hurt or killed by a buffalo in the Park. The largest quadruped on the continent, bigger than a moose, grizzly, Polar or Kodiak bear, it can outrun the fastest human sprinter in the world. My bison was on the loose over the plains of Nebraska, entirely unfenced and totally capable of making mashed potatoes out

How big can you go? This big. A bull bison meets a 54-caliber round ball. Don't try this at home.

Perhaps mounting a scope on a handsome custom Dale Storey side-hammer rifle was not the politically correct thing to do, but it proved the accuracy potential of a patched round ball out of a fine muzzleloader.

of our hides. My brother and I hunted the bull on foot, me with Number 47, my 54-caliber round ball rifle, Nick armed with a motor-driven Nikon loaded with 35mm film.

To make a long story short, the bull came our way instead of going the other way when we got close to him on the plains. Our plan was for me to signal when I was going to shoot. Nick was to photograph the action. I recalled the advice given by the fellow running the hunt. He said, "Don't aim dead center between the eyes. The ball could bounce off." I held to one side. The round ball penetrated the skull, passing completely through the right hemisphere of the brain. The bull rose to full height and toppled over. My code is to reload immediately after every shot, no matter how good the shot, and it's a good thing I did. The bull got up and a second shot was required to put him down for good, in spite of the placement of the first ball.

Muzzleloader Characteristics

Things have definitely changed in relatively recent times with regard to muzzleloader rifle weight, length, and overall shape. Realistically, the upswing of modern blackpowder shooting did not bring with it the true Pennsylvania/Kentucky long rifle, or the plains rifle. Sure, examples of both exist, and in no small number, but by far the non-replica muzzleloader leads the pack. It is shorter, lighter, and shaped differently than the originals. For example, the ever-famous Thompson/Center Hawken is a rifle made for today's shooter. It does not look like an original Hawken, nor is it as long or heavy as that rifle from the past. Custom gunmakers have created many copies of old-time Pennsylvania long guns, but custom guns are not prevalent either, nor nearly as widespread as non-replicas.

Although I have no intention of giving up on replicas of the past, there is no denying that the "average" blackpowder shooter today prefers modern renditions over copies of original designs. Sales prove it.

This bothers some shooters a great deal. They're the purists who feel that the whole thrust of blackpowder shooting is emulation of the past, especially the guns. They have a point. However, the fact remains that thousands of marksmen wouldn't shoot blackpowder guns if they were long and heavy, or if they had a lot of drop at comb, just as thousands of archers wouldn't use old-style bows.

A brief point on rifle weight, and we'll pace forward on the trail. This is personal, but I don't like super-light hunting rifles, old-time or modern design. I've had unhappy experiences with flyweights of

either category. As Jack O'Connor, the popular gunwriter, used to say, ideally a rifle would weight no more than a wristwatch when you carried it, but it would turn into a bench gun when it came time to shoot. Since that's impossible, the best bet is a rifle that isn't so heavy that it bends your back to carry it, but is certainly not a flyweight when it comes to steady holding for that one all-important shot. This is why I suggest starting out with a rifle that feels good in the hand, but holds great in the field, too. Also remember that blackpowder big game rifles use a heavy powder charge to gain power, and this means more recoil. Real light muzzleloaders kick real hard!

This aspect of getting started in blackpowder shooting is absolutely vital to success, and that's why I've devoted extra space to it. The newcomer will be grossly disappointed if his rifle—and we're dealing mainly with the big-bore rifle here—is uncomfortable to carry, unstable in the field, or so darn cumbersome that it just doesn't come to shoulder smoothly. So pick a rifle that feels right for you, but don't avoid replicas because of their length, weight, or configuration. What may seem bulky at first could well prove to be one of the most stable shooting machines of your career. Give yourself time to adjust. More drop at the comb of the stock than you are used to will probably

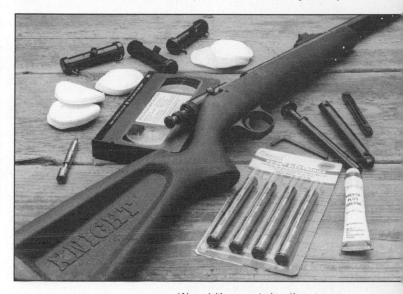

(Above) It's your choice. If you want to go modern, do it. This Knight LK93 Wolverine Value Pack comes with the accessories needed to get started in the game.

Rifle weight is very important to success. A very light rifle may not be ideal for the field, while a too-heavy rifle can be a burden to carry. Medium weight offers stability that can be packed along without too much trouble, and when it comes time to shoot, a good steady hold is possible.

The blackpowder revolver is for the shooter who wants multiple shots in a handgun. This Navy Arms replica of a 19th-century Colt is 44-caliber, six-shot.

The author with two very different types of shotgun, short and regular barrel lengths. The short one is at home on the trail where compactness is handy.

become an off-hand delight if you give the rifle a chance. The rifle-style buttplate with its crescent shape may seem foreign at first, but give it a chance. You may even like it better than the more common modern shotgun style or flat buttplate or recoil pad.

Getting Started with Blackpowder Triggers

Triggers are so vital to successful shooting that I feel obligated to ruminate in that field a good while. The double-set trigger is especially important to understand. When blackpowder "got going good" again, this trigger system was very popular. It's not as widely used today, but still exists on a number of firearms. Modern muzzleloaders don't use them as a rule, and many non-replicas have also gone to the single trigger system of contemporary design. In fact, many modern muzzleloaders have triggers that were designed for bolt-action cartridge rifles.

I like the double-set trigger, and feel that it deserves special mention. It allows for a very crisp and light let-off. This two-trigger setup, with a set trigger and a hair trigger, allows the shooter to touch off the instant the sight picture is correct, and with the least disturbance of aim. To each his own on triggers, and there is no argument that the single trigger is by far more familiar to modern shooters, but a trigger with a pull of only a half-pound or so is fantastic and deserving of attention. So how does it work?

To use the set trigger, the rear trigger is pulled fully rearward. Rarely, this is reversed (as on the Thompson/Center Patriot pistol) in which the front trigger is the set, while the rear trigger is the hair. You can hear it set with a click, and you can feel it set. Now the front hair trigger is ready. Depending upon how it has been adjusted, the front trigger can be tripped with only a few ounces of pressure. Incidentally, some gunmakers distinguished between double-set trigger and the multiple lever trigger designs. For these fellows, a double-set trigger means it can only be fired in the set position. In other words, the rifle will not fire until the trigger is set. However, the multiple lever design allows the front trigger to be pulled without having the rear trigger set, an advantage in getting off a fast shot.

The single-set trigger has one trigger, which is set by pushing *forward* on it until it clicks. When set, the rifle "touches off" with just a few ounces of pressure, depending upon specific adjustment. As with the double-set, this type is like breaking a taut spider web with your

forefinger. It provides excellent control, while the light let-off disturbs aim very little.

George Larsson, writing in a March, 1949 issue of *The American Rifleman* magazine, said this of the double trigger: "The object of a set trigger is to allow a rifle or pistol to have a firing mechanism with a heavy engagement between the sear and the firing pin, for purposes of safety. Then, by the process of pulling or pushing the 'setting trigger,' a bar is cocked under heavy spring pressure and because of a leverage advantage that is gained it may be released by a light touch on the 'firing trigger.'" Mr. Larson continues: "Practically all set triggers have a provision whereby the gun can be fired by pulling the firing trigger without first having set the mechanism. Most of them also have an adjustment screw which controlls the amount of contact surface at the crucial point of engagement." This set-screw appears on many muzzleloaders. The double-set trigger should not be set too light. I prefer

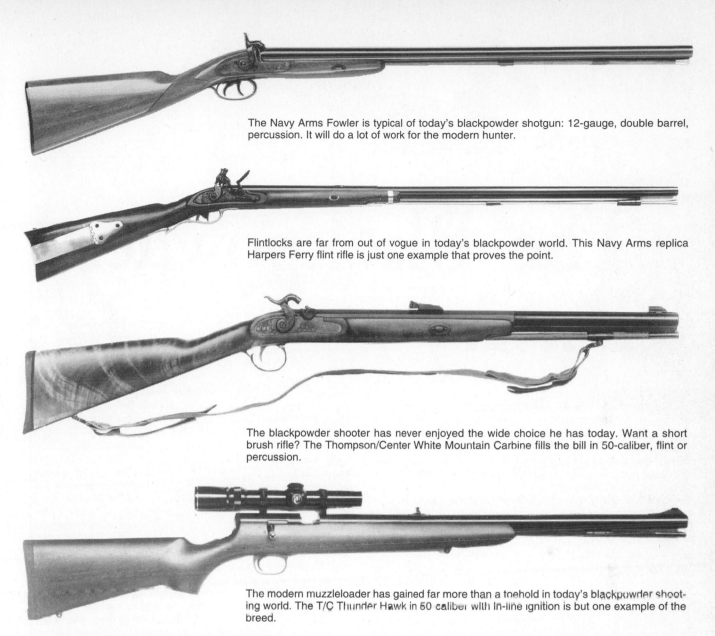

The Navy Arms Fowler is typical of today's blackpowder shotgun: 12-gauge, double barrel, percussion. It will do a lot of work for the modern hunter.

Flintlocks are far from out of vogue in today's blackpowder world. This Navy Arms replica Harpers Ferry flint rifle is just one example that proves the point.

The blackpowder shooter has never enjoyed the wide choice he has today. Want a short brush rifle? The Thompson/Center White Mountain Carbine fills the bill in 50-caliber, flint or percussion.

The modern muzzleloader has gained far more than a toehold in today's blackpowder shooting world. The T/C Thunder Hawk in 50 caliber with In-line ignition is but one example of the breed.

having mine adjusted, or at least checked, by a blackpowder gunsmith. Of course, set your trigger only when the rifle is put into action. Don't walk around with the double trigger in the set position.

To Buckskin or Not to Buckskin

Getting started in muzzle-loading also means making a decision on the style of shooting a person prefers to engage in. Whether 'tis nobler to go forth in old-time clothing and live in a tipi, or stick to everything modern, must be decided. Buckskinning is much more than blackpowder shooting—it's a way of life. The dedicated buckskinner is a lover of American history, a student of the past, and a devotee of an era. He knows not only the ways of the past, he also emulates them. He can start a fire with flint 'n' steel, make a capote, rig a tipi so that the smoke curls up almost magically out of the center hole, and he may be able to thrive in the backwoods with 19th-century regalia about as handily as some of us get by with modern garments and gear. Are these things appealing to you?

The best part of buckskinning, I think, is the association with like-minded people. The modern rendezvous is a trade ground not only for a great deal of hand-made accoutrements, but also for ideas. The satisfaction lies in reliving a piece of the past, doing things in accord with the way they were done in early America. There is comradeship. There is fun. Buckskinners are unique. About the only drawback to this sidesport of blackpowder shooting is the crowd factor for those who don't like group activities. The rendezvous is definitely a social gathering. Futhermore, it can attract outsiders who want to look on.

So there are many phases of our sport, but to keep things simple, let's break them down into three major categories: informal shooting, competition, and hunting. Buckskinning, by the way, may contain all three. The shooter can elect to partake of one or more of these categories. Because ours is a world of specialization, and because of the depth of each phase of blackpowder shooting, most marksmen gravitate to one aspect of the sport or another. But that need not be.

So start out wisely. Adopt the blackpowder attitude, and you'll probably find this sport to be one of the best things you ever did. And if you're already shooting the smokepole, keep it up. It isn't at all odd that in the midst of the space age, with personal computers, genetic engineering and microwave cooking, many eyes have turned rearward for a brief glance at a quieter time, an era when the mountain man roamed the American West and Canada, and leatherstocking trailblazers explored the eastern seaboard. There's real advantage in burning blackpowder, whether for the pure enjoyment of it, the relaxation, its challenge, rendezvous fun, the buckskinner lifestyle, or doing an exciting blackpowder-only big game hunt. So if you don't already own a muzzleloader, or you plan to buy another one, turn now to Chapter 4, dealing with the blackpowder battery. ●

chapter 4

The Blackpowder Battery

IN THE PREVIOUS chapter, getting started was broached. Now let's look at the purchase of the most important part of the blackpowder outfit—the gun. A complete blackpowder battery includes handgun (pistol or revolver), shotgun and rifle. Since we're a nation of riflemen, it is that category that receives the most attention. There are many different types of blackpowder rifles, but for the sake of clarity, only seven are listed here: originals, replicas, non-replicas, rifled muskets, modern muzzleloaders, customs and cartridge rifles. Wise firearm choice, be it handgun, shotgun or rifle, is vital to overall blackpowder shooting enjoyment and success. Make the wrong choices for your intended purposes and you won't touch them after the charm of making smoke and sparks wears off.

Initially, just shooting muzzleloaders is enough. You don't even have to hit the target all the time, or bring game home. But it doesn't take long for this attitude to wear thin. Bullseyes are made for hitting. Game departments issue tags for filling. And these goals require the right frontloader. Of course, there's nothing wrong with buying blackpowder guns just for collecting or informal plinking. But most are purchased for hunting, some for rendezvous and target work. And, of course, in order to participate in a broad way, more than one sootburner is needed. It's a "match the gun to the game" quest. The game can be the kind you eat or the kind you play. Here's a look at the modern blackpowder battery, starting with the sidearm. *Note*: in order to fairly assess many of these blackpowder guns, ballistics are used to make clear and definite separations.

The Revolver

The caplock revolver is one of the most enjoyable of the blackpowder shooting irons—if you get the right one for what you're trying to do. These guns worked surprisingly well for old-time law officers as well as lawbreakers. But today, the caplock revolver is mainly for fun. There are too many better sidearms available for anyone to rely on a caplock six-shooter (or five-shooter for that matter) to save his bacon, or bring the bacon home from the field in the form of wild game. For example, the 31-caliber revolver is a handsome little gentleman, just right for plinking. The Baby Dragoon, a 31-caliber wheelgun, shoots a little .319-inch 50-grain ball at around 800 feet per second (fps) for a muzzle energy of 71 foot pounds. This is sufficient for rabbits and other small edibles if you can hit 'em with the little handgun, but the majority of 31s don't have target sights, so counting on putting meat in the pot with them is usually only wishful thinking.

Going up a notch, the 36-caliber revolver shoots a round ball that weighs about 80 grains. I've chronographed this little lead pill at around

1100 fps for a muzzle energy of 215 foot pounds. Lots of good and bad guys were laid in Boot Hill by a 36-caliber ball. But in spite of that fact, it's no powerhouse. The 44 is much stronger. My Ruger Old Army revolver, 44-caliber, shoots a .457-inch 143-grain round ball at up to 1050 fps for a muzzle energy of 350 foot pounds. As a matter of quick comparison, the 357 Magnum handgun cartridge with a 158-grain bullet at 1300 fps gains close to 600 foot pounds of muzzle energy, considerably more poop than the 44 caplock. Yet most modern gunners no longer consider the 357 a powerhouse in light of the 44 Magnum and other big boys on the block. But with good sights, the 44-caliber caplock revolver can put a lot of edible small game meat on the table, and some big game has also been taken with the 44, but only by experienced handgunners. Ballistically, the 44 blackpowder revolver is too small for big game and is disallowed in most areas. See Chapter 5 for a discussion on some hunting rules pertaining to blackpowder guns.

It boils down to this: The replica blackpowder revolver is a lot of fun to shoot and a lot cheaper to collect than original blackpowder wheelguns. Small caliber revolvers are adequate for small game, while the 44 is good for small game, varmints and even deer—if an expert is behind the sights. Original-style revolvers, which generally have crude sights, are useful for close-range fun, mainly tin-can rolling. However, it's important to point out that there are formal and informal shooting matches that do feature the caplock revolver at blackpowder gatherings. Target-type sights are allowed for some, but not all, of these contests. And by the way, when thinking about purchasing your blackpowder revolver, if you really want to hit something with it, go with adjustable sights. Replica revolvers with topstrap notches for rear sights are great fun, but when it comes to hitting the mark, target sights are the ticket.

The Pistol

The blackpowder pistol is generally a single shot, but there are a few double-barrel pistols offered, too. Replicas in flint or percussion are usually best for the fun of owning and pure enjoyment of shooting, including plinking. Some blackpowder matches have pistol events that call for a flintlock or caplock single shot with "primitive" sights. If that's your game, go for it. But if you like clustering holes on the target, or putting a rabbit in the pot with a head shot, buy your pistol with sights. An example is the Thompson/Center Patriot. This 45-caliber pistol has double-set triggers and adjustable sights. I'm no pistoleer, but with the Patriot I can put five shots into an inch center-to-center at 25 yards from a rest. I have enjoyed a lot of camp meat supplied by the Patriot. If your goal is hitting the target, think sights for consistency.

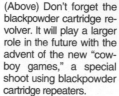

(Below) The author has run across a number of original revolvers, especially from the Civil War era, as this Remington is. They are not impossible to locate.

(Above) Don't forget the blackpowder cartridge revolver. It will play a larger role in the future with the advent of the new "cowboy games," a special shoot using blackpowder cartridge repeaters.

Some collectors buy replica pistols to fill in until the real thing comes along. Others collect replicas—period—considering them ideal for historical reasons, yet inexpensive compared with the "real thing." Blackpowder pistols are fun to own and show. Buckskinners need pistols to make their outfits complete for rendezvous, because trappers of the Far West often carried one tucked in a belt. These were for more than show, of course. They were deadly at very close range.

Then there's hunting. Today, there are a number of blackpowder single shot pistols that carry sufficient power to hunt big game, provided the shooter is an expert with both his gun and his hunting skills. Lyman's 54-caliber Plains Pistol is one. It shoots a .530-inch 225-grain round ball at around 925 fps for a muzzle energy close to 430 foot pounds. The muzzle energy is not exciting, but at close range with good ball placement the big 54-caliber hunk of lead does good work. Navy Arms Company has a 58 caliber Harper's Ferry Pistol that is even stronger. I've gotten as high as 800 fps with a 500-grain Minie from this gun for around 750 foot pounds of energy. Up close, that's big game power. Also, Thompson/Center's Scout pistol, calibers 50 or 54, is truly a blackpowder big game one-hander. Near the muzzle, it carries the punch of a 30-30 rifle.

The Shotgun

The blackpowder shotgun is a pure joy to own and shoot, and unlike handguns or rifles, this sootburner can come fairly close to modern performance. Being prescriptive is dangerous, and sometimes downright unfair, but I'm forced to advise today's blackpowder fan to go with a 12- or 10- gauge caplock scattergun. Smaller gauges are available, and by no means poor in power or pattern, but 12s and 10s throw sufficient shot to really get the job done, whether the target is a clay pigeon or a real pheasant. The 12 is entirely adequate for all upland game, although if wild turkeys are a major interest, a 10 is better yet. Waterfowl hunters who have geese on their list should consider a 10-gauge for pass shooting, while a 12 is enough for decoy or pit shooting.

Four other criteria are highly important to blackpowder shotgun selection: percussion versus flint, single or double barrel, steel shot capability, and choke.

Make my smokepole shotgun a percussion. Make yours a percussion shotgun, too—most, if not all, of the time. Flintlock shotguns are available, and darn good ones, too. They are, in my opinion, more fun to shoot than caplocks, as well as more challenging. If a shotgunner really wants to step back in time, he should consider the fowler with those wonderful little plumes of smoke rising from its flintlock pans. But from the practical side, the percussion shotgun wins hands down.

Not all originals of the past were well-made, handsome or worthwhile. However, some were, such as this engraved frontloader. Note the high fence (under the hammer arm), cleverly designed to prevent cap debris from coming back at the shooter.

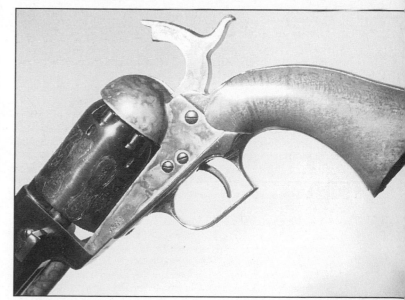

Is the real thing out of reach? Then how about a copy of a Colt cap 'n' ball revolver built by Colt today? This is Colt's 3rd Dragoon model.

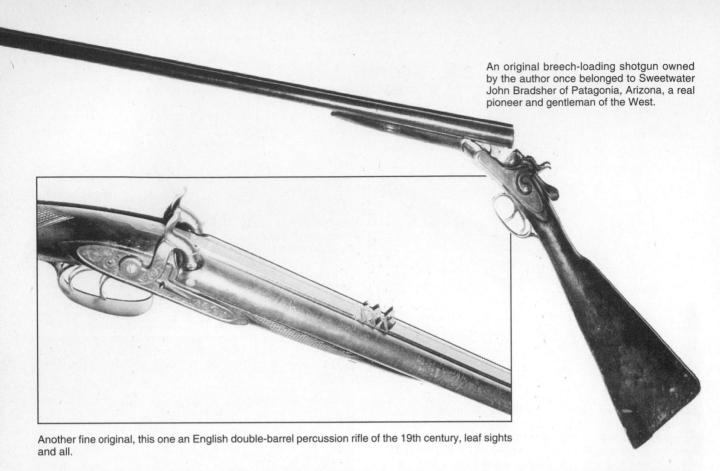

An original breech-loading shotgun owned by the author once belonged to Sweetwater John Bradsher of Patagonia, Arizona, a real pioneer and gentleman of the West.

Another fine original, this one an English double-barrel percussion rifle of the 19th century, leaf sights and all.

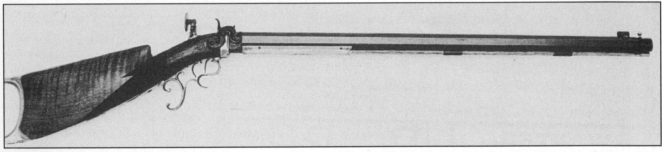

Only one shooter in thousands will have a beautiful original like this Schalk Schuetzen rifle built in the 19th century. But they are available for those who can find and afford them.

It comes in more models, is easier to shoot because of faster lock time, and reloads quicker.

Should you buy a single or double barrel gun? A single-barrel blackpowder shotgun will put plenty of game in the pot, but a double gun makes more sense in most situations. It's a very simple matter of two shots versus one. Since the muzzleloading shotgun takes a while to reload, the single shot shooter is decidedly handicapped. Picture this: It's early morning. The ducks are flying. Your decoys are out and your call is bringing a response. You get your chance. Boom! With a single shot gun, that's it, unless you happen to have a second shotgun in the blind with you. But with a double, it's Boom! Boom! And that can be a lot better if duck soup is your goal.

In many hunting situations, shooting non-lead shot is no longer a matter of choice. It's a matter of the law. If your blackpowder shotgun cannot handle steel shot, you cannot hunt waterfowl with it, and in many areas, you simply cannot hunt with it at all, because only steel shot is allowed. However, this situation is changing, since bismuth is beginning to be accepted as an alternative to steel. Down the road, there may be other allowances made, but all in all, steel shot is the rule these days in many areas. Fortunately, blackpowder shotgun makers have answered the call. There now are a number of excellent muzzle-loading

guns that are made to shoot steel shot. Warning: If you shoot steel in a gun not intended for this hard shot, be prepared to watch the ruin of your bores. If you don't need to shoot steel shot, because you plan only to break clay pigeons or hunt game which does not command, by law, steel pellets, fine. But it's obvious that if you plan to hunt where steel only is allowed, you'd best have a blackpowder scattergun that can handle this hard shot. See more about steel shot in Chapter 35.

The Rifle

Finally, this brings us to the blackpowder firearm with the greatest diversity. The rifle is known for its various types and evolution, most represented here.

The Original Rifle

The original muzzleloader is *not* defunct, impossible to find or always too expensive to buy. These concerns are simply untrue. Originals are available; some are entirely safe to shoot; and they don't always cost a fortune to purchase. I didn't truly understand these things myself until a few years ago when I made a visit to the East. In one private collection alone I saw more than fifty shootable originals. The friend I was with purchased one for $300. In fact, we fired the

This flintlock shooter did fine at a rendezvous shoot, proving that the old sparktosser is not dead yet. While caplocks rule, flint 'n' steel guns remain widely available, too.

rifle the next day. It was a caplock of unknown maker, plain as a city sidewalk, and in perfect repair. Having said all of this, let me now backtrack. Although there are originals around that still work—recall that I mentioned being introduced to one belonging to University of Alaska Professor Charles Keim—most of these guns should not be fired, for two reasons. First, unless carefully checked by a gunsmith and given a clean bill of health, old muzzleloaders should be left on the rack or in the closet. They may not be safe to shoot. And you don't know for sure until it's been checked out. Second, some originals are simply too valuable to fire. They should remain collector's items only. But it's nice to know that if a modern gun lover has a burning desire to own, and to shoot, an original frontloader, the possibility does exist.

The Replica

Considering the last two points made above—that originals may be unsafe to shoot or too valuable—the idea of a replica is about as smart as ideas get. Replicas did a great deal to get the blackpowder ball truly rolling again. Mainly, it was the replica handgun that got things started. A modern shooter could buy an old-time gun that looked like, and shot like, "the real thing," without risk, and at a modest price. Not all replicas were, or are, well-made or worthy of purchase, but there are many

excellent models. Those who want a very close copy of an original-style gun must know what to look for. Just because a firearm bears an original name does not mean it copies an original. But many do authentically replicate. For example, Colts from that company's blackpowder division certainly match the real thing, and the Navy Arms Ithaca Hawken was patterned after an original half-stock plains rifle. In short, there are simply dozens of replica blackpowder guns around, and they make infinite good sense for the shooter who wants to emulate the past, not only in collecting, but also in shooting.

The Non-Replica Rifle

Still the most popular blackpowder rifle of the hour, the non-replica leads the pack because it is practical. Thousands of shooters really don't want to search out, or pay for, an original that they may or may not be able to shoot. They want to make smoke. They want to put holes in paper. They want to take game home. The replica will certainly do all of these things, but the practical blackpowder shooter lives in the present and simply does not care about history or copying anything out of the past—except for the shooting part. The non-replica is ideal for these people. It copies nothing, but it loads and shoots, in every respect, the way Jim Bridger's rifle loaded and shot. It is also legal for small game and big game, and it is allowed in most shooting games, even at rendezvous. True, the dedicated buckskinner wants as close to the real thing as he can get, but there remain plenty of non-replicas at these gatherings all the same. Non-replicas come in all major calibers, from 32 to 58, and they are generally well-made, good-handling and totally effective.

The Rifled Musket

To begin with, the name is a bit odd. Musket generally referred to a military shoulder arm with a smooth bore. But when that same firearm was given rifling, it became a rifled musket—still looking like its former self, in the main, but no longer a smoothbore. I'm not entirely sure why the rifled musket is not more popular today, except for its military-like appearance, which may tend to keep some shooters away. In fact—and I realize the risk of making so bold a statement—the rifled musket could well be the best choice for the hunter who wants a rugged, powerful, accurate and dependable rifle. Of course, I'm thinking more of the rifled musket that was designed for target work than the pure military model.

Rugged: The rifled musket was originally intended for war. It was designed to take a beating. Powerful: My tests show the Volunteer with an allowed 130 grains *volume* of Pyrodex behind a 490 grain bullet doing 1514 feet per second for almost 2500 foot pounds of muzzle energy. At 100 yards, the bullet still carries over 1800 foot pounds of punch. (See *The Gun Digest Black Powder Loading Manual*, from DBI Books, for more information on loading the rifled musket.) Accurate: The common rifled musket is sufficiently accurate for big game hunting, while the target-type rifled muskets are more accurate yet. Dependable: I cannot recall a misfire with a properly loaded rifled musket that was kept clean. The rifled musket is also dependable because it's tough. A little whack in the field won't break it.

Two come to mind, both sold by Navy Arms Company. One is the Whitworth, a replica of the famous target rifle that Joseph Whitworth built in the 19th century. The Whitworth cleaned up at target matches, but it also produced good hunting ballistics, and still does. The other is the Volunteer, which in a way is even more favorable in the field than the Whitworth, because Navy Arms allows stouter loads in it.

The Modern Muzzleloader

Actually, there's nothing new about the modern muzzleloader concept. For decades, clever gunmakers have created many non-original, modern-like frontloaders. However, the modern muzzleloader concept came into wide acceptability only a comparatively short time ago. What is a modern muzzleloader? It is impossible to give a perfect list of criteria, because not all models "play the game" the same way. But here are some criteria that apply most of the time to most guns: in-line ignition; removable breech plug; modern trigger; modern configuration; adjustable sights; scope-ready; modern safety; fast rifling twist.

In-line ignition and direct ignition are two different things. Under-hammers and sidehammers (mule ear locks) of the past used direct ignition. Fire went directly from the percussion cap into the breech. In fact, in both cases, the nipple was screwed directly into the breech. In-line ignition is different in that fire from the percussion cap goes directly into the breech, but *from the back*. The percussion cap is, as the name implies, in a direct line with the breech. The cap rests, as it were, behind the powder charge, just as the primer in a cartridge lies behind the powder charge.

Most modern muzzleloaders can be cleaned from the breech end, just like a modern rifle. This is accomplished with a removable breech plug. The plug screws out, which opens the passageway into the bore. In this manner, a shooter can come home with his fired rifle, unscrew the breech plug, and clean the gun from the "back end," just as he would take the bolt from his Remington Model 700 rifle and clean it from the breech, rather than the muzzle end.

Triggers on the modern muzzleloader are generally of the same design as triggers found on today's bolt-action rifle. In fact, sometimes these triggers are the very same ones used in modern rifles. They can be adjustable, and there is absolutely nothing unfamiliar about them in function.

The modern muzzleloader does not look like a muzzleloader. Its configuration or style resembles a modern rifle. However, there is more to this than meets the eye. It not only looks like a contemporary rifle; it *handles* like one. It is not muzzle-heavy, for example, nor does it weigh a great deal, nor is it long-barreled. The modern muzzleloader picks up, carries and comes to the shoulder just like rifles that are already familiar to today's shooters. This factor plays a big role in selection. For those who simply want to shoot blackpowder—emphasis on shoot—with no regard for history, the modern muzzleloader can be the ticket. It even makes the non-replica frontloader look old-timey.

Modern muzzleloaders have sights that we would expect to find on any contemporary rifle. That's because they are often the very same sights in the first place. Adjustable open iron sights are common. Peep sights are less prevalent, but available, and are also adjustable.

Every modern muzzleloader I can think of comes scope-ready. That is, the receiver is drilled and tapped for a scope mount. It's a rather simple job to attach a scope. As noted, scopes are not always legal on primitive hunts; however, they are legal for small game hunting everywhere, as far as I know, and scopes are also OK on *some* blackpowder-only hunts. Scopes are also ideal for target work or for testing a rifle's potential accuracy. However, as far as I know, they are not acceptable at rendezvous shoots. But modern muzzleloaders in general don't make the rendezvous scene anyway.

Sliding safeties are found on the majority of modern muzzleloaders, as opposed to the hammer system with half-cock notch. As noted earlier, however, not all modern muzzleloaders have all of the features listed here. Some guns have a double safety system. Along with the sliding safety, the plunger acts as a safety, too. The gun won't fire unless the plunger is "cocked."

The vast majority of modern muzzleloaders have a rapid rate of twist, comparatively, whereas a round ball rifle carries a very slow twist. For example, a 50-caliber modern muzzleloader may have a rate of twist as "tight" as 1:20, one turn in 20 inches of projectile travel, while a 50-caliber ball-shooter, such as the Navy Arms Ithaca Hawken replica, will have a 1:66 rate of twist, a single revolution of the projec-

Sure, you'd like an original, but you don't want to spend that much, plus your real interest is shooting. A replica is the answer. This Dixie Gun Works Sharps New Model 1859 Military Rifle is a faithful copy of the real thing in 54-caliber.

Replicas still abound, even though the non-replica and modern muzzleloader generally rule these days. This Navy Arms Brown Bess 75-caliber musket is a copy true to the original and to history.

If your cup of tea tends toward utility, the non-replica may suit you just right. The T/C Pennsylvania Hunter Carbine is a handy frontloader that copies nothing out of the past.

While big game hunting is the number one reason for buying a muzzleloader today, it is not the only reason. More popular than ever is shooting muzzleloaders (and blackpowder cartridge guns) for the relaxation and pure enjoyment of the sport. This Navy Arms Mortimer Match Rifle in 54-caliber is an example of a frontloader that you can hunt with if you wish, but it also serves well on the target range.

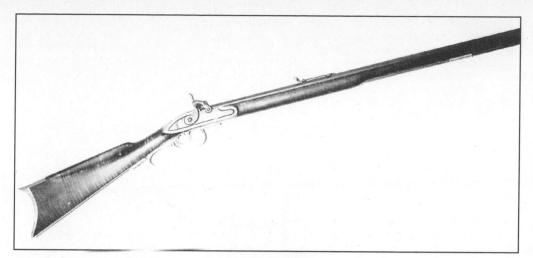

Handsome, straight-shooting, and very much like a long rifle of old, this Mountain State Mountaineer 50-caliber longrifle is not an exact copy of an old-timer, but it comes close enough for most of us.

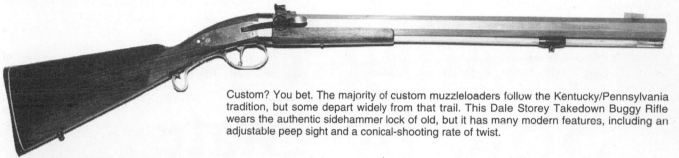

Custom? You bet. The majority of custom muzzleloaders follow the Kentucky/Pennsylvania tradition, but some depart widely from that trail. This Dale Storey Takedown Buggy Rifle wears the authentic sidehammer lock of old, but it has many modern features, including an adjustable peep sight and a conical-shooting rate of twist.

This beautiful custom-made pistol, the artwork of Kennedy Firearms, can be fired, of course, but its true value lies in historical beauty and interesting history.

tile in 66 inches of travel. For more on twist, see Chapter 27. The faster rate of twist is owed to the fact that modern muzzleloaders are meant to shoot conicals, including the pistol bullet, jacketed or lead, with a sabot (sah-bow). Chapter 17 has more on blackpowder conicals.

The Custom Muzzleloader

Then there's the custom gun. If you haven't seen a good handmade frontloader, you've missed something. There are modern gunmakers who create longarms (and pistols) just as handsome as their forebears made. There is only one thing wrong with a custom longarm. Bucks. Nobody can expect a man to handmake a rifle for pennies. Not in today's economy. A thousand dollars gets you a fairly plain custom longrifle. The price goes up from there, depending on quality and quantity of extras. If you're not heavily into muzzleloading, the custom's not for you. But if blackpowder shooting has become a way of life, then you deserve a custom rifle.

The Cartridge Rifle

Not a muzzleloader at all, but still a blackpowder rifle, this soot-burner comes in two general styles, lever action and single shot. Original leverguns, such as the Model 1886 Winchester, make some hunters happier than school kids on Saturday. But today it's mainly the single shot design that is used in blackpowder shooting with original fuel. The two most popular models are the same ones that were most popular in buffalo hunting days, the Remington rolling block sold through Navy Arms Company and the Sharps sold through C. Sharps in Big Timber, Montana. The former is in caliber 45-70 Government. The latter is offered in many of the old-time blackpowder cartridge chamberings, including the famous 45-120 Sharps. Shooters who want to stay with a cartridge, but still desire to be a part of the old days and old ways, owe it to themselves to look at the blackpowder cartridge rifle. The single shot cartridge rifle is gaining in popularity, mainly because of the interesting silhouette shoots designed for these guns. Hunters are also finding that the blackpowder cartridge, loaded with blackpowder or Pyrodex, makes for an interesting hunting tool.

That's a look at what is available gun-wise in the blackpowder shooting game. There is no doubt in my mind that the modern muzzleloader will continue gaining in popularity; however, I see no reason for the other branches of the blackpowder tree to fall away. ●

5

Blackpowder Hunting Rules State by State

THE AMERICAN HUNTER is more mobile than ever, flying from state to state to enjoy different hunts in different places. These can be extensive and expensive jaunts, or downright cheap vacations. Some hunters take their families along, or at least their spouses. Everyone may hunt, or, while the hunting member enjoys his outing, the remainder of the group sees the sights in the area. Also, hunting partners may go in groups of two, four, even more, driving for economy, where gasoline, motels and other expenses are shared. Admittedly, this chapter is a teaser, designed to inform the reader about state-to-state blackpowder rules, but also to let him know that many blackpowder hunting opportunities do actually exist. Of course, no single roundup can be absolutely up to date. Rules and regs change. But the following gives the reader a darn good idea of what he currently will find in his, and other states, pertaining to blackpowder gun laws for the sportsman.

Not all of the regulations, by the way, make sense. When blackpowder shooting returned full-force to the modern scene, new laws had to be formulated. Unfortunately, they were not always based on ballistic fact, and so we are left with some real dandies. For example, caliber 40 was chosen as a minimum for deer-sized game in many states. A good hunter/marksman shooting a 40-caliber rifle with a 93-grain lead ball can certainly take deer cleanly. But missile placement better be right. A 40-caliber round ball is pretty small potatoes: Even at a starting velocity of 2000 feet per second, muzzle energy is only 826 foot pounds. There are blackpowder pistols that beat that. What's worse, energy is down to around 200 foot pounds at 100 yards, which means hunters with 40s must refrain from taking anything remotely like a long shot. Of course, the same size conical is quite a different story. The 40-caliber White modern muzzleloader shoots a 410-grain bullet at over 1300 fps for a muzzle energy of around 1600 foot pounds, with over 1200 foot pounds left at 100 yards.

State-by-State Blackpowder Restrictions

Ballistic rules were not the only ones made by game departments. As frontloaders changed, departments had to consider new regulations to manage the new guns. For example, should scopes be allowed on primitive hunts? Where and when should they be used? Where and when should they not be? What about modern muzzleloaders? Should they be allowed? Some game management officials, as well as hunters, fear that the high-tech muzzleloader may threaten special blackpowder-only seasons. If these guns get good enough, they reason, why should there be special seasons for them? Of course, officials don't always study the facts. Are modern muzzleloaders really more effective than non-replicas, replicas or other blackpowder guns? Are they more accurate? Do they shoot farther? Are they truly more powerful than older-style charcoal burners?

In a sense, this chapter compares state game agencies' attitudes toward blackpowder arms, with a look at departmental positions on muzzleloader ballistics. The blackpowder shotgun is not addressed because game departments across the nation impose few restrictions on the frontloading scattergun. But keep in mind that steel shot laws do apply to all shotguns. Here is a look at regulations pertaining to blackpowder hunting arms, but remember the warning sounded above: rules change. Be sure to check local laws before using a muzzleloader in any state or, for that matter, any special hunt within any state. The following does not pertain to game laws, only the use of blackpowder firearms. Here are some things to consider: Are blackpowder arms allowed for small and large game? Are there caliber restrictions or powder charge (weight) restrictions? Are sabots and modern pistol bullets allowed? Any terminal energy restrictions? Are scope sights allowed during special muzzleloader only *big game seasons*? Or during regular big game seasons? Are peep sights admissible for big game hunting with a muzzleloader? May modern muzzleloaders be used for big game? Are there any special restrictions on muzzleloaders, such as barrel length, flintlock use, number of barrels allowed?

Alabama

The Cotton State allows muzzloaders for both small and big game with a 40-caliber restriction for deer and no powder weight rule. Smoothbores are legal, as are rifled barrels. Sabots with modern jacketed pistol bullets are OK. There is no foot pounds of remaining energy rule in Alabama. In-line ignition modern muzzleloaders are in, but no scopes on blackpowder-only hunts. Peep sights are allowed. Black-

Is this rifle legal on a blackpowder-only "primitive" hunt? Most likely not, because it's a blackpowder cartridge rifle, and special hunts are set up for muzzleloaders.

(Below) Hunting bears with a muzzleloader over a bait may be totally acceptable in some areas and absolutely illegal in other areas. Every hunter is obligated to check local laws before hunting.

powder pistols over 40-caliber are also allowed for deer. Alabama has what it calls "Stalk Hunting," a primitive weapon hunt where *only* muzzle-loading rifles, muzzle-loading shotguns, or bowhunting are allowed.

Alaska

The Last Frontier allows blackpowder guns for small and big game, with restrictions for the latter. While there is no powder weight limitation, there is a caliber/bullet weight minimum for brown/grizzly bear, black bear, moose, bison, elk, muskox and mountain goat. For these species, the frontloader must be 54-caliber or larger, or at least 45-caliber shooting a 250-grain bullet or heavier. Smoothbores are OK. Modern bullets/sabots are legal. Modern muzzleloaders are all right, as are peep sights. There are no specific firearm restrictions, such as barrel length requirements. No scopes allowed on special blackpowder-only hunts. Blackpowder handguns are legal, but subject to minimum caliber rules.

Arizona

Muzzleloaders are legal for small and big game in the Grand Canyon State with no caliber restrictions for either, nor any powder weight regulation. Sabots and modern bullets are legal. There is no terminal energy law, and scopes are OK for both special blackpowder-only seasons and regular hunting seasons. In-line ignition modern muzzleloaders are also fine, as are peep sights. In Arizona, the muzzleloader must have a single barrel. Multiple-barrel frontloaders are taboo. The game department's policy is to manage game more than firearms, and the modern muzzleloader is fully acceptable in Arizona at this time. The state does have special blackpowder-only hunts, along with its HAM hunt, which allows handguns, archery tackle, or muzzleloaders for javelina. No modern arms are allowed on this hunt.

Arkansas

The Wonder State allows muzzleloaders for small and big game, with some restrictions. For small game, the muzzleloader may be 45-caliber *or smaller*, but not larger. For big game, the blackpowder rifle has to be 45-caliber *or larger*. There is no powder weight restriction.

Sometimes rules pertain directly to the design of a muzzleloader. For example, a regulation may call for an exposed nipple, one that can be reached by the elements, such as rain or snow, while a closed system is illegal.

Modern bullets/sabots are legal. No terminal energy rule applies in Arkansas. Scopes are allowed, but there is a catch. Only 1:1 scopes are legal. Scopes that magnify the target are against the rules for special or general season hunting with muzzleloaders. Modern muzzleloaders are fine, as are peep sights. Multiple-barrel blackpowder rifles are acceptable, but each barrel must contain no more than one projectile. Also, barrels must be 18 inches or longer. Any caliber muzzle-loading pistol or percussion revolver is allowed in the field when the hunter also has with him a legal longarm as his primary hunting tool. When this information was collected, Arkansas had no departmental position on modern muzzleloaders.

California

The Golden State allows small and big game blackpowder hunting with 40-caliber or larger demanded for big game, but no caliber limitation for small game. There is no energy rating requirement or powder weight minimum. The issue of sabots and jacketed bullets was not specifically addressed in the statutes when this information was gathered. The rule reads "single ball or bullet loaded from the muzzle," which qualifies the jacketed bullet/sabot. Only iron sights may be used on muzzleloaders, and peep sights are included. No scopes. This rule applies to small game as well as big game. Barrels must be at least 16 inches long. An older rule that muzzleloader barrels must be at least 26 inches long was dropped.

Colorado

The Centennial State allows small game hunting with a muzzleloader with no caliber restrictions. However, the rifle must be 40-caliber or larger for deer, antelope and other big game, while moose and elk hunting demand a 50-caliber rifle or larger. There is no powder charge rule or remaining energy limit. Sabots are not allowed during the muzzleloader-only season, nor are scope sights, but peep sights are acceptable. Colorado has been considering a ruling that could limit modern-style muzzleloaders. The legal muzzleloader must be a firearm fired from the shoulder (not a handgun), with a single barrel shooting a single patched ball or a single conical. The conical must not exceed twice the length of its caliber. In other words, a 50-caliber projectile must not be over an inch long.

Connecticut

The Constitution State allows small and big game hunting with a smokepole. There is a 45-caliber minimum caliber restriction for big game, but not over 12-gauge. No powder weight or energy rule. Sabots and modern jacketed bullets, however, are not legal for big game hunting, although both are all right for small game. Scope sights are not allowed on frontloaders for regular or special seasons. Modern muzzleloaders and peep sights are legal.

Delaware

The Diamond State allows smokepole small and big game hunting, with restrictions. Caliber 42 is *minimum* for big game; caliber 36 is *maximum* for small game. A legal powder charge must weigh at least 62 grains (for big game). Bullets must be all lead. Jacketed bullets are not legal in Delaware. No minimum energy rating required. Peep sights are fine, but scopes are not currently allowed. Modern muzzleloaders are accceptable. Other restrictions include barrels that must be 20 inches or longer, no multiple barrels—single shots only—one projectile only. The bore must be rifled. Caliber 42 or larger in a pistol is OK for finishing shots, but not to hunt with. Muzzleloaders in general are *not* legal during the regular big game season. Legislation on this matter is pending.

Florida

The Sunshine State allows blackpowder guns for both big and small game with no caliber limitation on small game. The frontloader must be 40-caliber or 20-gauge for hunting deer or bear. Jacketed bullets and sabots are OK. No terminal energy restriction. Scopes and peep sights are legal for special or general season hunting, as are in-line modern muzzleloaders. Blackpowder pistols are legal for hunting. Be sure to check current regs.

Georgia

The Peach State allows blackpowder guns for small and large game with a 44-caliber restriction for big game frontloaders, none for small game arms. No powder weight restriction. Sabots/jacketed bullets are legal. No energy rule. Scopes are not permitted on muzzleloaders used in wildlife management area hunts, but scopes are allowed during general season. Peep sights accepted. Georgia's interest is to keep the rules simple, understandable, and in accord with the best interests of wildlife management. Modern muzzleloaders are welcomed. Muzzleloaders may be used during the regular firearms season and on special primitive weapons hunts on Wildlife Management Areas (WMAs) in Georgia. Jacketed bullets/sabots are OK, but no blackpowder sidearms for hunting.

Hawaii

The Aloha State allows muzzleloaders for big game hunting, but not for small game hunting. A 45-caliber limitation applies with no powder weight restriction. At present, there is no foot pounds of energy ruling. No scopes, but peep sights are OK on frontloaders. In-line ignition modern muzzleloaders are allowed, but no jacketed bullets. Sabots are legal with lead bullets. The breech must be open.

Idaho

The Gem State has been involved in building a set of rules for hunting with a muzzleloader. Currently, small game and big game can be hunted with a smokepole—no caliber restriction on small game, but a 45-caliber limit for deer, antelope and mountain lion, and a 50-caliber limit for elk, moose, big horn sheep, mountain goat and black bear. There is no foot pound rating requirement nor powder charge minimum. Sabots with jacketed bullets are legal. Peep sights are OK. Scopes are allowed on frontloaders during the regular season, but not on special primitive hunts. Special restrictions include no breechloaders, open or peep sights only, blackpowder or Pyrodex only, no more than two barrels. The gun must be equipped with a flintlock or a percussion lock that is "directly exposed to the weather." The last rule exempts some modern muzzleloaders whose percussion caps are not *directly exposed to the weather.* Idaho offers its *Idaho Hunter's Guide to Muzzleloader Performance and Limitations* for prospective hunters to study.

Illinois

The Prairie State allows big game hunting with blackpowder arms, but not small game hunting. No powder weight restrictions, but sabot/jacketed bullets are not legal. The muzzleloader must be at least 45-caliber, but foot pounds of energy is not a limitation. Scopes, peep sights and modern muzzleloaders are allowed. The barrel of the muzzleloader must be at least 16 inches long. For purposes of legality, an unloaded muzzleloader is one with powder removed from the pan, cap removed from the nipple, hammers fully forward. Maximum caliber for smoothbore is 10-gauge, minumum is 20-gauge.

Indiana

The Hoosier State has blackpowder big game and small game hunting with no caliber restriction for small game. Big game arms must be 44-caliber or larger. No powder charge restrictions nor terminal energy ratings. Jacketed bullets/sabots are legal, but only if the bullet itself meets minimal caliber restrictions. Scope sights and peep sights are allowed. So are in-line ignition muzzleloaders. No breechloaders allowed for blackpowder seasons. Indiana has no written position on modern muzzleloaderes. They are legal for hunting, as noted. The muzzleloader must fire a single projectile only—no multiple barrel arms. Blackpowder handguns are illegal for hunting.

Iowa

The Hawkeye State approves of frontloaders for big and small game with a 44-caliber minimum and a .775-inch maximum projectile diameter for rifles or smoothbores. Powder charge weight is not restricted, and sabots and jacketed bullets are allowed. No minimum energy figure is imposed. Scope sights and peep sights are legal. Modern muzzleloaders are also legal and "considered appropriate" by the state. However, the muzzleloader must shoot only a single projectile.

Kansas

The Sunflower State approves of both small and big game hunting with the muzzleloader, caliber 49 or larger for elk, 39 or larger for deer and antelope. Small game hunting demands no minimum caliber. Sabots/modern bullets are not acceptable. There is no terminal energy law, but scopes are not legal on frontloaders. Peep sights are. Modern muzzleloaders are legal. The blackpowder firearm must have a single barrel. Kansas has no stated position on modern muzzleloaders at this time.

Kentucky

The Bluegrass State allows blackpowder small and big game hunting with a 40-caliber minimum for the latter. Scopes are OK during regular and blackpowder seasons. Peeps also legal. No minimum powder charge or foot pounds rating. Sabots/jacketed bullets are allowed, as are modern muzzleloaders. Note: A patched bullet that is no more than twice the length of its diameter is legal—longer not allowed. Shotguns must be at least 20-gauge, but not over 10-gauge. There is no limitation on number of barrels allowed.

Louisana

The Creole State has small and big game blackpowder hunting with a minimum 44-caliber rule for the latter. Modern muzzleloaders are allowed, but no matchlocks and no wheelocks. Peep sights are OK, and sabots and jacketed bullets are legal. Scopes must be of 1x with no magnification. No minimum powder charge or foot pound law. Multiple barrels legal. Exposed ignition only; no covered breeches.

Maine

The Pine Tree State has blackpowder hunting for both small and big game. Caliber 40 is a minimum for big game. While there is no specific caliber limitation for small game, during special blackpowder hunts carrying a firearm under 40-caliber is illegal. Therefore, hunting small game under those circumstances with less than 40-caliber is illegal. No powder charge restriction. No terminal energy demands. Scopes are legal during regular season, but not for special blackpowder hunts. Peep sights are always legal. By regulation, modern muzzleloaders are legal. The law reads: "Ignited by a percussion cap or priming charge of a flint, match or wheel lock mechanism." Only single projectiles are allowed. The barrel must be 20 inches or longer. No breechloaders. Smoothbores and rifles are both legal.

Maryland

The Old Line State allows small and big game hunting with a muzzleloader. The big game arm must be at least 40-caliber. Maryland has an energy ruling: The frontloader must produce a muzzle energy of at least 1200 foot pounds for rifle and 700 foot pounds for sidearm. There is also a minumum powder level: not less than 60 grains for a rifle, 40 grains for pistol, blackpowder or equivalent Pyrodex. Sabots with jacketed bullets are allowed. Scopes are allowed only for a person with an opththalmic disfunction—by special permit. Modern muzzleloaders are legal, but all frontloaders must be ignited by a percussion cap or flint—no fusils with modern primers. No peep sights allowed, according to source contacted. Handguns must have a barrel length of at least 6 inches.

Massachusetts

Small and big game hunting is legal with blackpowder guns in the Bay State, but with a number of rules. Legal arms must be at least 44-caliber, but not over .775-inch bullet diameter. Smoothbores cannot be over 10-gauge. In-line ignition rifles and covered breeches are OK during the regular season, but not for special blackpowder hunts. No rifled barrels—smoothbores only—during regular season, but smoothbores and rifled arms are allowed for special seasons. Wheelocks and matchlocks not allowed during blackpowder season. Scopes are allowed for regular season and also for the blackpowder-only hunt when mounted on smoothbores only. Peep sights are OK, either hunt. Projectiles must be made of lead, and sabots are legal. Jacketed bullets are not legal. A minimum barrel length of 18 inches is needed for blackpowder hunts.

We must never forget that general rules also apply to muzzleloaders. These geese cannot be hunted legally with lead shot, even if that lead shot comes from a muzzle-loading shotgun.

Michigan

The Wolverine State allows small game and big game hunting with a muzzleloader—no caliber restriction for small game, 45-caliber projectile or larger for big game. No powder weight or foot pounds of energy law. Sabots/jacketed bullets legal. Scope sights are allowed during regular and blackpowder-only seasons, as are peep sights. Modern muzzleloaders are legal. No closed breech.

Minnesota

The Gopher State allows small and big game hunting with a muzzleloader, with caliber restrictions for the latter. The big game frontloader must be a smoothbore of at least 45-caliber or a rifled arm of at least 40-caliber. No minimum caliber for small game. No powder charge limitation. No energy rating limit. Sabots and jacketed bullets OK. Peep sights are legal, but scopes are not during the special blackpowder season. Scopes are acceptable otherwise. Muzzleloading pistols are legal only in the "all firearms zone." The caplock revolver is illegal on the special muzzleloader hunt in Minnesota. Minnesota accepts the modern muzzleloader with no immediate plans for altering that posture.

Mississippi

The Bayou State allows blackpowder hunting with a minimum 38-caliber smoothbore or rifled firearm. In-line ignition modern muzzleloaders are allowed. An enclosed breech is OK, as are peep sights. Scope sights are not. No matchlocks or wheellocks. As the rules currently read, it seems that Pyrodex is not allowed. Only blackpowder. No minimum powder charge or foot pound rating. Sabots and jacketed bullets are legal. No muzzleloading handguns.

Missouri

The Show Me State has small game and big game blackpowder hunting. The small game firearm is not restricted by caliber, but a big game frontloader must be at least 40-caliber. There is no powder charge limitation. Sabots and jacketed bullets are legal provided the bullet is over 40-caliber diameter. The smoothbore minimumm is 20-gauge, maximum 10-gauge. No energy rating. Scopes are not legal during special blackpowder-only seasons, but they are legal otherwise. In-line ignition and other moderen muzzleloaders are legal, as are peep sights. A special rule in Missouri states that while a person is "pursuing deer on a resident muzzle-loading firearms deer hunting permit" he may have and use more than one muzzleloading firearm, but shall have no other firearm, longbow or crossbow in his or her possession.

Montana

The Treasure State limits small game hunting with a muzzleloader to certain species only, so be sure to research the most recent, but caliber is not specified. No closed breech. No powder charge restriction. Sabots with jacketed bullets are OK except in specific Wildlife Management areas where only the round ball is allowed. No energy requirement. Scopes are legal for frontloaders, as are peep sights. In-line and other modern muzzleloaders are legal during the regular hunting season. In certain hunting areas restricted to archery equipment, shotguns and muzzleloaders are also allowed (all three simultaneously), the latter being restricted to a "round ball .45 caliber or greater."

Nebraska

The Cornhusker State denies scope sights for blackpowder arms used during a special muzzleloader-only hunt. Peep sights are legal. Modern muzzleloaders are also legal. Small game and big game may be hunted in Nebraska with no caliber limitiation for the former. Big game rifles (pistols not allowed) must be caliber 40 or larger. There is no powder charge rule, nor is there an energy level regulation.

Nevada

The Silver State has blackpowder hunting for small game and big

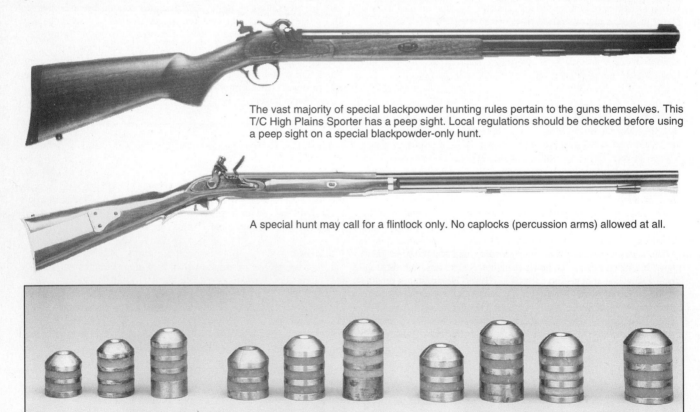

The vast majority of special blackpowder hunting rules pertain to the guns themselves. This T/C High Plains Sporter has a peep sight. Local regulations should be checked before using a peep sight on a special blackpowder-only hunt.

A special hunt may call for a flintlock only. No caplocks (percussion arms) allowed at all.

Are all of these fine Thompson/Center bullets legal for special blackpowder-only hunts? Not where the rules call for a bullet no longer than twice its caliber, as some of these are.

Game departments consider lethality when they come up with blackpowder ballistic rules. Unfortunately, many of the rules have no basis in fact. Is a 40-caliber round ball enough for mule deer like these? Sure, for the expert who gets very close and places his shot precisely. Otherwise, no. And yet, 40-caliber round balls are legal in most states for big game.

game with no caliber restriction on the former, but the big game rifle must be 44-caliber or larger. There is no powder weight or foot pounds of energy rule in Nevada. Sabots with jacketed bullets are not legal. Scope sights are not allowed during special or regular seasons. Peep sights (and of course open irons) are fine. A single-barrel rifle is the rule for big game, flint or percussion. Modern muzzleloaders are acceptable, since they function as standard muzzleloaders. A closed breech is not allowed.

New Hampshire

The Granite State offers the hunter both small and big game blackpowder hunting, with no caliber restriction on the small game rifle, but a minimum 40-caliber for deer and all other big game—except moose, in which case the muzzleloader must be 50-caliber or larger. There is no foot pounds of energy rule. There is no powder charge rule. Scopes are allowed, as are in-line ignition modern muzzleloaders. Peep sights are also considered all right. However, it is unlawful to use a muzzleloader that was not manufactured to be a single shot.

New Jersey

The Garden State permits blackpowder hunting for large and small game with a 44-caliber minimum restriction on the big game firearm and a 36-caliber or smaller rule for the small game rifle. No powder weight rule. No minimum foot pounds rating. Sabots/jacketed bullets are allowed. No scopes, but certain handicapped persons may use one of 1.5x maximum magnification (be sure to check current restrictions). Modern muzzleloaders and peep sights are acceptable. Rifles must be a single barrel with single missile. The blackpowder shotgun carries the same restrictions as the modern shotgun in New Jersey.

New Mexico

The Land of Enchantment is blessed with wildlife officials who are always studying their own rules with regard to making them more applicable to the wildlife resource. Small game and big game are legally huntable with blackpowder. There is no caliber restriction on small game; big game arms must be rifles caliber 45 or larger, or shotguns firing a single projectile—*no buckshot*. There is no regulation on how much powder must be used in a load. There is no energy rule. Scope sights on sootburners are out, but peep sights are OK. Modern muzzleloaders are legal as are sabots with jacketed bullets. Rules regarding the muzzleloader in New Mexico are generally relaxed, with the state depending upon the good sense of the hunter to regulate himself.

New York

The Empire State allows small and big game hunting with blackpowder guns, no caliber restriction on the former, but the big game rifle must be 44-caliber or larger. No smoothbore rifles allowed. No powder weight restriction. Only a patched round ball is legal for use during the muzzle-loading season—no conicals, no jacketed bullets with sabots. No minimal energy rule and no scopes during special blackpowder-only seasons, but scopes are OK for the general season. Peep sights OK. Double-barrel frontloaders are not legal during special muzzleloader season, nor blackpowder revolvers. Blackpowder handguns must be registered in New York for use. Modern muzzleloaders are legal, but may not be used in certain special blackpowder-only seasons. Check latest rules.

North Carolina

The Tarheel State regulates frontloaders very little, relying on hunters to choose those firearms best suited for the task. Both small and large game are legal fare with no caliber restrictions for either as long as 10-gauge is not exceeded. There is no minimum load or terminal energy requirement. Sabots with jacketed bullets are legal. So are modern muzzleloaders, scope sights and peep sights.

North Dakota

The Flickertail State allows small and big game hunting with the smokepole. The big game rifle must be 45-caliber or larger; no restrictions on the small game rifle. No powder charge rule. No jacketed bullets, but sabots are legal with lead bullets. Modern muzzleloaders are OK. No energy demands. Scopes are not allowed during special blackpowder season, but are legal for general season. Peep sights yes. Modern muzzleloaders are acceptable for the regular season, but not for special blackpowder-only hunts at the time this information was gathered.

Ohio

The Buckeye State permits small and big game hunting with the muzzleloader—38-caliber minimum for big game, no prescription for small game. No powder weight rule. No energy rule. Sabots with jacketed bullets are allowed. So are peep sights and scopes, as well as modern muzzleloaders. Rifled bores and smoothbores are both legal, but must shoot a single projectile only. Smoothbores must be 20, 16, 12 or 10 gauges only. Multiple-barrel arms OK.

Oklahoma

The Sooner State permits small and big game hunting with no caliber restriction on small game. The big game rifle must meet a minimum caliber rule of 40 or larger with a projectile of at least 38-caliber; 20-gauge minimum for smoothbore, 10-gauge maximum. No powder charge rule nor foot pounds rating. Sabots and jacketed bullets OK. Peep sights are allowed, as well as scopes. Modern muzzleloaders permitted. No blackpowder handguns, but sabot and jacketed bullets OK.

Oregon

The Beaver State sanctions small and big game hunting with the frontloader; no restriction on caliber for small game, but the big game rifle must be 40-caliber minimum for deer, bear, antelope and cougar, with a 50-caliber minimum for elk and bighorn sheep. No powder weight rule. No energy law. Sabots/jacketed bullets are OK, but modern muzzleloaders with a covered ignition are not allowed. Scopes not legal; peeps OK. Oregon's law states that the ignition system must be exposed to the weather, demanding an open ignition system only. Iron sights only. Peep sights are in. Scopes are out. While some modern muzzleloaders (with open ignition systems) are allowed, the rules of the game are up for review in Oregon and there may be some changes. Currently, the "traditional experience is balanced with modern technology" and that posture may well remain in effect.

Pennsylvania

Small and big game hunting are allowed in the Keystone State, but there are regulations that apply to both regular and special blackpowder-only seasons. In-line ignition frontloaders are OK during regular season, but not allowed on blackpowder-only hunts. Only flintlocks of 44-caliber or larger (may be rifled) are permissible on blackpowder-only hunts. Maximum for smoothbores is 10-gauge. Scopes and peep sights allowed during regular season. Neither is legal on blackpowder-only hunts. Sabots/jacketed bullets OK for regular season, not allowed for blackpowder-only hunt. The firearm must be a "long gun." (See Point-by-Point Notes at the end of the chapter.)

Rhode Island

Little Rhody permits blackpowder hunting for small and large game. The latest information calls for caliber 45 in rifles or smoothbores; however, other caliber restrictions may apply. Be certain to check before hunting. A solid ball in a blackpowder shotgun is prohibited at all times other than during the pursuit of deer. Deer hunting with sabots/jacketed bullets is forbidden in Rhode Island. There is no charge weight limit or energy requirement, and scopes are not allowed during the special blackpowder season and are limited to shotguns only during the regular season. Peep sights are OK. But laser sights on any hunting instrument are unlawful. In-line modern muzzleloaders are legal (but not with sabots). The projectile must be of lead and of approximate bore size, not undersized, which requires a sabot. The modern muzzleloader is accepted. Rhode Island recognizes in-line ignition design as pre-1865.

South Carolina

The Palmetto State has blackpowder hunting for small and big game with caliber restrictions for both: The big game rifle must be at least 36-caliber, while the small game rifle must not be over 40-caliber. The minimum smoothbore, 20-gauge. The state has no powder charge rule nor energy rating. Sabots with jacketed bullets are OK. Peep sights are lawful for big game hunting; scopes are not during special blackpowder hunts, but they are allowed for regular seasons. Modern muzzleloaders are allowed. Muzzle-loading handguns are OK, but are subject to caliber restrictions.

South Dakota

The Sunshine State has both small and big game blackpowder hunting with no caliber limitation on the small game rifle. The big game muzzleloader must be at least 44-caliber, and it must fire a bullet that is at least 44-caliber. No powder charge weight nor energy restrictions. The sabot/jacketed bullet is all right, but the bullet must be at least 44-caliber. Peep sights are acceptable at all times. Scopes are limited by the fact that "only rifles with open iron sights may be used" on the December muzzleloader deer season. This rule does not apply to peep sights. Scopes are allowed during general season. Check the rules for each specific hunt in the state. Modern muzzleloaders are permitted in South Dakota.

Tennessee

The Volunteer State has made it a point to level as few restrictions on blackpowder hunting as possible and practical. Small game and big game are open to muzzleloaders, with a 40-caliber minimum for the latter. No caliber restriction on small game. No energy ruling. No powder weight law. Sabots are allowed, but not jacketed bullets. Scopes OK, as are modern muzzleloaders and peep sights. Blackpowder revolvers and pistols are also legal as long as the barrel is at least 4 inches long.

Some states have called for energy rulings for muzzleloaders. How many foot pounds remaining energy (at 100 yards) would be right for this bull moose? It's a tough decision to make, and that's why caliber restrictions, sometimes with minimum powder charges, are generally used.

Texas

The Lone Star State calls for a caliber 45 minimum, rifled or smoothbore frontloader, for big game, with no maximum restriction. Both large and small game are open season. Texas has no powder charge limit nor terminal energy ruling. Sabots with jacketed pistol bullets are allowed. Peep sights are fine, as are scopes. Modern muzzleloaders are legal.

Utah

The Beehive State allows small and big game hunting with a muzzleloader. All big game hunts are open to blackpowder rifles, except special archery hunts. The big game rifle must be at least 40-caliber or larger. For elk, moose, bison, sheep and goats, the projectile must weigh at least 210 grains. If a jacketed bullet with sabot is used to hunt these species, that bullet must weigh at least 240 grains. The rifle barrel must be at least 21 inches long and capable of being fired only once without reloading. Iron sights only, which allows the peep, but of course excludes scopes.

Vermont

The Green Mountain State allows small and big game hunting with a frontloader. A blackpowder arm can be used wherever a modern firearm is legal. No small game caliber minimum, but 43-caliber is the smallest allowed for big game. No powder charge nor foot pounds of energy limit. Sabots/jacketed bullets OK. Scopes and peep sights also allowed, as are in-line modern muzzleloaders. The muzzleloader must be a single barrel, single shot, smoothbore or rifled, with a 20-inch or longer barrel for a rifle, 10-inch for pistol.

Virginia

The Old Dominion State has both small and big game blackpowder hunting. A general caliber ruling calls for 45-caliber or larger muzzleloaders. Virginia does have a powder charge limit. At least 50 grains of blackpowder or equivalant Pyrodex constitutes a minimum legal load—no energy restriction. Sabots with jacketed bullets not allowed, but sabots with lead bullets OK. Scopes are permitted during regular season, but scopes on frontloaders for blackpowder-only hunts are not allowed. Peep sights are all right. The rifle must have a single barrel only, flintlock or percussion only (which rules out wheellocks and similar devices). However, modern muzzleloaders are OK.

Washington

The Evergreen State allows small and big game hunting with the muzzleloader with specific restrictions. The blackpowder rifle must be at least 40-caliber for deer, and it must fire a single, lead, non-jacketed projectile, with the exception that buckshot may be used in a muzzleloading smoothbore provided the shot is at least #1 Buck size. The smoothbore used for deer must be at least 60-caliber. All other big game in the state must be taken with a frontloader of at least 50-caliber shooting a single, non-jacketed bullet of at least 170 grains weight. There is more. The rifle must have a barrel 20 inches or longer. It may be a double-barrel rifle, but during special blackpowder-only primitive hunts, only one barrel may be loaded; the other must be left unloaded. There are specific hunts with specific gun rules that must be checked before hunting. The fusil with modern primers is not allowed. Ignition must be flint or percussion, with wheelocks allowed. But the ignition system must be exposed to the elements and not covered. Scopes are not allowed in the blackpowder-only season, but are OK in the general season. Only metallic sights, which includes the peep, are legal during blackpowder-only hunts. Modern muzzleloaders are allowed if their ignition systems are exposed to the elements. No powder weight or energy restriction.

West Virginia

The Mountain State has small game and big game hunting with a minimum 38-caliber rating, rifled and smoothbore arms OK. In-line modern muzzleloaders are allowed on both special blackpowder-only and regular hunts, as are covered breeches. Scopes are OK for regular season, not allowed for blackpowder hunts. Peep sights are legal all the time. Sabots and jacketed bullets are legal.

Wisconsin

The Badger State allows small game and big game hunting with the frontloader with no caliber limitation on the small game rifle, but the big game rifle must be 45-caliber or larger if a smoothbore, 40-caliber or larger if rifled. No powder weight law nor energy rule is in effect in Wisconsin. Sabots with jacketed bullets are not legal, but sabots with lead bullets are. Modern muzzleloaders are not legal either during the muzzleloader-only season, but are OK during the regular big game season. Scopes fit the same ruling—not legal for blackpowder-only hunts, but legal otherwise. Peep sights are always acceptable. In-line ignition is now allowed, according to latest data.

Wyoming

The Equality State allows both small and big game blackpowder hunting with relatively relaxed rules. The big game muzzleloader must be at least 40-caliber. No small game minimum. While there is no foot pounds rule, Wyoming law requires a load of at least 50 grains of powder. Scopes and peep sights are OK. Modern muzzleloaders, sabots/jacketed bullets, multiple barrels are all allowed. Smoothbores are not legal for big game.

Canadian Blackpowder Restrictions

Briefly, here are a few rules pertaining to blackpowder gun hunts in the Canadian provinces. Blackpowder handguns are not legal at all.

British Columbia

Currently, there are no restrictions on muzzleloaders in BC. There are no special blackpowder hunts, but a muzzleloader may be used during the regular big game season.

Manitoba

Muzzleloaders are legal during regular hunting season, as well as during special blackpowder deer season for residents only. Minimum rifled barrel for moose, elk and bear, 50-caliber. For deer, 45-caliber. No restrictions on sights. In-line modern muzzleloaders are legal with closed breech design. Round ball, conicals and sabots are legal.

New Brunswick

No restrictions on muzzleloaders, no special seasons. Blackpowder rifles may be used during regular big game seasons.

Northwest Territories

No special muzzleloader seasons, but frontloaders are OK during regular big game seasons. There are no special regulations for blackpowder hunting rifles.

Nova Scotia

There is a special blackpowder-only hunt in a wildlife management unit. Minimum caliber is 43. There are no special rules pertaining to muzzleloaders used during the regular big game season, which is legal.

Ontario

It is legal to use muzzleloaders during the regular big game hunting season. There are no restrictions on sights, but barrels must be 18¼-inches or longer. Sabots and in-line modern muzzleloaders are legal.

Saskatchewan

Muzzleloaders are legal during big game hunting season, and there are special muzzleloader seasons as well. Check current regs for species and dates. Minimum caliber is 23. All sights are legal. Round balls, conicals and sabots with jacketed bullets are legal. Modern muzzleloaders with a closed breech design are legal.

Point-By-Point Notes

It's interesting and useful to see how states treat blackpowder, and this at-a-glance listing is designed to impart general knowledge along

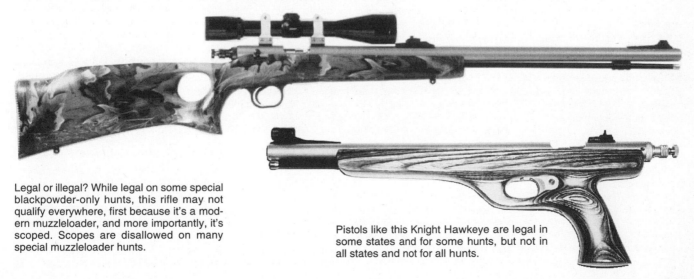

Legal or illegal? While legal on some special blackpowder-only hunts, this rifle may not qualify everywhere, first because it's a modern muzzleloader, and more importantly, it's scoped. Scopes are disallowed on many special muzzleloader hunts.

Pistols like this Knight Hawkeye are legal in some states and for some hunts, but not in all states and not for all hunts.

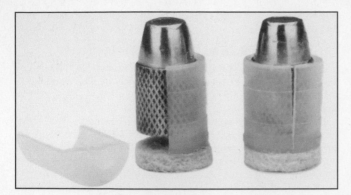

Jacketed pistol bullets in sabots are illegal for certain blackpowder-only hunts, while lead pistol bullets in sabots may be legal. Each state has its own rules, and even within a state specific hunts may have special regulations.

Quite a number of blackpowder rules were set down by people who did not understand muzzleloader ballistics. That's why the 40-caliber round ball at under 100 grains weight is legal for big game in many areas.

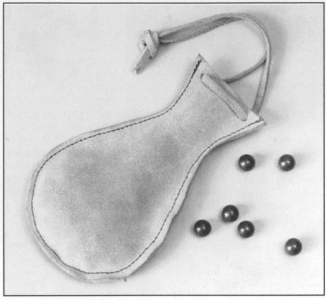

those lines. However, states are always adjusting their rules, so, as always, a study of current regulations for each state you want to hunt is imperative. Someday, we may see across-the-board rulings that make sense and apply, if not for every state, at least for most states. Special thanks to the IBHA (International Blackpowder Hunting Association) and its fine publication, *Blackpowder Hunting*, for the following data.

Special Muzzleloader Seasons

Georgia and Montana are the only two states to say that they have no special blackpowder-only hunting seasons, although Alaska's hunts are limited at this time.

Rifled Barrels

All states responded that rifled barrels were legal for hunting with a muzzleloading firearm.

Smoothbores

Delaware, Nebraska, and New York declare that smoothbores are not legal for muzzleloader hunting. The rest of the states allow them.

Minimum Rifle Calibers

Caliber 40 is the legal minimum in all states except the following: AK (see reference above); CO (50 for elk); CT (45); DE (42); GA (44); HI (45); ID (45 for antelope, deer, mountain lion/50 for elk, moose, sheep, goat, bear); IL (45); IN (44); IA (44); KS (39 for deer, antelope/49 for elk); LA (44); MA (44); MI (45); MN (40 for rifled barrel, but 45 for smoothbore); MS (38); MO (40 for rifled barrel, but 20-gauge for smoothbore); MT (no caliber restriction); NE (40 for rifle, but none for smoothbore); NV (44 for rifled and 20-gauge for smoothbore); NJ (44 and 20-gauge); NM (45 and 28-gauge); NY (45 and 20-gauge, but remember that smoothbores are not legal for big game hunting); NC (no restrictions); ND (45); OH (38 and 20-gauge); OK (20-gauge for smoothbore) OR (20-gauge for smoothbore) PA (44 and none for smoothbore); RI (45); SC (36 and 20-gauge); SD (44); TN (none for smoothbore); TX (45); VT (43); VA (45); WA (60 for smoothbore); WV (38); WI (45 for smoothbore).

Minimum Barrel Lengths

No restriction, or no response to the question, marks most of the states. However, the following states have barrel length rules: AR (18″); DE (20″); ID and IL (16/18″ for rifled and smoothbore); IN (18″); NV and ME (20″); MD and MA (18″); MS (10″); MO and NV (18″); NY, NC and OH (18″); PA (16/18″); UT (21″); VT (20″); VA (18″); WA (20″); WV and WI (18″).

Legal Number of Barrels

Most states limit the blackpowder hunting rifle to only one barrel, or they say nothing about the number of barrels. The follow states allow two barrels: ID, IL, MS, MO and WA.

Lock Types

Flintlocks are legal in all states. Percussion guns are legal in all states, except PA, which calls for an "original muzzleloading single-barrel long gun manufactured prior to 1800 or a similar reproduction of an original muzzleloader." In-line ignition is now legal in all states, according to the IBAH study, with the exception of PA. The following states do *not* allow a closed breech blackpowder firearm for hunting: HI, ID, LA, MI, MT, NV, NJ, OR, PA and WA.

Sights

All states allow adjustable iron sights. All states except PA allow peep sights. Individual states vary widely on the use of scope sights. See information above, state by state.

Projectiles

The patched round ball is legal in all states for hunting. The following states do not allow the conical bullet: CT and NY. Sabots are *not* legal in the following states: CO, CT, MA, NV, NY, PA, SC and RI. Some states are still trying to decide on the legality of the jacketed bullet for big game hunting with a muzzleloader. The following states said "no" to jacketed bullets in muzzleloaders when this information was gathered: CA, DE, HI, MN, NV, NY, ND, PA, TN, VT, WA and WI.

Muzzleloading Handguns

The states are fairly well split on the use of muzzle-loading sidearms for hunting, and there are various rules that apply for those states that do allow the sidearm. The following states permit blackpowder handgun hunting in some form. Check local regulations for details: AL, AK, AZ, AR, CT, DE, FL, GA, KY, MD, MI, MN, MO, MT, NH, NY, ND, OK, SC, SD, TN, TN, TX, VA and WV.

Conclusion

It's extremely obvious that the states do not share the same ideas about muzzleloader rules and regulations. The vast majority of rules seem to have scant ballistic or scientific basis, and appear to have been compiled by persons who had never studied blackpowder shooting. What's needed is a meeting of the minds based on ballistic facts and not whims. Those states which govern least are probably as well off as those with rules that cannot be defended logically. These no-rule states assume that the blackpowder hunter will decide for himself what is best for hunting. ●

chapter 6

Outfit for The Blackpowder Shooter

MUZZLELOADER TYPES WERE addressed in an earlier chapter with rifle, shotgun, and sidearm making up the battery. We need to touch on the guns again, while also looking at some other gear for the blackpowder shooter. The best advice I can give here is to go for quality. Top-notch goods last a long time. Cheap stuff is no bargain. Buy the best, enjoy using it, and have it for a long time.

What did the Old-timers Have?

Not as many accouterments as we do, that's for sure, but we enjoy more different types of guns today than great grandpa did. Plus we are more gadget-oriented than shooters of the past. Of course, the explorers of early America may have carried a lot more if they had the means. After all, gear to the trailblazers meant survival. Let's take a look at the kit of the old-time wilderness hunter for an idea of what yesteryear's blackpowder shooter considered essential. In Charles Hanson's book, *The Plains Rifle*, the author quotes 19th century writer and adventurer Ruskin, who listed mountain man Bill Williams' 1840 outfit as follows:

> In the shoulder-belt, which sustained his powder horn and bullet-pouch, were fastened the various instruments essential to one pursuing his mode of life. An awl [for sewing], with deer-horn handle, and the point defended by a case of cherry-wood carved by his own hand, hung at the back of the belt, side by side with a worm for cleaning the rifle; and under this was a squat and quaint-looking bullet mould, the handles guarded by strips of buckskin to save his fingers from burning when running balls, having for its companion a little bottle made from the point of an antelope's horn scraped transparent, which contained the 'medicine' used in baiting the traps.

The mountain man also had his trusty rifle, of course, probably a half-stock plains model, although that surmise does not carry across the board. Lead would be required for "running balls," which referred to casting projectiles, and blackpowder had to be carried (in a horn generally, as noted by Ruskin) to prime the flinter, and to provide the main charge in the breech. Something to shoot was needed (lead bullets), plus patches, flints, or percussion caps. The flintlock shooter used a vent pick, a small wire tool, to clear the touchhole of his gun so that

sparks from the pan could reach the main charge in the breech. Caplock carriers might use a similar tool to keep the vent of the nipple clear. Some carried a charger, a simple, non-adjustable but effective powder measure made of brass, iron, horn, antler, bone or other material.

Chargers were often supplied by the gunmaker, who, as part of the gunmaking bargain, devised a good load for the smokethrower and also sighted the rifle in for his customer. Supposedly, the charger threw an optimum powder charge that provided power and accuracy. As for the paper cartridge supposedly employed by the mountain men, Hanson suggests that there is precious little evidence of its existence during that period of time. I've not yet seen paper cartridges listed in scholarly texts on the mountain men, so Hanson is probably correct.

Powder was transported from the mouth of the powder horn into the charger. The latter provided the proper volume of fuel. There was also double-charging, which was the practice of dumping two scoops downbore instead of one (*not* recommended) from the charger. Some historians say pre-cut patches were used. Others say they were not. Patch knives found among the gear of the mountain men indicate that at least some shooters placed a hunk of patch material over the muzzle, followed by a ball, which was bumped downward past the crown of the muzzle with the snub portion of the short starter. Then the excess patch material resting above the ball was sliced away with the knife, after which the patched ball was run home on top of the powder charge in the breech. However, Ned Roberts had evidence that at least some old-time shooters used pre-cut patches.

The *worm*, a corkscrew affair, was run downbore with coarse cloth attached called "tow" to swab out blackpowder fouling. It could also be used to withdraw a stuck patch. We don't usually employ a worm and tow as a cleaning device today, but the mountain men were not known for the fastidious rifle maintenance we modern shooters are famous for. If a ball stuck in the bore, a screw was used to retrieve it. Conditions were tough. Before long, the bore of the mountain man's rifle was pitted. When it became too eroded to shoot well, it was *freshed out*, or bored to a larger caliber, with re-cut rifling.

The old-time kit included a hunting bag, also called a shooting pouch, and erroneously referred to today as a "possibles bag." This could be slung over the shoulder via a strap, and was a repository for the vital instruments of shooting, from spare lead balls to cap containers for the percussion rifle, extra flints for the flinter, nipple wrench, and other necessary tools. Vent or nipple picks, small screwdrivers, combination tools, and other devices have also been found in various original shooting pouches. However, to give an exact account of the early hunting kit is impossible. A "straight starter" could be there. This short version of the ramrod was used to start the patched ball downbore a short distance. Today, we call it a short starter.

Along with the shooting bag or pouch there was also a container some historians call a "possibles bag." In theory the name alludes to the chance of finding almost anything in it. The possibles bag was a larger pouch, and in it could be a bit of tobacco, spare gunpowder, flints, patching materials, a fire-steel for starting the well-known flint 'n' steel engendered fire, and according to Hanson's *The Plains Rifle*, this bag might also have a unit labeled a "fire-bag," which was a combination of fire-steel and tinder, sometimes carried on the mountain man's belt. In short, the possibles bag was a catchall.

Along with the regular powder horn, which might hold about a pound of powder, the flintlock carrier may have had a much smaller priming horn. Some historians argue that the old-time shooter didn't have time to fool with a priming horn. He simply poured a little powder from his regular horn into the pan. Maybe so. But priming horns were probably used by some to carry and disperse a fine-grain powder into the pan. A little capbox would be toted by a user of the percussion rifle. There were cappers, too, that stored caps and dispensed them directly onto the nipple of the firearm. There was the in-line capper, which held caps in a row, and the magazine capper, a larger-capacity unit which held many caps in the body section, gravity-feeding one

cap at a time through a spring-loaded opening. Luckily, both styles remain available to modern shooters, including a handsome magazine capper from Tedd Cash.

Rifles were, of course, the most important part of the outfit. Shotguns were popular with settlers, less popular with trappers. Some trappers carried pistols stowed in belt or sash, sometimes via an integral clip built onto the gun. Later, pistols were replaced with revolvers, but not for increased ballistics. The mountain man's pistol was often of large caliber. It didn't burn a lot of powder, but it shot a big ball. The revolver, however, had firepower—five or six fast shots in succession. Incidentally, the fur trade era was effectively over by the time revolvers were in wide use out west. After all, Civil War revolvers were twenty years away when most of the fur hunting was over with.

Although the items are many, most are small because compactness was important. Many mountain men were left afoot for one reason or another, as we learn from Osborne Russell's journal and other documented writings. What a man did not want was bulky gear. He might have to carry most of his lifesaving supplies on his own back. Of course, our western trailblazers allowed their trusty steeds to pack the goods, including the possibles bag, the necessary beaver traps, the plews of fur, and even spare clothing. If you are going to temporarily recreate part of the old-time lifestyle of the mountain man many of the above-mentioned items will be part of your outfit.

Where to Buy Supplies

Five easy-buy places for guns and gear are blackpowder direct mail supply houses, general direct mail catalogs, gun shops, and general department stores. Many of the larger companies like Navy Arms, CVA, Dixie Gun Works, Mountain State Muzzleloading Supplies, and

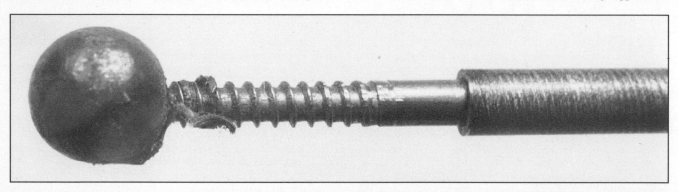

(Above) Simple, yes, but the screw can be a vital tool when a ball is stuck downbore.

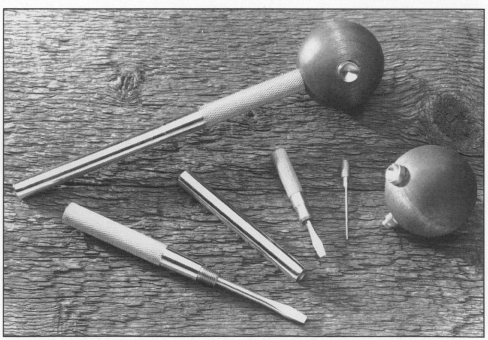

The Ball Buster short starter is much more. Take it apart and there are several screwdrivers, a nipple pick and other tools. Compactness is the byword here.

The shooting bag, erroneously known as a possibles bag these days, contains the small but vital essentials to muzzle-loading. This bag has it all: in-line capper, small powder flask, ball bag marked for 54-caliber, short starter, adjustable powder measure, hornet nesting material (in sandwich bag), extra nipples, worm, screw, cap box, readyloads, even solvent in plastic squeeze bottles, along with shooting patches and cleaning patches, plus a knife, pipe cleaners, and more.

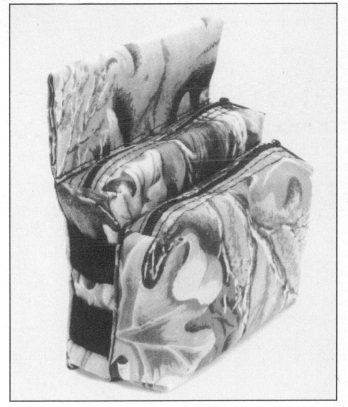

(Below) Totally modern, Thompson/Center's Hunter's Field Pouch is camouflaged in the patented Advantage pattern.

others sell blackpowder guns and gear directly by catalog order. Cabela's, Bass Pro Shops, and Gander Mountain are general mail-order houses, but they have a large array of blackpowder guns and supplies for sale. Meanwhile, I can't think of a gun shop that wouldn't order a muzzleloader for a customer, and general department stores, such as K-Mart and Wal-Mart, also sell blackpowder guns and supplies.

Build a System

The system goes like this: Have a shooting bag for each major blackpowder firearm. When you head for the field to target shoot or hunt, all you have to do is grab the shooting iron in question, the appropriate pouch to match that gun, and you'll have all of the essentials required for that specific firearm. Second, if the exercise spans two days or more, consider a blackpowder shooting box. Mine is a simple wooden box that holds a multitude of shooting supplies, from powder and flints to nipple wrenches, projectiles and patching material, a mini-warehouse of blackpowder "stuff."

The Shooting Bag and its General Contents

No two shooting bags will contain exactly the same items, but here is a list to consider. Blackpowder shooting does demand quite a few accessories, and many of these can be kept in a shooting bag, permanently. Here they are:

bullet bag
charger or powder measure
short starter
hornet nesting material
worm and screw
nipple wrench
combination tool and/or screwdriver
small tin box with pre-cut, pre-lubed patches
small (but ample) powder flask
priming horn
cleaning rag
solvent (in plastic eyewash bottle)
readyloads (2)
pipe cleaners
bristle bore brush

jag
capper
cap box (for percussion caps)
extra flints
cleaning patches
spare nipple
knife (which may attach to strap of shooting bag)

Modern muzzleloaders prompted modern shooting bags, like this CVA Deluxe Hunters Set with readyloads, flask, in-line capper, short starter, etc.

The following important items are not generally kept in the shooting bag, but they are important gear for the blackpowder shooter and should be owned and used:

    wiping stick or loading rod
    shot pouch or shot horn
    cleaning kit
    bullet moulding outfit
    patch cutter
    shooting glasses
    ear plugs
    textbooks and notebooks

    Tap-O-Cap (for making percussion caps)
    holsters, scabbard, gun cases
    clothing (buckskinning or standard)
    knives and tomahawks

General Shooting Supplies

Some of these items are repeated because they are supplies directly involved in shooting, such as blackpowder, as well as items that should be kept on hand to replace worn or depleted necessaries.

    flints
    powder

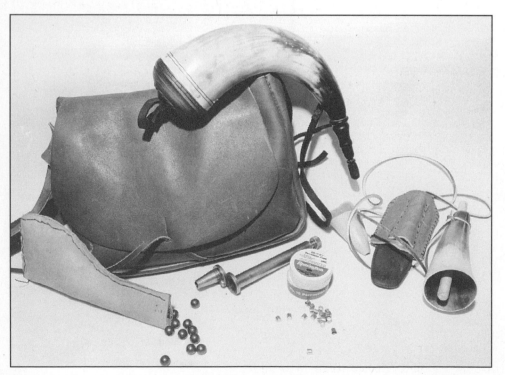

There are numerous shooting bag options. Add your powder horn if you like, a ball bag if desired, caps in original containers, as shown here, or in a metal cap box—even a turkey call in this case, on the far right.

42

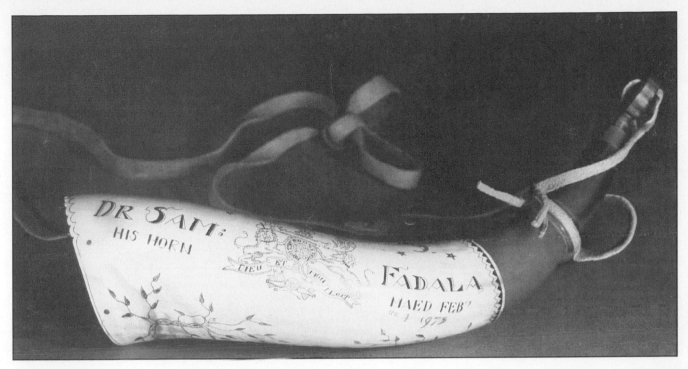

Some powder horns are as much art as function, and Fadala's favorite carries considerable decoration.

percussion caps
cleaning patches and cloths
solvents and other chemicals (including lubes)
pre-cut shooting patches and patch material
extra nipples
extra ramrod
home-cast or commercial round balls
conicals
shotgun wads
shot
cleaning gear

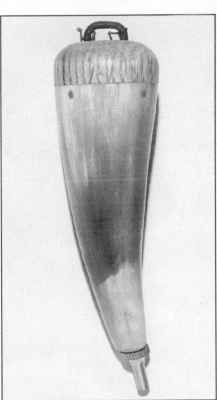

(Above) This fine pan charger built by Bill Knight is spring-loaded. The spout is forced against the bottom of the pan and one push on the body of the little horn dispenses a small charge of FFFFg blackpowder.

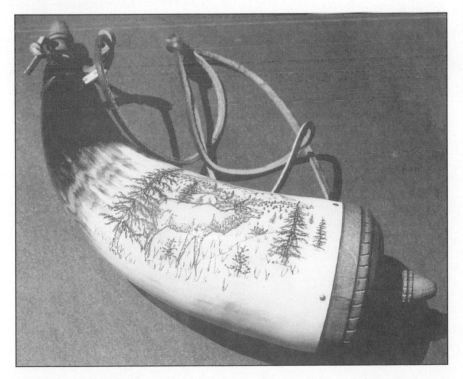

The powder horn is much more than decorative, although it can be that, too. It is a safe way to carry blackpowder.

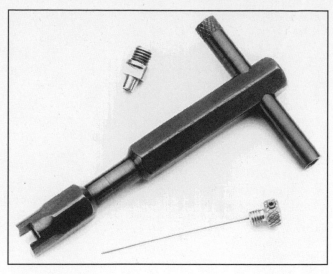

Luckily, companies have seen fit to provide specialty items for black-powder shooters. This CVA Walker Nipple Wrench is a perfect example, a tool designed specifically for the Walker cap 'n' ball revolver.

Thompson/Center's Deluxe Nipple Wrench includes a tiny item that has big application: the nipple pick (at the bottom), which is no more than a strong, fine wire to reach into the vent of the nipple or the touchhole of the flintlock gun.

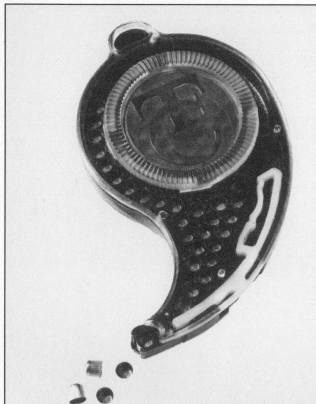

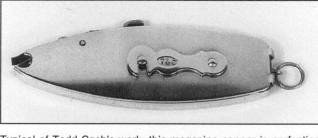

Typical of Tedd Cash's work, this magazine capper is perfection itself, compact, made to last a lifetime, and yet historical and func-

A Look at the Blackpowder Rifle

Chapter 4 deals with different types of blackpowder guns, but here we'll define them succinctly, like a buyer's reference guide.

The Original

Practical shooters will not have an original rifle, nor should they. Some of these guns are in perfect working order, others are not, and

A modern magazine-type capper is Thompson/Center's U-View model.

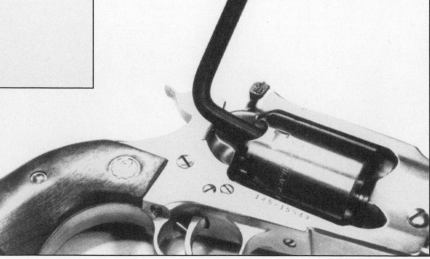

The correct nipple wrench is not just handy; it is vital to proper maintenance. An ill-fitting nipple wrench could damage this Ruger revolver.

there could be danger in shooting the latter. Originals may also carry too much value to be fired.

The Replica

This rifle style makes absolute sense for the person who wants to copy the past. Most replicas are also hard working, practical, accurate, and totally reliable. Replica rifled muskets, as detailed in Chapter 4, are a perfect example of guns from the past that still work today. So are rifles like Browning's version of the Winchester Model 1886 chambered for the venerable 45-70 cartridge. Replicas are a good choice. That goes double for the many replica handguns and pistols.

Rifled Muskets

Already mentioned above, the rifled musket is a good buy for shooters who want total ruggedness, good-enough hunting accuracy in most models, and target accuracy in a few—all in a dependable and powerful rifle.

The Non-Replica

There are several reasons the non-replica remains number one in sales among blackpowder guns. First, most shooters don't care about copying the past; they simply want to shoot blackpowder. Second, non-replicas generally cost a bit less than replicas. Third, non-replicas work well. They also have a bit more familiar look, and handle a lot like modern arms, although they are not nearly as modern as modern muzzleloaders.

The Modern Muzzleloader

Simply ideal for the shooter who has no interest whatsoever in buying a rifle that looks like a frontloader of yesteryear. He only wants to shoot blackpowder in the most convenient way possible. The majority of these guns are well-designed and built of top grade materials.

The Custom Rifle

The only negative I can come up with for the custom muzzleloader is cost. You won't get a good one without a fairly thick roll of cash. Other than that, a custom blackpowder rifle is a wonderful purchase, a one-of-a-kind, personal gun meant to be kept for life—when it's built right.

The Blackpowder Cartridge Rifle

In recent times, the blackpowder cartridge rifle has gained a large following, mainly due to growing interest in the silhouette game that demands a single-shot smokethrower. While old-time blackpowder cartridge repeaters are also nifty guns, the single-shot is more popular.

Buying the Handgun

As in selecting a rifle, the intended use is the key to picking the right blackpowder sidearm. The choices are many, but the major question is, Is it for show or shooting? Do you want to tote the handgun around, perhaps as part of your rendezvous regalia? Then a single-shot pistol of the period is the thing. Maybe you want to collect blackpowder sidearms. There are many pistol and revolver replicas to choose from. Informal plinking can be accomplished with any blackpowder handgun, but formal target work demands sights. Hunting with a muzzleloading handgun generally means a big-bore pistol, as discussed in Chapter 4. Incidentally, there are also beautiful custom pistols to be had from gunmakers who specialize in the work, William Kennedy being one such artisan. The custom pistol has the same problem associated with the custom rifle—cost. But it is available for those willing to part with bucks for the best.

Buying the Blackpowder Shotgun

Chapter 4 discussed the major features and differences of the various shotguns: flint vs. percussion, bore sizes, choke, and so forth. It's a very simple matter to choose the right gun to do the job, be it clay pigeon shooting, upland birds, waterfowl, wild turkey, or other applications. A choked 12-gauge double designed to shoot steel shot just about fills the bill.

The Blackpowder Knife

A case can be made for a blackpowder knife. Certainly, the patch knife is muzzleloader oriented. Although useful for peeling an apple, too, the patch knife was designed to cut patches on the muzzle of the frontloader. Other knives fit into blackpowder niches, too. The modern knife is carried by thousands of hunters, but it would be all wrong at rendezvous, where other knives are more appropriate, including the trade knife, Bowie, and Green River. Those interested in more on the beaver trapper's old-time knife should consult *Firearms, Traps & Tools of the Mountain Man*, a book by Carl P. Russell.

The trade knife was for trading, but the term may also have been used to mean "line of work," as in "What's your trade?" It was generally of simple blade, and often referred to as a butcher knife which was used for many purposes, from skinning to cutting meat for camp. I envision the trade knife as a tool and not much more, but correct for those reliving the time frame when such knives were in vogue.

The Bowie gained fame more from the man than from the knife itself. Here was a large knife, probably designed by Jim Bowie himself, and possibly altered by the blacksmith who built it, James Black. If anything made the Bowie famous, it's the way Jim died in 1836 at the Alamo. It is doubtful that the Bowie was all that popular in its time.

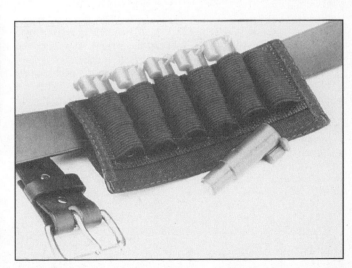

Thompson/Center's Quick Shot Butt Stock Carrier offers readyloads carried right on the firearm.

Readyloads are convenient and fast. These Thompson/Center readyloads are in the company's Quick Shot Belt Carriers.

The mountain man's knife was always handy, oftentimes carried in a sheath as illustrated here by Jim "Bear Claw" O'Meara. Note two knives on this outfit, both on the belt. Also, there's a tomahawk, making three cutting edges.

A hard-working butcher knife might cost as much as $2.50, plenty to spend on a knife out west. The Bowie had to cost a lot more. There have been quite a number of articles and books on the Bowie knife. Check your library.

The Green River knife was, and still is, tied in with the mountain man. Stories vary on the Green River. Apparently, John Russell built his knife factory in Green River, Massachusetts. It may not have been a knifemaking operation to begin with, but in time knives came from the plant. They were marked "Green River" because of where they were made. The Green River that flows through the Far West, however, was a mountain man's treasure. Probably, the Green River trademark symbolized that favorite body of water to him, but we don't know that much about it, except that the name of the knife was used in conversation. For example, "Hand me that Green River" simply meant "Give me that knife." Someone might say, "He gave it to him right up to the Green River," referring to stabbing an enemy. "Give 'em Green River, boys!" is self-explanatory. In my estimation, the Green River was a good working knife, and it makes a fine addition to the blackpowder outfit today.

Obviously, the patch knife is for cutting patches. For rendezvous, however, it's not so simple. A Bowie knife, whether or not this type was truly used by the mountain man, fits the picture, but so does the Green River, trade knives of various description, and a host of simple-bladed models of the period. Knives at rendezvous are used for throwing as well as showing. Not all knives are throwers, however. It's a matter of balance and blade style, and special throwing knives are available. Of course, the blackpowder blade is used for every cutting chore imaginable, from campside duties to cooking at the tipi.

Special cutlery shops grace every part of the country, offering a multitude of different knife styles, including models that definitely fit the blackpowder scene. Blackpowder shops usually have special knives on hand, too, as do mail order houses. There are knife kits as well as finished products, too, so a shooter can hand-craft his own blackpowder blade. Various knife companies continue to offer old-style knives, like Lynn Thompson's Special Projects company. As with the custom rifle or pistol, the hand-made knife rides at the head of the procession. Certain custom knifemakers, like Martin Kruse, will build you a patch knife, for example, or a Bowie. Naturally, custom knives require more folding green than factory models. Green River knives are available, too, and Atlanta Cutlery and others sell them.

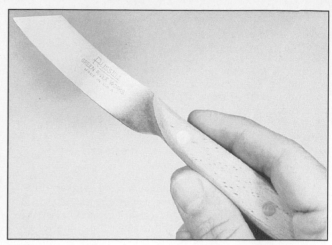

Russell Knife Co. offers this patch knife for the blackpowder shooter who likes to cut his patches from a larger piece of cloth at the muzzle.

CVA's Trade Knife is of the simple design often found in the 19th-century kit. While it is historical in style, it also functions well for camp duties.

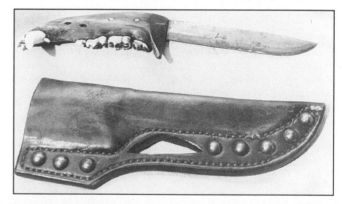

Dennis Mulford hand-made this knife and its sheath. The buckskinner wants a knife that relates to times of old, and this one certainly does.

Blackpowder Software

Tandy and other leather stores often sell kits for making old-timey knife scabbards and other accessories associated with blackpowder days, and the handy do-it-yourself shooter can make many of his own soft goods, including an entire buckskin outfit, shooting bags and other items, such as capotes. A few buckskinners have even built their own tipis. *Muzzleloader Magazine* is loaded with ads for various companies offering just about any make-it-yourself buckskinning item you could imagine, as well as dozens of wonderful ready-made blackpowder goods for the trail and camp.

This has been a little look at "toolin' up for big doins" in the world of blackpowder shooting.

chapter 7

Blackpowder Accoutrements

WHETHER YOU SPELL it accouterments or accoutrements, the word means the same thing: accessories. They're not very expensive, although they can be when custom-made, and they're generally not very big. Most of them fit into a shooting bag. But they certainly are important in blackpowder shooting. In fact, without some of them, you won't shoot at all. In short, a sootburner is not much better than a doorstop without its accoutrements.

The Necessary Items

The items listed here are much more than simply convenient things to have around because without them the gun can't be loaded, cleaned, or worked on. These little things of muzzle-loading can make or break the game. Many were mentioned in Chapter 6 on outfitting, but in this chapter, each is described and we explain why it's necessary.

The Powder Horn

The powder horn is more than romantic. It's one of the best blackpowder storage and dispenser containers ever. Despite the age of its origin, the horn has many superior attributes. With a tight stopper, it can be made waterproof. It's rugged and won't break easily; therefore, it's good for carrying powder into the hunting field or onto the range. Should a spark or ash hit the stoppered powder horn, the cargo inside won't go boom! Horn of itself is non-sparking, which is another good attribute for a powder-carrying vessel. Horn can be shaped to fit the shooter's needs, too. When softened in hot water it can be bent to shape. All in all, the old-time powder horn is a hard-working tool with few equals, and it comes in various sizes to match specific shooting bags, or for that matter, fit right inside them. Finally, the powder horn is also authentic, a true implement of the past.

Priming Horn

A miniature powder horn, generally with a spring-loaded spout protruding from its end, this little fellow is used as its name promises, to prime the pan of the flintlock rifle, pistol, or shotgun. It works with ease. The spout is pressed down on the hard surface of the flintlock pan, thus depressing the spring, which allows a trickle of FFFFg powder to flow. The priming horn dispenses just the right amount of powder right where you want it, toward the outside of the pan, which is conducive to more certain ignition.

Priming Tool

This is a priming horn that's not a horn. It is more like a small met-al flask, but it serves the same purpose as the priming horn, containing and dispensing FFFFg pan powder.

The Adjustable Powder Measure

There are several variations on this theme; however, the adjustable powder measure is easy to pigeon-hole because it is adjustable, and measures out powder by volume. Blackpowder and Pyrodex are best loaded by volume. It is the correct and safe way to go. Adjustable measures are accurate, and therefore wonderful tools for the job. The adjustable measure may or may not have a swing-out funnel. When it does, the measure can be correctly set, loaded with powder to slightly overfull, then the funnel can be swung into place, which tops off the charge by swiping away a few granules of powder from the top of the barrel. A sliding rod or tube alters the volume of the funnel. The adjustable powder measure is calibrated with markers that clearly show how much powder is in the funnel, in what we should refer to as "grains volume," since the adjustable measure does not, directly, tell the weight of the charge.

Powder measures normally vary a little from brand to brand. Most I've tested correspond with FFg blackpowder. In other words, with the setting on "100 volume," the measure throws 100 grains weight of FFg blackpowder. The shooter can check his own powder measure by gauging it with a modern powder/bullet scale. Simply toss your charge into the pan of the scale and see how much it weighs in grains. This little test shows how closely volumetric charges correspond by weight. For Pyrodex, set the measure for the charge, such as 100 grains, and the measure will provide a proper load in grains weight. Set at 100, Pyrodex RS will yield about 71.5 grains weight. This will produce, for all practical purposes, the energy represented by 100 grains weight Goex FFg blackpowder.

The Charger

The charger is a non-adjustable powder measure. It may be as simple as a metal tube, drilled-out antler tip, or just about any other non-sparking material. Once the proper volume is settled on, the tube is cut to length, or the antler tip is drilled no more. A shooter need do no more than pour powder from a horn into the charger and he has a volumetric load ready to go. As with the adjustable powder measure, the charger is used to directly pour powder down the muzzle, rather than pouring powder from a horn. At least in theory, a lingering spark downbore could ignite the powder charge. While this is unlikely, it's better to have a rather small amount of powder affected in the charger or measure, rather than a horn full of powder.

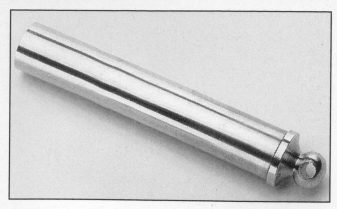

Whether in a shooting bag or range kit, the non-adjustable powder measure is useful. The CVA Brass Fixed Powder Measure comes in either 90-grain-only or 120-grain-only sizes.

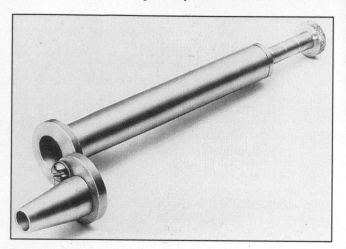

The adjustable powder measure is also workable in either a shooting bag or range kit. This is Uncle Mike's adjustable measure, which throws up to 120 volume.

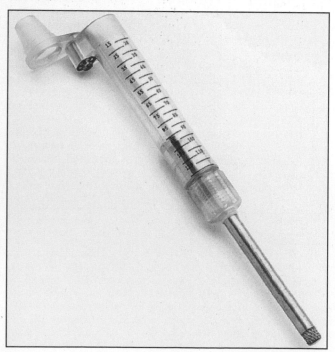

Modern muzzleloaders prompted modern accoutrements, like this Thompson/Center U-View Powder Measure, which is calibrated in 5-grain increments.

Uncle Mike's pan primer, made of metal, not horn, operates by placing the nose of the spout against the bottom of the pan and pushing on the body of the primer. A small dose of FFFFg powder is expelled directly into the pan without spilling.

The Flask

Yet another means of metering out powder for a proper charge, the flask can be made of metal, hard leather, or even cloth. It incorporates a metering device of one type or another, perhaps as simple as a tube (spout) varying in length to provide different powder charges. I carry a small powder flask in my big game shooting bag. Though compact, it holds enough powder for several shots from a big game rifle or shotgun, and a number of shots for a small game rifle or sidearm. I use flasks when hunting small game, and like those with interchangeable spouts that alter the powder charge. Want a light charge? Use a short spout. Want more powder? Use a longer tube. A flask is rugged and certain. With multiple tubes or spouts, it's also versatile. Flasks are useful for rifle, shotgun, pistol, and revolver.

The Readyload

The readyload is another way to carry a proper powder charge. The charge is still produced by volume, but it is pre-measured into the readyload. There are so many possible variations that it is difficult to list all different types of readyloads. For shotgunning, my readyloads are no more than plastic 35mm film containers, some for powder, some for shot. Powder and shot are premeasured and loaded into these small plastic containers. In the field, I simply pop a top off the appropriate container and pour the contents down the barrel. Just about any non-sparking tube can be worked into a readyload. Commercial models include plastic tubes with partitions to separate powder from bullet. These even have slots to hold percussion caps. Check with your blackpowder shop or catalog for the readyload of your choice.

The Ramrod

The ramrod rests in the pipes of the gun, usually beneath the barrel, always ready for use. In the field, the ramrod is used to force a projectile downbore firmly onto the powder charge in the breech. It can also be used to clean the gun in camp. A worthy ramrod has a threaded end that accepts various accessories, including worms, screws, jags, and other maintenance tools. The Uncle Mike's Thread Adapter Set allows cleaning accessories of different thread to be used on one rod. Wood-

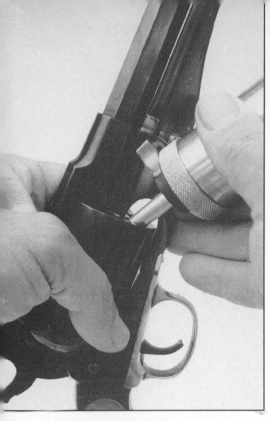

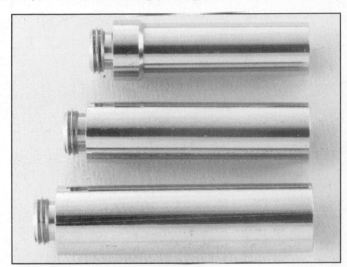

Also useful in the field or on the range is a flask, which not only carries powder, but also dispenses it. This CVA flask is patterned after a Remington design.

Different spouts are used to vary the powder charge thrown by a flask. These simply screw in and out of the flask.

On the shooting range or in the field, a flask is quick and easy to use. Here it charges a cylinder chamber on a Remington-style revolver.

en ramrods work fine and dress up the firearm. Check the grain, however, to ensure straightness. Curly grain ramrods are more likely to break than straight-grain types. Ramrods can be made of synthetics or steel as well as wood. Today's synthetic ramrod is safe to use because it won't lap the muzzle of the rifle. A steel rod is ok, too, provided it is undersized and used with a muzzle protector, which keeps it centered in the bore. This prevents it from harming the rifling.

The Loading Rod (or Wiping Stick)

In the old days they were called wiping sticks. They are essentially cleaning rods and are also used to load the rifle at the range. The usual loading/cleaning rod is longer than the ramrod, and generally made of steel, whereas wiping sticks were, and still are, made of wood. These rods are great for cleaning, not only at home, but in camp and at the range. The metal loading/cleaning rod is perfect for its dual jobs of seating projectiles and delivering brushes and cleaning patches and other tools downbore. Those with muzzle protectors are especially nice.

Palm Protector

Uncle Mike's makes a wooden palm protector that slips over the end of a ramrod. It's a good device for protecting the palm of the hand when loading a stubborn patched ball downbore. It keeps the end of the rod from working into the hand, and allows a little more pressure on the projectile.

The Short Starter

Accuracy calls for a patched ball that fits fairly closely to the bore. That means that pressure is required to seat the projectile. The longer wooden ramrod may break if you try to start a tight ball with it (or even a concical), while the even longer loading rod is unwieldy for this operation. That's where the short starter comes in. It generally has two rods. One is the stub, the other is the starter rod. The stub is just that—very short. It fits over the nose of the projectile and with pressure on the ball (handle portion) the missile is pushed past the crown of the muzzle. But that's not good enough because the ramrod or loading rod still

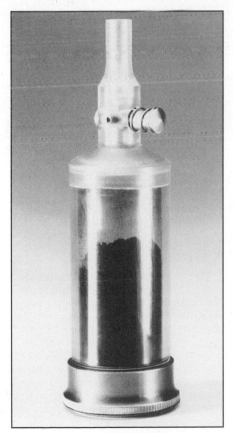

Matching the high-tech approach of the modern muzzle-loader, this Thompson/Center U-View Powder Flask reveals its cargo of powder visually.

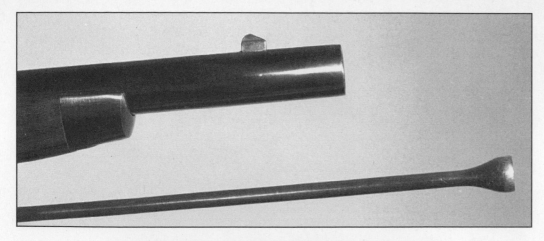

Ramrods come in many different styles. This one is steel. Note the small diameter, which helps to keep the rod from scraping the bore and damaging the rifling at the crown of the muzzle.

Uncle Mike's palm protector is called the Palm Saver, which is what it does when placed on the end of a loading rod or ramrod.

Another item necessary in the field or on the range is the short starter. This is Uncle Mike's Oregon Ball/Bullet Starter with built-in muzzle protector.

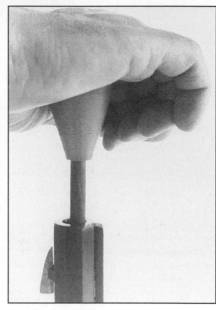

won't seat the ball easily from this position. So the longer rod on the starter is used to push the projectile still deeper into the barrel. Now the patched ball or conical is ready to be driven fully down on the powder charge with either a ramrod or loading rod. A short starter is a must-have tool.

Bullet Starters

A variation on the short starter, these tools are used to drive all manner of bullets downbore, including conicals of various styles. An example is the Buffalo Bullet Co. Universal Bullet Starter. It's made of super-tough synthetic material and is especially useful for starting conicals and sabots. It protects the nose of the projectile from deformation. Although designed to seat conicals, I find the unit excellent for round balls too.

The Combo Tool

Over the years there have been many combination tools on the market. One current model is a perfect example of the spirit of the combo tool. This is the "Ball Buster," a trademarked unit from Cousin Bob's Mountain Products of Ben Avon, Pennsylvania. In this one tool a shooter has screwdrivers, three powder measures, a light-duty hammer, signal whistles, short starter and more. The combination tool spirit lives in various other tools, including the October Country Adjustable Powder Measure with Funnel and Nipple Pick. This beautifully-made tool has a threaded vent pick in the base of the sliding measure bar. Such a pick is perfect for clearing a fouled nipple or touchhole and also serves to block the touchhole when loading the flintlock. There are many different combo tools available today.

The Vent Pick

Also called a nipple pick, this simple bit of wire is useful for clearing a clogged nipple vent or flintlock touchhole, as well as blocking the flintlock's touchhole prior to pouring the powder charge downbore. This step ensures a clear touchhole for the passage of flame from the pan into the breech of the firearm.

The Nipple Wrench

Just make sure it fits, that's all, and the nipple wrench will install or remove a percussion nipple without marring it in any way. An essential tool for the caplock.

The Pipe Cleaner

This is not a muzzle-loading tool, of course. It's a wire device with a fuzzy wrapping meant to clean the smoking pipe. However, it just happens to work great for blackpowder guns, blocking the touchhole, as mentioned above, or swabbing oil or dirt from any part of the firearm. My blackpowder shooting kit is never without pipe cleaners.

The Capper

There are two basic kinds of cappers. One is the magazine model. It holds many caps and gravity-feeds them one at a time for seating on the cone of the nipple. The other kind is the in-line model, where caps are held, as the name implies, one after another, in a line. One cap at a time is placed on the nipple. Cappers offer a handy way to carry percussion caps, and they are a safety device as well. Instead of introducing a cap with the fingers, the cap is placed on the nipple with this tool.

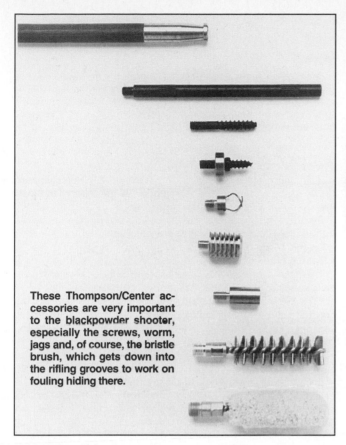

These Thompson/Center accessories are very important to the blackpowder shooter, especially the screws, worm, jags and, of course, the bristle brush, which gets down into the rifling grooves to work on fouling hiding there.

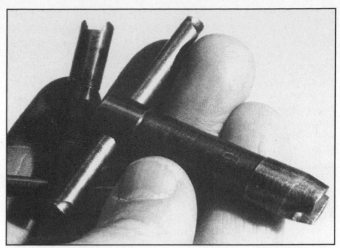

Some tools are less likely found in a shooting bag or hunting camp, and more likely encountered on the shooting range or in the home workshop. These large nipple wrenches are meant for heavy-duty work, very much at home in the shooting box, but not ideal in the shooting pouch.

The lowly pipe cleaner gains considerable status when you have to clean hard-to-reach places, such as the nipple seat of a muzzle-loader.

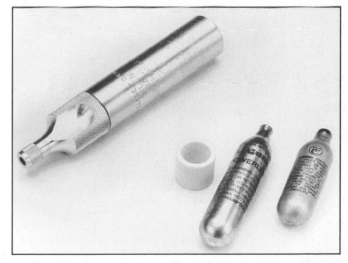

Another interesting product, the Thomson/Center Magnum Ball Discharger forces a load out of the gun without having to shoot the firearm. It's handy when you *can't* shoot.

If the million-to-one chance comes up and a cap was detonated by squeezing it on the nipple, fingers would be burned or worse.

The Screw

A screw is just that. It looks like a wood screw with a threaded shank, and it installs in the ramrod or loading rod tip. The screw is used to remove a stuck ball. It is centered on the lead ball, screwed into it, and then the ball is pulled out, not always easily. The N&W loading rod has a special knocker unit integral to the shank of the rod. This knocker makes easy work of using the screw for removing a stuck ball. Use the screw with a muzzle protector so that the screw does not make contact with the rifling.

Jags

These are small metal devices, generally with an hourglass shape. They screw into the tip of the loading rod or ramrod and are designed to grip a cleaning patch. They usually have a concave nose so they can be left in place to seat the round ball, as well as the conical projectile. It's important to match the jag not only by caliber size, but also by style. A specific breech shape may require a special jag shape. Other jags may get stuck (seriously) downbore.

The Worm

The worm is also a threaded device that fits the end of a ramrod or wiping stick. But instead of a screw, it's fitted with tiny protruding wires or arms. The worm is made to withdraw a patch that may be stuck in the bore. It is not made for pulling a stuck ball.

The Fusil

The fusil is made to hold a modern metallic cartridge primer in place of a percussion cap. It's a real boon to stubborn caplock muzzle-loaders that refuse to fire on command. A small pistol primer, for example, is placed in the fusil, and if that won't set off the main charge in the breech, see your gunsmith because something is very wrong. Fusils of various types can be ordered from your blackpowder dealer. Another good source is the big Dixie Gun Works catalog.

Kap Kover

A weatherproofing device, the Kap Kover goes over the top of the nipple. A gasket keeps water out, but the firearm will not go off until the Kap Kover is removed, making it a safety device, too.

The Possibles Bag

My research indicates that this is a large container apparently used by the mountain men to hold a great deal of gear, from traps and extra smoking tobacco to heaven knows what. The array of plunder owned by a mountain man, I suppose from knife to flints, was his "possibles," and so the bag to contain them is a possibles bag. I've seen leather kits in museums that may have been possibles bags. Jim O'Meara, an expert on the mountain man era, has a large basket-like container that serves the same capacity as the possibles bag.

The Shooting Bag

This is the smaller bag, often called a possibles bag today, that is used to contain all of the little devices necessary to blackpowder shooting. I have one shooting bag for each of my most-used firearms. Each bag contains the ball, patches, powder measure, cleaning gear, powder flask with appropriate spouts, and all items needed to shoot that one rifle. In this manner, I can grab a given rifle and the matching bag and take to the field knowing I have everything with me to shoot that one rifle.

Ball Bag

Some original ball bags were made from the tanned scrotum of buffalo or other large bovines. Today, there are various commercial models, usually made of tanned leather, but many other materials would also work. They can also be handmade, and are generally used to hold a number of lead round balls, but of course can be used for conicals as well.

The Flash Cup

This is a great little device for the caplock rifle. It's a metal unit that fits on the nipple seat and is held in place by the screwed-in nipple itself. It diverts flame away from the wood parts of the rifle, thereby saving them from being burned and marred. Tedd Cash makes beautiful flash cups in various sizes to fit different guns.

Cleaning Gear

Cleaning gear includes the wiping stick (cleaning/loading rod) mentioned already, plus patches, solvents and especially well-fitted bore brushes. No matter what method of cleaning a muzzleloader is used, bore brushes are ideal for getting down into the grooves of the rifling and removing fouling. Cleaning equipment also includes everything from rags and toothbrushes to pipe cleaners.

This is a brief look at some of the major blackpowder accessories that have come down the pike in the past centuries as well as in recent times. A full list would make a book. Blackpowder shooters, like trappers, are industrious people with a lot of do-it-yourself in their souls and plenty of imagination. They have come up with some of the most fanciful, outlandish, interesting as well as useful (and now and then useless) gear you ever saw in your life. But without accoutrements, shooting would be no more possible than keeping your car running without any tools. ●

Here, the Kap Kover is shown along with a Tedd Cash flash cup, both in place on a Mulford custom rifle. The Kap Kover prevents accidental discharge while helping to waterproof the gun, and the flash cup keeps sparks from damaging the fine wood of the rifle.

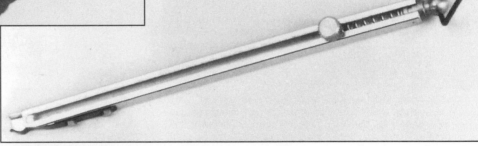

(Above) Uncle Mike's One-Handed Capper is of the straight-line type. Caps line up in a row, one behind the other.

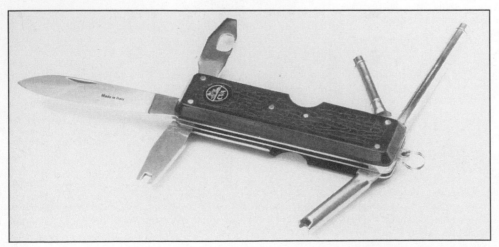

The CVA Possibles Tool and Sheath is a combination tool with multiple functions. It includes the essential knife blade, as well as screwdrivers and a nipple wrench, and it can be carried on the belt.

chapter 8

Safe and Sane Blackpowder Shooting

CLICHES GOT THAT way because they're true. The old one about familiarity breeding contempt is no exception, and it applies well to shooting. A person can get so used to handling firearms that he may forget the most basic concepts of safety. Blackpowder shooting is a very safe sport, when done safely, but it's prone to the same dangers associated with any other activity where potentially dangerous tools are used. So never lose caution. Always consider your guns tools that, when properly handled, are totally safe, but when mishandled can be dangerous. Muzzle-loading is as safe as crossing the street. I mean that literally. Look both ways, and you're okay. Close your eyes and you're done. The only worthwhile shooting is safe shooting. I like what Mark Twain said about guns and safety a long time ago. He wrote:

> Never meddle with old unloaded firearms; they are the most deadly and unerring things that have ever been created by man. You don't have to have a rest, you don't have to have any sights on the gun; you don't have to take aim, even. You just pick out a relative and bang away at him. A youth, who can't hit a cathedral at thirty yards with a cannon in three-quarters of an hour, can take up an old empty musket and bag his grandmother every time, at a hundred.
>
> Mark Twain

Every shooter has seen a warning printed on 22 rimfire ammunition, like "Range 1¼ Miles," or "Range One Mile—Be Careful!" The maximum range of bullets is indeed far. Sharp-profiled big game bullets fired from modern cartridges can fly three miles when fired at about a 35-degree angle. Blackpowder guns have a much shorter extreme

Are you sure of your target? Do you have a good backstop? The general rules of firearms safety pertain just as much to muzzleloaders as they do to modern arms.

How far will it shoot? Today, the conical-shooting rifle like this Thompson/Center Scout with its quicker-twist rifling is very popular. Conicals shoot farther than round balls because they retain their velocity better. In other words, know the approximate extreme range of your frontloader. Even rather stubby conicals may travel more than a mile when the muzzle of the rifle is at the optimum angle.

53

Muzzleloaders load from the front. That's why they're called muzzleloaders. The shooter must be certain that the firearm is truly unloaded before he begins to put down another load, and it must also be unprimed (flintlock) or uncapped (percussion) during the entire loading process up to the point of shooting.

Although repeat proof is hard to come by, the short-started blackpowder load has definitely caused trouble. In some cases, barrels were completely destroyed, and in other cases, like this one, a bulge occurred. This barrel was ruined with a target load, not a hunting load, the bulge occurring at the base of the short-started patched round ball.

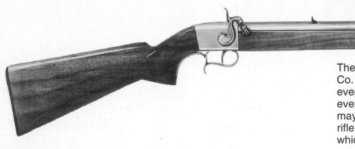

The author was sent a Morse rifle like this one from the Navy Arms Co. for experimentation. The rifle was so well made that it did not fail even with outlandish powder charges and multiple projectiles. However, that does not mean damage wasn't done. Because the tests may have caused unseen problems that could cause failure later, the rifle was purposely fired with a charge of W-748 smokeless powder, which destroyed it.

range, of course, especially when they shoot the round ball. That's what makes them so great in certain areas where longer-range guns are unacceptable. But this fact does not excuse us from being careful. Make sure of your backstop, and never shoot over a hill.

Another warning often found on ammunition boxes reads "Keep out of Reach of Children." That certainly goes not only for blackpowder guns, but for the powder and percussion caps as well.

The Short-Started Load

Short-starting can ruin a barrel. I know this from tests, and yet it's difficult to prove, because it doesn't happen all the time. The short-started load means that the projectile—any bullet or shot charge—is not pushed all the way downbore firmly upon the powder charge. Considerable testing on this phenomenon has yet to provide a certain answer as to why short-started loads *sometimes* ruin barrels. But all the same, I am convinced that they can. My most startling tests were conducted several years ago. Dale Storey, the Casper, Wyoming gunsmith, built devices consisting of breeched barrels. Many vicious overloads were fed into these test guns, including multiple projectiles and dramatic overcharges of FFFFg blackpowder. The barrels were of top quality, and they held up to initial abuse, but that doesn't mean they weren't damaged. Only a metalurgical test could prove the condition of those barrels following the destructive loads they were subjected to. However, they did not burst.

They survived until subjected to short-starting. We had no idea what we were dealing with. Backed away by 50 feet, Dale and I felt safe. Should a barrel crack, we would be all right—we thought. However, after the first major blowup, all subsequent tests were conducted with a greater respect and a substantial barrier between firearm and experi-

Every year someone refuses to obey the cardinal rule of loading blackpowder guns: Never use smokeless powder. The warning is clearly stamped directly on many muzzleloaders, as seen here on this White rifle, which states "Black Powder Only. Read Instructions Before Use."

mentors. We became true believers when a piece of barrel steel blew so far from one short-started test gun that it was never recovered. It was stopped in flight by the high-speed shutter of a 35mm camera, however, proving that the missing section of barrel skyrocketed well above the test site. Instead of the usual roar when the gun was fired, there was a thunderous boom. We knew the barrel had blown. The motor drive on my camera provided film sequence evidence of the event. That little test was followed by several more. For a long while, we had only three barrel failures. Then a couple more ruptured. All were destroyed or damaged only when the condition of short-starting prevailed.

In a later test, I built several test devices using copper pipe. I found that 54-caliber patched balls and certain 54-caliber conicals fit just right in these pipes. They were loaded in the usual fashion, just as if they were muzzleloaders. *These devices were built for test purposes*

The results of a smokeless powder charge in a muzzleloader clearly stand out here. This firearm was purposely fired with smokeless powder (by remote control) to destroy it following severe tests with blackpowder that may have weakened the gun.

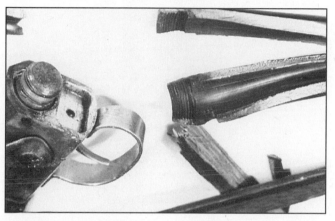

How did the gun blow up? The author has been called upon to examine exploded muzzleloaders. Total destruction at the breech suggests that this one probably blew because of smokeless powder—which it actually did because it was a test gun purposely destroyed with a charge of smokeless powder. Note that the entire breech section fragmented and parted from the rest of the gun.

only and must be considered very dangerous. All testing was done behind a barrier and from a distance. A telephoto lens was used to record results. Remarkably, the thin pipes did not break, even with charges up to 120 grains of FFg and a single patched ball.

However, when the pipes were short-started with normal loads in the 100 grains volume range, there was trouble. They either blew apart or they split wide open. With light loads the familiar bulge or walnut appeared. The bulge was always situated exactly at the base of the projectile where it had rested in the "bore" of the pipe. Damage was also incurred by using a conical in place of a patched ball, indicating that the greater mass and bearing surface made a difference. In short, *don't short-start projectiles; seat them firmly on the powder charge.* That is not new advice. Elisha Lewis, in *The American Sportsman,* published in 1885, said "We are consequently forced to adhere to the ancient doctrine of explosion, and still believe that a fowling-piece is more apt to burst with a wad or a ball far up the barrel than if pushed home upon the shot or powder." He also added, "This phenomenon we cannot account for." Neither can we.

Multiple Projectiles

Part of the reason for the startling blowup of one test rifle was the use of two projectiles, although I must quickly add that other short-started rifles with a single projectile also failed when short-started. Multiple projectiles are not a good idea, in spite of recommendations by some shooters who have apparently gotten away with the process.

Firing two round balls simultaneously may not stress a sturdy rifle if a modest powder charge is used. However, accuracy goes straight downhill when two balls are fired together. Two conicals could cause real trouble, as we'll see later on. Forget multiple projectiles; they can be unsafe, are inaccurate, and there's no reason to use them.

Duplex Loads, Bore Obstructions, Dirty Bores and Pressure

Certain smokeless bulk powder/blackpowder duplex loads have worked when very light charges of bulk powder were combined with the main blackpowder charge, the object being more complete combustion of the blackpowder. However, fouling still occurred, and there was no increase in muzzle velocity. Furthermore, it's difficult to locate bulk smokeless powder, and when some is found, it may be very old. Therefore, the warning *never use smokeless powder in a muzzleloader* must be taken to mean NO smokeless powder whatsoever, no matter how little, or how it is used.

Some shooters believe that short-starting is related to the bore obstruction. As everyone knows, a barrel can be burst if it is clogged with mud or other solids. This is true of modern or blackpowder firearms. *Be certain that the bore of any firearm is clear. Never allow an obstruction of any sort to exist.*

Increased pressure without benefit of additional velocity can occur when too much powder is used. The law of diminishing returns takes over, and instead of the extra powder working to push harder on the projectile, it seems to spend itself getting rid of the ejecta from the bore. Ejecta is the total mass fired from the bore, including the bullet, but also all unburned powder as well as fouling.

Increased pressure can also result from a dirty bore. Partly, this is due to the fact that the effective volume of a bore is decreased by caked-on fouling, allowing less room for gas expansion. Tests have shown that dirty bores can raise pressures while velocity either remains the same, or in some cases, drops. So a dirty bore may cause trouble. Swab the bore when appropriate. *Do not allow fouling to build up.*

Managing the Short-Starter and the Ramrod

Of course it is wise to recognize that when a projectile of any type is seated downbore with a ramrod, there is powder underneath that charge. That is why some shooters call for pinching the ramrod. Then if the gun goes off, your fingers will be saved—in theory. I was contacted by a reader who saw a magazine photograph of mine showing a fellow grasping his ramrod fully, rather than pinching the rod with thumb and forefinger. "Never show that again!" he reprimanded. "The rifle could go off and the ramrod could go through your hand. You should only pinch the ramrod with your thumb and forefinger." In fact, accurate rifle and pistol loads generally require a tight conical or ball/patch fit to the bore, which is not likely achieved by seating with only thumb and forefinger on the ramrod. The safe practice is to ensure that there is no cap or pan powder in the firearm, rather than worrying about the pinched ramrod. Ballistically acceptable loads can be produced in the shotgun with a well-fitted, *but not tight* wad column. In fact, as we'll learn later, an overly tight wad column can be dangerous in the shotgun.

I wrote to the man who advocated the pinched ramrod technique. Did he use a short-starter? I asked. Of course he did, he replied. Did he center his palm over the ball of the short-starter? Of course he did. How else would you use it? He got my message. The rifle was just as loaded when he used a short-starter with his hand over the ball as it was when he fully seated the patched ball with a ramrod. Here is the way to be safe with the short-starter, ramrod or loading rod: Never load a firearm that could possibly have a lingering spark downbore. Don't hurry. Wait for any possible sparks to die out, or swab the bore to extinguish them. And make darn sure there is no cap on the percussion gun, or pan powder in a flintlock. That's the way to be safe using a short starter.

Blackpowder is considered pressure-sensitive. A lot of force may set off a charge. On the other hand, it is supposed to burn a bit better when compacted. What to do? Two reasons given for compacting powder

This double gun is both shotgun and rifle, and it poses two special safety factors. First, the shotgun load demands the correct wad column, one that will stay where is belongs down in the breech. Second, with any double-barrel frontloader, when one barrel is fired, especially several times, and the other is not fired, the unfired barrel should be checked for the possibility of the load riding upbore, creating a separated charge/projectile condition.

Can blackpowder generate enough pressure to blow up a gun? Yes it can, as has been proved many times. Do not overload any powder, and that includes blackpowder substitutes as well as the original propellant itself.

are a slowed burn rate and uniformity of combustion. The tightly fitted patched ball in the muzzleloader aids in maintaining constant pressure upon a compacted powder charge in the breech, thereby producing improved burning uniformity.

Blackpowder Pressures

This one is related to the overload noted earlier, but it deals directly with blackpowder pressure, and the reader should be armed with this information. When I was a beginner in the muzzle-loading game, I was told by several "blackpowder experts" that you could not overload a blackpowder gun. The overcharge of powder just "blew out the bore." That is not true. The notion that it is impossible to achieve greater than 25,000 psi (pounds per square inch pressure) with blackpowder, no matter how much powder you install down the muzzle, is false, in spite of its popularity. Captain Noble, along with his partner Mr. Abel, generated blackpowder pressures of about 100,000 psi in the late 19th century. Simply be aware that blackpowder is very powerful, and never treat it as anything else.

Pressures and Conicals

In my own demonstrations—I can't call them scientific tests—I have shown that elongated projectiles can bring damage with loads that are OK with round balls. This does not imply that Minies or Maxies are dangerous. They are excellent projectiles, and I will continue using them. However, the added mass of the conical does alter the manner in which pressure is built behind it. *Don't exceed the maximum conical load allowed by the manufacturer of the firearm in question.* That's the way to be safe.

Proofing

Now and then I still hear about proofing a muzzleloader. It's a bad idea. The process includes securing the firearm so that it can be fired

Respect for the elements is part of any safe hunt. Hypothermia, a word almost unheard in the not-too-distant past, is now mentioned often as the cause of trouble for hunters every year. Dress for the weather. It's part of your safety program. Also, in wilderness settings, think about carrying some form of shelter, such as a bivy (small one-man tent often weighing under 2 pounds). Then if a storm does catch you, shelter will be yours almost instantly.

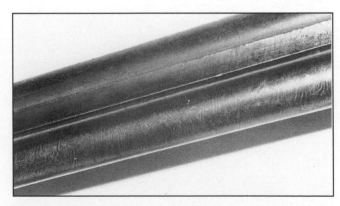

Damascus or "twist steel" is handsome and historical, but it may also be unsafe. Experts these days recommend not shooting Damascus-barreled guns.

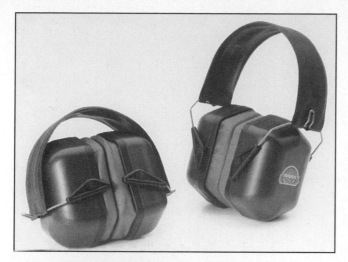

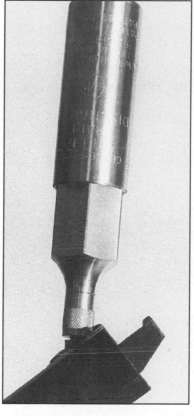

Once upon a time, muffs like these were not nearly so prevalent. Today, they're for sale in just about any gun-shop. These Silencio muffs have deep cups for good protection, but they fold up small to fit in a shooting kit.

Everyone knows that the supposedly un-loaded firearms is a menace. These days, you don't have to fire a load away at the end of the day's hunt if you don't want to. The Thompson/Center Dis-charger will "blow it" out of the muzzle. Even this, of course, demands caution. The CO_2 gas used to push the load out can do so at a pretty good clip, so aim the muzzle in a safe direction before expelling the load.

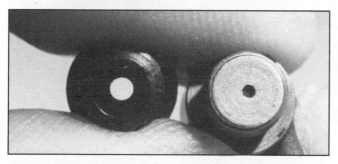

While it may be of comparatively small concern, the nipple should have a flat base with a pinhole, as on the right. The large orifice on the other nipple is uncalled for and can allow considerable blow-back.

remotely with heavy powder charges, and sometimes multiple projectiles. The idea is that if the gun can handle such loads without coming apart, it's well made. However, the very act of proofing can cause incipient damage. Later on, the damage suffered during proofing could get somebody—you—hurt.

Double-Barrel Shotgun Management

I've seen shooters recharge a fired barrel of a double-barreled blackpowder shotgun without uncapping the remaining loaded barrel. Don't do this. Always first remove the percussion cap from the loaded barrel before recharging the fired barrel. Throughout this process, the muzzles of the gun are to be pointed away from the shooter. You may ask, how can a shooter be hurt by not uncapping the loaded barrel? How can that barrel go off? The shooter can be hurt because of Murphy's Law, which suggests that anything can go wrong, and at the most inopportune moment. Should the hammer come down somehow on that capped barrel, a shot charge is on its way. *So unncap it—always.*

Check for Load Shift

It is possible for the charge in one barrel of a double-barreled muzzle-loading shotgun to move upbore during the firing of the other barrel, creating what amounts to a short-started load. *After decapping the loaded barrel, run a ramrod down to ensure that the entire load—powder, wads and shot—is still firmly seated in the breech of the gun,* thereby avoiding the unwanted short-started condition.

The Shotgun Wad Column

Don't use hunks of old undershirt, newspaper, or diaper cloth as part of a wad column in the muzzle-loading shotgun. I tried these in my romantic pre-testing days of blackpowder shooting and found that bits

of cloth and/or paper could remain smoldering in the bore. That's bad because subsequent powder charges may be ignited by these embers. There is far more chance of smoldering with paper or cloth than with proper shotgun wads. Today, we have no excuse for using foreign objects as wad columns because so many companies offer proper materials. All the same, *take your time in putting that second charge downbore. Give bits of unconsumed powder a chance to burn out.*

Restricted Wad Column

More energy is required to put an object into motion than to keep it in motion once it is moving. In modern handloading, great care is taken to prevent brass cartridge cases from becoming overly long. They are trimmed to correct dimensions because extended case necks may pinch the bullet firmly in the leade or throat of the chamber. Impeding the initial progress of the bullet in this manner can seriously raise pressures. In modern shotshell reloading, careful shooters make certain that wads do not bind in the hull for the same reason—impeding their progress downbore can raise pressures. While the blackpowder situation is not identical to the above examples, it is important that restricted wad columns be avoided. The wad system for the muzzle-loading shotgun must do its work, that of containing and guiding the shot charge, but it must also escape upbore in front of expanding gases. *The wad must not hang up in the bore; if it does, high pressures can result.*

Nipples

This one is a nitpicker, but the shooter should be aware that there are nipples with a huge orifice through the cone. They are essentially without a base as well. I've found as high as a 200 fps muzzle velocity loss with such nipples, indicating gas diversion. The venturi principle is at work here, so gas is not jetting out of the cone like a volcano. To explain the venturi, it is the foundation for the ramjet, which has a

Even the quick-to-run-away javelina can be dangerous under certain circumstances. Any wild animal, even a mouse, may cause a problem. All hunters, including those carrying charcoal burners, must play it safe in the field, especially when approaching a downed animal that may still have the ability to kick or bite.

"vent" in either end, but of different diameters. It is propelled forward because more gas flows through the larger hole than through the smaller hole. The touchhole of the flintlock works this way, too.

Some gas escapes through it; however, a properly-built touchhole does not allow a super jet of gas to blast through, because by far more gas is expelled through the much larger hole called the bore's muzzle. However, there is definitely a gas diversion with "straight-through" nipples, and I question how much flying cap debris the large orifice promotes. My advice is to *check the nipple on your new muzzleloader and exchange straight-through large-hole nipples for those with a base and tiny orifice.* Nipples with flat bases and pinholes serve better in containing expanding gases.

The Patch

The patch is not a true gasket because it does not seal hot gases behind the ball. Many tests and demonstrations have revealed this. However, patches are vital not only in taking up windage in the bore, *but as safety devices.* They help hold the missile on the powder charge in the breech, thus preventing a short-started condition. Patches also hold lube, which can be important from the standpoint of keeping the bore free of heavily caked fouling that may raise pressure. And patches, by maintaining force on the powder charge, indirectly aid fuel combustion. So the patch offers several safety features. Further discussion of this topic appears in Chapter 18.

Granulations

This is another nitpicker, but it is wise for muzzleloader fans to understand that the various granulations do make a difference in pres-

sures. Blackpowder is composed of saltpeter, charcoal and sulfur. Granulations are of the same mixture, be they Fg, or FFFFg. However, alteration of the kernel size (and shape to some degree) changes burning characteristics dramatically. Therefore, there is a big pyrotechnic difference between Fg and FFFFg, in spite of their identical chemical makeup. For example, my favorite 54-caliber rifle does not use Fg powder because it's too coarse. It needs a very large charge of Fg to produce any sort of reasonable muzzle velocity.

On the other hand, that gun doesn't use FFFg either, not for heavy hunting loads. For target loads, sure, even in this big bore, because a mild charge of FFFg gives a reasonable muzzle velocity. But the best balance between muzzle velocity *and pressure* in that 54 comes with FFg powder or RS grade Pyrodex. When a manufacturer notes a particular maximum safe charge for his product, he should state the granulation. If he doesn't, the information isn't complete. Of course, you also need the projectile weight and type as well. However, "100 grains of powder maximum" is not enough data.

In a Lyman test, 100 grains of FFFg gave a 54-caliber ball about the same muzzle velocity as 140 grains of FFg (these were test loads in both cases). However, the 100-grain FFFg charge gave a pressure rating over twenty-five percent higher than the FFg load. Whenever you get a twenty-five percent pressure increase without so much as one extra foot-per-second velocity increase, that's the wrong ballistic direction to travel. I am not suggesting that FFFg is wrong in your rifle. I am saying that in a 50- to 58-caliber long rifle, chances of gaining greater velocity with less pressure *for heavy hunting loads* reside with FFg granulation instead of FFFg granulation. It takes more FFg than FFFg to gain that velocity, but I prefer the trade-off. My own cus-

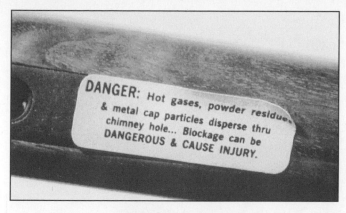

Read the instructions that come with your muzzleloader. They are your avenue to safe operation of that particular firearm. This Gonic GA 87 wears a clear warning concerning its chimney hole, which is an avenue for cap debris in front of the trigger guard.

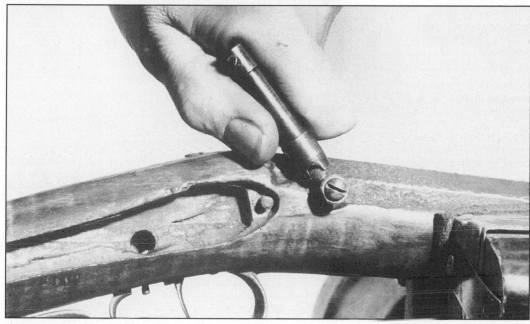

Firearm condition is vital to safety. This particular original is intact, but is it safe? A professional gunsmith should check it out for wood as well as metal damage.

tom 54 long rifle burns 120 grains of FFg behind a .535-inch patched round ball for a muzzle velocity slightly under 2000 fps, and with a good pressure-to-velocity ratio.

The Capper

Cappers are safety-oriented from at least two standpoints. First, they keep percussion caps under control, rather than being carried loosely in pockets. This is good because there is less liklihood of the caps getting into trouble—in the campfire, children's hands, or anyplace else. Second, the capper precludes having to force a cap on the cone of the nipple with fingers. On the outside chance of a cap going off during this application, it won't be pinched between thumb and forefinger when it does. I use a straight-line capper, as well as the Tedd Cash magazine capper. The capper is a convenience tool, but it is also a safety tool.

The Powder Measure and Charger

Adjustable powder measures and non-adjustable chargers are safety devices in two ways. First, they preclude pouring powder directly from a powder horn into the muzzle of a recently fired gun. Should a spark linger downbore, there is a minute chance that the powder may ignite. But with the measure or charger, only a modest amount of fuel would go off, and there would be no container to make a bomb. Second, the measure or charge prepares a correct, safe amount of powder, rather than some haphazard dose.

The Right Propellant

I know this advice seems as silly as telling a man not to grab a strange pit bull by the ears; however, there are still cases every year of shooters trying various fuels in muzzleloaders—with disastrous consequences. For this book, only blackpowder and Pyrodex were used, both of which are formulated to shoot in muzzleloaders. Just because some other powder is the color black does not make it blackpowder. Plenty of smokeless powders are black in color, too, and they can blow your muzzleloaders to smithereens.

Blackpowder Management

Already mentioned is the fact that under sufficient pressure, blackpowder may ignite. Treat it that way. Don't put super pressure on a loading rod, for example. Also, store it not only away from kids, but also separated from sources of combustion, such as heaters. Keep it in the original containers, not in glass jars. Blackpowder is also sensitive to sparks. Many mills of early days went up in smoke because an external spark ignited the product during manufacture. Tray drying of blackpowder was abandoned in the 1820s for a process using rotating wooden drums, which resulted in some explosions. The problem was prevented by adding graphite to the rotating drum. Graphite apparently dissipated static electricity charges.

But what about atmospheric sparks? In hauling blackpowder, the Army found that static sparks were directed to the metal container, not the powder, and the metal would have to get very hot in order to ignite the powder inside. Lightning would do it, of course. Having said all of this, the way to reduce the odds of an outside source of ignition setting off a charge of powder is to cap the horn or can after use, preventing the invasion of any spark. This is especially directed toward that hot little bit of potential disaster that comes from the lighted end of a cigarette—*don't smoke around blackpowder.*

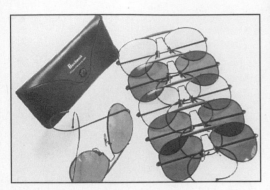
There are many fine shooting glasses on the market. They protect against flying cap debris as well as blackpowder smoke.

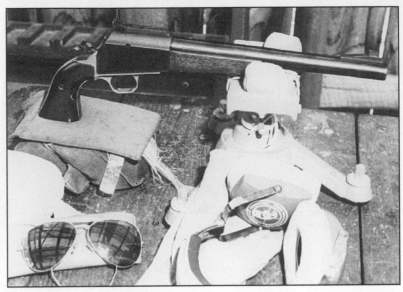
Eye and ear protection are absolutely necessary for all shooting, including blackpowder guns. Make it a habit to wear them.

As for concocting your own blackpowder, forget it. It's not worth the risk. Buy your powder factory-made. Phil Sharpe said it well in his book, *Complete Guide to Handloading*, when he warned that the manufacture of blackpowder was a dangerous practice, and he tells of various powder plants exploding. He warns: *"Home-made black powder is extremely dangerous both to make and to use!"* That's Sharpe's last word on the subject. Mine, too.

Big Game & Muzzleloaders

The blackpowder hunter should remind himself that though the odds of getting knocked around by a big game animal are small, they do exist. The hunter generally has one shot, so he must not start something he cannot finish. Mountain lions, black bears, bison, moose, grizzlies, even deer have clobbered people in recent times. Don't put yourself in a compromising situation. For example, many blackpowder hunters dream of taking a bison someday, hopefully in fair chase. While considered by some to be a barnyard fixture, the bison is not that kind of animal. Be careful. Old-time buffalo runners didn't get hurt because no big game animal is dangerous at long range. *Fear no big game, but respect it all.*

The Blackpowder Cannon

Blackpowder cannons can be fun, and safe, when used correctly. However, they are guns, and they demand respect. Proof of this fact occurred at a football game when a blackpowder cannon exploded, inflicting serious wounds. If you have a cannon, or are involved in the use of one (such as at a sporting event), treat it as a gun, not a toy. If you do load a ball, be sure to use only the prescribed powder and charge for that projectile and fire it only on a proper range. Miniature cannon balls have been known to travel over a mile.

Muzzleloader Condition

Be absolutely certain that your frontloader is in good operating condition at all times. Push forward on the cocked hammer spur with your thumb. If the hammer falls, the half-cock notch is worn. Don't use that gun again until it is repaired by an expert. Retire worn-out firearms of any kind. If in doubt concerning the condition of any gun, take it to a competent gunsmith for a checkup. And don't shoot junk. Some blackpowder guns were sold as wallhangers, but their owners insist on firing them, which can be dangerous.

Bullet Casting

Casting lead bullets, round balls or conicals, is a fine hobby in itself, but you're dealing with molten lead. Wear protective clothing to prevent burns, and use eye protection, too. Don't let water get near hot lead because it can cause an eruption. Work in a safe place. More on casting safety is presented in Chapter 24.

Accidents

Most blackpowder accidents I've studied were caused by the shooter, not the equipment. One fellow claimed that a loaded rifle he was leaning against suddenly fired because of some element in the air. Unlikely. Guns usually "go off" because the trigger is pulled. Another fellow claimed that his rifle blew to bits because there was a gap between stock and barrel and that gap collected powder, which then exploded the rifle all over the landscape. The most powder that could have been collected in the gap was 10 grains weight. Ten grains, 20, even 30 grains of blackpowder cannot blow a firearm into confetti. Another fellow said that the barrel of his frontloader was obviously of poor steel because he put down a proper load and the rifle blew up. It turned out that the ball was short-started by a foot. The old addage "accidents can happen" is true, but they are usually *caused*.

Use commerical ear plugs or ear muffs when shooting, and don't depend on anything less than the real thing. And make sure your shooting glass lenses are made of break-resistant material.

Rules of the Road

The modern Ten Commandments of shooting safety apply fully to muzzleloaders, but they fall short because of the many unique aspects of the charcoal burner. We have not, by any means, covered all aspects of blackpowder safety here; however, the above are sound rules. Sometimes frontloaders are given less respect than modern arms. I don't know why. I have seen people resting with their long rifle muzzles pressed into their bellies. The rifles were unloaded, of course, but see Twain's remarks about unloaded guns at the beginning of this chapter. Habits are funny things. They don't know when to go away. Bad shooting habits can show up any time. The sport of blackpowder shooting is too great to mar with accidents, so let's do our best to prevent them.

●

chapter 9

Replica Blackpowder Guns

SUPPOSE YOU OWNED Old Gabe's favorite rifle. I mean the real thing. Would you shoot the original plains caplock that belonged to Jim Bridger? Probably not. Nor Kit Carson's rifle, nor Daniel Boone's nor Davy Crockett's. But you certainly would like to take that oldtimer to the shooting range or attempt to bring game to the table with it. Replicas make this possible. They also serve collectors, too. For example, if you wanted to display—for study, historical interest or just plain enjoyment—guns of the Civil War, you could spend a lot of time and money rounding up originals, or you could use copycats. Naturally, the dedicated collector wants the real thing, but even the most avid may fill a niche with a replica until he can find the real McCoy.

What is a Replica?

If you insist on absolute correctness, *a true replica is a copy made by the original artisan.* In other words, a replica of the Mona Lisa would be painted by Leonardo Da Vinci himself. But we can forget this nitpicking definition. For our purposes, a replica is a look-alike that embodies the intended spirit of the original. I realize the word "spirit" doesn't really fit the picture, but it's important all the same. Spirit means that you don't have to have a screw-for-screw, bolt-for-bolt, piece-for-piece blackpowder matchup in order to have a replica. True, you can find such copies, but they're rare.

Normally, a close study of a blackpowder replica gun reveals a few differences compared with the original. It might be a sight or the curve of a buttplate, or an extra screw or bolt. Certainly, finishes rarely match up. Old-time finishes, either metal or wood, and modern finishes are not the same, although a replica can be finished in the old-fashioned way, and some are. We still have the formulas for both metal and wood finishes if we want to use them. In short, close is darn good enough when you're talking about replica blackpowder rifles, shotguns, pistols or revolvers, as well as cartridge guns, such as the Sharps line, or copies of old-time Winchesters. Close is good enough because of that

Technically, a replica is, by dictionary definition, "A duplicate of a picture, etc., executed by the original artist." However, for our purposes, a replica is a copy or near-copy of an original muzzleloader or blackpowder cartridge gun not made by the original builder, for he's long gone. Here, replica lever-action blackpowder cartridge rifles are being assembled.

word used above—spirit. The spirit of the original is intact in a reasonable copy of the "real thing."

Several companies deal in replicas, or at least in firearms that embrace the heritage and flavor of originals. One specific company that deserves special mention is Navy Arms of Ridgefield, New Jersey.

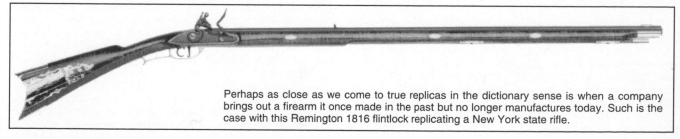

Perhaps as close as we come to true replicas in the dictionary sense is when a company brings out a firearm it once made in the past but no longer manufactures today. Such is the case with this Remington 1816 flintlock replicating a New York state rifle.

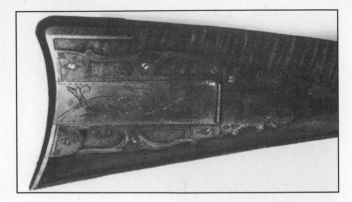

Is it an original or a replica? The modern blackpowder gunsmith is capable of producing firearms that very closely resemble the "real thing," especially after the piece has been fired and carried into the field to get that used appearance.

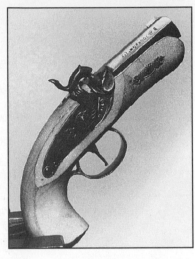

You don't have to shoot a replica to enjoy owning it, nor does it have to be an exact copy of the real thing. This little Philadelphia Derringer, built from a Traditions kit, is destined for a desktop.

Colt is famous for bringing back its own guns. Here's the Colt 1860 Army, the cap 'n' ball 44-caliber revolver of the Civil War era. There were various versions of this handgun, but all in all it was just what this replica represents today, including the engraved cylinder.

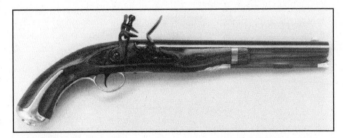

This Navy Arms Harper's Ferry 58-caliber pistol can be fired, but it also serves as a conversation piece or decoration because it is, for all intents and purposes, a copy of an original.

Navy Arms has gone well out of its way to offer a long line of replica blackpowder guns, not only muzzle-loading rifles (and rifled muskets), revolvers, pistols and shotguns, but also lever-action and single shot blackpowder cartridge rifles. In fact, the company provides more replicas than non-replicas. The president, Val Forgett, wanted it that way. He harbored a strong interest in military blackpowder firearms, and thanks to this interest we now have a copy of the great Whitworth rifle, as well as Civil War exotics, such as the 1841 Mississipi Rifle in calibers 54 and 58.

Replicas represent blocks of real history. The Civil War period is a perfect example. Modern shooters can concentrate on this era if they so choose, selecting the guns that match the time period. The Western movement, sometimes referred to as the fur trade in the Far West, is another clear-cut time period that had its own guns, especially the half-stock plains rifle. The golden age of American arms is often considered that time when immigrants from abroad plied their trade in the New World. There were many specific types of long guns in those days. We refer to these as "schools" now, such as the Lancaster School, or York School. But we're generally speaking of the Pennsylvania/Kentucky rifle in this niche, recognized by their slender stocks and long barrels. The modern blackpowder shooter needs to understand that replicas always represent a certain era. There are no exceptions, because the original guns being replicated fit into a specific period of history.

Pistols

Replica pistols can be smoothbores or rifled. Smoothbores are often represented by big single shots capable of firing a heavy projectile. Some were used as boarding pistols for the purpose of swarming upon the deck of a ship at sea. These flintlocks, generally without sights, were used at very close range. Although they fired big bullets, usually round balls, powder charges were normally quite modest. After all, low velocity provided plenty of power for such close work. Today, there are quite a number of replica blackpowder pistols, far too many to list here. A perfect example, however, is the Navy Arms 54-caliber

Harper's Ferry Flintlock Pistol, Model of 1806. It was the first U.S. military pistol manufactured by a national armory, the Harper's Ferry national armory, to be exact. There are also dueling pistols from the past, as well as target pistols. We have even had double barrel pistols reproduced, as well as the the old-time pepperbox and duckfoot pistols, which had multiple barrels.

Revolvers

Replica revolvers had significant impact on the current rebirth of blackpowder shooting, which began sometime in the 1950s and is still running strong today. Two Civil War revolvers have made the greatest impact: the Remington Model of 1858 and Colt's Model 1860 Army. Both were produced originally with many variations, and both have been reproduced with many minor variations. But they are entirely recognizable. Along with these two revolvers, dozens of other replica models have come forth. The 1851 Colt Navy is but one example. Remington's long line of revolvers has also been reproduced. Even the big Colt Walker has come back, and speaking of Colt, that company now operates a blackpowder division devoted to faithful replicas of past revolvers.

While Colts and Remingtons dominate, we're also lucky to have other great revolvers from the past born again in the present. The Rogers and Spencer 44-caliber six-shooter is a perfect example. It was one of

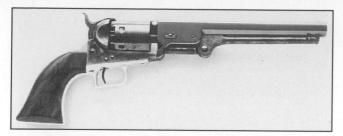

Colt has replicated a large number of its original blackpowder firearms, including the 1851 Navy caplock revolver.

Replicas serve the important function of providing a look-alike and shoot-alike firearm that either cannot be found as an original or would cost too much to risk in the field. This interesting LeMat Navy Model revolver from Navy Arms, caliber 44 plus a single 65-caliber barrel, is one of those guns.

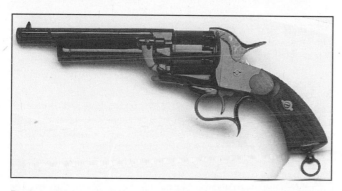

Replicas often represent a very specific firearm of the past. This LeMat is the Cavalry model, which differs from the Navy model, also sold by Navy Arms.

the finest models of the American Civil War era, and it saw service into the Spanish-American War. The Spiller and Burr 36-caliber revolver has also been reproduced. It was originally built in Atlanta, Georgia, between 1862 and 1864. In brief, there is no lack of fine blackpowder caplock revolvers copied from the past. While these generally lack the authority for big game hunting, they are enjoyable simply to own and shoot, and they make interesting target handguns as well.

Muskets

The various models of the Brown Bess Musket, caliber 75, came to America in the hands of the British soldier. Today, the Brown Bess is offered in replica for those shooters who want to experience a piece of 18th-century history. Many other muskets have also been replicated, including the Harper's Ferry Flint Rifle of 1803 in 54-caliber. The

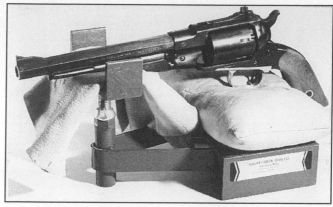

Is it a replica? Mostly. But the target sights on this Remington cap 'n' ball revolver do represent an alteration. The shooter who wants a more exacting replica of the original Civil War-era Remington must opt for the model without target sights.

Lewis and Clark Expedition was known to employ this model. Perhaps we should never have been corraled into allowing the term to survive, but "rifled muskets" were army-like firearms that resembled the military musket, but in fact were not smoothbores. We have a host of these today, including the famous and accurate Whitworth.

Pennsylvania Long Rifle

These great blackpowder firearms are replicated by custom gunsmiths today, and they usually adhere strongly to the exact style of the past. Sometimes you have to look twice to tell the difference between a contemporary Pennsylvania long rifle and the "real thing." This famous style of firearm is also replicated more loosely in many commercial models. The spirit is there, while duplication is not. It's fairly easy to distinguish between these commercial long rifles and originals. However, they do embody the shooting idiosyncracies of the old-time rifle, and many of today's marksmen choose these over customs for the obvious reason of cost. Commercial replicas are far less expensive than customs.

Plains Rifles

True replicas of original plains rifles are less prevalent than one might imagine. The Ithaca Gun Co. decided to offer a close replica of a Hawken rifle, and so they did. Later, Navy Arms took over the Ithaca Hawken, and that company continues to offer the rifle to this day, in calibers 50 and 54, both in kit and finished form. Lyman also offers a close-enough replica of an original plains rifle, appropriately named the Lyman Great Plains Percussion Rifle. In calibers 50 and 54, this rifle generates the spirit of the old-time half-stock rifle, although it is not an exact replica of an original. It's offered in a left-hand version for the southpaw, too, as well as a flintlock for those who want or need a sparktosser plains rifle.

In recent times, Mountain State Muzzleloading Supplies has taken on a line of plains rifles represented by several different models, such as the Leman Prairie Rifle. This is a modestly priced percussion in kit or finished, calibers 36, 40, 45 and 50. Rendezvous target shooters may wish to look into the 40 or 45, since this rifle does embody the general appearance and function of the original. Mountain State also offers a Mountaineer Rifle (formerly from Cheny Rifle Works) and an Elkhunter that speak of the 19th century.

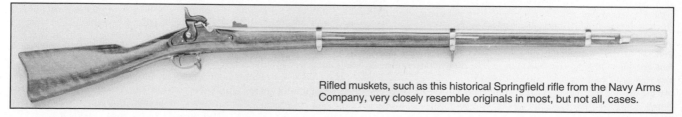

Rifled muskets, such as this historical Springfield rifle from the Navy Arms Company, very closely resemble originals in most, but not all, cases.

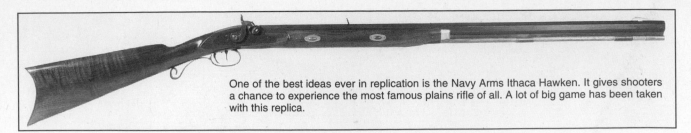

One of the best ideas ever in replication is the Navy Arms Ithaca Hawken. It gives shooters a chance to experience the most famous plains rifle of all. A lot of big game has been taken with this replica.

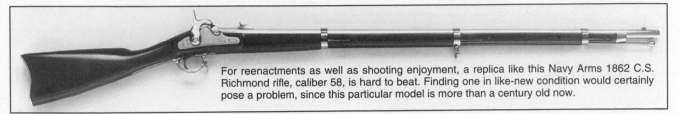

For reenactments as well as shooting enjoyment, a replica like this Navy Arms 1862 C.S. Richmond rifle, caliber 58, is hard to beat. Finding one in like-new condition would certainly pose a problem, since this particular model is more than a century old now.

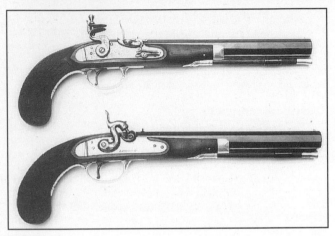

Two William Moore replicas, one in percussion, one in flintlock. Quite a pair, and they cost a lot less than originals.

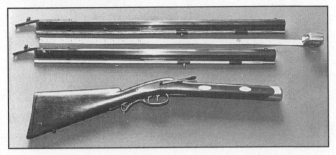

Was there ever a sidehammer rifle exactly like this one? Probably not. However, the handsome "mule ear" rifle with two barrels does represent all traits of the original era.

Mowrey Rifles

Standing in its own niche is the Mowrey replica sold through Mountain State and built by Mowrey Gun Works in the U.S.A. This rifle replicates the Model of 1835 originally associated with Ethan Allen. It's half-stock in style, but with a unique design. The lock is especially interesting because of its simplicity with so few moving parts. Today, the Mowrey percussion blackpowder rifle is offered in calibers 32, 40, 50 and 54, with either brass or steel furniture. One turn in 66 inches makes the Mowrey a ball-shooter; however, a demand for conicals prompted the factory to build a 45, 50 and 54 Mowrey with a 1:30 twist. This very different replica has been around for a long time.

Backwoods One-Man-Shop Rifles

It's difficult, if not impossible, to pigeon-hole the trustworthy muzzleloader of the past that was built primarily in the one-man gunshop, often located in the hinterlands. These were truly American rifles. I say that because they were designed in a no-nonsense fashion to suit no-nonsense Early American hunters. They were accurate, reliable and sturdy, while not boasting a lot of dash and splendor. Today, we have a few rifles that could be squeezed into this rather slender domain. One is Dixie Gun Works' Tennessee Mountain Rifle. This is a true long tom with a barrel that measures a full 41.5 inches long. Offered in caliber 32 for small game and wild turkey hunting, this rifle is about as plain as a brown paper bag, but it shoots fine and it produces over 2000 feet per second muzzle velocity with only 30 grains volume of FFFg blackpowder. It is also produced in caliber 50, flint or percussion, for the big game hunter.

There are other one-man-shop replicas around today, some built in small shops not entirely unlike those of the past. The J.P. McCoy, sold

by Mountain State, is a good example. Calibers are 32 and 45, either percussion or flint. The barrel is 42 inches long on this double-set-trigger long rifle, and the stock is made of curly maple. It's a perfect example of its genre. It wears an L&S Manton lock and a Green River barrel, and the furniture is finished in traditional brown.

Miscellaneous Historical Arms

These types of replicas can be fired, but they exist primarily because they were important in firearms history. Consider the Black Watch 577-caliber pistol from Dixie Gun Works. This 1½-pound single shot smoothbore is a faithful reproduction of an original military flintlock pistol. An interesting brass-frame beauty, yes, but not really suitable for today's target shooting or hunting. It's value is historical. Navy Arms Co.'s Colt Paterson revolver fits into the same category. This was Colt's original, as manufactured in Paterson, New Jersey, by the Patent Arms Manufacturing Co. It was the first truly successful revolver, carving a very important niche in arms history. The replica is faithful right down to the hidden trigger and five-shot cylinder.

Dixie Gun Works' English Matchlock Musket, caliber 72, is another firearm that falls directly into the historical realm. With its 44-inch barrel, this gun is quite a conversation and study piece. It replicates English matchlock firearms of the 1600s, which were prevalent prior to flintlocks. The modern version is a true matchlock, which means it is fired not by spark or flame, but from a type of fuse, as it were. The 8-pound replica is almost 58 inches long. While it may be legal to hunt with in some regions, its obvious reason for being is historical interest. The aforementioned Brown Bess, along with many other replicas, meets the criteria of historically oriented arms.

Blackpowder Cartridge Pistols

Very few of these have been offered to the shooting public in replica form. However, single shot rolling block pistols reminiscent of the past have been produced in modern times.

64

You can commission a custom gunmaker to build a rifle that closely resembles an original without making an exact replica, if that's what you want.

Blackpowder Cartridge Revolvers

Quite the opposite position is held by the blackpowder cartridge revolver. Many models have been, and continue to be, provided in replica form. The list is amazingly long. Most obvious, of course, is the Colt single-action revolver, which began life as a blackpowder cartridge sidearm. Today, there are dozens of different models, including the famous 1873 Colt. These are usually offered in 44-40 Winchester or 45 Colt chamberings. The latter is often referred to, incidentally, as the "45 Long Colt." However, the correct name is simply 45 Colt. Navy Arms recently introduced the 1873 U.S. Cavalry Model, which is an exact replication of the U.S. government issue of the Colt Single Action Army revolver. This beauty has a 7.5-inch barrel and weighs 2 pounds, 13 ounces. And it comes in 45 Colt.

But it's not all Colts—not by a long shot. The Remington blackpowder cartridge revolver of 1875 is back in a modern-made replica.

The Remington competed with Colts from about 1875 through 1889. It was, and is, a well-designed single-action revolver, now chambered in either 45 Colt or 44-40 Winchester. The Navy Arms version weighs 2 pounds, 9 ounces, and it has a 7.5-inch barrel. The list goes on and on, including the interesting Schofield revolver of 1873, a favorite of George Armstrong Custer. Jesse James carried one, too. Legend has it that Bob Ford took Jesse's Schofield after shooting Jesse in the back. This is a breaktop Smith & Wesson-type revolver modified by Major Schofield for the U.S. Cavalry.

Blackpowder Cartridge Single Shot Rifles

Montana Armory has founded a company on blackpowder single shot rifles, mainly replicas of different Sharps models, but others, such as Winchesters, as well. Today, the two major blackpowder cartridge single shots in replication are the Sharps, in one form or another, and

Replicas work. They're new in manufacture, old in style, and functional. This Navy Arms Mortimer Match rifle, caliber 54, is a good example of a working replica. It shoots well.

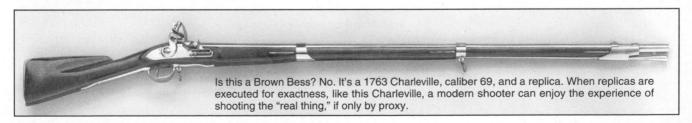

Is this a Brown Bess? No. It's a 1763 Charleville, caliber 69, and a replica. When replicas are executed for exactness, like this Charleville, a modern shooter can enjoy the experience of shooting the "real thing," if only by proxy.

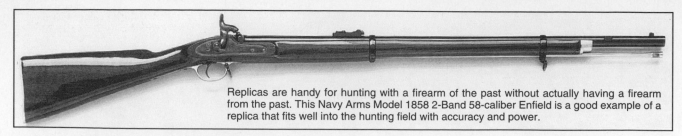

Replicas are handy for hunting with a firearm of the past without actually having a firearm from the past. This Navy Arms Model 1858 2-Band 58-caliber Enfield is a good example of a replica that fits well into the hunting field with accuracy and power.

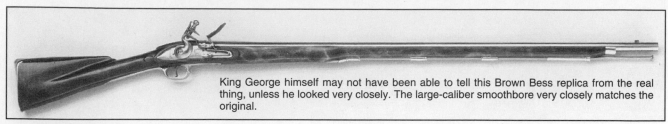

King George himself may not have been able to tell this Brown Bess replica from the real thing, unless he looked very closely. The large-caliber smoothbore very closely matches the original.

the Remington Rolling Block. Caliber 45-70 Government is most popular in these rifles and is by far the number one choice of blackpowder silhouette shooters. However, many other chamberings are also available, especially from Montana Armory. These include the original Sharps cartridge line, such as the famous 45-120-550 Sharps round. Browning continues to offer its handsome replica of the Model 1885 single shot blackpowder cartridge rifle, now in caliber 40-65 Winchester, as well as 45-70 Government. Browning admits that the growing popularity of blackpowder silhoutte matches prompted the offering of their New Model 1885 BPCR (Black Powder Cartridge Rifle). A Vernier tang sight makes this blackpowder single shot a long-range shooter.

Blackpowder Cartridge Repeaters

Mainly, these are Winchesters from the past, including the Model 1866 Yellow Boy, Oliver Winchester's improved version of the trend-setting Henry repeater. Winchester's even more famous Model of 1873 has also been replicated. For the most part, these lever-action repeaters are chambered for the 44-40 Winchester cartridge; however, the 45 Colt has also been chambered in these replicas. Not to be forgotten are Browning's modern-made replicas of past blackpowder cartridge rifles, including the Model 1886 and Model 1892.

Blackpowder Shotguns

I've left the blackpowder shotgun for last because it is a bit of an enignma as far as replicas go. The double-barrel blackpowder shotgun sold today is very much like its original brother, so much so that it fully embodies the spirit of that old-time hammergun. And yet, it does not always replicate the original. However, as far as I am concerned, the hunter with a side-by-side double-barrel caplock shotgun is experiencing the same shooting as our forefathers of the past. Furthermore, there are a few true replica blackpowder shotguns offered today. Two examples I can think of are the Mortimer, a flintlock, and the Mowrey, a single shot caplock. The Mortimer is a replica of the original English flintlock shotgun. Navy Arms offers this shotgun in 12-gauge with a 36-inch barrel. The Mowrey is a single shot based on the Ethan Allen design with 32-inch barrel. It is also available in 12-gauge, with a 28-gauge model as well.

These are some of the replicas available to the modern blackpowder fan. These guns allow today's shooter to enjoy yesterday's experience. But remember, sometimes a replica is close enough in spirit to fill the bill; it does not have to be an exact copy of a past firearm. I think this spirit business is even more important than a screw-for-screw copy of an old blackpowder firearm, because it is the spirit that supplies the reliving of a long-ago time. ●

chapter 10

Non-Replica Frontloaders

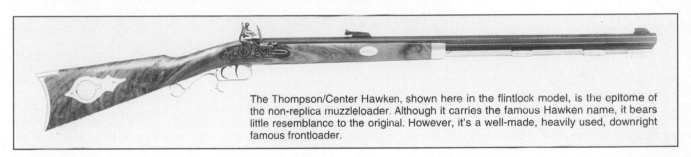

The Thompson/Center Hawken, shown here in the flintlock model, is the epitome of the non-replica muzzleloader. Although it carries the famous Hawken name, it bears little resemblance to the original. However, it's a well-made, heavily used, downright famous frontloader.

WHILE IT MAY be logical to assume that muzzleloaders disappeared the moment the metallic cartridge came along, it just didn't happen that way, nor did blackpowder cartridge firearms die out when smokeless powder became a commercial commodity. Original muzzleloaders remained in service in their own small way, and a few gunsmiths continued to build custom frontloaders. Eventually, certain arms sellers saw the light: "Hey, modern shooters want to keep on toting old-time guns!" they said. That's when replicas took hold. As catalogs prove, a shooter in the 1940s and 1950s could buy muzzleloaders, albeit they were often of rather questionable manufacture. Men such as Dixie Gun Works' Turner Kirkland and Navy Arms Co.'s Val Forgett figured it was time to give the public something better.

Why Non-Replicas

Without a doubt, replicas first stirred the imagination of modern shooters, but it wasn't long before the non-replica was off the drawing board and into the gun store. Actually, the progression was rather natural: Non-replicas represented a way to build charcoal burners using modern manufacturing methods and materials. If the shooter of the day didn't like the newly designed blackpowder guns, all he had to do was leave them on the shelf. Ultimately, that didn't happen; thousands bought and shot them instead. While the historical value of the non-replica was very low to nil, the romance of shooting was there, even though the romance of the past was not. Purveyors eventually learned something: A whole lot of people wanted to shoot blackpowder, and they didn't give a hoot about the geneology of the firearm they used to make smoke with.

This factor was a source of consternation to the purist, who believed then, and still maintains, that shooting muzzleloaders is a means of emulating the past. Certainly, old-style guns offer a greater challenge than smokeless-powder cartridge arms, and they often supply a greater reward when that challenge is met either at the target range or in the hunting field. No matter. The numbers of shooters interested in charcoal burners swelled, but their interest was not history. It was simply shooting. And good non-replicas served the purpose at reasonable prices, so the non-replica was here, and here to stay. To this hour, the non-replica blackpowder firearm remains number one in popularity among modern blackpowder marksmen and hunters, although the modern muzzleloader is certainly catching up.

Buckskinners still cling to replicas and customs when and where they can, but the "average" frontloader fan shoots non-replicas (and modern muzzleloaders). Sales prove it. Although "just plain shooting" is the thrust behind the non-replica's success, there is another reason for shooting these look-like-nothing-from-the-past guns that is even more powerful: opportunity. Today's blackpowder hunter has a special opportunity originally called the "primitive hunt," although it now goes by many different titles. While it's described by different names, it always means the same thing: a chance to hunt either before or after the regular season, often in a special place where modern long-range firearms are not allowed. Just about every state offers these blackpowder-only seasons, as proved by a quick walk-through of Chapter 5. But if you want to do the blackpowder-only hunt, you need a charcoal burner. Most of today's hunters figure that should be a rather simple, not-too-expensive, reliable frontloader, generally shorter and lighter in weight than an original or replica.

The Non-Replica "Hawken"

If purists were disappointed with non-replicas in general, you should have heard them howl when some of these guns picked up the Hawken name. Sam and Jake Hawken built the epitome of the plains rifle, and the so-called modern Hawken resembled the old-time rifle as much as Popeye looks like Batman. There was no doubt about it: Anyone with a

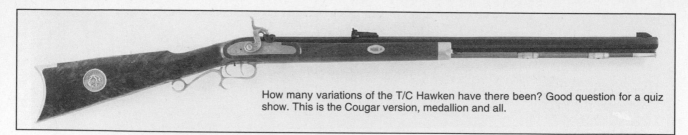
How many variations of the T/C Hawken have there been? Good question for a quiz show. This is the Cougar version, medallion and all.

sense of history had to despise the newfangled Hawken. Maybe so, but here is what happened: The non-replica Hawken took over, lock, stock and barrel, as the old saying goes. It appealed to the very shooters mentioned above—the men and women who didn't give two hoots about Jim Bridger and his plains rifle. They just wanted to shoot blackpowder. What made the newfangled Hawken so popular when it first came out? The same factors that continue to propel the non-replica to the top of the sales list today: ruggedness with manageability at a good price.

The modern Hawken non-replica was strongly made, quite reliable, more than sufficiently accurate for general target work and hunting, and yet it did not weigh 10 pounds, nor did it have a 32-inch barrel. In short, the 20th-century Hawken suited the 20th-century shooter, while the 19th-century Hawken was, it appears, more gun than most modern marksmen wanted to bother with. And don't leave out the cost factor. Build a copy of an original Hawken and I guarantee it will have to sell for more than most non-replicas. And so the non-replica Hawken made its niche in the modern shooting world. To the best of my knowledge, the well-made Thompson/Center Hawken non-replica still outsells just about every other blackpowder firearm because it fills the bill. It does everything the average blackpowder fan demands of a smokepole, and you don't have to get an extra night job to pay for one.

Of course, the non-replica Hawken didn't end with the T/C. Just about every company that went into making or importing frontloaders attached the Hawken name to at least one of their models. For example, the current Navy Arms Hawken does not resemble the same company's Ithaca/Navy Hawken Rifle. The former is their non-replica; the latter is the company's replica. On the face of it, the two rifles are extremely alike—just as a compact car is extremely like a Rolls Royce. Both cars have a frame, a body, four wheels with rubber tires, doors, seats, windows, an engine, plus a host of other appointments that are similar, such as radios and heaters. Both types of Hawken rifles also share the same general features: locks, hammers, stocks, barrels and sights. They even load the same way—from the muzzle. However, you could tell the copycat Hawken from the non-replica in the dark, by feel alone.

The bottom line is that the non-replica remains king of the hill, and the Hawken-type non-replica is at the very top of today's muzzleloader sales. Strongly made and reliable, non-replica Hawkens will be popular with shooters when we are writing "2000 and something" for a date on our letters. Here are a few currently made non-replica Hawkens that are examples of the genre: the Armoury R140 Hawken, calibers 45, 50 and 54 with 29-inch octagon barrel, weighs 8.25 pounds with adjustable sights (unlike the original Hawken, which had fixed sights—see Chapter 29). The popular CVA Hawken is available in calibers 50 and 54. It has a 28-inch octagon barrel, 8 pounds of heft, an adjustable open rear sight, a flintlock or caplock action, and brass patch box. The Thompson/Center Hawken was already mentioned. It's offered in a great variety of sub-models in calibers 45, 50 and 54. Sights are adjustable. Weight runs around 8 pounds. Percussion or flintlock choice is available. Double-set triggers are the rule. There's a left-hand model to go with the right-hand. Traditions sells their Hawken Match rifle, caliber 451 and 32-inch barrel with conical-shooting 1:20 rate of twist. It weighs close to 10 pounds and is 50 inches long overall. It has double-set triggers, blade front sight and buckhorn rear sight. The same company also offers a Hawken Woodsman Rifle in calibers 50 and 54. This rifle weighs only 7 pounds and wears adjustable hunting sights.

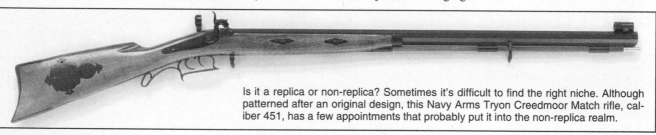
Is it a replica or non-replica? Sometimes it's difficult to find the right niche. Although patterned after an original design, this Navy Arms Tryon Creedmoor Match rifle, caliber 451, has a few appointments that probably put it into the non-replica realm.

Keeping a good thing going, Thompson/Center has added model after model. This is the T/C Hawken Silver Elite, named for its silver-colored metal.

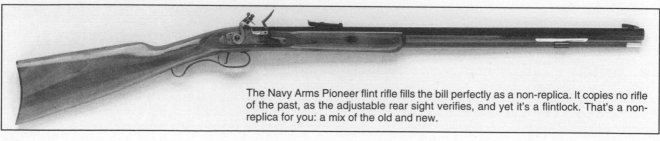
The Navy Arms Pioneer flint rifle fills the bill perfectly as a non-replica. It copies no rifle of the past, as the adjustable rear sight verifies, and yet it's a flintlock. That's a non-replica for you: a mix of the old and new.

Today, the modern muzzleloader influence has pushed itself into the non-replica arena. This Traditions Deerhunter is not a modern muzzleloader. It's a non-replica, but it's scoped, has a rubber buttpad, PVC ramrod, sling swivels, and other features associated with modern muzzleloaders.

The Compact Non-Replica

"Oh, I like this one!" That's what a friend of mine said when he picked up a little non-replica rifle I'd been testing at the range. He had just put down a beautiful custom long rifle that he felt was "pretty, but just too darn long and heavy." One of the greatest selling points of the non-replica was, and still is, compactness, along with modest weight. Remember that there were many lightweight originals in the past. Mind you, not Hawkens, which often fell into the 10-pound-plus realm, but originals of other types. For example, I own a replica of a Pennsylvania flintlock long rifle. It's a very close copy of an old rifle because the gunmaker had the original on hand when he built mine. This long tom only weighs 7.5 pounds, but is long—real long—and hardly compact. Meanwhile, take a look at the Navy Arms Hawken Hunter Carbine. It weighs only 6 pounds, 12 ounces, but more to the point, it is only 39 inches overall with a barrel length of 22.5 inches. That is compact, and yet this shorty comes in calibers 50, 54 and 58, with big game capability.

The compact aspect of non-replica blackpowder rifles is so important that further examples are in order. Let's take a look at the CVA Timberwolf non-replica rifle with its 26-inch barrel and weight of 6.5 pounds. Overall length is a mere 40 inches, a far cry from my Pennsylvania flintlock at 54 inches total length. Thompson/Center's White Mountain Carbine is shorter yet, only 38 inches overall. This 6.5-pound 50- or 54-caliber non-replica muzzleloader has a 21-inch barrel, half-octagon/half-round, and is offered in both flint and percussion. The Traditions Buckskinner Carbine is also compact, a mere 36.25

inches long with a 21-inch barrel. In calibers 50 or 54, and with a weight of 6 pounds, this adjustable-sight shorty is downright small. The buyer also has a choice of 1:20, 1:40 or 1:66 twist to match his bullet preference.

While there are many other non-replica compact muzzleloaders around today, I want to close this portion of the chapter with another Traditions model because it fits the bill so well. It's the Traditions Whitetail Series, designed for light weight. The barrels in this series run 24 inches, and this percussion rifle goes a few ounces under 6 pounds with an overall length of 39.25 inches. This rifle proves that compactness in a non-replica is a reality. While it is possible to find replica and custom muzzleloaders that are also compact, they are not nearly as prevalent as compact non-replicas.

The Non-Replica Handgun

The next chapter gets into the real rage of the hour—modern muzzleloaders, including a few handguns. Still, there are, and have been, quite a few non-replica blackpowder handguns offered to the shooting public. To be exact, pistols such as the CVA Hawken are non-replicas. They do not truly copy the old-time gun, and that is not a black mark at all. It's simply a fact. Lyman's Plains Pistol falls into the same category. CVA's Hawken, offered in 50-caliber and fitted with a barrel 9.25 inches long, possesses pretty good power capability. Lyman's pistol is made in 50 or 54 calibers with an 8-inch barrel, also capable of good power. The Traditions Trapper Pistol, 50-caliber, is another non-replica single shot blackpowder sidearm. While the majority of

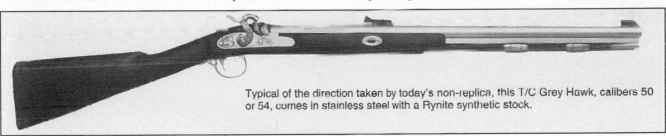

Typical of the direction taken by today's non-replica, this T/C Grey Hawk, calibers 50 or 54, comes in stainless steel with a Rynite synthetic stock.

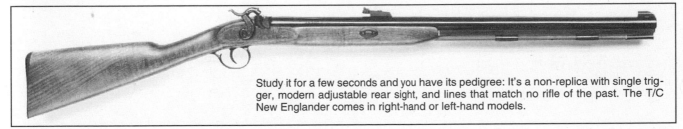

Study it for a few seconds and you have its pedigree: It's a non-replica with single trigger, modern adjustable rear sight, and lines that match no rifle of the past. The T/C New Englander comes in right-hand or left-hand models.

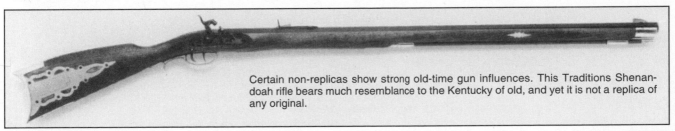

Certain non-replicas show strong old-time gun influences. This Traditions Shenandoah rifle bears much resemblance to the Kentucky of old, and yet it is not a replica of any original.

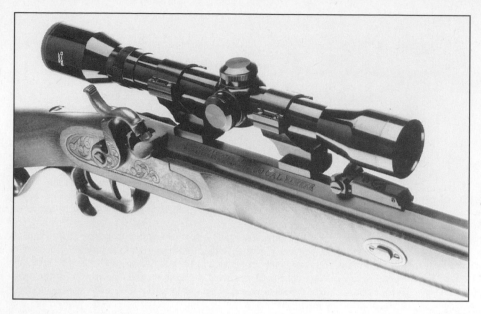

The modern muzzleloader influence is at least partly responsible for scopes mounted on non-replicas today.

blackpowder cap 'n' ball revolvers make at least an attempt to replicate originals, many actually do not faithfully copy the oldtimers, so we have blackpowder revolvers, too, that are truly non-replicas. Ultimately, the non-replica blackpowder handgun fulfills the same criteria enjoyed by the non-replica rifle—shootin' a smokepole sidearm for the fun of it, without regard for history.

The Non-Replica Shotgun

Noted in the previous chapter, blackpowder shotguns are a bit of a problem when it comes to pigeon-holing them. There are some true replicas, as well as a host of side-by-sides that are darn close enough to the old-time gun to give full spirit to the shooting. Then there are downright non-replica blackpowder shotguns that make no bones about it. The Thompson/Center New Englander is one of these. It looks like the non-replica Hawken the company already sells. It's a 12-gauge single shot with 28-inch barrel, and it only weighs 5 pounds and 2 ounces. This percussion shotgun also accepts a T/C-compatible rifle barrel. Traditions' Fowler shotgun is a non-replica, too, a 12-gauge single shot with 32-inch barrel at only 5 pounds and 6 ounces.

The Black Powder Cartridge Gun

There is no doubt about the fact that many of our blackpowder cartridge six-guns of the hour are not true replicas of the past. They are very much like the original, but do not truly duplicate it. Likewise with some of our cartridge rifles, which closely resemble originals, but do not copy them exactly. The same can be said of certain blackpowder cartridge repeating rifles. If the shooter demands a replica blackpowder cartridge gun, rifle or sidearm, he should look into the matter carefully, because he may find that his choice is more closely aligned to the non-replica than a copy of an original.

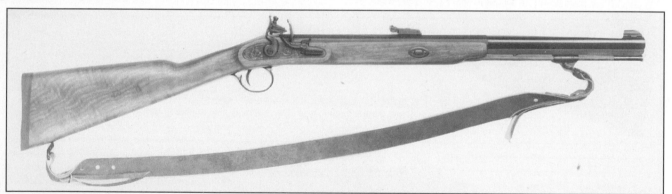

Rugged and handy. Those two features have been part of the non-replica code from the beginning. This T/C White Mountain carbine, in flintlock to meet the rules for certain primitive hunts, is built to take it. Handy to carry, this carbine has a barrel only 21 inches long with a 1:20 conical-shooting twist.

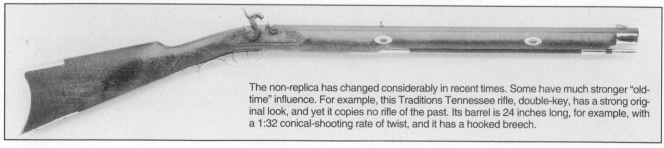

The non-replica has changed considerably in recent times. Some have much stronger "old-time" influence. For example, this Traditions Tennessee rifle, double-key, has a strong original look, and yet it copies no rifle of the past. Its barrel is 24 inches long, for example, with a 1:32 conical-shooting rate of twist, and it has a hooked breech.

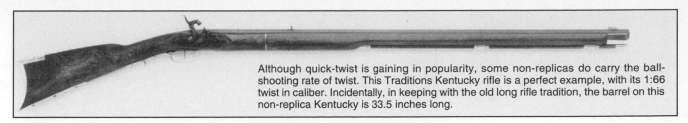

Although quick-twist is gaining in popularity, some non-replicas do carry the ball-shooting rate of twist. This Traditions Kentucky rifle is a perfect example, with its 1:66 twist in caliber. Incidentally, in keeping with the old long rifle tradition, the barrel on this non-replica Kentucky is 33.5 inches long.

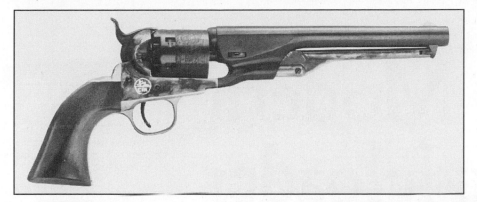

Copycat or not? This is another of the guns that rides the fence. But the CVA logo on the frame tells us that the company had no intention of making this 1851 Colt Navy a replica, even though it's a definite look-alike to the real thing.

It's not all rifles. This pistol is also represented in the non-replica class. This Traditions Kentucky pistol is a perfect example. It's not a replica, but it closely resembles the oldtimer. Rate of twist, Incidentally, is 1:20. It'll shoot conicals.

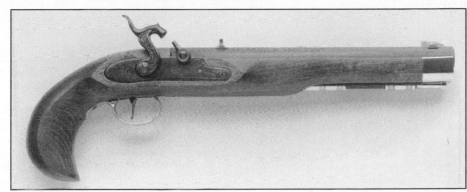

Should I Buy a Non-Replica?

The object of a book like this is to tell the truth, and the truth is, very few shooters who bought non-replicas when they first appeared actually had much experience with blackpowder. It's my opinion that only a few knew for a fact that the "Hawken" they bought had nothing to do with the Hawken carried by the mountain men of the Far West Fur Trade era of the 19th century. These shooters were not duped. They simply didn't have a point of comparison, unless they were students of muzzle-loading arms, which the vast majority were not. Eventually, most of us became familiar with the original muzzleloader, not only in the Hawken style, but in many other types, such as the rifled musket.

At that point, a few shooters said, "Hey, my blackpowder rifle is nothing like those true old-time originals." Some then opted to look into replicas or even customs, but it's obvious that the vast majority did not. They simply went on buying and shooting non-replicas. Why? I think the reasons were outlined in the beginning of this chapter. That is exactly the way things stand today: The 20th-century blackpowder

shooter, on the threshold of the 21st century, has spoken. He continues to buy, shoot targets with and hunt with the non-replica form of muzzleloader. And when he switches, he's more likely to go for a modern muzzleloader than he is a replica of an original or a custom. That's just the way things are. But the question has not been answered. Should you buy a non-replica?

If you don't care about the history of the blackpowder firearm, if you are not a collector who wants to fill a niche temporarily or permanently with a copycat, if you are not interested in experiencing the guns of yesteryear or taking on the full challenge of working with designs of the past; if you simply want to shoot blackpowder in an expedient manner, using well-balanced, well-made guns that do not copy firearms of days gone by; if you prefer lighter, shorter, easier-to-carry and a bit handier-to-clean blackpowder guns over originals, then go for non-replicas by all means, especially if you just want to take part in a blackpowder-only big game hunt, but don't want to spend a lot of money on your rifle. If these things are true, buy a non-replica. It's the blackpowder gun for you.

●

The Modern Muzzleloader

Davey Crockett would never have recognized this Knight LK-93 Wolverine Thumb-hole Stainless with Mossy Oak Treestand camouflaged stock as a rifle that shoots blackpowder. But it is, because it's a modern muzzleloader.

WHEN THE REMINGTON Arms Company offered a special limited-edition flintlock rifle, I was not surprised. After all, Remington was there from the beginning, making flintlocks since 1816. The firm's famous name on a flintlock made sense. But when the reknowned manufacturer came out with a modern muzzleloader, I was surprised. Still, there it was all the same, a Model 700 bolt-action rifle if ever I saw one—built to be a muzzleloader. The Remington Model 700 ML and Model 700 MLS blackpowder rifles were announced as "a marriage between the past and the present." They verified what we all knew—blackpowder shooting was "hot stuff" in America, and the modern muzzleloader was definitely on fire.

Why a Modern Muzzleloader?

The modern muzzleloader is a natural progression from the non-replica blackpowder firearm. The many thousands of hunters and marksmen in America, Canada and across the world who desired to simply shoot blackpowder in the most expedient way got more of what they wanted. The theme of the modern muzzleloader is to come as close to a contemporary firearm as possible, while maintaining the basic function of a blackpowder gun. In short, the modern muzzleloader is a rifle, handgun, or shotgun that loads from the front end, just like any other charcoal burner, but with appointments that make it look and handle like a smokeless-powder gun. The specific aspects of the modern muzzleloader are attended to individually below, which is important to understanding the phenomenon that has taken blackpowder shooting by storm in the past decade.

History is a great teacher. In looking at the modern muzzleloader, it's useful to glance back in time to see how we arrived at today. Did you know that some of America's best hunters did not rush to the smokeless-powder gun when it was readily and widely available? Well, it's true. A stampede to new things is the American way, but jogging at the periphery of the crowd there are always a few stalwarts who don't like newfangled gadgets and futuristic ideas. What has worked for them over the years is not only good enough, but probably better than new, not to mention morally and aesthetically correct. A perfect example of a well-known (for his time) hunter/writer who stayed with blackpowder guns when smokeless cartridge guns became available was the great woodsman George W. Sears, known as Nessmuk to his many readers.

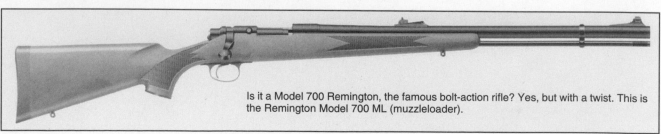

Is it a Model 700 Remington, the famous bolt-action rifle? Yes, but with a twist. This is the Remington Model 700 ML (muzzleloader).

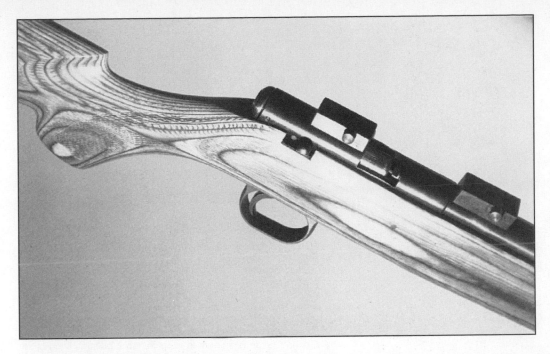

This Gonic Magnum shows the breech style so prominent on modern muzzleloaders these days, with in-line ignition. That is, the percussion cap flame is in line with the bore, so the spark darts straight into the back of the charge as the charge rests in the breech.

Nessmuk continued happily with his muzzleloader. In discussing a particularly excellent deer hunt in *Woodcraft and Camping*, he pointed out that his rifle was a "neat, hair-triggered Billinghurst, carrying sixty round balls to the pound, a muzzleloader of course, and a nail-driver." E.N. Woodcock, another respected outdoorsman of his era, embraced the cartridge rifle, but did not find smokeless powder appealing. "I would not hesitate to say that I prefer the 38-40 and blackpowder," he said in *Fifty Years a Hunter and Trapper*. He found his old 38-40 with blackpowder plenty fine, thank you.

I think it's important to recognize that many great shooters stayed with blackpowder guns when the "newfangled" firearms came along, because it helps us understand why the modern muzzleloader enjoys a stampede of success on one hand with firm opposition on the other. At the close of this chapter, there's the same question asked of the non-replica muzzleloader: Is the modern muzzleloader for you? Maybe. Maybe not. You have to do blackpowder shooting your way. But consider all points before deciding. If you end up sticking with your "old-fashioned" frontloader, you'll be like Nessmuk and Woodcock, who were perfectly happy with their "outdated" shooting instruments. But you just may want to go modern, too.

But before you decide, you may as well know it before we go into the details concerning the modern muzzleloader—the new gun has caused concern among some wildlife managers. It seems that they are as split on the subject as shooters. Today's blackpowder enthusiast should be aware that not all game department leaders are happy about the modernizing of the blackpowder firearm. Some have gone so far as to suggest severe limitations on these updated smokepoles. They consider their use as cheating. One game manager stated that the modern muzzleloader had no historical value. It was entirely untraditional, simply a blackpowder shooting gadget, not a "real muzzleloader," a way of getting around the challenge of muzzleloading, a means of skirting the rules. "Blackpowder hunters asked for special seasons because they were handicapped. Now after getting their hunts, they're doing everything possible to erase the handicap," he announced.

A comparison of modern muzzleloaders to compound bows has been presented by several officials. Archers, these game managers complain, got early, late and extra-long seasons in special places because they were hunting with a primitive tool. However, once they had these unique hunts, they set out to turn the ancient bow into a high-tech tool with sight pins, scopes, releases, relaxation factors and every gimmick that can be imagined. The same people contend that the modern muzzleloader is like the compound bow—just a way of getting around the rules and the challenge. Meanwhile, the modern muzzleloader continues to grow in popularity, and its legal use on hunts has been denied in relatively few places.

Properties of the Modern Muzzleloader

There is nothing new about introducing the flame of the percussion cap into the breech of the frontloader by tapping a hole directly into the barrel and inserting a nipple. Both the underhammer and side-hammer accomplished this many years ago. Today, the understriker and mule ear—two different names for the underhammer and side-hammer—are not very popular, but both remain in the lineup. The sidehammer, in fact, is the forte of custom gunmaker Dale Storey, as noted in the next chapter. The modern muzzleloader enjoys direct ignition, but not by tapping a nipple into the barrel. Rather, the nipple becomes a part of the breech.

In-Line Ignition

It is difficult to find entirely new ideas in the world of firearms. So it is with in-line ignition, which is a very old concept. For example, the Paczelt flintlock in-line rifle existed long before the telephone. (See *1996 Gun Digest*, "The In-Line Muzzle-Loader," by Doc Carlson.) Today, in-liners are not flintlocks. They are percussion firearms using a percussion cap or modern primer. There are many variations, but most modern muzzleloaders that employ in-line ignition use some form of plunger. The plunger system, whereby a striker of some sort lies on the same plane as the nipple, the percussion cap and the breech, is also old hat, but the specific design of the modern muzzleloader's in-line system is up-to-date: A nipple is tapped into the breech plug, putting the percussion cap or primer directly behind the breech. A striker-hammer flies forward to detonate the cap or primer, with fire darting directly into the breech to ignite the powder charge. This creates a relatively quick lock time (the time elapsed from the pulling of the trigger to the gun going off). There are modern muzzleloaders that do not employ in-line ignition, but most do, and so it is a major trait of this new type of blackpowder firearm.

Loading Style

The modern muzzleloader is charged from the "front end," just like any other blackpowder muzzleloader. This aspect has allowed the firearm to qualify where blackpowder guns only are allowed on special hunts. While speedloaders are at home with modern muzzleloaders, they are equally functional with other guns, so we can safely say that loading a modern muzzleloader is not unlike loading any other type of blackpowder firearm.

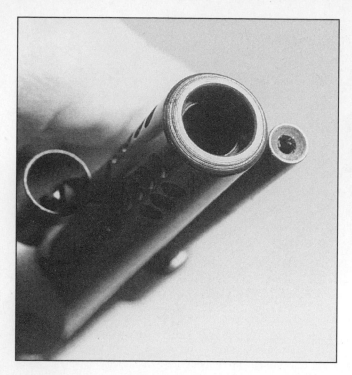

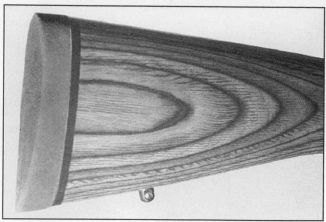

This modern muzzleloader stock is, first of all, laminated wood. Then it wears a full-scale recoil pad. And notice the sling swivel eye—all traits of modern rifles.

The Gonic Magnum modern muzzleloader uses a recessed muzzle for two reasons. One is good bullet alignment to the bore. The recessed muzzle allows the bullet to enter perpendicular to the bore. The second reason is ease of loading. When a bullet is lined up, it slides downbore easier.

While it should go without saying, remember that the modern muzzleloader shoots *only* blackpowder, Pyrodex or other blackpowder substitute for muzzleloaders. The use of any other type of propellant can constitute great danger to the gun and the shooter.

Rate of Twist and Projectiles

There remain a few modern muzzleloaders with what might be called a rather slow rate of twist, but they are waning. It is safe to say that a fast rate of twist is an important feature of today's modern muzzleloader. Twist is, of course, a comparative feature (see Chapter 27). Compared with a ball-shooter that may have a 1:66 rate of twist, a modern style may have a 1:20 to 1:40 rate of twist. The significance of this quicker twist is conical stabilization. Of course, the old-fashioned round ball may also be shot from a quick-twist gun, especially if depth of groove is deep, thereby providing a fairly tall land.

The tall land of a fast-twist barrel may help grip the ball, controlling the patched lead sphere in the bore. But there is always a chance that the ball will, as they said once upon a time, "strip the bore" or "trip on the rifling," meaning it rides over the lands rather than being guided by them. A full-throttle powder charge may cause this situation in a quick-twist bore, whereas a light charge may not, due to lower velocity. When a ball does in fact "trip on the rifling," the gun is essentially a smoothbore as far as accuracy is concerned. So the modern muzzleloader, with its comparatively fast rate of twist, is a conical-shooter by nature, and that is important to recognize. Except for low velocity plinking and perhaps small game shooting, there's no reason to put a round ball down the bore of a modern muzzleloader.

Because of the fast-twist nature of the modern muzzleloader, many newly designed blackpowder projectiles have surfaced. One is the modern jacketed pistol bullet installed in a plastic sabot. The idea of a sabot (pronounced sah-bow) is hardly new. The term originally meant "wooden shoe," or "wooden work shoe," which paints the picture of a unit that surrounds the projectile the way a shoe surrounds the foot of its wearer. In modern times, sabots have been used to shoot sub-caliber projectiles. Remington's Accelerator ammunition does just that. It allows the use of a 22-caliber bullet in a larger bore by encasing the smaller bullet in a plastic sabot. Thereby, the 30-30 Winchester, 308 Winchester, and 30-06 Springfield cartridges, as examples, can shoot 22-caliber bullets.

Shooting a sub-caliber projectile in a muzzleloader that is capable of launching a heavyweight projectile does not always make sense, but the fact remains that a great many hunters find the lead or jacketed pistol bullet accurate and effective when fired with a sabot from a modern muzzleloader. For example, the current world-record elk was taken with such an arrangment using a Knight modern muzzleloader and a Sierra jacketed pistol bullet with sabot. Many other bullet designs have followed the advent and popularity of the modern muzzleloader, including a host of conicals. (See Chapter 17 for more on conicals.)

Choosing the correct bullet for the modern muzzleloader is no different from finding the right bullet for any other blackpowder gun—trial and error. The individual marksman is encouraged to head for the range, more than once, with several prospective missiles, shooting for accuracy while keeping careful records of the resulting target groups. A pattern will emerge. Hopefully, different bullets will fire with accuracy at good velocity. However, guns are individual in nature, as anyone who has handloaded for modern guns knows, and muzzleloaders are no exception. Bench-tests with some modern muzzleloaders have revealed best accuracy with jacketed bullet/sabot combos, while other modern frontloaders like lead conicals better. So test your gun to find out what it shoots best.

Modern Muzzleloader Ballistics

It's wise to get this extremely important aspect of the modern muzzleloader out of the way right now, because there's some confusion about how much punch these new guns really have. They have a lot. For example, Gonic's Magnum is allowed a big powder charge behind a heavy bullet, and that spells ballistic authority in no uncertain terms. Most modern muzzleloaders are capable of shooting 50- and 54-caliber conicals at reasonable blackpowder speeds. But we must not think of the modern muzzleloader as modern in power. It is not a 338 Winchester! Furthermore, it is wrong to consider modern muzzleloaders as necessarily more powerful than traditional muzzleloaders.

The most powerful muzzleloader I ever owned was a Hawken-styled heavy-barrel 54-caliber, a strong-breeched rifle capable of withstanding big charges of FFg or Pyrodex RS behind a heavy lead bullet, such as the 460-grain Buffalo Bullet round nose. A velocity of 1700-1800 fps was achieved with that bullet from this traditional-style rifle for close to 3000 foot pounds of muzzle energy. But the aforementioned Gonic Magnum is no slouch, either. Shooting the heavy 586-grain Gonic bullet, the 50-caliber version earned 1510 feet per second for a muzzle energy of 2968 foot pounds. Recall that the great 30-06 Springfield with a 180-grain bullet at 2700 feet per second muzzle velocity gains 2914 foot pounds of muzzle energy.

The Prairie River Arms PRA 50 is a bullpup muzzleloader. The bullpup design, with breech way back in the stock, makes for a very short overall length.

Another trait of most modern muzzleloaders is cleaning ease. When the plunger-bolt is removed from this one, it can be cleaned from the breech, which is very handy.

Of course, the modern, jacketed, streamlined, high-ballistic-coefficient bullet retains its velocity/energy better than the blunter black-powder bullet, but within normal frontloader ranges, the modern muzzleloader is capable of plenty of big game power. Warning: A heavy blackpowder bullet may in fact pass right through the chest cavity of deer-sized game without telling effect. If bone is hit, it's quite another matter. Recommended: On truly big game, shoot more into the shoulder (scapular region) than behind the shoulder and always follow up on shots to make certain that you truly missed, should the animal run off after you fire at it.

Thin-jacketed pistol bullets designed for normal handgun velocities tend to open up well on thin-skinned game and therefore may provide an apparently more pronounced effect on deer-sized game compared to heavy lead conicals. This is the reason that some hunters report top results with the pistol bullet/sabot, even though the greater force rests with the larger lead conical projectile. The pistol bullet, as well as lighter lead missiles with hollow points, tends to shed more of its energy in a light big game animal than a heavy bullet whistling through and delivering a knockout blow to a tree in the background.

Overall Configuration

We've spoken of in-line ignition and fast twist as traits of the modern muzzleloader, but overall configuration cannot be ignored. The Remington Model 700 ML and 700 MLS are perfect examples of overall configuration—the shape or lines of a firearm. The modern muzzleloader not only looks up-to-date, it handles that way. Mounting one to the shoulder gives the feeling of familiarity. In fact, that's the

whole point of the design—for the shooter to feel that he is firing a contemporary gun when he handles a modern muzzleloader. It looks modern. It handles modern. Every feature from stock fit to safety is modern. The overall design of the modern muzzleloader is so contemporary that sometimes you have to look twice to see proof that the gun is a muzzleloader.

Along with modern-style stocks, recoil pads are at home on modern muzzleloaders. With heavy loads, recoil pads are a welcome addition to these guns.

Sling swivels are commonplace on modern muzzleloaders. In other words, slings on these guns are the norm. This feature lends another aspect to the modernness of the modern muzzleloader, for most contemporary bolt-action cartridge rifles wear slings for carrying, as well as shooting steadiness.

The modern muzzleloader is also very easy to waterproof for most models. Some are darn near waterproof without adding much more than a rubber muzzle cap. This is an important feature for anyone who hunts in wet weather.

Removable Breech Plug

Modern bolt-action rifles are readily cleaned from the breech end. Muzzleloaders are normally cleaned from the muzzle end. Breech-cleaning is faster and easier than muzzle-cleaning. Almost all of today's modern muzzleloaders have a removable breech plug, allowing cleaning from the back end, just like a contemporary bolt-action rifle. I feel this is one of the modern frontloader's best features, a real maintenance advantage.

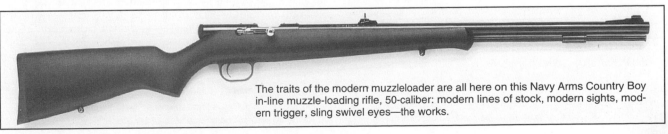

The traits of the modern muzzleloader are all here on this Navy Arms Country Boy in-line muzzle-loading rifle, 50-caliber: modern lines of stock, modern sights, modern trigger, sling swivel eyes—the works.

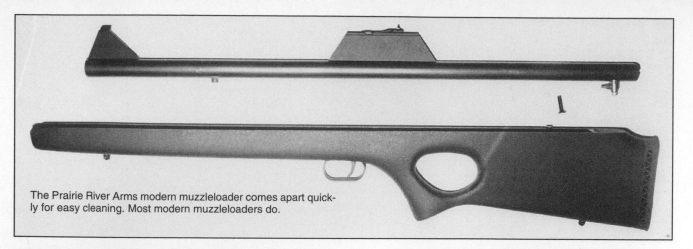

The Prairie River Arms modern muzzleloader comes apart quickly for easy cleaning. Most modern muzzleloaders do.

The idea behind the modern muzzleloader is familiarity. This Knight MK-85 exhibits a multitude of traits found on modern arms, including the shape of the stock, the sights and the trigger.

Scoping the modern muzzleloader is matter-of-fact. This scoped Thompson/Center ThunderHawk rifle has a Weaver-style, detachable mount with See-Through rings for use with iron sights.

Handling Qualities

Broached briefly in "Overall Configuration," the handling qualities of the modern muzzleloader are just that—modern. This aspect shows across the board for the ilk. Rather than trying to describe what a modern muzzleloader handles like, it's easier to say what it does not handle like. It is not a Pennsylvania/Kentucky long rifle, or a rifled musket, or a plains rifle, or any other style from the past. Stocks appear and "feel" modern. Drop at comb is modest. The stock is rather straight. Pistol grips are common. So are cheekpieces. Also, modern muzzleloaders are seldom "nose heavy." They have good balance. Albeit the so-called nose-heavy long tom of yesteryear, with plenty of drop at comb of stock, was one fine offhand shooter, the modern muzzleloader is much more familiar in how it hangs for the shooter.

Modern Materials

Synthetic stocks are at home on modern muzzleloaders, certainly not on replicas. Other modern materials are incorporated in these guns as well. Remington's Model 700 ML, for example, has a blued carbon steel barreled action. The Remington Model 700 MLS has a receiver and barrel made of 416 stainless steel. Each of these guns wears a fiberglass-reinforced synthetic stock.

Sights

Modern muzzleloaders generally have modern adjustable iron sights, usually of the open rear sight style, but also aperture or peep sight options. Folding rear sights so often found on modern cartridge rifles are often used on modern muzzleloaders.

On the other hand, most modern muzzleloaders are drilled and tapped for scope mounts as a rule and are expected to wear glass sights. Scopes work well on these guns because of the modern-style stock, which normally puts the eye in line with the ocular lens of the scope. This factor is true of some non-replica frontloaders, too, but it is not true of older-style stocks with plenty of drop at comb, where the eye sees exactly where the iron sight rests low on the barrel. The fact that modern muzzleloaders are expected to wear scope sights was one aspect that bothered some game managers, who saw the use of a scope as overstepping the original intention of the primitive hunt.

Of course, scope sights are very old, so it's not a simple matter of using history as a criterion. Some states allow scopes on blackpowder guns for special primitive hunts; other states do not. (See Chapter 5 for more information.) Some hunters thoroughly enjoy scoping the blackpowder rifle to gain the obvious advantage of a magnified target and better bullet placement, not only on paper, but also in the hunting field. In spite of the fact that scopes were used on 19th-century muzzleloaders, today's blackpowder traditionalists do not, and I can safely say never will, accept glass sights on smokepoles. Peep sights were also found on old-style smokethrowers, yet these, too, are basically unacceptable to traditionalists.

Workmanship

Good workmanship attends the modern muzzleloader. Naturally, this attribute is found on any good blackpowder firearm, original, repli-

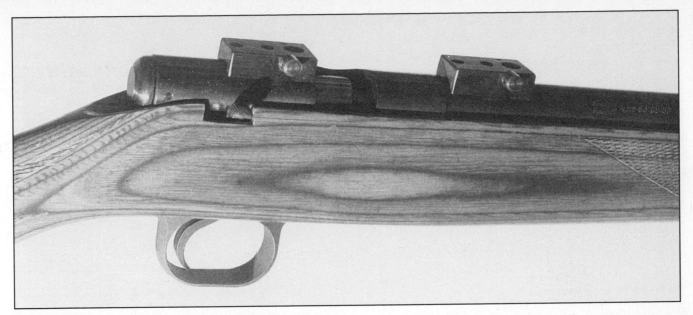

The Gonic Magnum uses scope blocks for mounts. Note, too, the single adjustable trigger, as well as the strong laminated stock. Also see the modern-style sliding safety. It's easy to see why the modern muzzleloader is called modern.

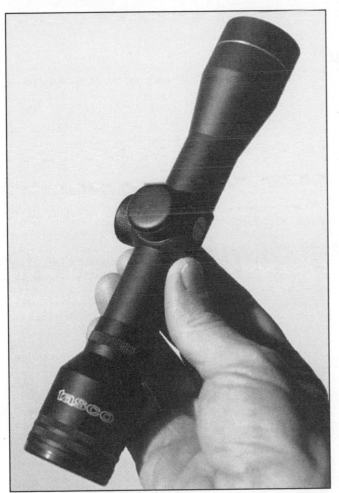

Sometimes smaller compact scopes are used on modern muzzleloaders. This Tasco compact is a good example. Some hunters feel that they still have a comparatively short-range rifle even in the modern muzzleloader, and they don't need super-power scopes.

ca, rifled musket—the entire clan. But it was soon obvious to me that modern muzzleloaders as a rule are extremely well-made. And that's a good trait of this new type of frontloader. Combined with top-grade materials, the well-constructed modern muzzleloader is rugged and reliable in the field.

Operational Features

The White Systems Model 91 is typical of a modern muzzleloader with a handful of special features right down to the sliding safety. Modern safeties abound on modern muzzleloaders. The usual safety is of the sliding variety, but others can be found. Some have double safeties, where the plunger can be locked up as well as the trigger.

Modern triggers are found on the "newfangled" old-time blackpowder firearm. In fact, some of these triggers are taken directly from contemporary bolt-action rifles.

While most modern muzzleloaders employ the standard No. 11 percussion cap, some have gone to modern primers. For example, Gonic's Magnum can use a No. 209 shotgun primer, as can certain Knight firearms.

Most modern muzzleloaders break down easily with the proper tools, which are provided with the gun. This is a nice feature, since removal of the breech plug allows breech-end cleaning.

Barrels

The octagonal barrel so often seen on replica, and even non-replica, charcoal burners is rarer than a heat wave in January in Nome. Round barrels prevail. So do short barrels—short, that is, by usual blackpowder standards. While many non-replicas also wear shorter barrels, as opposed to originals and replicas, the modern muzzleloader's trademark is compactness via a short barrel. Ballistic tests proved that good big game power could be realized from shorter blackpowder barrels, and that did it. Barrels sometimes get real short on the newfangled guns. For example, there's a Knight modern muzzleloader with a 20-inch barrel. I tested one in 50-caliber and it got 1541 feet per second at the muzzle using 120 volume Pyrodex RS with a 350-grain Buffalo Bullet conical for over 1800 foot pounds of muzzle energy.

It almost goes without saying that compactness with shorter barrels equals modest overall weight. The modern muzzleloader is seldom as heavy as most originals or replicas, and is lighter in weight than many non-replicas as well.

The Modern Muzzleloader Pistol

Look for more of these—shooters like them. The Thompson/Center Scout Pistol is a good example of the type. It's a true blackpowder muzzleloader in every sense, but it's modern in design, a takeoff on

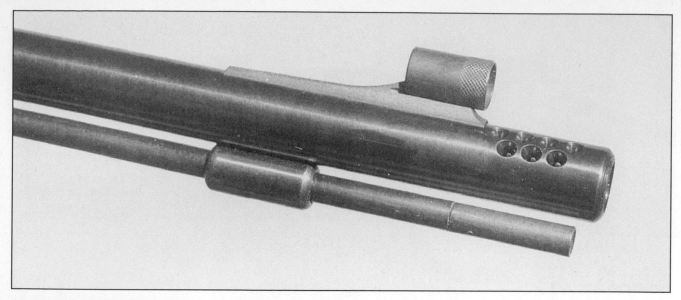

Gases diverted through these holes in the muzzle of the Gonic Magnum modern muzzleloader do a great deal to control muzzle arc. The Gonic Magnum is allowed a big powder charge, so there's plenty of gas to work with.

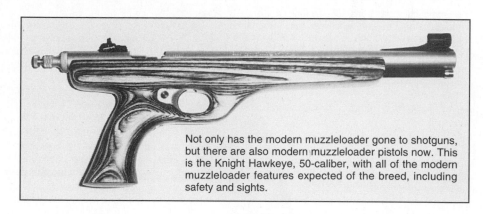

Not only has the modern muzzleloader gone to shotguns, but there are also modern muzzleloader pistols now. This is the Knight Hawkeye, 50-caliber, with all of the modern muzzleloader features expected of the breed, including safety and sights.

the Scout Carbine. In calibers 45, 50 and 54, this is a very powerful handgun, capable of taking deer-sized and even larger game in the hands of a good hunter and marksman. It has a 10-inch barrel, is 15-inches overall and weighs 40 ounces. In-line ignition coupled with adjustable sights lend to its modern style. The White Shooting Systems Javelina is another modern muzzle-loading pistol. It a stainless steel beauty with Insta-Fire ignition and adjustable sights. Calibers 45 and 50, 14-inch barrel, drilled and tapped for scope sight, 5.2 pounds weight, match-grade trigger, too. There's also the Knight Hawkeye pistol, another fine entry into the domain of modern muzzleloaders. This is a target-shooting blackpowder handgun, with a 12-inch 50-caliber barrel and 1:20 twist for pistol bullet/sabot shooting. It has in-line ignition and a double safety system. The breech plug is removable for cleaning, and the Hawkeye has a fully adjustable modern trigger. It's also drilled and tapped for a scope.

The Modern Muzzlelaoder Shotgun

Not to be left out, the blackpowder shotgun has also gone modern. Knight's 12-gauge is allowed a heavy powder charge. It also has a Full-choke insert for tight patterns. Put the two together, and turkey/goose hunting comes to mind. This is a true modern muzzleloader with all the features normally associated with the well-known Knight rifle. Also in the hunt is the White Shooting Systems White Thunder Shotgun—a modern muzzleloader all the way. It's a 12-gauge shotgun with Insta-Fire ignition and 26-inch vent-rib barrel. It only weighs 5.25 pounds. Choke tubes include Improved Cylinder, Modified and Full. White also has a "Tominator" modern muzzleloading shotgun. It's 12-gauge, too, but it also comes with a Super Full Turkey choke.

Should I Buy a Modern Muzzleloader?

As the threads of the modern muzzleloader story weave into a fabric, it is clear that this new way to shoot blackpowder does not necessarily surpass other frontloaders in range, rapidity of fire, out and out power, or even accuracy. Practical shooters, however, love it, while purists hate it. It's a personal matter, of that there is no doubt. But one thing the modern muzzleloader is for sure is a muzzleloader. It shoots no farther than traditional frontloaders using the same bullets. It has to be loaded from up front, one shot at a time, and cleaned after shooting, just like any other muzzleloader. It simply handles a heck of a lot like the familiar cartridge guns we all grew up with.

If the modern muzzleloader has a problem, it is one of image. It is perceived by some as a super shooting mechanism that is not truly a muzzleloader, but an excuse for one—a cheat, if you will. As mentioned earlier, some game department officials also feel that way, as proved by the rifle's exclusion from certain blackpowder-only hunts. Oddly enough, this perception is promoted by the very people who have the most to lose—the manufacturers, who are following in the footsteps of compound bowmakers, each one trying to outdo the other with a list of special features longer than a 20-mile detour. Modern muzzleloader makers would be better-served by "telling it like it is." The updated frontstoker is a well-made shooting iron with contemporary features, a different form of blackpowder thunderstick with a bright present and a solid future—as long as it isn't excluded from special blackpowder hunts because game officials take modern muzzleloader advertisements too seriously. ●

chapter 12

The Custom Muzzleloader

DURING AMERICA'S "GOLDEN Age of Firearms," every one-man, one-room gunsmithing operation in the country was a custom shop. A gun shop had to be self-contained. There was no way for a smith to airmail for a spare part, because there was no airmail, so he often had to fashion a piece himself. But there were parts to be bought and even the best-known armsmakers purchased them. The Hawken brothers, for example, ordered many parts for their rifles. But the local gun shop remained the star of hope for those who needed new firearms or the repair of old ones. The star, yes, but its planets were many: locksmiths, spring makers, hardware manufacturers, and more. Sideplates to nosecaps were made as specialty items. Expert barrel makers surfaced early in our country, too, and they're still at work. However, in spite of parts makers being on the job, there remained a body of do-it-all-yourself gunsmiths, who could pound a barrel out of a hunk of steel, rifle it, stock it, and even engrave the finished product.

These intrepid craftsmen could hand build a powder horn and shooting bag as well as construct a complete rifle. They did it all without electricity, too, the bulk of their tools powered by hand or perhaps the wheel of a watermill. Ah, you ask, where did they all go, those marvelous gunmakers who could create with their skills a rifle, even a sidearm, of grace, beauty, and exceptional function? They didn't go anywhere. Chances are, there's a custom muzzleloader gunmaker in your town. I live in a relatively out-of-the-way part of America, but my city has an expert gunmaker named Dale Storey. The modern blackpowder riflemaker has power-driven tools and he happily purchases many gun parts from individual experts. But he still embodies more than the Emersonian self-reliance of the old smith. Today's custom muzzleloader makers, the good ones, are doing work which is on a par with, and sometimes a cut above, the best gunmakers of the past.

Nobody has to have a custom-made muzzleloader. Over-the-counter models are good enough for any shooting function, and there are enough replicas around to satisfy the needs of most historically-minded shooters. So "need" is the wrong word. However, I feel that every

This fine custom dueling pistol from Kennedy Firearms has a 10-inch full octagon swamped barrel that goes from .910-inch to .755-inch, and then back up to .780-inch diameter. Barrel making in the old days was sometimes a singular business. It still is with some makers.

dedicated blackpowder shooter *deserves* just one special-made front-loader, either rifle or pistol. Truly, there is nothing quite so fine as a properly handmade firearm. It's like owning a great painting, a hand-finished myrtlewood coffee table, or a handmade antique chair. The custom long rifle or pistol earns its keep because it is in itself a functional work of art, a joy to own and to shoot. Unfortunately, not every handmade long rifle or pistol is a true custom.

A True Custom Blackpowder Gun

A "real" custom is not a *customized* rifle. There's a big difference. Just about any woodworker can take an existing rifle and alter it in some way, sometimes to the detriment of the original. The true custom rifle or pistol, on the other hand, is unique. It is *your rifle*, and there is not another exactly like it anywhere, although the work of the gunmaker may be recognized as his individual *style*. Furthermore, a custom rifle is, if not perfect, at least a powerful attempt toward the goal of perfection. There

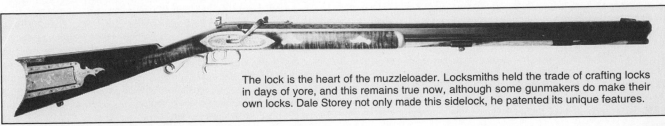

The lock is the heart of the muzzleloader. Locksmiths held the trade of crafting locks in days of yore, and this remains true now, although some gunmakers do make their own locks. Dale Storey not only made this sidelock, he patented its unique features.

History is very much a part of custom blackpowder gunmaking. This fine French Type D Trade pistol is from the 1730 to 1765 period. Kennedy Firearms handcrafted it in 54-caliber smoothbore.

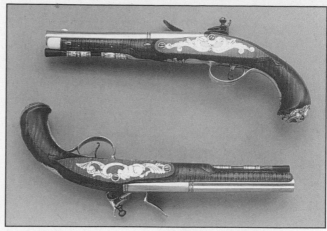

Why buy a custom? Sometimes for the pure joy of ownership. These are one-of-a-kind pistols built by Bill Kennedy of Kennedy Firearms. They show the pride of their maker.

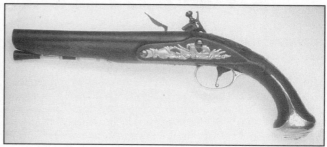

Superior craftsmanship shows in this Kennedy Firearms custom R. Wilson English Georgian Pistol.

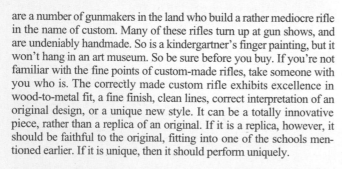

What makes a custom? Fine touches like this frog's-head butt on a Kennedy Firearms pistol gives part of the answer.

are a number of gunmakers in the land who build a rather mediocre rifle in the name of custom. Many of these rifles turn up at gun shows, and are undeniably handmade. So is a kindergartner's finger painting, but it won't hang in an art museum. So be sure before you buy. If you're not familiar with the fine points of custom-made rifles, take someone with you who is. The correctly made custom rifle exhibits excellence in wood-to-metal fit, a fine finish, clean lines, correct interpretation of an original design, or a unique new style. It can be a totally innovative piece, rather than a replica of an original. If it is a replica, however, it should be faithful to the original, fitting into one of the schools mentioned earlier. If it is unique, then it should perform uniquely.

The Custom Replica

This rifle or pistol should be the result of dedicated research. The gunmaker must know his history. I have a very plain flintlock long rifle given to me by Charles J. Keim, one-time dean at the University of Alaska, as well as a famed Alaskan Master Guide. Professor Keim had the rifle custom-built by R. Southgate to replicate "the real thing," and that it does. If we could turn the hands of time back, and I walked into an Early American settlement on the east coast carrying this flintlock, no one would ask where I got it. People would recognize it as a rifle style they had seen before, many times. Any rifle or pistol from the past can be duplicated fairly closely by a modern master. If this is what

you desire, choose your gunmaker wisely, based on his historical interest and knowledge, as well as his artistic hand.

The Unique Custom

The best example I can think of is another rifle I own. It was built by Dale Storey to copy nothing from the past, but it is a custom in every sense of the word. This professionally made rifle was designed for a specific purpose, with the following criteria: powerful; capable of shooting big bullets; compact; takedown for travel; quick ignition; adjustable sights; accurate; and not a modern muzzleloader in design, but a firearm based at least somewhat on an original. Storey's sidelock became the heart of the little powerhouse. The octagon barrel, a full $1\frac{1}{8}$ inches across the flats, is only 22 inches long. The entire rifle goes under 38 inches overall length.

The heavy barrel, the lock and all steel parts are blued, not browned. This is not a copy of a rifle from the past, remember. The two-piece walnut stock is plain in figure, straight-grained for strength, and deeply oil-finished to withstand the weather. The rugged little rifle wears a sling eye in the toe of the stock, with another sling eye integral to the front ramrod pipe. The front sight is of the Patridge style, while the rear sight is a Williams peep, adjustable for windage and elevation. There's an ordinary wooden ramrod riding beneath the barrel, and ignition is via a standard No. 11 percussion cap.

A custom rifle to match the occasion. This flintlock was designed to fit a special period of time and to represent an Eastern rifle that may well have gone West in the 1800s.

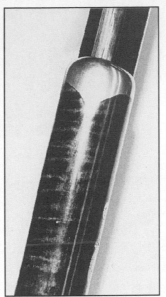

What makes a custom? Metal-to-wood fit must be close, but at the same time, the hand of the artisan is seen in the work. Machined accuracy it may not be, but handwork it is.

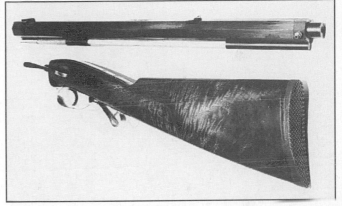

This look at the Dale Storey custom Buggy Rifle shows it taken down for travel. Custom it is; traditional it is not.

Storey calls his little masterpiece the "Storey Takedown Buggy Rifle." Takedown indeed. One bolt holds stock to barrel; remove this bolt with an Allen wrench and the rifle breaks in half, as it were. The barrel half is 22 inches long, the stock half a bit under 16 inches. The single-trigger, 50-caliber rifle has a 1:24-inch rate of twist. One of its favorite loads is a 600-grain conical with a strong charge of Pyrodex RS or blackpowder. This custom frontloader fits a specific need, my need for a compact, portable blackpowder gun with a lot of power. Did I mention how easy it is to clean? Well it is, because of the short barrel, which swamps out in a hurry. A few swipes with a solvent-soaked bristle brush, flushing water and cleaning patches, and the little Buggy Rifle is ready to put away.

Fitting a Need

I said above that my Storey Takedown Buggy Rifle fit a special need. The custom frontloader should serve a function, made to do something special. It may be a lightweight hunting rifle, or a longarm for rendezvous competition, or a collectible piece. Speaking of the later, Dennis Mulford built a number of Jaegers that were commissioned strictly for collecting and not shooting. The bad news about customs is cost. When you consider what goes into one, the price is fair and proper, but it's unlikely that you'll find a custom crafter who can build anything more than a very basic piece for fewer than 1000 greenbacks. It'll cost much more than that if your rifle wears many of the refinements, like patchbox, a touch of engraving, a line or two of carving. That's where dollars and *sense* come in. The shooter who wants to spend his custom rifle money wisely needs to plan before he buys. He has to decide what he wants and *what he wants it for* before contracting for the rifle.

Planning for the Custom

Communication, as with so many other things in life, is the key to success when planning a custom rifle or pistol. By all means, the

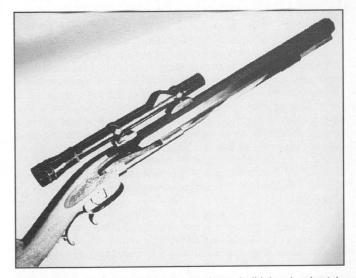

Today, the custom gunmaker may not always build the classic style of old tradition. This Storey Buggy Rifle is proof of that. It takes down in the middle and wears a scope sight.

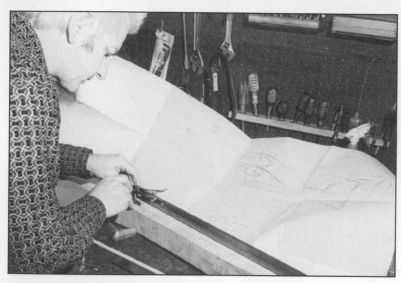

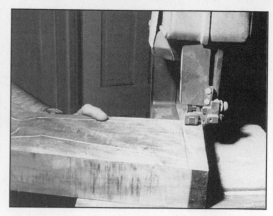

Although lock and barrel may come from specific makers, building the custom rifle remains a hands-on process, including the cutting of the stock from a blank of wood.

Dale Storey starts his custom rifle with a plan, clearly drawn so that the customer knows just what he's getting, and Dale knows just what he's building.

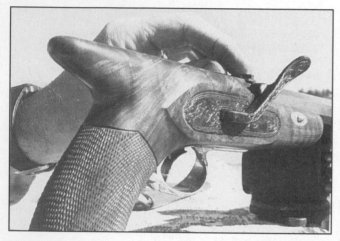

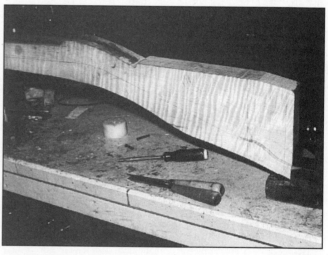

The lock on this custom pistol is wedded to the wood, not in an over-tight fit, but with a metal-to-wood joining that is professional in nature, and the mark of an expert gunmaker.

This is what a custom stock looks like in the early stages—literally hewn, as it were, from a blank of fine wood.

buyer and gunmaker must be on the same page if the venture is to succeed. The buyer has rights, but so does the gunmaker. I wanted a rugged, large bore, custom hunting rifle, hand built, and at least reminscent of the plains rifle genre. It would be tough, reliable, accurate—a true Hawken. That's what I wanted. I shopped for a gunmaker. John Baird, then publisher of the *Buckskin Report*, gave me the name of Dennis Mulford, whose work was highly praised by a number of satisfied customers. I laid my plan out before Dennis for his approval, but he balked. His interest in riflemaking leaned to the fine lines of the Pennsylvania rifle, rather than the sturdy Hawken. But maybe there was a way for me to have Hawken toughness with Lancaster looks, Dennis said. He came up with a solution.

He had seen, in a museum gun collection, an original Pennsylvania rifle of the Lancaster school, but the rifle had been extensively modified. The barrel had been cut from what was no doubt a very long length to around 34 inches. I could live with that. The bore had been "freshed out" from a smaller caliber to a larger one. I wanted a 54-caliber rifle. This old rifle had plenty of barrel for that much bore. The original flintlock had been replaced with a drum and nipple, turning the gun into a percussion piece. That was

my cup of tea. Although flintlocks fascinate me, and I shoot them, I wanted a percussion long rifle for big game. I could enjoy a rifle such as Dennis described, but why was the original Lancaster redesigned? Often when a settler hit the Oregon Trail, he had his graceful long rifle altered for the heavier game west of the Mississippi. That's probably what happened to the museum rifle Dennis studied.

Had I not listened to my gunmaker, my fine rifle would never have materialized. But I did communicate, and my custom gun was born a modified Lancaster with a 34-inch barrel, 1 inch across the flats, 54-caliber, made to shoot the patched round ball accurately with a twist of 1:79 inches. The barrel, made by Richard Hoch, was ordered based upon the following note: "We'd like a slow twist, something like a turn in 75 to 80 inches." Hoch cut her at one turn in 79 inches, which turned out to be excellent. My rifle was a relatively unadorned piece except for an engraved oval plate in the cheekpiece and a few relatively minor embellishments on the barrel. The answer to my dollars and sense question was a clear one—the money spent was worth every penny. Had I never fired the rifle, it would be worth the money it cost, as a beautiful painting of a mountain is worth its price even though you can't climb that mountain.

Fitting metal to wood is a process of careful forethought and measuring. When the custom gunmaker is finished, the brass furniture will be a part of the rifle, not an add-on.

Metalwork is very much a part of the professional gunmaker's art. He is not a stockmaker alone.

But use the rifle I did. Antelope to bull elk were tagged with this custom gun, which I called No. 47, the serial assigned by Mulford. All game fell to one shot, except for a far-off antelope buck that required two balls, thanks to my own greed which bade me shoot a bit farther than I should have. Accuracy was supreme. I once fitted a scope on a temporary mount and the rifle printed several three-shot inch-sized groups at 100 yards. Groups of 2 inches center-to-center spread were made with iron sights. Power was ample for medium range, a .535-inch 230-grain ball at about 1970 fps average muzzle velocity, and muzzle energy of close to a ton.

That rifle proved just right for me. Was it a custom? You bet it was. There wasn't another exactly like it on the planet, although you could see the Dennis Mulford touches in it. It fulfilled two personal needs of art and hunting. It was truly handmade, not turned out on an assembly line, and it showed top-grade craftsmanship. So we have discussed two custom rifles that couldn't be more different: the Storey Takedown Buggy Rifle, a practical, hardworking powerhouse well-suited to travel, and a handsome, yet functional long gun remindful of the past. Two different rifles with two different purposes in mind—both customs, and both joys to own.

Dollars and Sense

As admitted above, custom guns aren't cheap, especially when they're built by the masters. So weigh it out. Consider the dollars, but use your good sense, too. Don't buy less than you want or need, because if you do, chances are your custom rifle will be replaced down the line with another, destroying all original "savings." When you know what you want, find the right gunmaker for the job. There are many ways to locate that builder. The first is through your local blackpowder shooting club. If you can find a rifle that truly pleases you among the club members, you'll have a chance to see firsthand the work of the gunmaker. The rifle owner may even let you try his pet at the range. Gun shows also offer a good avenue to the right muzzleloader craftsman for you. I've seen some fine work displayed at shows.

Advertisements are another means of locating a gunmaker. Many excellent craftsmen advertise in blackpowder journals, such as *Muzzle Blasts* and *Muzzleloader Magazine*, as well as other arms periodicals, *Rifle* magazine for one. The annual *Gun Digest* lists many custom guncrafters, among them blackpowder rifle and pistol craftsmen. Once you do find the right artisan, then you have to strike an agreement with him. And don't forget that both of you have rights. When I asked Mulford for that Hawken rifle, he refused. I could have gone elsewhere, but Mulford's refusal made sense to me, and I ended up with a prettier piece which functioned perfectly for my needs. The gunmaker must have input, and he has to have some say over the style of the finished product. It may be your money and your rifle, but it's his name on the finished product.

Sometimes, however, the customer has to stand his ground. I wanted a very different sort of rifle for big game hunting, and I knew that it must be of the Hawken style, late model. The big rifle was built by two gunmakers with all metalwork by Dale Storey, and it was stocked by Dean Zollinger. It ended up just what I ordered, not graceful, and maybe not pretty, but one super powerhouse for elk and larger game. The high-quality barrel of this special rifle was a full 1 1/8 inches across the flats. It weighed 11 pounds, caliber 54, and with a twist of 1:34 inches, a conical-shooter. Not an exact replica, the rifle strongly resembled a very late model Hawken, with a slight pistol grip at the wrist of the stock and a Storey peep sight. Powerful? How about a 460-grain Buffalo Bullet Co. projectile at up to 1800 fps muzzle velocity, and a muzzle energy of around 3000 foot pounds.

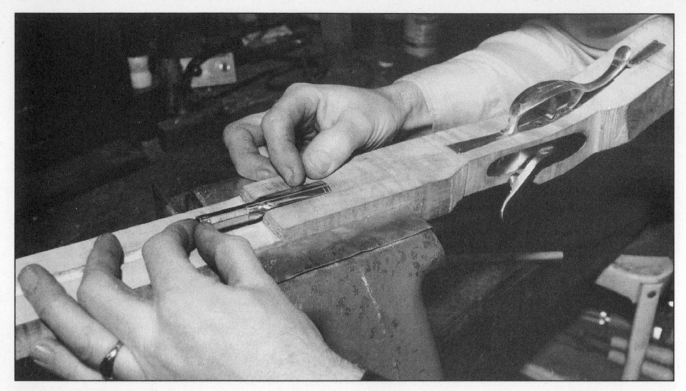

Parts are a big consideration in building a custom muzzleloader. This "store-bought" entry thimble is just what the gunmaker needed. Had it not been commercially available he probably would have fashioned his own.

The Contract

A contract should be drawn up, but often the customer and custom gunmaker don't do this. Not that a verbal agreement is no good. It is binding because the gunmaker generally demands half the cost of the gun up front. So both gunmaker and customer have a stake in the project before the first sliver of wood is sliced off the stock blank. I still prefer the idea of a contract spelling out the fine points of a specific custom gun to be built, its style, quality of wood, particular barrel and lock, patchbox or no patchbox, engraving or none, carving, sight configuration, finish of wood and metal, caliber, barrel length, trigger arrangement, and so forth. A contract should also include exact price and a target date for delivery. I am much in awe of the wonderful work accomplished by custom riflemakers and I appreciate their efforts, but even Michelangelo had a deadline when he painted the ceiling of the Sistine Chapel. The artist should not be rushed, but he should be held accountable for getting the work done somewhere within the lifetime of the customer.

Appreciation of artwork, be it fine taxidermy, sculpture, or handmade firearm, is not always forthcoming from the general public. Therefore, the riflemaker/rifleowner relationship is enhanced greatly, I think, if the owner has an idea of what goes into a custom gun. A friend looked at a custom rifle of mine and declared, "That much money for a hunk of iron and a piece of wood?" He had no concept of what went into that project or he wouldn't have been so critical of it. Custom armsmakers vary in their approach to the job, and nothing said here is meant to indicate that there is only one way to accomplish the task of turning metal and wood into a beautiful rifle or pistol. So the notes traced out below are generalized only. Individual makers do not all follow the same steps, by any means.

The Work Begins

Some custom gunsmiths work with a blueprint, creating a "paper gun," as it were, a model to work from. A blueprint can be very bene-

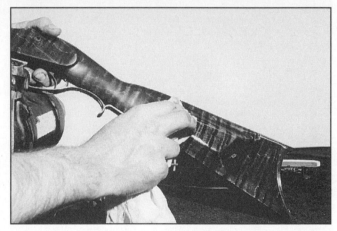

Finishes are very often unique to the individual custom gunmaker, who will have his own preferences and may consider his brew a secret.

ficial to both craftsman and customer, a plan committed to sketch on paper, a line-by-line pencil drawing. Sometimes the plan can be placed directly on the stock blank to show the way. Sometimes a gunmaker will provide a blueprint (at a fair price) for the prospective buyer to see before he plunks his money down.

Finding and buying parts has become more of a problem in recent times than in the past. I've listened to complaints from gunsmiths who have trouble finding the parts they really want to work with. So this step is definitely real. After all, parts that can't be located and purchased must be handmade by the smith, and that takes time. My own Mulford long rifle has many metal pieces handmade by the artist.

The smith generally cuts the barrel to length and crowns it, and cuts dovetails to accept sights and underlugs. Underlugs may be fitted. Contrary to a layman's thoughts, locks do not necessarily come ready to fit into the mortice in the stock. Often, they must be assembled first, and most definitely tuned. These steps require time, patience, knowledge and skill.

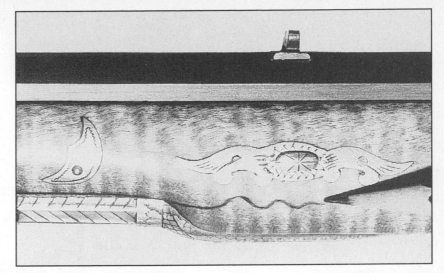

On a standard rifle, it's unlikely that the escutheon would bear engraving or have any particularly ornate styling to it. But on a custom, this touch is expected.

A machine can make perfect productions time and again. The human hand is less perfect, but preferred and even demanded on the true custom frontloader.

The stock is a huge undertaking. Not only must the wood be generally, and then specifically, contoured into a handsome stock, it must also be fitted with all parts, from barrel to lock and furniture. Inletting requires a trained hand, believe me. Wood-to-metal fit must be darn near perfect, and that requires skill. The lock has to fit perfectly into the stock mortice, and all working parts must have room to function. The ramrod channel in the stock is drilled—a mistake here, and the stock becomes firewood. The trigger is positioned in the stock through careful inletting. An upper tang screw hole may be drilled now. The buttplate must be fitted, along with integral parts, and with consideration for length of pull and drop at heel. The stock may be cast-off or cast-on, too.

The trigger guard has to be buffed and fitted. A vent hole may now be drilled and a vent liner installed for the flintlock, or the drum and nipple may be fitted here on that type of percussion rifle. An upper and lower tang may be crafted and fitted (the tang hole was previously drilled). Carving, if any, may be started now, first with an outline and then with relief. Other embellishments may also transpire at this point. The patchbox may be fitted. Inlays can be set. Much sanding is going to be in order, for the wood is still in a relatively rough state. Metalwork might need to be engraved. Finally, the stock can be stained. Metal is polished. The barrel is browned. The stock is rubbed down and finished. All parts are now assembled and mechanical refinements are made.

Thimbles may be purchased, or hand-made from sheet metal. Buttplates, likewise. The sideplate may require fitting or even complete manufacture. A nose cap may have to be cast and then worked into finished shape. Ramrods are either handmade by the smith or purchased, and if purchased, they still demand fitting. Patchboxes and cap boxes are whole jobs in themselves, either handmade or purchased and then fitted. This is a mere glance at the furniture that can be installed, not to mention engraved name plates, escutcheons, wire inlays, other metal inlays, and so much more.

These steps are not necessarily in order, nor meant to be complete. They are given here only to apprise the prospective rifle buyer of what goes into a custom piece. The finished product is generally sighted-in by the gunsmith, too, or at the very least fired for mechanical function. The end result is a true custom rifle, not a homemade thunderstick, but a personal and unique chunk of hard-working art. While the blackpowder rifleman can easily go a lifetime without a custom long arm, the ownership of at least one such shooting iron can certainly bring a lot of pleasure and reward into a shooter's life. The best way to spend money on a custom rifle is with a lot of good sense and careful planning guiding the way—finding the right man for the job, setting up a firm agreement on price and product, knowing ahead of time exactly what the rifle will and will not be, and will or will not have, from patchbox to engraving, and determing a realistic delivery date. Getting all of the cards on the table before the game begins ensures that the end result will be a pleasurable and successful experience for both gunmaker and rifle owner. Custom long rifles don't come cheap, but they are worth every cent of the investment when built correctly by an expert.

●

chapter 13

Collecting Blackpowder Guns

FORTUNATELY, GUN COLLECTING is one of the oldest avocations in the world. Hobby, yes, but many have made careers of locating, preserving, identifying, and cataloging old firearms. This is fortunate for those scholars, for that is what they are—true scholars—because their lives have been enriched by the adventure of arms collecting. It's also fortunate for the rest of us. Guns, you see, relate history. Every year, I make it a point to visit, at least once, the Buffalo Bill Historical Society located in Cody, Wyoming, at the gateway to Yellowstone National Park. I have a great interest in this well-run museum because of the fantastic Plains Indian artifacts and wonderful paintings of the Old West by the masters. But I also gravitate each time to the Winchester arms display, as well as to the Remington and Browning presentations. They are, in and of themselves, an education, not only historically, but also for the knowledge of arms styles, workings, and idiosyncracies.

A long-range study of firearms often begins with a look at the atlatl, precursor of the bow and arrow, followed by the bow and its missiles, especially the interesting stone broadheads used by ancient man. Viewing the development of these tools brings a person finally to the dawning of the age of firearms. Gun collecting is, of course, far more than just a history lesson. But I venture to guess that we all see the story of man in front of us every time we examine old firearms. We cannot help but ask, "Who used a gun like this? When did he use it? What did he use it for?" The very way arms are made denotes an era. In Chapter 12 the one-man shop of Early America was mentioned, with its handmade rifles and pistols. By the time the Volcanic, Henry, early Winchesters, Remingtons, Marlins, and other interesting factory-made firearms came along, America was in an industrial revolution with tremendous social change.

Getting Started in Gun Collecting

How does one get started in gun collecting? He studies resource books and magazines. There are many of both widely available. A look into *Gun Digest* proves that. My 1996 copy of "The World's Greatest Gun Book," a fantastic 50th Annual Edition, lists Periodical Publications on pages 529 and 530. Many magazines listed on these pages are dedicated in some way to gun collecting. I'll leave it up to the reader to investigate this fact for himself, but it's nice to know that currently published gun magazines do, in one way or another, treat firearms historically and functionally, which is always of interest to collectors. Some of these magazines are printed abroad, but many, such as *The American Rifleman* and *Rifle* magazine, are published in America.

Along with periodicals, there are literally hundreds of gun books that help the blackpowder arms collector. Each of these titles is of high interest to those who want to study firearms. Once again, *Gun Digest* is valuable. On pages 532 to 536 of the 1996 edition, titles of gun collector books appear, including *Flayderman's Guide to American Antique Firearms*, a well-established tome on the subject. There are also many books that deal with specific firearms, such as the Sharps rifle, Winchesters, Remingtons, Brownings, Hawkens. There are well-researched books. For example, *A History of the Colt Revolver*, a study of the famous sidearm from 1836 to 1940, is loaded with blackpowder collecting information. After all, the first Colts were cap 'n' ball smokethrowin' sidearms.

A list of magazines and books on guns would fill many pages of this book. *Gun Digest* is a great starting point, but the local library is another place to begin looking. Nowadays, listings are on computers, and inter-library loans are prominent. This feature of the modern library is ideal for students who want information on specific firearms. If you find a book you want to study, but your library doesn't have the title, that library can get it for you through inter-library loan. My local college library, as well as my local public library, charge a couple bucks

Peter Hasserick (left), director of the Buffalo Bill Historical Center in Cody, Wyoming, along with Howard Madaus (center), curator of the Cody Firearms Museum, accept a check from Leon Weir, president of the Remington Society of America. The check is for a display that houses the Remington collection of firearms at the Buffalo Bill Historical Center. Collecting, on a public or private level, has never been more exciting.

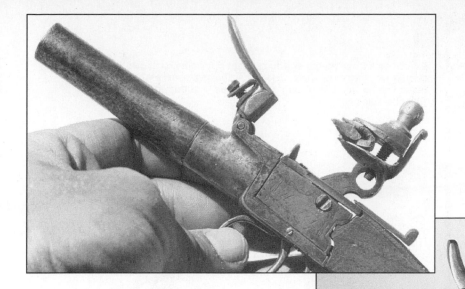

Condition is very important to gun value, although certain arms carry a high dollar figure simply because of their rarity or historical interest. This very old original flintlock pistol shows about Good condition.

Colt's 1861 Musket is once again in production. This single shot 58-caliber is true to the original. Plus, bayonet and accessories are available to complete the package.

for the service, but it's well worth it. I've had titles shipped from many states away, and in only a few days these books are on my desk. So don't overlook inter-library loans when your local sources don't have a title you wish to look at.

Guidebooks to Blackpowder Guns

Books about "old guns" act as encyclopedias for collectors, showing an interested person what was out there in blackpowder arms once upon a time. Specific books on collecting give price listings, answering the question: "What should I pay for an original?" I mentioned Flayderman's fine title above, available from DBI Books. Another excellent source is the *Standard Catalog of Firearms* from the publishers of *Gun List* newspaper. This big book of guns is published by Krause Publications. What do you find in such a title? Some downright interesting guns from the past, along with values. Values are vital to collecting. A collector must know how much he should pay for a specific firearm.

For example, let's take a look at a James Warner belt revolver, a double-action percussion in 31-caliber that was available with 4-, 5-, or 6-inch barrel with an etched non-fluted cylinder. Perhaps a collector might find the Warner revolver fitting for his holdings. But what should he expect to pay for one, if he can find one? The *Standard Catalogue of Firearms* lists this blackpowder sidearm at $650 in Excellent condition, all the way down to $300 in Poor condition. Since shooting is probably of no concern to a collector, the condition of the firearm may not be all that important, although of course any collector prefers Excellent over Poor if he can get Excellent at the right price. But what does Excellent mean? What does Poor mean? Fortunately, standards have been set so that the collector is able to place faith in the degrees of condition.

Two major American arms newspapers are available for those who collect seriously. These papers are filled with used collectibles, as well as used shooters. *The Shotgun News* from Snell Publishing Co. has been around for many years. It's loaded with arms from the past. *The Gun List* is another excellent arms newspaper. It's available from

Standing at the gateway of the Buffalo Bill Historical Center is a likeness of its namesake. Along with the huge Winchester collection, the museum now holds 120 famous Remington guns. Gun collecting is almost as old as gun manufacturing, it seems. Examples of even the distant past still exist because someone saved them.

Krause Publications. Serious collectors subscribe to one or both of these papers, faithfully reading each issue every month.

Condition Standards

Before going into the ratings, remember that the term "modern," which is used often below, does not necessarily mean a repeating smokeless powder firearm. Nor does the term have to mean a gun built in very recent times. But the modern niche does not include antiques. Antiques are older firearms, no longer manufactured. Furthermore, it is

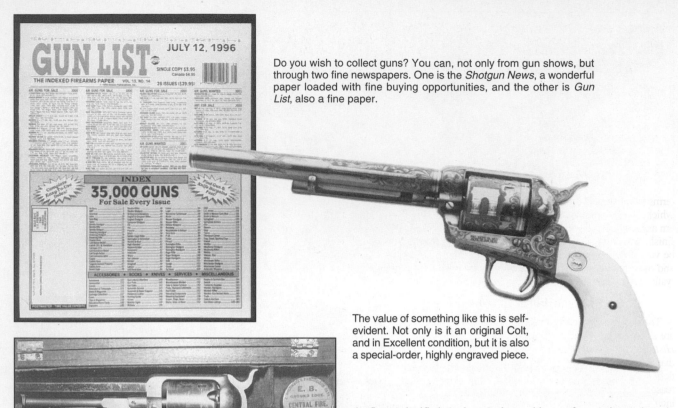

Do you wish to collect guns? You can, not only from gun shows, but through two fine newspapers. One is the *Shotgun News*, a wonderful paper loaded with fine buying opportunities, and the other is *Gun List*, also a fine paper.

The value of something like this is self-evident. Not only is it an original Colt, and in Excellent condition, but it is also a special-order, highly engraved piece.

What a find! This cap 'n' ball revolver is in at least Very Good condition, plus it has its special accoutrements with it, including flask and mould, bullets (conicals) as well as percussion caps.

almost impossible to pigeonhole some firearms. They seem to stand in two worlds simultaneously. There are "modern" Colt revolvers, for example, that didn't truly change their shape or function over the years. These guns began in the days of the blackpowder cartridge. Are they antiques or modern? It all depends upon who is placing them, and what his criteria are.

There are different grading systems. The following is only one way to categorize firarms according to their condition. However, it is a good one.

NIB

NIB means New in the Box. Usually, the box actually comes with this firearm. It's rather unlikely that a blackpowder firearm from yesteryear will have a box with it, if indeed it ever had one to begin with. But you get the picture: NIB means sparkling new.

Excellent

Excellent refers to firearms that retain 100 percent factory originality. This means that no one has refinished the piece or added parts to it that were not put there by the original maker. Nor is anything missing. Excellent also means superb working order. Modern firearms must retain 95 percent of their original finish in order to qualify as Excellent, while old-time guns must show 80 percent of their original finish. Even on the antiques, however, bores should show sharp and clean rifling, if

the firearm is rifled. And remember, evidence of tampering and repair mean that the firearm is not qualified for the Excellent rating.

Very Good

Very Good condition may not sound all that great on the surface, but this rank is quite high. It means that a product is in fine working order, and modern arms must retain between 85 to 95 percent original finish. Antiques are allowed to show as little as 25 percent of their original finish. Bores must be in good shape, and there must be no evidence of repairs. Perhaps "evidence" is a giveaway word here, as there are probably some professionally repaired firearms of the past that don't show it.

Good

This rating is given only to used firearms that are still in working order, and definitely not broken. Modern firearms must show from about 60 to 85 percent of their original finish, but antique arms in the Good category can be refinished or considerably cleaned up. There is one big difference between modern arms and antiques in the Good niche. Modern guns must have reasonably decent bores, whereas antiques can rank Good regardless of bore condition. In fact, with a Good rating, antique bores are not even considered important. This is understandable when you consider that the Good antique or blackpowder gun is without doubt a collector's item, much more than a shooter.

Fair

While Fair is not nearly as stringent as the higher ratings, a Fair gun must still "show" reasonably well and it should function. In short, a Fair rating does not mean broken. Evidence of repair, however, is all right, not only for modern arms, but antiques as well. Modern guns should retain about 25 to 60 percent of their original finish, while antiques can be very slim on original finish. Modern bores should be shootable, rather than entirely pitted and eaten away by corrosion, while antique bores are of no consequence in the Fair niche.

Poor

Poor is the lowest rating. A firearm noted as Poor may not even be collectible, unless, of course, it happens to be a very rare model. Poor firearms are often inoperable, and can even be dangerous to fire. The

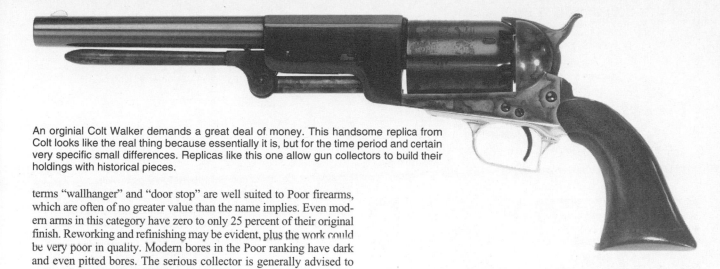

An orginial Colt Walker demands a great deal of money. This handsome replica from Colt looks like the real thing because essentially it is, but for the time period and certain very specific small differences. Replicas like this one allow gun collectors to build their holdings with historical pieces.

terms "wallhanger" and "door stop" are well suited to Poor firearms, which are often of no greater value than the name implies. Even modern arms in this category have zero to only 25 percent of their original finish. Reworking and refinishing may be evident, plus the work could be very poor in quality. Modern bores in the Poor ranking have dark and even pitted bores. The serious collector is generally advised to walk right by guns listed in Poor condition.

Some Criteria for Assessing Value

The collector of blackpowder firearms needs to know how values are arrived at so that he will understand what he's asked to pay for. *Historical significance* is definitely important. Suppose, for example, that an original Colt Walker shows up. This cap 'n' ball six-gun has definite historical significance, and part of its value is assessed for that reason. The massive Walker model, at 4 pounds, 9 ounces, with its 9-inch barrel, was indeed unique. Plus, most of these guns saw action, which means they were probably involved in some historical event or another. If you came across a Walker Colt in Excellent condition, it could run a cool $50,000. Even in Poor condition, an original Walker may run $20,000. There was a civilian Walker identical to the military model with serial numbers running 1001 to 1100. This one is worth a bit less than the military model, but it still commands about $45,000 in Excellent shape, down to $18,500 in Poor condition. Why less than the military model? Because the civilian piece is not considered as important historically.

Another obvious value criterion is *rarity*. You can bet if they made a few million Colt Walkers that model wouldn't command quite so high a figure today. Engraving, carving, and other embellishments also raise values, but only when properly executed, of course. For example, an old-time Winchester engraved professionally at the factory carries a higher price tag than the same gun engraved by a semi-professional. To take this one step further, poorly executed embellishments can *reduce* the value of a collectible firearm. Naturally *condition* means a lot to value, too. This fact was well established with the ratings given above. *Serial numbers* may also make a difference. For example, you can well imagine that a very low serial number may be desirable to a collector, especially if it happens to be number 1! Specific *models* make a big difference, too. Certain models are worth a lot more than others of very similar design. This factor was touched on above concerning the military and civilian models of the Walker Colt. Obviously, arms are priced according to their condition, too, as outlined above in the rankings.

Who Made It?

Blackpowder guns are also valued with regard to the manufacturer. There were a number of great gunmakers in the past who created arms that today are highly valuable. Let's look at just one example, of which there are many. Since the Hawken brothers, Sam and Jake, are widely known to modern shooters, we'll pick on them. A true Hawken rifle is worth a lot of money. The Hawkens are credited with building the epitome of the plains rifle carried by Kit Carson, Jim Bridger, and other mountain men. They started out in St. Louis, Missouri, and moved to Colorado in later years. Their guns are generally lodged in the period from 1815 to about 1870. Jacob and Samuel Hawken may not have made the very best plains rifles in the world.

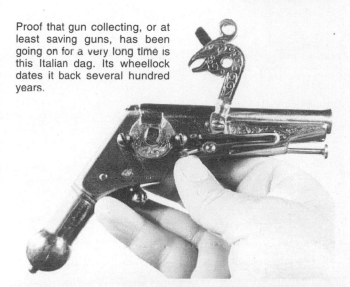

Proof that gun collecting, or at least saving guns, has been going on for a very long time is this Italian dag. Its wheellock dates it back several hundred years.

One of the real thrills in gun collecting is enjoying masterpieces from the past, such as this 19th-century English double rifle with tight, neat engraving.

That factor is open to question. But they certainly built the most famous of the breed. Today, Hawkens are valued on an individual basis. We might say they bring what the market will bear, but they certainly carry value. Why? Because of their makers. Companies are also considered when collections are prepared. That's why we have the great Winchester collection at the Buffalo Bill Historical Society in

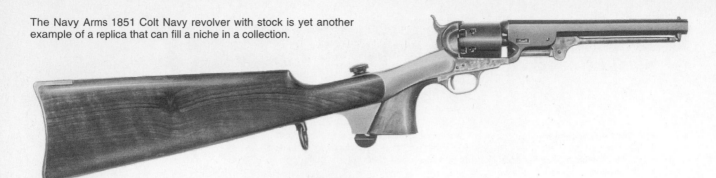

The Navy Arms 1851 Colt Navy revolver with stock is yet another example of a replica that can fill a niche in a collection.

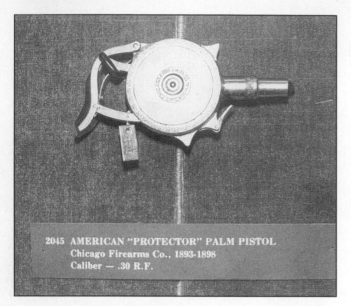

2045 AMERICAN "PROTECTOR" PALM PISTOL
Chicago Firearms Co., 1893-1898
Caliber — .30 R.F.

Fortunately, gun museums, such as the Winchester Collection, have many rare old guns for all to see, enjoy, and learn about. This American "Protector" Palm Pistol dates back to the late 1800s.

Pistols have not been left out at all. Replicas of fine old single shots like the William Parker Pistol continue to this hour. This one is from Traditions.

Cody, Wyoming. Obviously, the entire collection is dedicated to only one company, just as the Remington collection deals only with that brand name. Both of these arms factories turned out blackpowder guns, by the way, as did many other American and foreign gunmakers.

Blackpowder Guns in History

One of the most sensible plans in gun collecting is staying with a specific period in history. The problem, of course, is one of numbers. Every era had hundreds of different arms. The collector hoping to complete a specific period of time may be looking at a warehouse to hold them all. Of course, the practical means of dealing with a time in history is collecting either major pieces only, or being satisfied with a partial list of the guns. Above, it was noted that books are very important to gun collecting, because well-researched titles offer an ocean of information. Below, each period of history is treated only briefly, as space permits, with mention of one, two, or three book titles serving as reference. Naturally, there are many, many books devoted to each of the following periods. But this is not the forum for their inclusion. Furthermore, no attempt has been made to list all time periods because that would be far too extensive for our purpose. The following time periods are samples only, and not necessarily the most important, or in perfect order.

Very, Very Long Ago

While this initial category carries a tenuous title at best, it means guns from the beginning, including original "hand-guns," matchlocks, wheellocks, and eventually flintlocks. The cut-off point is arbitrary. The "long ago" niche was chosen only to get the ball rolling. Considering the fact that examples of extremely old guns have survived, even if only as broken items, it is possible to have collections that include these very early firearms. It is highly unlikely that the average collector will come upon truly ancient arms. Most of the existing samples are already in museums. But they do exist. Museum gun collections often have examples of matchlocks, wheellocks, snaphaunces, and of course flintlocks. The Winchester collection, mentioned earlier, contains several ancient arms.

So this strictly made-up category does, in fact, have merit because there really are collectible arms from very, very long ago. I said there would be mention of book titles for each of these categories. For this period, consider *Guns and Rifles of the World* by Howard L. Blackmore. The edition I have was printed by the Viking Press, New York, 1965. Also take a look at *The Flintlock: Its Origin and Development* by Torsten Link (translated from the Swedish by Urquhart and Hayward). Dr. Link's book begins with the snaphance and flintlock in literary reference, and moves forward from there. Obviously, these guns are not from the very, very long ago period—just the long ago, perhaps at the tail end of this time period that I invented.

The Kentucky Rifle

Skipping way ahead in time, past the Jaeger and other European firearms, there is the famed Kentucky rifle, which was born in

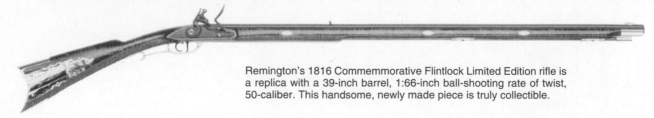

Remington's 1816 Commemmorative Flintlock Limited Edition rifle is a replica with a 39-inch barrel, 1:66-inch ball-shooting rate of twist, 50-caliber. This handsome, newly made piece is truly collectible.

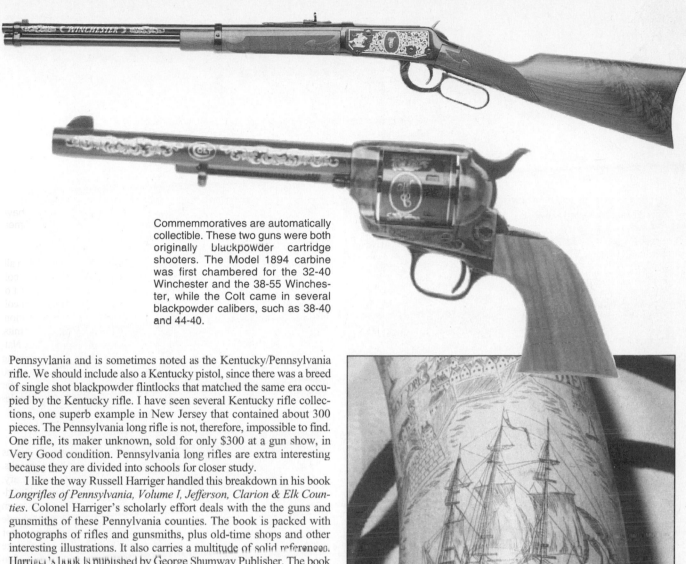

Commemmoratives are automatically collectible. These two guns were both originally blackpowder cartridge shooters. The Model 1894 carbine was first chambered for the 32-40 Winchester and the 38-55 Winchester, while the Colt came in several blackpowder calibers, such as 38-40 and 44-40.

Pennsyvlania and is sometimes noted as the Kentucky/Pennsylvania rifle. We should include also a Kentucky pistol, since there was a breed of single shot blackpowder flintlocks that matched the same era occupied by the Kentucky rifle. I have seen several Kentucky rifle collections, one superb example in New Jersey that contained about 300 pieces. The Pennsylvania long rifle is not, therefore, impossible to find. One rifle, its maker unknown, sold for only $300 at a gun show, in Very Good condition. Pennsylvania long rifles are extra interesting because they are divided into schools for closer study.

I like the way Russell Harriger handled this breakdown in his book *Longrifles of Pennsylvania, Volume I, Jefferson, Clarion & Elk Counties*. Colonel Harriger's scholarly effort deals with the the guns and gunsmiths of these Pennylvania counties. The book is packed with photographs of rifles and gunsmiths, plus old-time shops and other interesting illustrations. It also carries a multitude of solid references. Harriger's book is published by George Shumway Publisher. The book is a living history of the wonderful Pennsylvania long guns and the gunsmiths who made them. At the same time, it lends insight to an era and a segment of American geography. Business records from the gunsmiths themselves show not only the work they were commissioned to do, but the charges made for their labors.

Two more classics concerning the Kentucky rifle are both titled accordingly. There is *The Kentucky Rifle* by Captain John G.W. Dillin, a book printed in 1924 by The National Rifle Association of America. This title is long out of print, but can been located through inter-library loan. Also, it is possible to find a copy through a book search, which is provided by certain used book stores. Check book stores in your area. Dillin's book includes chapters such as "The Name Kentucky," which explains "Why the American pioneer rifle, first made in Pennsylvania, was known as 'The Kentucky Rifle,'" and "The Evolution of the American Rifle," relating the link between European and American gunmakers. It's a fine book. Another title worth studying is *The Pennsylvania-Kentucky Rifle* by Henry J. Kauffman, published by Stack-

Another look at a powder horn with scrimshaw. Some of the fine old horns of the past are artwork in themselves and well worth collecting.

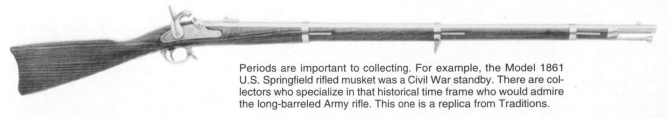

Periods are important to collecting. For example, the Model 1861 U.S. Springfield rifled musket was a Civil War standby. There are collectors who specialize in that historical time frame who would admire the long-barreled Army rifle. This one is a replica from Traditions.

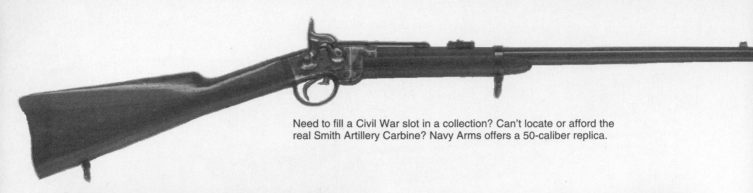
Need to fill a Civil War slot in a collection? Can't locate or afford the real Smith Artillery Carbine? Navy Arms offers a 50-caliber replica.

Powder horns are also collectible, especially when they are decorated with scrimshaw like this one.

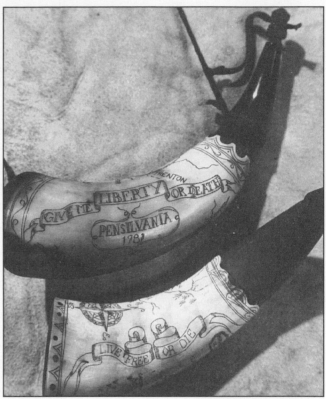
Some powder horns bear interesting words, as these horns relate.

pole Books. This book has been reprinted. Check with your bookstore or library for a copy.

Wars

Wars all over the world were fought with blackpowder arms before smokeless powder came along. The American Revolutionary War saw the use of muzzleloaders, for example. Collectors may wish to find and display much more than arms in this case, as there were many other interesting implements of the period used in this grim contest. *The Book of the Continental Soldier* by Harold L. Peterson does a fine job of explaining the smoothbore musket of the period, as well as uniforms and equipment. Peterson details not only the firearms, but how they were employed. The Brown Bess is given attention on page 27, for example. Recall that this firearm is available

in replica form today, allowing students of the period to enjoy owning—even shooting—a Brown Bess without having to find and purchase an original. British infantry muskets are explained, too, as well as a host of American rifles. The same treatment has been given the American Civil War and its guns. Naturally, after this time period, wars were fought with smokeless powder.

The Fur-Trade Era

This time capsule has captured the imagination of shooters everywhere, and many books have been written on the mountain men and that era. Individual heroes, such as Joe Meek and Jim Bridger, have been frozen in time on the pages of such books. The guns of the time also caught the imagination. We still shoot them today as replicas. Charles E. Hanson, Jr. wrote *The Plains Rifle*, a book dedicated to the gun of the same title. It was printed in 1960 by The Gun Room Press and can be found to this day. Hanson begins with early trade guns, then moves into early Leman flintlocks, and on to Samuel and Jacob Hawken and their arms.

John D. Baird singled out Hawkens and wrote two books on these rifles. One is *Hawken Rifles, The Mountain Man's Choice*. The edition

This original Colt is still very shootable and actually could be used in the fast-action cowboy shooting games of the day. One reason old guns remain not only in collections but in the field is how they were made and what they were made of. Many were made quite well, and of good materials.

The real thing? Yes, it is a real Colt, but not real old. Colt has, over the years, continued to make some of their great guns. A collector who can't find or afford an original may find this model just right, although it, too, carries a fairly heavy price tag.

I have is from 1976 as reprinted by The Gun Room Press. Baird begins with the Hawken customer, then goes into early Hawkens, full-stock Hawkens, specific Hawken rifles, and much more. The same author also wrote *Fifteen Years in the Hawken Lode*, published in the same year by the same press. This work includes a look at double guns on the frontier, the famous Hawken guns, and it goes on with buckskinning adventures in modern times.

Blackpowder Guns in General

Blackpowder gun collecting also exists on a very general level. An acquaintance of mine simply collects "blackpowder guns," for example, while a fellow I met in South Africa has concentrated on any muzzleloader used in his homeland at any time, including the ivory hunting era. Neither man looks for only one historical era, one company, or one gunmaker. Each has an interest only in blackpowder guns, period. The first person has originals of pepperbox pistols to Kentucky rifles. The second has many blackpowder big bores. There are also books generalized in gun collecting which include blackpowder arms. For example, *Famous Guns from the Smithsonian Collection*, a 1966 title published by Arco Publishing Co. includes "The story of firearm patents between the years 1836 and 1880." It includes drawings and photographs of models which were held in the Smithsonian Institution. The same writer did another book that included many blackpowder

arms of interest to the collector. This was *Antique Guns from the Stagecoach Collection*, and it dealt with a short history of small firearms, including at look at serpentine "hand guns," as well as various muskets and Kentucky rifles, a host of pistols, the Walker Colt, Colt Dragoons, the LeMat rifle, Sharps First Model, and much more. *Antique Guns* by Bowman and Lucian Cary is yet another title that shows many blackpowder arms, such as Le Page percussion duelling pistols and percussion Colts.

This chapter represents a single brushtroke on the canvas of gun collecting. The topic is vast and varied, and far too involved to do more than introduce it here. However, it is more than obvious that blackpowder gun collecting is viable, and happening right now, and it will continue to go on indefinitely. Even if a modern blackpowder marksman is not interested in collecting firearms himself, he can read about the ones that interest him, and he can see many of them in gun collections housed in museums all over the world. These "old guns" are fascinating to anyone interested in "smokepoles." ●

chapter 14

Blackpowder— The Propellant

BLACKPOWDER'S "...BURNING mechanics remain poorly understood because of their complexity," said Professor F.A. Williams, an expert in propellants and fuels. Those who have not traveled the blackpowder trail find it unbelievable that this ancient simple mixture could tax modern science. But it's true. Many of our instruments of study are too blunt to penetrate the shell of the problem. Blackpowder is a compound. It is a mixture with only three major ingredients: saltpeter, charcoal, and sulfur. For centuries, this mixture was simply referred to as "gunpowder." After all, there was no reason to distinguish it from other propellants, because there weren't any. Blackpowder was invented in China—maybe. Or it could have been India. Some students of the subject prefer Greece. The truth is, we don't know where it first surfaced, or even what it was used for in the beginning.

To be ultra clear on this important point, I repeat that blackpowder is an intimate mixture of three basic ingredients, saltpeter, charcoal, and sulfur. That is essentially it. No wonder modern shooters think of blackpowder as simple. Yet this fuel has so many properties that it is, as Professor Williams said, extremely complex in nature. Of course, there were many different brands of blackpowder over the years, each differing in burning characteristics. Some of these were Hazard's Kentucky Rifle & Sea Shooting powder, Loflin & Rand's Orange Extra, Lightning, Ducking, du Pont Diamond Grain, Eagle Sporting, Eagle Duck, Eagle Rifle, Oriental in America, Curtis & Harvey Diamond Grain, Col. Hawker's Duck Powder, Pigoo, Wilks and Laurence's from England. There were many others. Although all of these powders contained the same essential ingredients, they differed by precise amounts of each product in the mixture, and by exact burn rate, and composition (how the ingredients mixed together), plus granulation or kernel size, and other properties. The manufacturing process itself promoted differences from brand to brand, and while saltpeter, sulfur, and charcoal made up these different powders, the exact nature of each part often varied.

What we think of as very early blackpowder may have been something else altogether, by the way. Since some form of this explosive may have been around as early as the 7th century A.D., it's no wonder we're confused about its origin and original composition. *The Book of Fires for Burning the Enemy* by Marcus Graecus mentions "Greek Fire," which some scholars assumed to be early blackpowder. It may have been a mixture of potassium nitrate (saltpeter), sulfur, and oil. But this is mostly conjecture. Greek Fire probably did exist, but it may not have been an explosive. We just don't know. Nor are we so sure about Roger Bacon's 13th century accounts of some form of blackpowder. The Chinese cer-

Blackpowder differs markedly from one brand to the next. Elephant brand is a fine product, but it is not the same as other blackpowders. This is Elephant FFFg, which the author used with great success in a Storey Buggy Rifle for tests. Elephant brand comes in several sporting granulations, including Cannon grade on the large end and FFFFFg (5F) on the small end.

Over the years, there have been dozens of different blackpowders offered to shooters. These are but two: Green River and DuPont Superfine.

Currently, GOEX brand is the best-selling black-powder on the market. It comes in several useful granulations, including FFg, which is excellent in the majority of big bores shot today.

tainly had an explosive in the 1200s. And the Europeans were making smoke on the battlefield, we believe, at least by the 1400s.

Blackpowder Components

We will never know the truth concerning the advent of blackpow-der, but it is certain that the three ingredients noted above made up this propellant a very long time ago, and still do. Saltpeter is potassium nitrate, KNO_3, an oxidizer. Sodium nitrate was also used as an oxidiz-er for blackpowder. KNO_3 has a melting point of 334 degrees Centi-grade, but the ignition point of blackpowder, depending on the source, is noted as 300 to 350 degrees Centigrade. Then there is charcoal, which represents a form of carbon. Charcoal is more than the body of blackpowder. Different charcoals affect burn rate and combustion properties. Certain willows, for example, have been sought after for high-grade blackpowder charcoal. One opinion says sulfur, the third ingredient, is a binding agent for saltpeter and charcoal, serving to maintain the integrity of the powder. Another suggests that sulfur pro-motes both ignition and combustion. Make your choice. Mixture ratios of the three main ingredients changed over the years. A 1350 English gunpowder shows 66.6 percent saltpeter, 22.3 percent charcoal, 11.1 percent sulfur, while a 1650 French powder ran 75.6/13.6/10.8. A pop-ular mixture was and still is 75/15/10.

One gram of blackpowder yields 718 calories of heat, 270 cubic centimeters of permanent gas, and roughly a half-gram of solid residues. Some gases of combustion are: CO_2, CO, N_2, H_2S, H_2, K_2CO_3, K_2SO_4, and K_2S: carbon dioxide, carbon monoxide, nitrogen, hydrogen sulfide, hydrogen, potassium carbonate, potassium sulfate and potassium monosulfide. In one test, 82 grains weight of FFg black-powder left 42 grains of solids after combustion. The factor of remain-ing solids is related to powder efficiency and muzzleloader cleanup, two vital factors discussed below. *Granulation also determines how blackpowder behaves.* (See more on granulation below.) Blackpowder is a surface-burning compound and particle size and shape are in part responsible for burn rate. The kernels are polyhedral (many-faced). The shape of the powder charge is always cylindrical in the breech of the muzzleloader, and when properly seated, there is always 100 per-cent load density—no air space between powder and projectile. Black-

powder pressure is in part related to kernel dimensions. Per charge, fine kernels yield higher pressure than coarse granulations. Knowing this, the correct granulation can be selected for each application.

The Lyman Black Powder Handbook illustrates pressure per kernel size with a 54-caliber pressure gun and a patched round ball. A 100-grain charge of FFFg and a 140-grain charge of FFg, same brand of powder, achieved within 39 fps of each other, but the 100-grain charge of FFFg showed 3200 LUP greater pressure: 8500 LUP for the 140-grain charge of FFg and 11,700 LUP for the 100-grain charge of FFFg. Also, different projectiles alter pressure results. So do different test bar-rels. A 45-caliber long rifle with 28-inch barrel, also shooting a patched round ball, got about 1650 fps with 70 grains of GOEX FFg. It took only 50 grains of FFFg GOEX to achieve the same velocity, indicating more work accomplished for FFFg.

The difference here was not as pronounced as it was with the 54-caliber barrel, but it was there. Pressures were about 8000 LUP for FFg and 8500 LUP for FFFg, quite close, but remember the 20-grain pow-der difference in favor of FFg. It's clear that fine-grained blackpowder produces greater pressure than the coarse grains. In calibers 45 through 54, FFg makes sense for hunting loads because good velocity can be reached with modest pressure, while FFFg is ideal for target loads even in big bores because mid-range bullet speeds can be reached with a mild powder charge. Calibers under 45 do well with FFFg granulation all around. As usual, things are not that simple. For example, the Gonic company suggests FFFg loads for its sturdy Magnum muzzleloader. So it is, as usual, a matter of individuality. Nonetheless, it's good to understand that one property of blackpowder is pressure per granule size. It is a piece of "blackpowder knowledge."

The concept of granulation was so important to blackpowder pyrotechnics that the policy of developing powder with specific kernel size continued into smokeless powder development. Producing ker-nels, even of irregular size, had many advantages. Granulation pro-moted powder integrity. Rather than a dust-like product, the propellant remained intact as kernels. Powder that was not granulated tended to break down, whereas powder in kernels stored much better. Also, granules thwarted moisture better than a dust-like powder, partly because the kernels could be coated. As noted above, kernels tended to

Smoke is definitely a part of the charcoal-burning game, and while it's definitely there, it really is not the cause of too much trouble. Some shooters feel that without smoke, it just wouldn't be a muzzleloader.

break down less than a dust-like substance, which was already like flour in consistency. This, too, fought moisture. Blackpowder is hygroscopic, meaning it will absorb moisture from the atmosphere, so preparing the propellant as kernels helped. Also, granulated powder was easier to transport, especially onto a battlefield. It came ready to use. And burning rate was easier to control with granulated powder, as proved by the differences in kernel sizes noted above.

Barrel length is always a shooting consideration. How does blackpowder change in performance in relation to barrel length? This is a variable worth considering. Comparing a 32-caliber squirrel rifle with a 41½-inch barrel to another 32-caliber rifle with a 29-inch barrel, both burning 30 grains volume FFFg blackpowder behind a 45-grain round ball, the longer barrel gets close to 2100 fps muzzle velocity, while the shorter gets about 1900 fps. Two hundred feet per second is a statistically significant difference in velocity, indicating that blackpowder does not go *Whoosh!* in one fell swoop when ignited.

Over the years, some shooters have misunderstood that. I have read that blackpowder "just explodes," and that it does not "burn progressively in the bore." Certainly, blackpowder and smokeless burn differently, but blackpowder neither goes Boom! instantly in the bore, being used up in the first couple inches of barrel, nor does it burn so slowly that only long barrels are any good. Many factors pertain, including caliber, bullet weight, projectile style, and so forth. Conclusion: Shorter barrels on big game muzzleloaders are worthwhile. A shorter barrel means a more compact firearm, which is worth trading for a minor velocity loss. That is a practical way to look at it.

Blackpowder Pressure

Speaking of pressures, it is vital for the blackpowder student to understand that dangerous pressures are possible with this fuel. Unfortunately, there has been a long-standing dictum about blackpowder pressure that is wrong now, and always has been wrong. It goes like this: "You cannot overload blackpowder, because no matter what you do, there is a ceiling on pressure, which is 25,000 psi." Wrong! In the first place, 25,000 psi could be too high anyway in some guns, and in the second place, far higher pressures are possible with blackpowder. Noble and Abel, 19th century English experimenters, burned blackpowder in a closed vessel. Velocity of combustion is much higher under these circumstances than in open air; however, pressures up to 100,000 psi announced the possible dangers of misloading blackpowder in a firearm.

Barrel-bursting was common in the past and is not unheard of today with blackpowder. The problem is, how do we *prove* that overload dangers exist? Repeatability of test results has been difficult. I built some "barrels" using copper tubing of a size just right to accept 54-caliber missiles. When loaded properly, the thin tubing withstood charges

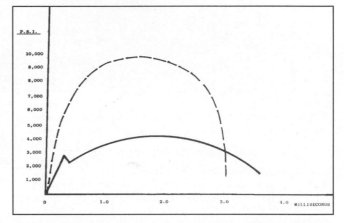

The blackpowder "burning curve" shows a sharp spike at ignition, followed by a long smooth line. The unbroken line is blackpowder; the broken line is for smokeless.

of around 100 grains volume FFg powder with one patched round ball. Conicals were another story. Heavy bullets caused the tubing to burst, especially at the meeting point of projectile and powder charge. My little pipe tests were indicators of blackpowder behavior, but could not be judged absolutely factual, with a set of provable laws.

Unfortunately, the problem of "short-starting," which is failure to seat the projectile firmly upon the powder charge, continues to arise. Proving the dangers of short-starting has also been difficult. I did blow up several gun barrels that were short-started. These barrels withstood heavy powder charges, and blew only when short-started. Proof? Not really, but certainly a powerful indication of trouble. I also tried proving the dangers of short-starting with my test pipes. When an air space existed between the powder charge and the round or conical bullet, the usual result was a bulge, known as a "walnut" in blackpowder circles. This bulge generally occurred where the ball rested in the bore. A ring in the bore (circular inner recess) also occurred. But obtaining identical results from test to test proved to be very difficult, and therefore I have yet to say that I have absolute *proof* of short-starting dangers.

I just know from experience, and from literature from the past, that blackpowder can cause trouble when a load is short-started. This whole business of short-starting and bore damage is coupled with blackpowder pressure and how the powder burns. Kernel size and shape were already mentioned. As these change, so does the air space around each kernel of powder—less air space around fine granulation,

What can blackpowder do? In this pipe test, a charge of blackpowder destroyed the test instrument. Blackpowder can achieve high pressures, in spite of statements to the contrary that have circulated in the past.

Blackpowder Traits at a Glance

1. Muzzleloaders have a maximum load rating. The maximum allowable charge of powder is provided by the manufacturer of the firearm. Do not exceed maximum loads because it has been proved that blackpowder can achieve high pressures.

2. Short-starting a muzzleloader may damage the bore and in severe cases split a barrel. This statement is based on a number of cited examples. However, to date no conclusive repeatable test of short-starting has been devised.

3. Blackpowder firearms seem to kick hard with full-power hunting loads in big-bore rifles and pistols because it takes heavy charges to generate reasonable muzzle velocities, and the weight of a powder charge is part of the recoil formula.

4. Volumetric loading (with a simple powder measure) is acceptable and safe with muzzleloaders, producing good accuracy, because blackpowder is not as efficient as smokeless powder.

5. The conical's base or skirt (Minie ball) can be damaged by the powder charge. This can cause inaccuracy. If a conical continues to provide inaccurate groups, a test can be run by lowering the powder charge to see if that improves grouping.

6. Blackpowder can cause patch burnout. A cure for this burnout is the use of hornet nesting material. A couple sheets of this between the patch and the powder charge will safeguard the patch from burnout. Likewise, a shotshell wad can be burned through by blackpowder, thereby destroying the pattern. A card wad on top of the powder charge can prevent this.

7. Extremely small powder charges may result in poor accuracy due to obturation failure. If a muzzleloader gives poor accuracy with ultra-small powder charges, carefully increase the powder charge (only to safe levels) and check accuracy results. Do not load higher than a maximum allowable charge in any muzzleloader.

8. Blackpowder remains vigorous at low temperatures, which is a helpful condition for those who hunt with muzzleloaders in cold weather.

9. Blackpowder has an excellent shelf life; however, it must be stored in a cool and dry place in a closed container or it may degrade.

10. A shotgun wad column must be loaded into the breech so that it can easily move upbore. A restricted wad column can cause high pressures with blackpowder.

11. Varying ramrod pressure may cause accuracy deterioration. Reasonably consistent pressure alleviates this possible problem.

12. Blackpowder reaches a point of diminishing returns, where adding more powder does nothing to achieve any higher velocity.

13. Longer barrels generally give a bit higher velocity than shorter barrels with blackpowder, but not enough to worry about. Shorter-barreled big game blackpowder rifles "carry well."

14. Projectile design and weight can make a difference in the generation of energy and in how blackpowder behaves in the bore during combustion.

15. Kernel size makes a difference in generated pressures. Use the appropriate granulation for a given blackpowder gun.

16. Muzzleloaders must be cleaned after use due to the properties of blackpowder, including the fact that about half of blackpowder is not converted from solid to gas during combustion, but rather it remains in a solid state after burning.

17. Blackpowder ignites readily. The can should always be closed after use. A spark could ignite the powder.

18. Blackpowder is an intimate mixture of charcoal, saltpeter and sulfur. Many other powders may also be black *in color*, but they are *not* blackpowder and must never be used in a muzzleloader for any reason.

19. Smoke from blackpowder results from several properties of the propellant, but mostly from the fact that about half of the powder remains a solid during combustion and is not converted to a gas.

20. Although blackpowder is a mixture of only three main ingredients, its burning properties are not entirely understood to this day.

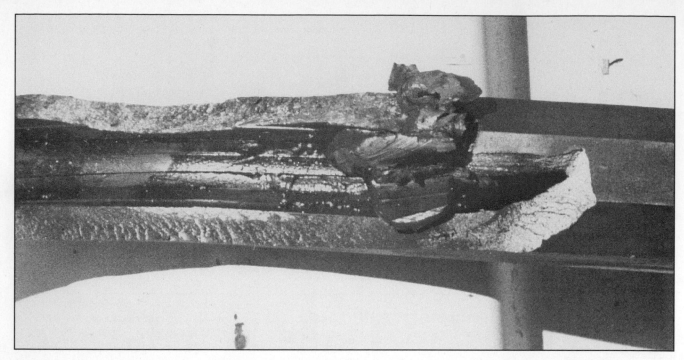
This piece of barrel shows the potential power of blackpowder. It was a short-started overload that destroyed the barrel.

more around coarse granulation. The full impact of air pockets in the powder charge is not entirely understood. We know that finer granulations generate more pressure for equal powder charges than coarser granulations, and my assessment is that this factor has to do with the way the smaller kernels are consumed.

Another blackpowder fact that bears considerable attention is residue in the bore from a fired charge. Recall that roughly half of a blackpowder charge remains in solid form after the gun is fired. What does this mean to consequent pressures, and to the problem of short-starting? There could be a link between leftover solids in the bore and short-started loads representing a bore obstruction. In a sense, part of the powder charge is a missile that must be expelled from the bore as ejecta (ejecta is the sum total of everything fired in the bore, including fouling itself). If part of the powder charge acts as a projectile and there is a real projectile lodged upbore, too, then there is a solid pushing on a solid—a little like loading two bullets in the breech—one well downbore, and the other part-way down the bore. The upshot of this whole business is not to short-start a bullet in the muzzleloader. It's a bad practice.

I was a so-called "expert witness" in a court case dealing with an exploded muzzleloader that resulted in an injured party. A study of the firearm revealed no mechanical faults. The steel was judged of proper quality and formulation by a metallurgist. The findings, however, did show that the projectile was not seated upon the powder charge, but was in fact about twelve inches off the charge. The barrel burst precisely at the point where the base of the patched ball rested. This is a common occurrence. The powder charge was 120 grains, which was acceptable in that particular firearm. The shooter swore that he had short-started that same rifle many times without a problem. He probably had. I have seen test barrels withstand several short-started loads with no overt sign of damage. One more try and Boom!—a burst barrel.

Other Pressure

Varying ramrod pressure on the projectile, and therefore on the powder charge, affects how powder burns in the bore. Varying ramrod pressure has been shown to result in a high standard deviation from the mean velocity. This indicates that a change in pressure on the powder charge alters burning characteristics. One supposition concerning this phenomenon suggests that with extremely high pressure

distributed by the ramrod to compact the powder charge, the melting point of potassium nitrate is lowered. Conversely, the melting point of potassium nitrate is higher when the powder charge is not so heavily compacted under pressure from the ramrod or loading rod. Perhaps.

The practical upshot of this is to maintain constant pressure on the powder charge from shot to shot by applying consistent force on the ramrod. When working up test loads, I use a device that slips over the end of the loading rod. The spring-loaded tool relaxes when 35 pounds of pressure is reached, so each charge receives the same amount of pressure. For everyday shooting, no such device is warranted. The shooter simply applies reasonably consistent pressure on the ramrod and all will be well. This does not mean leaning on the ramrod. That can actually crush a powder charge. Don't do it.

Barrel bursting is obviously a result of—if we can use the notion—"misguided pressure," because barrels have been damaged with reasonable powder charges. Therefore, it's not a simple matter of an overcharge of powder. For example, it is theorized that barrel blowup from a bore obstruction, such as a muzzle clogged with mud or snow, is due to the elasticity of the air that rests between the projectile and the bore obstruction. Naturally, a bore obstruction can also destroy a modern cartridge rifle using proper ammunition. But blackpowder barrels have been ruined when no bore obstruction existed and with appropriate powder charges. My friend Jim O'Meara noticed that one of his barrels was bulged during a shoot. The person who had borrowed Jim's rifle short-started it. The charge of powder was only 60 grains, which was far from excessive. Yet, there was a large bulge in the barrel precisely where the patched ball had been located.

Many surmises are made. Ignition spikes are blamed for certain barrel failures. Maybe? Blackpowder ignition is quick, that's for sure. Blackpowder ignites so readily and sends out a flame so rapidly that it has been used to ignite the huge charges of powder burned in the guns of a modern battleship. A burning curve, time vs. pressure, read on an oscilloscope, shows an instant peak at ignition, then a drop/rise followed by a smooth curved line. Another interesting fact: while some propellants burn slowly at low ambient temperatures, blackpowder remains vigorous in the cold. And as with all guns, the size of the bore determines in part what pressures will be. A 19th-century Naval Report on blackpowder translated from the French notes that, "In similar guns charged with the same powder, the maximum pressure is

Some people collect blackpowder cans, like this smallfry from the past, just one of many brands packaged in many different ways.

Elephant brand blackpowder has a long list of instructions concerning their propellant on the can, including disposal precautions.

BLACK POWDER
SUITABLE FOR MUSKETS, PISTOLS & SHOTGUNS
DANGER EXPLOSIVES!
YOU MUST READ WARNINGS ON SIDE OF CAN!
WARNING: This canister contains BLACK POWDER made from Potassium Nitrate, Sulfur and Charcoal. Black Powder is highly flammable and explosive and must be stored in such a way to avoid exposure to friction, heat, impact, sparks or open flame. ELEPHANT BLACK POWDER is to be used only with black powder firearms. Its use for any other purpose is hazardous and prohibited and can cause personal injury (including death). The use of black powder should be undertaken only by those who are familiar with and are extremely careful to observe, all safety precautions and practices.

HANDLING PRECAUTIONS:
· KEEP AWAY FROM CHILDREN
· Handle with care. Avoid impact, friction, heat, sparks, or open flame.
· Prevent contact with food, chewing material, and smoking material.
· Keep containers tightly closed when not in use.
· Clean up any spilled powders. USE A BRUSH AND DUSTPAN ONLY.
· Do not carry in clothing. If clothing is contaminated with black powder, do not approach heat or flame source.
· Wash hands thoroughly after handling black powder.
· Do not take internally. In case of ingestion, call a physician.
· Do not repackage. Do not purchase or accept any ELEPHANT BLACK POWDER not in its original, factory sealed container.
· Do not mix this powder with a powder of any other type.
· Do not dispense black powder directly from canister into firearm.

STORAGE PRECAUTIONS
· Obey all laws and regulations regarding quantities of explosive material and methods of storage.
· Always store in a cool, dry place.
· Do not store in same area with solvents, flammable gases, or highly combustible materials.
· Store powder in its original ELEPHANT BLACK POWDER approved container.
· Keep out of reach of children.
· Do not subject storage cabinets to close confinement.

DISPOSAL PRECAUTIONS
· Be sure this powder container is completely empty before discarding. Do not use this container to store other powders or materials, or for any other purpose.
· Do not keep old or salvaged powders.
· Check old powders for deterioration regularly. Disposal, if necessary, must be in accordance with all local, state, and Federal laws and regulations.
· Do not dispose of this container into fire.
· Consult manufacturer for more details on disposal.

MADE IN BRAZIL BY PERNAMBUCO POWDER FACTORY
IMPORTED BY PETRO-EXPLO, INC.
ELEPHANT BLACK POWDER DIVISION
7650 U.S. HWY. 287, #100
ARLINGTON, TEXAS 76017 U.S.A.

proportional to the caliber." This fact is clearly reflected in our maximum loads. Small bores, such as the 32, reach peak velocity of about 2000 fps with a modest powder charge, while true big bores, 64-caliber for example, never achieve 2000 fps muzzle velocity with anything short of a shovelful of powder, figuratively speaking.

There is also a theoretical maximum in the work that any powder can perform, with power representing the gas produced minus the energy consumed in pushing solids in the bore, and also minus heat loss. If we think of the firearm in terms of a heat engine, heat loss takes on serious proportions. Depending upon the specific firearm and load, the barrel may absorb 25 percent of the heat generated from the powder charge. This represents lost "power" in terms of work applied to a projectile. Cooling of the big barrel walls of the muzzleloader is greatest when the weight of the powder charge bears a small ratio to the interior surface of the bore. In short, a modest powder charge in a big bore allows faster barrel wall cooling than a heavy charge in a smaller bore size (less volume and also reduced bore wall surface). Another job caused by the force of the burning powder charge that is important in muzzleloaders is obturation or "base spread" of the missile. The ball moves upbore a distance before obturation because there must be sufficient pressure on the base of the projectile to promote the mild foreshortening of the bullet that takes place. But this factor can be overrated. In one test, 2200 psi was reached in less than .1 (one-tenth) millisecond after ignition of the powder charge, showing that indeed there was instantaneous force on the projectile.

Obturation

In extreme cases of tiny powder charges in large-bore firearms, accuracy may not be there, possibly due to failure of projectile obturation. Gas passes around the bullet and out through the grooves of the bore. The subject of gas acting upon the base of the projectile brings up two more salient points: the partial destruction of a conical's base and patch deterioration. The first is especially possible with a thin-skirted Minie ball. The skirt is literally gas-cut, and accuracy suffers. Patch destruction is not as simple. If the round ball obturates sufficiently in the bore to engage the rifling fully, then accuracy may be achieved in spite of a burned-out patch. All in all, however, it behooves the shooter to check fired patches. They are found lying downrange.

There are burned-out patches and cut patches. The first are called "blown patches" because they show blowouts in the cloth, which are holes of various size from huge to minor. On the other hand, a patch

Square Mesh Screen Granulation Sizes		
Granulation	Go	No Go
Fg	.0689-inch	.0582-inch
FFg	.0582-inch	.0376-inch
FFFg	.0376-inch	.0170-inch
FFFFg	.0170-inch	.0111-inch

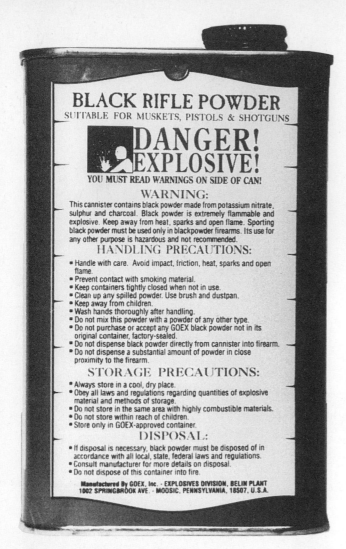

GOEX includes a whole list of warnings on the back of their blackpowder cans. Every item is important, including "Do not mix this powder with a powder of any other type."

may be cut, not blown. Patch cutting is caused by the lands of the rifling, not by the action of the powder charge. Here's how to check for patch cutting: Load no powder downbore. Seat a patched ball on the empty breech. Don't force the patched ball all the way to the bottom of the breech, or it may be difficult to extract. Withdraw the patched ball with a screw on the end of a loading rod. The patch should be intact. If it is cut by the rifling, there are a couple of cures: shoot more because shooting laps the bore, or have the bore lapped by a blackpowder gunsmith to smooth the sharp edges.

Blackpowder Traits

Blackpowder also presents a minor problem of smoke that the shooter must deal with. Since blackpowder does not efficiently alter from a solid to a gaseous state when burned, the result is plenty of smoke. That's not serious, but poor solid-to-gas transfer is the cause of an important aspect of blackpowder shooting: the need to clean the guns afterward. Failure to do so can ruin a firearm. The literature states that the bulk of the sulfur content of the powder does not truly burn. Sometimes tiny beads of sulfur remain after firing. The amount of the fouling left in the bore is in part determined by powder granulation. Since blackpowder is surface-burning, each kernel is consumed from exterior to interior. Smaller granulations yield not only greater energy per charge, but they are also better consumed as the outer portion of the kernel flakes off through burning. That's why I like modest charges of FFFg instead of

FFg for target shooting and plinking. The finer granulation leaves less residue to clean up than larger-granuled powder charges that deliver similar muzzle velocities.

Another trait of blackpowder is its efficiency level. Understanding its efficiency, or lack thereof, is vital to the serious shooter. Blackpowder is not nearly as efficient as smokeless. That's not altogether good, but it's not all bad, either. You see, blackpowder loads well by volume because it is not highly efficient. Certain smokeless powders can be volumetrically loaded, too, but the modern smokeless powder cartridge must be loaded by scale or precise measure.

No chronograph known to man can detect one iota of difference in a 45-caliber rifle shooting 55.0, 55.5 or 56.0 grains of FFg granulation blackpowder. Nor will the target group show the difference. In short, blackpowder inefficiency allows perfectly fine bulk loading, because a little bit more or less does not change muzzle velocity. Volumetric loads are not only efficient and accurate, they're the correct way to use blackpowder, which should never be run through a modern powder measure. There is no reason to scale-weigh each load because the powder is not efficient enough to "know the difference" in one grain weight per charge or so. I recently ran a test comparing scale-weighed and volumetric charges of FFg. Both created the same low standard deviations.

In short, blackpowder inefficiency allows accuracy with powder charges that are not exact. Anyone can prove this to himself by keeping a log or journal of group sizes fired with carefully scale-weighed and volumetric loads. I have found no difference in accuracy between the two, all other things being equal. This makes field-loading volumetrically with no loss of accuracy a pleasant experience. Over the years, heavy slug guns shooting elongated projectiles with blackpowder have provided extremely tight groups, further proof that volumetric blackpowder loads don't hurt accuracy.

But blackpowder inefficiency presents a bit of a problem with recoil. You see, formulas for recoil include the actual weight of the powder charge, so that burning more powder means more recoil. Since it takes a large blackpowder charge to gain reasonable muzzle velocity in a big-bore gun, the muzzleloader has a reputation for "kicking" more than a smokeless powder gun for the kinetic energy it produces. For example, a 50-caliber muzzleloader firing a 177-grain round ball may consume 110 grains volume FFg blackpowder to gain about 2000 fps muzzle velocity, while a 30-30 cartridge driving a 170-grain bullet at similar velocity may use less than 30 grains of smokeless powder. The blackpowder rifle burns almost four times the powder charge in this comparison—and as stated above, more powder consumption means more recoil.

Blackpowder Storage

Another important aspect of blackpowder is its storage nature. When kept in sealed containers in a cool, dry environment, it lasts indefinitely. Furthermore, it retains its energy. I have chronographed rifles with extremely old (1850s), but apparently well-stored, blackpowder. Velocities were on a par with fresh powder. But remember that this is for carefully stored powder in sealed containers. Blackpowder exposed to air, heat, and high humidity can "go bad." Blackpowder deterioration is not altogether well-studied, but experts think they know a few things. It's noted that potassium sulfide may explode if rapidly heated, or when in contact with powerful oxidizers. Also, potassium sulfite, when heated to a point of decomposition, emits sulfur dioxide, which may be corrosive to metals, such as gun barrels and locks. In other words, excessive heat is no good for blackpowder, so store it in a cool place. However, I do want to make something clear. Heat hurts blackpowder, but apparently not nearly as much as it degrades smokeless powder. That's what tests indicate at this time.

The ingredients associated with blackpowder can cause trouble if they break down. For example, the purity of potassium nitrate is suspect as powder deteriorates. Potassium nitrate and sulfur may convert to potassium sulfide, sulfite, and sulfate, which may lead to problems with power and possibly metal etching. All of these, and other negatives associated with deterioration, seem to be connected with poor

Granulation makes a big difference in performance, and that is why different types are offered. Elephant's FFg and FFFg are two of the most widely used "grinds."

Another excellent granulation from GOEX is their Cartridge Grade, which was developed for the blackpowder cartridge, but has proved itself more widely.

storage. "Keep your powder dry" was good advice in the old days, and it still is. And since oxygen may also attack blackpowder, the rule should be upgraded to "Keep your powder dry and in an airtight container." Blackpowder is hygroscopic by nature; it attracts moisture. As it breaks down, it becomes more hygroscopic, partly because of degraded kernel integrity.

Can a good whack, as with a hammer, set off blackpowder? Good question. I've never gotten an urge to lay out a pile of blackpowder to beat on it with a hammer. However, the data I have indicates that it may be impact sensitive. I have wrongly referred to this in the past as percussion sensitive, which is not a good choice of words. Impact implies a blow, and yes, it could be possible to set off a charge of blackpowder with a healthy blow. Also, the term "detonation" has been used by many to describe a charge of powder "going off," words I just used. Better to say "ignition." The notion that free-standing blackpowder blows up like a bomb is wrong. Combining the two key terms, "impact sensitive" and "ignition," it is safe to say that we should treat blackpowder as if it were impact sensitive, igniting if struck with a certain impact.

Blackpowder remains an absolutely fascinating study in itself. I opened this chapter with a statement on the complexity of this basic and ancient propellant because the educated modern muzzleloader fan simply must know that his major fuel is indeed a simple mixture of three major ingredients, but its pyrotechnics are far from easy to understand. This chapter is but a brief glance at the properties of this amazing propellant, just a peek. A 19th-century treatise on the subject of blackpowder ended up on my desk a few years ago. Translated by the U.S. Navy, this work was so complex that parts of it were turned over to physicists for deciphering, and a couple of them had trouble understanding the finer points. Because there are so many missing paragraphs in the story of blackpowder as a propellant, writing or speaking on the subject is a little frustrating, but it's better than wrongful thinking about this old-time fuel. Now we all know a bit more about the stuff that makes our muzzleloaders go.

Blackpowder Sizes and Types

Over the past years, the smallest granulation I have readily found for sale has been FFFFg (4F). However, that changed with Elephant Brand Black Powder, which offers an even finer pan powder in FFFFFg (5F). Pan powder is used in flintlocks for ignition purposes. FFFg (3F) is a great kernel size for small-bore rifles, small-bore pistols, cap 'n' ball revolvers, and light loads in just about any muzzle-loading rifle. FFg (2F) is a real hard worker. It's superb for big game power out of 45-caliber and larger rifles. It works fine in shotguns, too, and is appropriate for big-bore pistols. Fg (1F) is at home in the 12-gauge shotgun, and better yet in the 10-bore. It's a great powder granulation, especially in the blackpowder cartridge, but does not yield good velocity in the "average" blackpowder hunting rifle of 45- to 58-caliber. These are the granulations most noted today. In the past, there was Life Saving, a granulation larger than Fg. Larger still in kernel size was Whaling, and still bigger was Cannon.

Corning of Powder and Granulation Sizes

Corning of powder was a big step in the development of a successful blackpowder that could be turned into granulations, carried into the field by soldiers and hunters, and relied upon for decades when stored properly. Corned powder was prepared wet, then turned into kernels (granulated), unlike the previous serpentine powder, which was ground dry. The following is for the reader's reference only. In fact, various powder companies have always insisted upon their own kernel sizes. At this very hour, kernel sizes still vary among different brands.

Large differences in granulation size do exist. On the left is Black Canyon Powder. Note how large its kernels are compared with GOEX FFg on the right.

GOEX granulations are well defined. On the left is FFFg, with FFg in the middle, and FFFFg pan powder on the right.

Square Mesh Screen

The following granulations are determined by screening with various meshes. A "Go" and "No Go" method is used. Blackpowder emerges in various kernel sizes. For lack of a better term, it is "sifted" to size. FFFFg (4F) and FFFFFg (5F) are the smallest kernel sizes currently available on a wide scale. No screen data is given on FFFFFg (5F), but note that FFFFg (4F) will drop through a "Go" screen mesh of only .0170-inch. If powder falls through the "No Go" mesh at a mere .0111-inch, it is even smaller than FFFFg (4F).

Glazing

The glazing of blackpowder was another improvement. Glazed powder was coated, generally, with graphite. This process was carried out in the final phase of manufacture, where the powder was tumbled to coat it. Glazing could go back as far as the 1500s. Hints from gun literature suggest this. While glazing was not a part of every powder manufacturing process, it was considered by some as a safety measure, as powder so treated was supposed to be more spark-resistant.

Other Powders

The warning has been sounded so often that I know it must be boring by now, but the modern muzzleloader enthusiast must be knowledgeable in order to be safe. Sure, he knows smokeless powder will blow his gun from here to Aunt Matilda's barn. However, as unlikely as it may seem, shooters still run across other propellants from the past. Let's touch on these very briefly.

Brown Powder

Brown powder differed from blackpowder due to the wood used for charcoal, which was under-oxidized and therefore more brown than black in color. Furthermore, brown powder had a sulfur content of only three percent. Some records show brown powder as powerful as early smokeless. Odds of locating brown powder are as good as falling in a stream and coming up with a 10-pound trout in your hand, but just in case, don't mess with the stuff if you do run across some in a dark attic somewhere.

Schultze Powder

This powder surfaced about 1867. Some sources feel it was the first smokeless powder. As such, it obviously has no place in muzzleloaders.

King's Semi-Smokeless

The literature goes two ways on this stuff. I tried some and got identical velocities from the test rifle with FFg GOEX. However, there are patent papers that show King's Semi-Smokeless as a modified blackpowder. If you run across any, avoid its use.

Dense Powder

Cited in *Gun Week* newspaper, December 22, 1978, Dense Powder was "a modern smokeless powder, frequently combined with nitroglycerine, that gives ballistic results identical to those obtained with blackpowder." If you find any Dense Powder, which will be like discovering a million dollars buried in your back yard, avoid it at all cost. It could blow your muzzleloader to smithereens.

Ballistite Powder

Gun Week newspaper, same issue as above, said that "Ballistite powder, also known as Nobel powder, was the first of the modern smokeless powders. First made in 1887, it consists of 40 percent nitroglycerine and 60 percent nitrocellulose." This stuff will blow your muzzleloader beyond smithereens.

Bulk Powder

Credit *Gun Week* again for coming up with data on Bulk Powder, which is noted as obsolete smokeless powder with a nitrocellulose base. It would be deadly stuff in a muzzleloader.

Du Pont Bulk Smokeless

This powder carries with it great confusion, so the following is very important: Du Pont's Bulk Powder was marketed from 1893 into the 1960s. Heed the warning that came with this stuff: "Warning: While it [Bulk Smokeless] is intended for volumetric loading by drams it is not suitable for use as a replacement for blackpowder in the older guns." Definitely, it was never intended for muzzleloaders. ●

chapter 15

Pyrodex— The Replica Blackpowder

PYRODEX IS A viable, popular, powerful replica blackpowder. It is, in fact, a powder unto itself and, perhaps, should not be referred to as a replica of anything. However, the term applies because it signifies the fact that Pyrodex safely works in muzzleloaders. At the same time, it is not blackpowder. It shares certain characteristics with the older propellant, such as after-shooting cleanup, but it allows more shots in a row without cleaning in between. It also has more energy per grain weight than blackpowder, but it is *not* loaded weight for weight with blackpowder. This fact is amplified later. Pyrodex is, in short, good stuff for frontstuffers.

Pyrodex is classified as a flammable Class B solid propellant, while blackpowder is listed as a Class A explosive. Pyrodex and blackpowder are by far the most popular powders acceptable for muzzleloaders. Pyrodex is also an alternative for the blackpowder cartridge. But remember, Pyrodex is not blackpowder, and blackpowder is not Pyrodex, but the two can be used interchangeably *by volume*, although they do not share identical composition or pyrotechnic characteristics. Successful management of Pyrodex demands an understanding of its properties, and that is the goal of this chapter. Everything said here deals with the latest Pyrodex formulation.

Pyrodex Now

Pyrodex's patents include different formulations. The Hodgdon Powder Company is conducting ongoing research to steadily improve upon its product. Today's Pyrodex is not exactly the same as lots from the past. For example, modern Pyrodex ignites more readily than earlier samples. It also gives more shots per pound, volume for volume, as compared with original recipes. These positive aspects translate into two essential factors: more certain muzzleloader ignition and a little more economy per can.

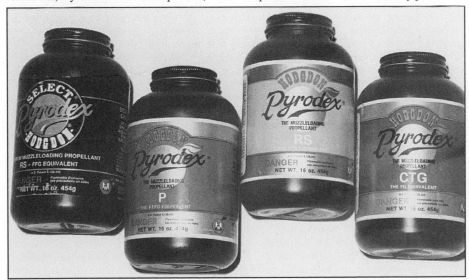

Pyrodex comes in various granulations, but not the same ones as blackpowder. Select RS (left to right), P, RS and CTG are all different. While RS and Select RS are quite similar, they are not identical, and shooters who want ultimate accuracy should try Select to see how it performs in their specific firearm. Pyrodex now comes in another form, a solid pellet.

The manufacture of Pyrodex is very sophisticated, including special buildings as part of the operation.

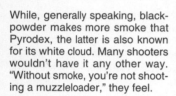

While, generally speaking, blackpowder makes more smoke that Pyrodex, the latter is also known for its white cloud. Many shooters wouldn't have it any other way. "Without smoke, you're not shooting a muzzleloader," they feel.

Cleaning Up

The best place to begin the Pyrodex story is at the end, not the beginning—how to clean up *after* a shooting session. While Hodgdon never suggested that Pyrodex was non-corrosive, the idea has flitted about from time to time. Actually, Pyrodex does demand after-shooting cleanup, so reject any notion to the contrary. Also reject any thought whatever about Pyrodex "eating up your blackpowder gun." This tale came my way a few times, so I decided to test for myself. The results of my extensive "Corrosion Tests" are located in Chapter 38. Five gun barrels were used over a period of time with three different powders and two different cleaning methods: water only and solvent only. See Chapter 38 for details. You'll see that Pyrodex did no damage to its two test barrels, even though barrels were sometimes left for a week between cleanings.

After shooting with Pyrodex, muzzleloaders and blackpowder cartridge guns can be maintained just as if the guns had fired blackpowder. The old-fashioned hot water method is fine. So is a no-water, solvent-only cleaning program. And for those who wish to combine the two—great. I have come to like that approach more and more. Chapter 38 deals with care and cleaning, so there is no point repeating that information here. However, it is very important to understand that

Pyrodex, while safe and excellent in frontloaders, is not a non-corrosive propellant, and the guns that shoot it must be cleaned afterwards.

You also may have heard that you don't have to clean between shots with Pyrodex. This time you heard right. I shoot twenty or thirty times before worrying about swabbing the bore when shooting Pyrodex at the target range. This process goes for the blackpowder cartridge gun as well as the muzzleloader. In my meager tests, cleaning in between shots with Pyrodex raised standard deviation instead of lowering it. In other words, variance between shots was greater with in-between-shot bore swabbing with Pyrodex than without swabbing between shots. This is not to imply that bore cleaning can be postponed *after* a shooting session. That point was clarified above.

Bore Prep with Pyrodex

Dressing the bore is an extremely old concept, also known as firing a fouling shot. In fact, it pertains to modern smokeless powder cartridges as well as muzzleloaders. Before I shoot for groups with, just as an example, a 30-06 rifle with a clean bore, I fire one shot downrange into the butts to dress the bore. This prepares the bore for the shots to follow. The idea is to create a condition of stability and uniformity. I used to advise firing four shots with Pyrodex in order to dress the bore.

Pyrodex definitely comes in distinct granulations. Currently, the two extremes are P (top) and CTG (bottom). The difference in granule size is clear and unmistakable. Over the past few years, rumor holds that CTG could be dropped, since RS works so well in blackpowder cartridges as well as shotguns.

Pyrodex RS (right) and Select are of the same granulation, but are not identical in formulation. RS appears a bit darker and has been known to develop superior accuracy in certain muzzleloaders.

That was wrong. I'd had good luck with four shots, arbitrarily, I might say, and so I stuck with that rule. Later, I tried three shots, then two. Now, I fire only one "prep shot" with Pyrodex to dress the bore. It should be sufficient.

Naturally, the amount of powder fired has much to do with this. If you're shooting light target loads in your 50-caliber muzzleloader, burning perhaps only 40 volume Pyrodex RS, you may wish to shoot twice in order to prepare the bore for grouping. That will put 80 volume RS through the bore, sufficient to burn off remaining metal preservers and so forth. What about hunting? Dressing the bore is mainly for target work. For hunting, I dry the bore with cleaning patches before loading up. And that's it. I do not dress the bore with Pyrodex before taking the rifle into the hunting field. However, one caution: The prep shot can also dry up any remaining oil in the vent (hole) of the nipple. You still don't have to fire a prep shot before loading Pyrodex for hunting, but do ensure that the nipple vent is clear. Fire a cap or two with the unloaded rifle, then use a pipe cleaner to swab out the vent, clearing away any possible debris left from firing the cap(s).

Pyrodex Granulations

Pyrodex is prepared in distinct granulation size. There are three "grinds" today; however, Hodgdon Powder Company is talking about

dropping CTG from the list. As this is written, P, RS and CTG granulations are still available. P stands for Pistol, and it is the finest granulation. While it's not popularly known, P is excellent for *light* loads in big bore rifles for target work, plinking, small game hunting or any other squib application. P is also at home in the smallbore. For example, I load 30 grains of P for excellent small game hunting results in 36-caliber long rifles. As its designation suggests, P is also useful in pistols. Naturally, it is equally workable in revolvers.

RS stands for Rifle/Shotgun. While P is like FFFg blackpowder in application, RS applies very much like FFg. It's highly useful for all big bore rifles and shotguns. RS is the king of the Pyrodex line, and no doubt about it. I've tried RS in all muzzleloaders from 32-caliber up, from small to large bore pistols, as well as shotguns from 20- to 10-gauge. RS can do it all. RS also comes in a Select Grade. More on this at the close of the chapter. While RS is the most important granulation in the Pyrodex lineup, I still like P for smallbores, light loads in bigger bores, and in revolvers.

CTG is similar to Fg blackpowder in application. On that score alone, you can see why CTG trembles on the threshold of discontinuance. Fg is not terribly popular or widely used. Neither is CTG, which stands for Cartridge, and that's where it shines, rather brightly, too. CTG deserves special mention as a blackpowder cartridge propellant

In-line ignition modern muzzleloaders like this Lyman Cougar—50- or 54-caliber, 1:24 rate of twist, 22-inch barrel—perfectly shoot Pyrodex.

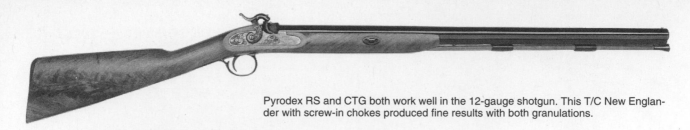

Pyrodex RS and CTG both work well in the 12-gauge shotgun. This T/C New Englander with screw-in chokes produced fine results with both granulations.

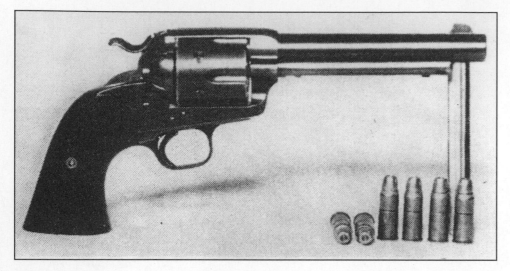

The blackpowder cartridge, revolver or rifle, does just fine with Pyrodex.

because it has a propensity to produce excellent accuracy. Tests indicate, if not prove, that Fg offers a general accuracy advantage over FFg or FFFg in a large-capacity blackpowder cartridge, so maybe it should be no surprise that CTG follows suit, providing high accuracy for the many old-time blackpowder rounds still shot around the country, from the popular 45-70 Government to the excellent Sharps rounds in many

Pyrodex loads work by volume, not weight, with blackpowder. For example, current lots of RS run 70.5 grains weight for 100 volume, while CTG goes 71.0 grains weight per 100 volume, and P runs 73.0 grains per 100 volume. That's why a powder flask like this T/C U-View can be used with Pyrodex.

different styles. Incidentally, CTG works fine in muzzleloading shotguns, too.

While RS shines in many smaller-capacity blackpowder cartridges, CTG is useful in larger-capacity rounds. Note: although no bore swabbing between shots is all right when using Pyrodex in the cartridge, it's like anything else—too much of a good thing is no good. For rounds that burn a lot of powder, such as the 45-120 Sharps, it's not a bad idea to clear the bore with Hoppe's No. 9 followed by swabbing with a blackpowder solvent after twenty shots or so. Let common sense prevail. If you look down the bore of your blackpowder single shot rifle and it appears "caked up," swab the darn thing!

To be fair, in certain instances, CTG is not as accurate as RS in the blackpowder cartridge. While I have not had a chance to study the situation, I suspect that bullet obturation (the projectile upsetting to fill the bore) is not as complete with CTG as it is with RS. I also wonder about bullet makeup, an all-lead projectile versus a bullet that is, in truth, an alloy of lead plus tin, antimony and/or other metals. The all-lead bullet may obturate to the bore just fine using CTG, while the alloy missile requires a swifter kick in the behind to upset it. A wise shooter will try both CTG and RS in his blackpowder cartridge rifle to determine which works best in his specific rifle shooting his specific bullets.

Finally, a word about Pyrodex Select Premium Muzzleloading Propellant, to state its full title. It is billed as "RS - FFg Equivalent," which it is. Select is also labelled "Specially Processed & Tested." Pyrodex Select is used identically to standard Pyrodex RS, because it is essen-

If it shoots blackpowder, it will shoot Pyrodex. This T/C 50-caliber White Mountain carbine was tested with Pyrodex RS. Results were sparkling: perfect ignition, top accuracy and full velocity.

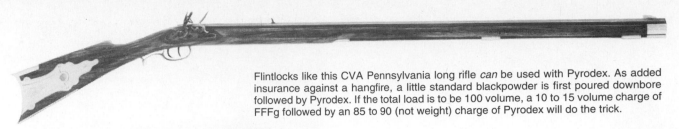

Flintlocks like this CVA Pennsylvania long rifle *can* be used with Pyrodex. As added insurance against a hangfire, a little standard blackpowder is first poured downbore followed by Pyrodex. If the total load is to be 100 volume, a 10 to 15 volume charge of FFFg followed by an 85 to 90 (not weight) charge of Pyrodex will do the trick.

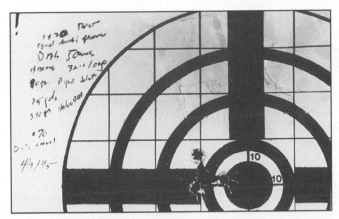

This group was produced with Pyrodex Select. It's .70-inch center to center from a Dale Jones barrel at 25 yards under poor range conditions (windy). Under more ideal conditions, this group repeated at 50 yards. Obviously, Pyrodex is capable of producing fine accuracy.

tially the same propellant; however, Select is not standard RS. The labelling provides a hint to this fact. The clue is in the words "Specially Processed." Select is, as advertised, a specially processed and tested powder that the perfectionist will want to try.

Loading Pyrodex

All correctly loaded muzzleloaders enjoy 100 percent load density, because the bottom of the bore is in fact the breech of the gun. Once powder is dropped downbore, with a projectile seated on the charge, you have to have 100 percent load density with no air space between the powder and the projectile (including a shotgun wad column). For cartridge guns, best accuracy is achieved when Pyrodex is also treated to 100 percent load density in the case, again leaving no air space. Otherwise, accuracy falls off. If lighter loads are desired, air space should be taken up with fillers or wads. Incidentally, don't forget that cartridge cases using Pyrodex must be cleaned after use, or they will corrode. The soapy water method is a good start, but finish up by polishing the interior of the case with modern solvent on a cotton swab. Cases must dry completely before reloading.

Pyrodex is loaded volume for volume with blackpowder—that's why it's called a blackpowder substitute. It truly does substitute for blackpowder, volumetric load for volumetric load, *but not weight for weight*. No need to worry about volumetric loads and accuracy, incidentally. In spite of information to the contrary, every test I ever ran proved that weighing Pyrodex charges to tenth-grain levels on a powder scale does *not* enhance accuracy. It cannot, because Pyrodex, like blackpowder, is not nearly as efficient as smokeless powder. In a way, this is an advantage because a grain weight more or less makes little difference in bullet velocity and, therefore, no significant difference at the target. As a test, I loaded rifles of 45, 50 and 54 calibers with carefully scale-weighed Pyrodex charges versus Pyrodex charges delivered from an adjustable blackpowder powder measure.

There was no difference in accuracy at the target. Groups were the same size with weighed charges and volumetric charges. Furthermore, my Oehler 35P chronograph showed identical standard deviations between scale-weighed Pyrodex charges and volumetric charges. So, load Pyrodex using a blackpowder measure, not a scale and never a modern powder measure. Incidentally, charge weight, volume for vol-

ume, will be lighter with Pyrodex. That is because Pyrodex is less dense than blackpowder. This factor is more pronounced today than in the past.

The latest lots of Pyrodex I tested showed the following: A powder measure set at 100 produced right on 100 grains weight of Goex brand FFg blackpowder. Leaving the powder measure at the same 100 setting, Pyrodex RS averaged 70.5 grains weight, CTG averaged 71.0 grains weight, and P averaged 73.0 grains weight. In the past, the ratio was closer to a 20 percent difference. In other words, the Pyrodex RS charge would have been 80 grains weight when the powder measure was set at 100. So Pyrodex is indeed less dense than blackpowder. However, it also yields more energy than blackpowder in like amounts. That is why Pyrodex delivers similar velocities when loaded volume for volume with blackpowder.

Powder Power

Ballistic results between the two powders are, for all practical purposes, the same when used volume for volume. Obviously, there are always individual gun to gun differences. Therefore, head-to-head velocity tests between blackpowder and Pyrodex vary. However, here are a couple actual figures that show the usual results of FFg versus RS in chronograph tests. A Jonathan Browning 50-caliber Mountain Rifle shooting 120 grains volume FFg earned 1974 feet per second muzzle velocity with a patched round ball. The same rifle using 120 volume Pyrodex RS, which in this test turned out to be 87.5 grains weight RS (an older lot of powder), yielded a muzzle velocity of 1929 feet per second with the same patched round ball. That's mighty close. A 58-caliber Zouave Carbine shooting 100 volume FFg drove a 460-grain bullet at 1088 feet per second. Using the same powder measure setting of 100, Pyrodex RS, at a weighed 72.5 grains (a later lot of Pyrodex), earned a muzzle velocity of 1064 feet per second with the same 460 grain bullet—no true difference because normal variations cause that much disparagement between two loads in the same rifle.

Pyrodex can be overloaded, which generally results in a clash with the law of diminishing returns. More powder than practical is loaded downbore, the result being either no increase in velocity whatsoever or so very little velocity increase that there is no practical gain. Meanwhile, pressures can rise. So the overload is counterproductive. The overload may also prove dangerous. While the gun may not show immediate effects of too much powder, wear through gas cutting could show up later. Furthermore, adding more powder than called for means more recoil—all for naught. The way to load Pyrodex is by the rules. Do not exceed the gun manufacturer's recommended maximum volumetric load at any time.

Pyrodex in Flintlocks

These two do not make a perfect combination, although the latest Pyrodex works much better in flinters than earlier formulations. Ignition is the primary problem because Pyrodex is harder to ignite. Due to its design, the flintlock does best with FFFFg blackpowder in the pan and an easy-to-ignite powder as the main charge in the breech. Blackpowder is easy to ignite. For those who wish to shoot Pyrodex in flintlocks, a duplex load can be used. For example, if the flintlock load calls for 100 volume RS, put down 10 volume FFg first, followed by 90 volume RS. This lays 10 grains of blackpowder first into the breech and next to the touchhole for quick ignition. Try your flintlock muzzleloader with the new Pyrodex first to see how ignition goes, and if there is a hangfire (slow ignition), then use the duplex method to promote faster ignition.

Proof that Pyrodex is at home in any muzzleloader, the Kentucky pistol was tested with RS and the results were perfect.

Some shooters like Pyrodex in the caplock revolver because it tends to cake up less than blackpowder. It's a personal matter.

Percussion Caps and Pyrodex

No problems were experienced with standard No. 11 percussion caps of high quality during the testing of Pyrodex loads in all types of frontloaders, including modern models such as the Knight MK-85 and the White rifle. There were no ignition or accuracy problems with Pyrodex in the blackpowder revolver, either, where once again No. 11 percussion caps were employed. Large rifle primers were used in blackpowder cartridges for all Pyrodex testing. Again, there was perfect ignition.

Lubes and Pyrodex

Various standard blackpowder lubes also caused no problem with Pyrodex, although it must be pointed out that Hodgdon now offers a special Pyrodex lube. It is billed as "Pyrodex Lube, patch or bullet lube & cleaner, especially designed by Hodgdon for use with Pyrodex propellant. All natural, biodegradable, environmentally safe." The back of the can bears instructions for use of this special lube. It is also sold as prepared cleaning patches called Pyro Patch.

Pyrodex in the Future

I doubt that Pyrodex was ever intended as a complete replacement for blackpowder. Replica blackpowder, yes. Substitute blackpowder, yes. But most shooters are going to use both blackpowder and Pyrodex, because both are readily available. The advantages of Pyrodex fall mainly in the realm of field shooting and target work, where cleaning between shots is not critical for long strings. Pyrodex is also convenient in trap, Skeet and other shotgunning games where multiple firing is the rule and cleaning between shots is inconvenient. Pyrodex is valuable simply as another available powder for any firearm that normally uses blackpowder. While it is a substitute, Pyrodex is a unique propellant in its own right, and highly worthy of our attention. It is here and will continue to be here. And don't worry. Pyrodex does make smoke, and some shooters consider smoke an aesthetic necessity with frontloaders and blackpowder cartridge guns. Pyrodex is now legal in all fifty states for blackpowder-only primitive hunts—because it works reliably with all blackpowder firearms, from small to big bore rifles, in all types of shotguns and cartridge guns, as well as in sidearms. ●

Testing the Pyrodex Pellet

WHAT IF YOU didn't have to carry powder in a horn, flask or other container? What if you had a solid pellet made up of an exact amount of propellant? Drop one downbore for target shooting, two for most big game loads, three for muzzleloaders allowed a big powder charge, and you're set. That would be convenient. Well, the Pyrodex Pellet does just this. Each one runs 37 grains *weight*, or 50 volume.

While examining this new product, one word came to mind: consistency. A random sample of five Pyrodex Pellets weighed 36.8, 36.7, 36.7, 36.9 and 36.9 grains, with an average of 36.8 grains. The same test performed on five two-pellet sets yielded 73.2, 73.6, 73.4, 73.8 and 73.3 grains, for an average of 73.46.

That's close enough to 71.5 grains weight, considering muzzleloader propellant efficiency (or lack thereof), to call it a 100 volume charge. And this size charge with proper bullets is considered a big game load for most muzzleloaders, while some frontloaders are allowed even more powder.

Weighing five three-pellet sets, these figures were produced: 110.2, 110.0, 110.6, 110.8 and 110.6 grains, averaging 110.44.

That's about 140 volume roughly, but as stated above, some gunmakers build rifles that allow this much fuel. However, for the most part, the 50-caliber Pyrodex Pellet is used singly for target shooting, with two at a time for normal hunting applications.

The Pyrodex Pellet is solid Pyrodex RS in a .450-inch diameter pellet, .750-inch long, intended for 50-caliber muzzleloaders. And length measurements of five random pellets, again, proved remarkably consistent. But there's more. Although current Pyrodex formulations enjoy excellent ignition, the Pyrodex Pellet, to ensure super ignition, has a base impregnated with blackpowder. Ignition was 100 percent in tests. Incidentally, laboratory tests of the pellet without the blackpowder base still revealed excellent ignition.

But it is at the range where a propellant's effectiveness is truly measured. So I ran five sets of three-shot chronograph tests using the Pyrodex Pellet. The test gun was a 50-caliber Storey Buggy Rifle with 22-inch barrel. In the first three tests, a 50-caliber 385-grain Buffalo Bullet containing T/C Natural Lube 1000 in the base cavity was seated with 60 pounds of pressure atop two Pyrodex Pellets. The final two tests used a 260-grain lead pistol bullet with sabot, also seated with 60 pounds of pressure. Remington No. 11 percussion caps were used throughout the testing.

In test one, the two pellets were loaded with the bases down. The three shots averaged 1426 fps, with a high of 1432 and low of 1422, giving an extreme spread of 10 and a standard deviation of 5. Test two was run under the same conditions and resulted in an average of 1433 fps, with a 1454 high, 1411 low, 43 extreme spread and 21 standard deviation. For test three, the Pyrodex Pellets were loaded inverted, with the bases up. The three shots averaged 1411, with a high and low of 1449 and 1375, ending up with an extreme spread of 74 and standard deviation of 37.

Tests four and five used the sabotted pistol bullets, the former with two pellets seated base down and the latter with three pellets base down. The double-pellet charge averaged 1802 fps—high of 1816, low of 1789, spread of 27, standard deviation of 13. The triple-pellet charge gave an average of 2135, with the other figures running, 2165/2106/59/29. Ignition with all tests was perfect.

As the numbers show, the Pyrodex Pellet, when loaded properly, can be very consistent. Add to this the convenience of the prepackaged load, and this product is sure to find a niche. ●

The Pyrodex Pellet has functioned well in testing and makes loading your rifle much easier.

chapter 16

The Surprising Round Ball

ANYONE WHO STUDIES ballistics, even a little bit, knows that the round ball is designed all wrong. It's too darn short and fat to "carry up" well downrange, so it loses velocity/energy rapidly. It shouldn't penetrate well, either. On paper, the pumpkin-like bullet is downright pitiful. Gun writers have railed against it for decades. It has no charisma. The only projectile I can think of that might fare worse ballistically would be a disc of lead. As it is, the lead pill is the lightest bullet for its caliber. For example, a .375-inch round ball only weighs 80 grains. Meanwhile, a .375-inch conical might weigh 300 grains or more. Also, the round ball has less bearing surface than any other bullet. In fact, a patch is used to take up windage (space between the ball and the interior of the barrel), so the strange lead pill doesn't even make direct contact with the rifling. In spite of all this, there are thousands of blackpowder shooters all over the world who won't quit the round ball.

Round Ball Ballistics

Of course a ball is round. Ever see a square one? So the very term seems redundant. But it isn't, because the first word means round, but the second word does not mean ball. Ball means bullet. So "round ball"

translates into "round bullet." Therefore, the reference to a round ball is correct and properly descriptive. There is a conical ball as well as a round ball. Military ammunition, with pointed jacketed projectile, is referred to as "ball type." R.A. Steindler writes in *The Firearms Dictionary* that "Ball ammunition is jacketed, military ammunition."

Of course, it is the round ball's globular shape which is responsible for its ballistic character. In flight, the round ball can lose about half of its initial speed over a range of only 100 yards. It is simply not aerodynamically sound. Over the same 100-yard span, a 140-grain .264-inch spitzer bullet loses only 15 percent of its starting velocity. Poor round ball. It's lousy in three important ballistic areas: sectional density, ballistic coefficient and kinetic energy.

Sectional Density

This is the relationship of a projectile's diameter to its length *plus the inclusion of mass proportion*. This vital ballistic factor is computed as the bullet's weight in pounds divided by the square of the bullet's diameter in inches. The resulting math provides a decimal figure, and that figure is used in comparing the sectional density of various mis-

The simple round lead ball has low sectional density and low ballistic coefficient, yet it works within its limits. "Give me a 12-gauge round ball," says one expert blackpowder hunter with a lot of game to his credit, "and I'll take anything on this continent with one shot, as long as it's close enough."

The rate of rifling twist dictates whether a firearm is a ball-shooter or not. Slow is for round ball. In 50-caliber, a turn in 66 inches is quite popular for round ball shooting, and as caliber goes up, twist can be even slower for ball stabilization. One turn in 12 feet was noted on some English big bores.

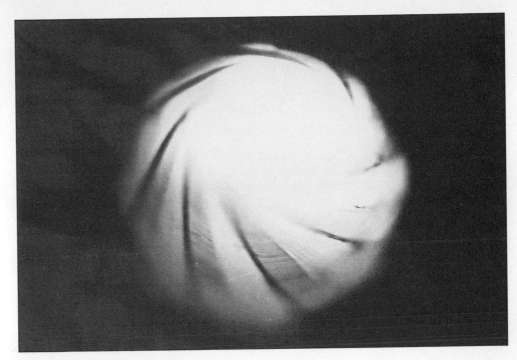

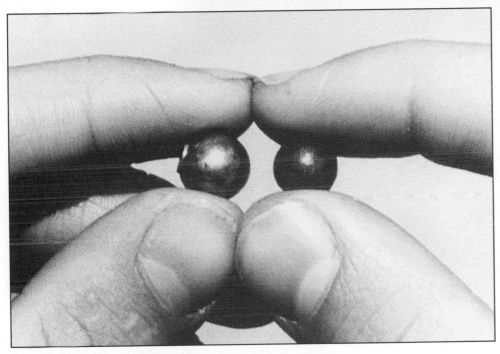

A round lead ball can only gain mass through larger caliber—it gets no longer than it already is. The ball grows in mass out of proportion to its caliber. Make it bigger and the ball gets heavier right away.

siles. Computing for a 30-caliber 150-grain bullet, the number is .226 (the result of dividing 150 grains by 7000 to reduce to pounds weight, which is .0214285; squaring the bullet diameter, .308-inch, the result is .094864; dividing the last figure into the first leaves .2258865, which rounds off to .226). By itself, that figure doesn't mean much.

But compare it with a 30-caliber 180-grain bullet, which has a sectional density of .271, and a meaningful relationship appears. It's clear that the 180-grain bullet has greater sectional density than the 150-grain bullet of 30-caliber. Carrying this one step further, a 175-grain 7mm (.284-inch) bullet has a sectional density of .310. The long (for its caliber) 7mm bullet with 175 grains of weight whips both of the 30s. It's not merely a matter of long and pretty, however. The 500-grain 45-caliber bullet, as loaded in the 458 Winchester, has a good sectional density, too—.341, better than the sleek 175-grain 7mm bullet. Surprised? Meanwhile, a 133-grain 45-caliber ball (.445-inch diameter) has a sectional density of only .096. That's not a very exciting figure.

Ballistic Coefficient

Sectional density is a useful guide, but it does not say enough about the projectile's ability to buck the atmosphere, because there is nothing in the sectional density formula that includes the shape of the missile. That's where "C," or ballistic coefficient, comes in. A number representing coefficient of form is derived from a "standard" model. That figure is coupled with the sectional density number as the ratio of the bullet's sectional density to its coefficient of form. In terms of sectional density, a 150-grain round-nose 30-caliber bullet and a 150-grain spitzer 30-caliber bullet are identical at .226. But C tells the truth about these two bullets in flight. The pointed one is better in retaining velocity (therefore, energy) downrange, and this shows up in C figures. The streamlined 150-grain bullet earns a C of .409. The round-nosed 150-grain 30-caliber bullet carries a C of .205. That same 45-caliber round ball referred to previously has a C of .063—that is blunt!

111

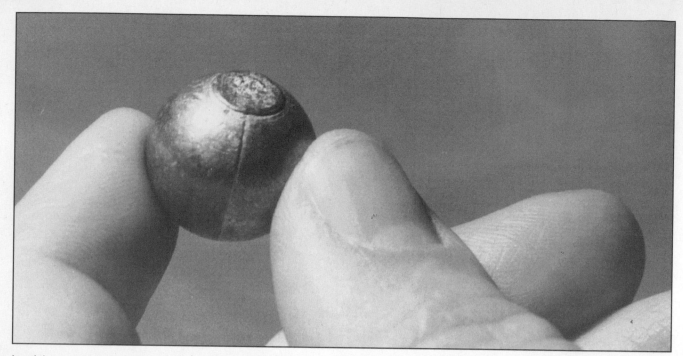

Load the sprue muzzle-out for accuracy, the old story goes, but it's not entirely true, nor false. A well-cast round ball can be loaded with its sprue down, too, for accuracy. But if you load sprue-up, you visually can center it, which you cannot do if you load it downward.

Kinetic Energy

Its ballistic coefficient is so poor because the round ball is not aerodynamically inclined. It rapidly loses velocity, which means it also rapidly loses energy. Newton's child, kinetic energy (KE), squares the velocity of the bullet in order to derive energy. Since the round ball's impact velocity at only 100 yards is comparatively low, the ball does not have exciting terminal paper ballistics. While KE does not tell everything about harvesting power, it is a useful barometer. When KE shows a 7mm Magnum as more powerful than a 54-caliber round ball rifle, that's only because it's true. A 7mm Magnum firing a 162-grain bullet at a muzzle velocity of 3165 fps, as chronographed with RWS factory ammo, has a muzzle energy over 3600 foot pounds. A 54-caliber ball-shooter firing a 230-grain missile at 2000 fps MV carries a muzzle energy of a bit over 2000 foot pounds. At 100 yards from the muzzle, the 7mm Magnum arrives with more than 3200 foot pounds of remaining energy. The ball reaches the century mark with less than 700 foot pounds of force remaining.

Round Ball Calibers and Characteristics

As a projectile approaches the speed of sound, drag greatly increases. The atmosphere is the number one cause for projectile velocity loss, and drag escalates with bullet speed. A high C helps to thwart drag, and since the round ball does not enjoy good ballistic coefficient, the atmosphere severely retards its flight. However, there is a way to pick up downrange ballistic punch with a ball. Since high velocity is out of the question in blackpowder shooting—1500 to 2000 fps MV is extremely fast, but compared with the 3000+ fps MV enjoyed by modern cartridges, such velocities are modest—the only way to make up for the lack of speed is increase of mass. That means a big caliber. Cal-

iber is all to the ball. In a conical-shooting rifle, if you want more mass, simply make a longer bullet. But the round ball doesn't get any longer, so if you want to make it heavier, you have to make it bigger. As the ball grows in diameter, its weight increases out of proportion. A little formula helps clarify this important fact: First, compute for the volume of a sphere, which is diameter to the third power times .5236; the resulting number is multiplied times 2873.5, the weight of a cubic inch of pure lead. Consider a round ball of .530-inch diameter (54-caliber). Its diameter to the third power is .148877; times .5236 (a constant) = .077952; times 2873.5 = 223.99, which rounds to 224 grains weight, the theoretical perfect weight of a .530-inch ball in pure lead.

This formula is also valuable for testing round ball lead purity. Remember it when casting your own round ball. Compare the actual weight of your product with the theoretical perfect weight of that size ball and you can determine its lead purity. Of course, you must mike the ball. You cannot rely on the general mould size. If you mike a ball and it comes out to .530-inch diameter, but it weighs 220 grains instead of 224 grains, chances are your lead contains tin, antimony or some other element of lower atomic weight than lead.

Of course, you should use the purest lead you can find when making round balls (described in Chapter 24). Lead cleaning is discussed in that chapter, the idea being to remove all other elements and foreign matter from the lead. I have tried lead alloys, which work well for cartridge bullets, and found them poor in the muzzleloader. Obturation is one reason for using "pure" lead. The bullet, round or conical, obturates or fills out somewhat in the bore as it is smacked in the behind by expanding powder gases from the blackpowder charge. Any bullet, due to inertia, is predisposed to remain at rest. Of course, expanding gases win, furiously driving the missile upbore. But, there is a tiny delay in takeoff that fat-

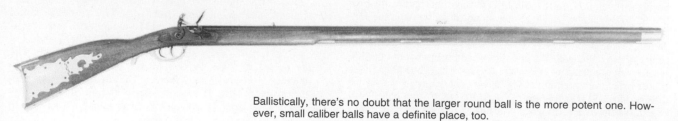

Ballistically, there's no doubt that the larger round ball is the more potent one. However, small caliber balls have a definite place, too.

tens the ball just a little bit (if a normal powder charge is used). And pure lead obturates better than an alloy. So go with the purest lead you can buy, or clean your lead supply before running ball.

Another way to guarantee that a lead ball is free of antimony, tin or other non-lead agents is to buy it from Speer, Hornady or Buffalo Bullet. I've tested these projectiles for lead purity, and you can, too: Mike the ball to be certain of its exact diameter, then use the previously mentioned formula to see how close your lead pill comes to the theoretical perfect figure for a pure lead ball. Another reason to use pure lead, especially for a hunting ball, is integrity. Pure lead round balls have high molecular cohesion. That is, they tend to remain intact rather than fragmenting, and round ball integrity is in part responsible for good penetration. If a projectile fragments, it gets lighter in weight, thereby losing its ability to penetrate deeply and make a long wound channel.

Choosing a Caliber

The best choice of round ball size depends upon the application. If you're interested in small game hunting, a large caliber is destructive and unnecessary, as well as more expensive to shoot than a smaller caliber. My favorite small game/plinking caliber is 32; however, this size is not always available. Therefore, my Hatfield Squirrel Rifle is a 36-caliber. On the other hand, the 32 can be cheaper to shoot than a 22 rimfire if you cast your own projectiles and make percussion caps with

the Forster Tap O Cap tool. A 32 rifle shooting the 45-grain .310-inch round ball attains 22 Long Rifle muzzle velocity, generally speaking, with only 10 grains of FFFg blackpowder. That allows 700 shots for 1 pound of fuel. My 36-caliber rifles have done better with about 15 grains of powder. As the ball size increases, more powder is necessary, of course, to propel the missile at useful velocities.

The middle-sized calibers, 40 through 50, are as good as the shooter behind the rifle, and we're speaking mostly of rifles now. I've taken antelope with a 40, and many deer have been dropped with the 50. My son, Bill, has taken elk with his 50-caliber ball-shooter. Ultimately, my favorite big game ball size is 54. It's not too large for deer and antelope, and at close range, it works all right on elk, although it's my opinon that shooting elk at more than 75 or 100 yards with a ball in this weight class is inviting trouble. Elk are big and tough, so get close, put the ball on the money, and the bull is yours. But when the power of the lead pill has dropped off, it's asking a lot of it to put a tenacious wapiti to the earth with one shot. I like the 54 because it's not so large that it requires huge charges of blackpowder to gain a decent muzzle velocity, yet it's not so small that the projectile is light. A .530-inch ball weighs around 224 grains, depending upon actual diameter and lead composition, while a .535-inch ball goes 230 grains. That's good heft. And in a well-made muzzleloader, the 54 ball can scoot at around 1900 to as high as 2000 fps MV.

The author considers the 54-caliber round ball "balanced." In .535-inch diameter, it weighs 230 grains, and yet it can achieve 1900 to 2000 feet per second with heavy but safe charges of FFg or RS powders. This antelope buck was taken with a 54-caliber round ball.

A single 54-caliber round ball instantly put this buck down. No second shot was required. That's the norm with a well-placed hit from close range with a hefty round ball.

The old round ball really works. Just remember that its diameter determines its ability to drop big game. There is a neat formula that can be used to illustrate this point. Consider two projectiles, one exactly 50-caliber (.50-inch ball) and the other 60-caliber (.60-inch ball). Work a little math on both, cubing each figure. The first, $.6^3$, comes out to .216; the second, $.5^3$, results in .125. Now divide .216 by .125 and you will get 1.728, rounded to 1.73. The theory of this little math game: the 60-caliber round ball has the potential of gaining 1.73 times the energy of the 50-caliber round ball. This is just a game, not a law, but let's give it a try. We'll propel both balls to 2000 fps MV.

The result is 2887 foot pounds of energy for the 60-caliber ball, which weighs 325 grains in pure lead, and 1670 foot pounds for the 50 ball, weighing 188 grains. Now prove the arithmetical point by multiplying 1670 times 1.73. If the little formula hasn't lied, the resulting figure should be pretty close to 2887. It is 2889. And that's pretty close. So the potential kinetic energy of the 60-caliber ball, which is only .10-inch greater in diameter than the 50-caliber, is about 1.73 times superior. This little formula is not a law to live by, but it shows that as diameter goes, so goes round ball power potential.

Make no mistake, the real big boys are great, too. But a 58-caliber ball, for example, requires a large dose of powder in order to gain a reasonable trajectory. The 60 and 62 are even more demanding. In testing one 62-caliber ball-shooting rifle, a huge powder charge was required to gain only 1600 fps MV. Nonetheless, if you want ultimate power from the round ball, the big missile is the only way to go. Round ball guns have dropped all manner of ponderous beasts, including elephants. In fact, several of the old-time elephant hunters of Africa chose the ball for ivory hunting. Major Shakespear and Captain Forsyth, both well-known shooters of the 19th century, liked the round ball very much. The famous explorer of the Nile country and great

In tests, damage to the nose of the ball proved detrimental to accuracy, even more than base damage. However, there were not enough repeat tests to make this a law. Ideally, both ends of the ball should be without damage.

The round ball should not be able to do what it does. On paper, it's a pitiful thing. But a 54 caliber round ball, which is an excellent size for big game, achieving good speed with decent weight, can flatten out like this and still penetrate the full breadth of a deer's chest cavity.

hunter of Ceylon, Sir Samuel Baker, preferred the ball over the conical, even for pachyderms. Here is how Sir Baker felt about it.

> I strongly vote against conical balls for dangerous game; they make too neat a wound, and are very apt to glance on striking a bone. . . . In giving an opinion against conical balls for dangerous game, I do so from practical proofs of their inferiority. I had at one time a two-groove single rifle, weighing 21 lbs., carrying a 3 oz. belted ball, with a charge of 12 drachms powder. This was a kind of 'devil stopper,' and never failed in flooring a charging elephant, although, if not struck in the brain, he might recover his legs. I had a conical mould made for this rifle, the ball of which weighed 4 oz., but instead of rendering it more invincible, it entirely destroyed its efficacy, and brought me into such scrapes that I at length gave up the conical ball as useless.

Sir Samuel stayed by his convictions. He went on hunting, espe-

cially in 19th-century Ceylon, with the round ball, even for elephants and water buffalo. He was known to shoot the latter from ranges of 300 yards or more with the round ball. How could the lead sphere be so effective at long range, especially on such heavy animals? Ball size. Baker's rifles were big bores, and I don't mean 58-caliber. A smaller Baker rifle would be in the 12-bore range, meaning it would require twelve round balls to make 1 pound. The 4-bore was also used, and with only four projectiles to the pound, each one weighed 1750 grains. Even the 458 Winchester, which is considered an elephant rifle, drives a bullet of "only" 500 or possibly 600 grains weight. The old-time hunter of dangerous game knew that the only way to gain true power from a round ball was to use a large one. The same fact holds true today, although we are not hunting elephants as a rule, so our large bores are petite by comparison with Baker's.

Contrary to the opinions of some, a round ball that hits faster has more energy for penetration. True, it may dispense of that energy quicker, flattening out more than a slow-moving ball, but all in all, go for speed with the round ball, just as you would for any other rifle. The laws of ballistics do not change with muzzleloaders.

Modern shooters had to rediscover what yesteryear's marksmen already knew: The patched round ball can be very accurate, even at more than 100 yards.

In tests, round balls penetrated considerable medium, creating long wound channels. They also flattened out considerably in the process. The ball on the left started out just like the ball on the right.

Penetration

I am sure that laws of physics explain why a lead round ball often drives entirely through the breadth of a buck mule deer standing 100 yards from the muzzle of the rifle, and yet such performance seems to stand outside the boundary of scientific explanation. To be sure, I've collected a good many lead pills from various game animals, especially mule deer and elk. However, just as many or more have whistled through non-stop. Major Shakespear said of round ball penetration: "I have Minie bullet-moulds for my rifles; but so long as the spherical bullets go through and through large game, I do not see the use of running the risk of shaking the stock of the gun, and of extra recoil, by using heavier balls." The Major, author of *Wild Sports in India*, remained a devotee of the round ball. The ball can penetrate game well; however, don't fall into the "low-velocity trap." There is a theory which goes, "Why load for higher velocity when the ball loses its speed so rapidly anyway?" The logic behind this notion hinges on a statement made earlier: As a projectile enters the domain of the speed of sound, the atmosphere plays particular havoc on its flight. That's true. But, if you want to end up with something, you still have to begin with something. Penetration is, in part, reliant on striking velocity. I have this to say of the low-velocity advocates who insist that big game should be hunted with modest powder charges only—if velocity is not important, then why not dispense with the rifle? Carry a bag full of round balls and throw them at the game.

Round Balls in the Wind

Several factors govern the performance of a projectile in the wind, and I'm afraid these criteria don't favor the round ball. Its time of flight is not that good, and its ballistic shape leaves a lot to be desired, too. Again, as the ball size goes, so goes its drift off course in the wind. A 36-caliber squirrel rifle shooting a .350-inch diameter round ball will miss the mark by a drift-over of almost 4 feet at 100 yards in a crosswind of 30 miles per hour. In a 10-mile-per-hour crosswind, this little 65-grain job drifts about a foot, and these figures are for a starting velocity of 1800 fps, not too far from the initial speed of a 22 WMR bullet. A 54-caliber rifle does a bit better, but even its 230-grain .535-inch round ball gets bullied by the wind. Again, let's use a starting velocity of 1800 fps for the sake of comparison with the above 36-caliber ball. A 30-mile-per-hour crosswind will shove the 54 ball over 2 feet to the side at 100 yards, and a 10-mile-per-hour crosswind will force the 230-grain lead bullet over about 10 inches. The moral of the story is simple: If you shoot a round ball in the wind, either know how to judge drift or get ready to miss the target.

Round Balls in Brush

This may seem like "pick on the round ball" day, but facts are facts. The ball is no great shakes in the foliage, not to say that any projectile is wonderful in the weeds. I've tested for brushbucking and found that no bullet, not even a 500-grain missile from a 458 Winchester rifle, made its way through a lot of brush in a straight line. Although, the ball is even worse. I once missed a deer at 50 yards. The ball deflected through a modest screen of brush that stood between my muzzle and his neck. He moved away slowly, having never seen me, and by the time he walked into a little clear area, now about 100 yards out, I was ready to shoot again. This time the buck was harvested for the freezer. But there was no brush barrier in the way to deflect the shot.

How Far is Too Far?

Long range for the muzzleloader isn't very long by modern standards. I am aware that Baker and others dropped game at 300 yards

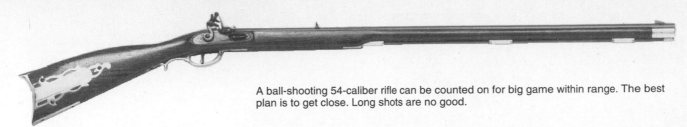

A ball-shooting 54-caliber rifle can be counted on for big game within range. The best plan is to get close. Long shots are no good.

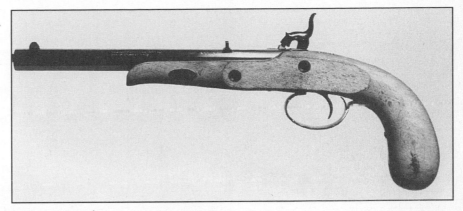

The patched round ball is also at home in the blackpowder pistol, although modern muzzleloader pistols, like this Traditions Pioneer, do gain more punch with conicals.

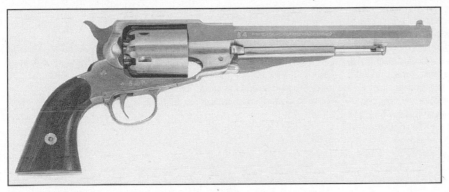

The cap 'n' ball revolver shoots a round ball with excellent accuracy. In fact, most caplock revolver shooting these days is done with the round ball rather than the conical.

with round-ball rifles. However, as pointed out, these rifles fired round balls that were measured by the pound. Even our 58-caliber ball weighs less than 300 grains. Back in the days of the mountain man, a long shot may have been in the domain of 200 yards. I'm not sure. There's some confusion on the issue. Joe Rose, who was applauded for dropping a buck antelope at rendezvous at the paced-off distance of 125 long steps, was considered a fine marksman for his effort. Now, was the buck truly 125 steps out there, or was he farther? A reader, Mr. Art Belding, explained the term "pace" in its more archaic meaning. A pace used to stand for 2 yards, not 1. A step was 1 yard; two steps equaled one pace. So in reverse, one pace equaled two steps or about 2 yards. The "about" is important. Belding studied the issue further and arrived at a pace equaling 5 feet. So 125 paces times five divided by three is 208 yards for Joe's antelope. I believe he, and the better marksmen of the day, could shoot 200 yards and hit a mark. There are many downwind shooters who can do that today.

Although I get my practice, I'm going to call my personal limit 125 yards with the muzzleloader, and that's round ball or conical, because either travels about the same trajectory when the Minie or Maxi is started at 1500 fps MV and the ball is started at 2000 fps MV. You can see why. The conical hangs onto its initial velocity better than the ball, but the ball starts out faster than the conical. Sighted dead on for 75 yards or so, depending on the caliber of the projectile, either missile, ball or conical, will drop about a half-foot out to 125 yards. That's as much Arkansas elevation as I care to deal with. Even sighted for 100 yards, a guessing game ensues past 125 yards. Therefore, my limit is 125 yards. Have I exceeded it? Yes, and I'm not so proud of the fact. Greed made me shoot at an antelope at what turned out to be about 200 yards. I held high, knowing the prairie goat was beyond my usual 125-yard limit, and

fired. The buck fell. I haven't done that since and don't intend to. Blackpowder hunting is a game of sighting, stalking, getting close, and putting one ball or conical right where it belongs.

Velocity is important to the issue of round ball shooting because the slow ball takes on a looping trajectory that makes hitting at reasonable and prudent blackpowder hunting ranges improbable if not impossible. My favorite 54-caliber ball-shooter puts the 230-grain .535-inch pill out of the muzzle at 2000 fps. If I sight the old gal for a flat hundred yards, I'm a few inches low at 125 yards, which remains my outer limit to the best of my field-judging ability. If the same ball starts at 1500 fps MV, and is again sighted for 100 yards, it must rise several inches at mid-range in order to achieve a reasonable below-line-of-sight drop at 125 yards. All of you who have shot modern firearms know that if your 30-06 150-grain bullet starts at 3000 fps it shoots flatter than if it starts at 2000 fps. The laws of shooting don't change just because the rifle is loaded from the front instead of the back.

Forgive an old story, but it applies to the round ball. Scientifically, the bumble bee cannot fly. It has too much mass for its minimal wing area. Liftoff is impossible. But the little insect doesn't know it cannot fly, so it goes about its business merrily buzzing from flower to flower. That's like the round ball. Its shape is ridiculous for fighting the atmosphere, and it shouldn't be able to penetrate a box of oatmeal, let alone the breadth of a buck. But the little ball merrily goes about its business just like the bumble bee, defying the laws of nature and successfully doing its job. The shooter must do his part, of course. First, he must select the proper round ball size for the job—small ball for small game, large ball for large game. Then the lead pill must be loaded to its safe potential for good energy and trajectory. Finally, the round ball must be well-placed. But that goes for any bullet. ●

chapter 17

The Blackpowder Conical

BALLISTICALLY, THE MODERN jacketed streamlined conical projectile retains its velocity and energy better than any other bullet of the hour. On the other end of the spectrum is the popular round ball, which loses roughly half of its muzzle velocity over a range of only 100 yards. In between the shapely modern bullet and ancient lead globe is the blackpowder conical, its profile more blunt than the space-age elongated bullet, but with much higher sectional density and ballistic coefficient than the round ball. In the past decade, literally dozens of different blackpowder conicals have emerged from the drawing boards of designers everywhere. "Build a better mousetrap, and the world will come to your door," the old saying goes. That's what today's blackpowder conical bulletmakers believe. They want to make better lead bullets for the old sparkthrower.

But be warned, it's only fair to point out that this chapter is not a diatribe against the old round ball. After all, the lead pill has proved it can do it all, provided it has a large diameter, is used within a reasonable range and is properly placed on target. That first criterion is very important, as pointed out in Chapter 16. Ball diameter is vital to mass, which in turn is paramount to round ball performance on big game. On the other hand, the conical does not depend entirely on caliber for its authority, because it does not require a tremendously large caliber in order to gain mass. The conical gains mass through elongation.

For example, a 45-caliber .445-inch round ball only weighs 133 grains, while a 45-caliber conical may go well over 500, even 600+ grains weight. Of course, this factor has not stopped gunmakers from providing huge-bore conical-shooters. There are many 58-caliber rifled muskets sold today to prove the point. These big bore conical-shooters do not, as a rule, fire super-heavy bullets, as proved by the fact that many 58-caliber lead bullets weigh no more than conicals for 45-caliber muzzleloaders. A conical I've used often in my own 58-caliber rifled musket is Lyman's 500, which in fact weighs 525 grains in "pure" lead.

What are Blackpowder Conicals?

The conicals of our interest are the Minie, Maxi, "standard lead bullet," pistol bullet/sabot, and what I'll call the "modern blackpowder conical," meaning the new breed of projectile that has made such a big splash in the pond of blackpowder bullets. There are many different types of conicals, and these five are presented as major examples only. They can differ dramatically in style, but each shares a commonality—all are elongated or conoidal in shape. Many are of the round nose configuration, although this is certainly not a criterion shared by all blackpowder conicals.

The Minie

This type surfaced in wartime because it met the need for a conical projectile that could be rather easily rammed home during the heat of battle, offering fast repeat shots, while retaining adequate accuracy. A hollow-base bullet seemed to be the answer. Many inventors stepped forth to claim the design as their own. Delvigne, a Frenchman, is often

These Gonic 50-caliber bullets are typical of the standard conical, fairly long for their caliber, round-nosed and heavy in weight. These go 565 grains each, with high penetration ability.

This is a good view of the Minie's hollow base. The depth of the base depends upon the specific projectile. Variation is common.

(Below) The Minie ball's most notable feature is its hollow base, which creates what is called a skirt. The lead skirt, as shown here, can be flared from the blast of the powder charge.

These Black Belt bullets from Big Bore Express Co. are typical of conicals with respect to weight for caliber. The round ball can only gain mass by getting larger, while a conical can also gain mass by getting longer. These bullets use a plastic base cup, by the way, as a bore seal.

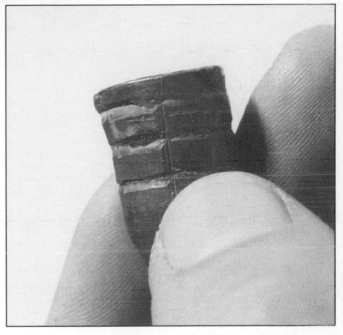

credited with the original hollow-base bullet. However, he's more accurately associated, I believe, with the 1828 invention of a round ball which did not require a patch (though it could still use one). It was undersized for easy ramming, but fattened out to meet the walls of the bore by power-stroking the ramrod downbore, which forced the ball to spread out via a wedge-like piece embedded in the missile.

Delvigne may have had a good idea, but it was Captain C.E. Minie of the French army whose name became attached to a conical with a hollow base. His hollow-based bullet came along in the 1840s, as far as my research goes. But wait a minute. You cannot leave W.W. Greener out of the picture, because he also claimed invention of the hollow-based projectile, and an English court agreed with the famous gunmaker, granting him a government settlement on the matter. In short, Greener satisfied the English courts of his day that his hollow-based bullet was indeed first, and that the government owed him not only credit, but also cash, which had already been granted Minie. The English government paid off, which leads us to believe that Greener's case was pretty darn convincing.

For our purposes, it doesn't matter who invented the Minie. What

does matter is that it works. Used in the American Civil War, the Minie, usually in 58-caliber and of about 500 grains weight, was a terrible killer, responsible for felling many of the more than 600,000 men who were casualties on the battlefield. The Minie is generally a caliber undersized—and I mean generally. The exact size is not important. The concept is. The undersized bullet could be loaded in a fouled bore, at least a moderately fouled one. Although the bullet went down undersized, it came out bore-sized. Upon firing, the hollow base created a lead skirt, and it was that skirt that flared out via gas pressure to engage the rifling—our old friend obturation at work. The Minie concept is alive and well today. It is incorporated into many different blackpowder conicals.

The Hornady Great Plains Maxi-Bullet is a cross between the Minie and Maxi design, so it fits in here. Since it has a hollow base, we'll classify it as a Minie-type bullet. It carries multiple grease grooves, a maximum bearing surface, a hollow base, and a hollow point. A tapered base eases muzzle entry, and it comes pre-lubricated. Hornady's Great Plains bullet comes in calibers 45, 50 and 54, weighing respectively 285, 385 and 425 grains. This design is indicative of a

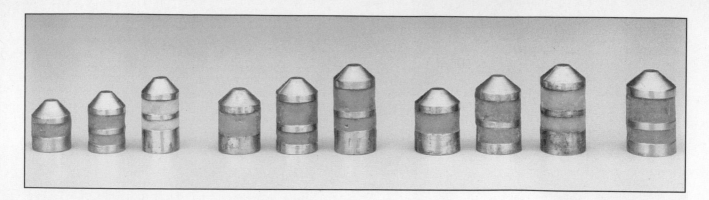

(Above) Thompson/Center's Maxi-Ball remains a fine bullet. These are pre-lubed for multiple shooting, and they come in many weights, up to a 58-caliber 560-grain missile. Maxi-Balls are known for their deep penetration.

The Maxi-Ball has always been known for penetration, but in fact it can also expand, provided the target offers considerable resistance. The Maxi on the left was fired into a bullet box test device, where it penetrated very well, but also expanded.

trend toward well-balanced, carefully thought-out elongated black-powder bullets. This cross-referenced bullet also fits into a third niche—the modern blackpowder conical, which is discussed later.

The Buffalo Bullet also falls into three classes. Some are flat-based, making them much like Maxi projectiles. At the same time, other Buffalo bullets have hollow bases, which clearly earn them a place in the Minie ball category. And to thicken the broth, the designs are mainly modern, so the Buffalo Bullet also earns the badge of modern blackpowder conical. It's entered here, however, because of the Minie-like hollow base on some examples, which forms a skirt and flares out to engage the rifling when fired. This bullet comes prelubed and has an interesting knurled effect on its shank, along with a driving band. In 45-caliber, a hollow-based/hollowpoint 285-grain bullet is offered, or a flat-base/hollowpoint 325-grain projectile. In caliber 50, there's a hollow-base/hollowpoint 385-grain bullet and a solid-base/hollowpoint 410-grain model. The 54 comes in three weights: 425-grain hollow-base/hollowpoint, 435-grain hollow-base/round-nose and 460-grain flat-base/round-nose bullet—the latter is a real hellbender in a strong long rifle built to handle heavy powder charges.

The Maxi

The conoidal shape was appreciated quite early in the shooting game—it didn't take gunners long to realize that the elongated missile retained its velocity/energy better than a round bullet. The Minie design was good, but it didn't satisfy everyone. A solid-base bullet of bore size, or almost bore size, would also shoot with accuracy—good accuracy—and it could be readily introduced to the barrel if the shooter maintained the bore in a reasonably foul-free condition. Today, with so many modern lubes that attack blackpowder fouling, the close-fitting projectile works better than ever. Here's the key: The Maxi-Ball is usually engraved as it is driven downbore, so even when slightly under bore size, the Maxi functions because it obturates to the bore.

The Maxi and standard solid-base bullets are closely related in function, but do not enjoy the same configuration. While Hornady and Buffalo Bullets both have flat-based bullets in their catalogs, the T/C Maxi-Ball is more like a true Maxi. It has a spire point, a good grease groove, and a flat base. Calibers 32 through 54 are available in many different weights, including a 103-grain projectile in 32-caliber. It comes unlubricated or prelubed.

There are a number of flat-based blackpowder bullets offered today, and we'll call these Maxi Offshoots. The T/C Maxi-Hunter is but one example. It is an advanced-design Maxi projectile with multiple grease grooves, a flat base with an engraving band and a slightly hollowed point. It's offered in three calibers as this is written—45, 50 and 54—and in a number of bullet weights to suit a big game hunter's personal preferences and needs.

The Standard Lead Bullet

This is also a flat-based projectile (although a few may have slightly concave bases). This is a very familiar bullet style, for there is no significant difference between these bullets and cast lead projectiles used in cartridges, such as the 30-30, 30-06 or, for that matter, any other metallic. This is the bullet fired in the blackpowder cartridge today. Shooters can buy them ready-made or cast these projectiles for themselves. They are at home in muzzleloaders as well as blackpowder breechloaders. Although most of these bullets are of the round-nose configuration, some are flat-nosed, and others are pointed.

There are quite a few standard lead bullets around these days, and many companies offer moulds to cast these projectiles. A blackpowder riflemaker who has long preferred the standard lead missile is Dr. Gary White. Dr. White's company, makers of the excellent White Muzzle-loading Systems modern muzzle-loading rifle, offers a complete lineup of conicals. Dr. White was a fan of the "long lead bullet" years before designing his own muzzleloaders. He used such bullets in Alaska and all over. Today, the White Superslug is well known. This bullet slides downbore with ease, and it is intended for quick-twist barrels only. It will not stabilize in slow-twist barrels. Accuracy is excellent. An example of a White Superslug that has proved itself in the game field is the 50-caliber 430-grainer. From the White rifle, a muzzle velocity of about 1600 feet per second is no trick with this bullet. At 100 yards, velocity is still close to 1300 fps. Chapter 33 on rifled muskets includes a few more words on standard lead conicals.

Many fine standard lead bullets for blackpowder arms are offered to the homecaster. Lyman alone has dozens of different calibers and configurations to choose from. Other companies add to the list. For example, the Champion mould from Rapine Associates comes in many Minie and Maxi styles, as well as calibers. There is an interesting 50-caliber Minie-Hawkins missile that weighs 325 grains, a short 250-grain 50-caliber Minie-Hawkins (could be good in the somewhat slower twist rifle), plus several bullets for Sharps and Spencer fans, an International conical in 58-caliber, and so forth. NEI also has a great many excellent moulds for conicals. This fine company specializes in custom moulds, too.

Gas checks are metal bases that attach to cast lead bullets. I have fired gas check bullets in muzzleloaders with success. On the other hand, the best reason to use gas checks is to safeguard the base of the bullet from gas damage, so if there is no gas damage to begin with, then the need for a gas check is not met. More testing is required; however, at this time I can safely say that most blackpowder lead conical projectiles do not require gas checks.

The Pistol Bullet/Sabot

Since this combination is described elsewhere in the book, suffice it to say this is a lead or jacketed pistol bullet that fits into a bore-sized plastic sabot (sah-bow), allowing the undersized missile to fly accurately from both muzzle-loading rifles and pistols. This theme has undergone many variations in the past few years, such as the Thompson/Center Break-O-Way Sabot, which is a modern jacketed bullet

held in a two-piece sabot with a "doughnut" for a base. This sabot does just what the name implies—it breaks away, not only the two plastic halves, but also the doughnut. All three parts fall away due to lack of momentum, while the bullet, spinning, travels toward the target. Incidentally, the term "pistol bullet" is widely used; however, the exact bullets employed in muzzleloaders are more often fired in the modern revolver, such as the 44 Magnum. The advent of the modern muzzleloader has much to do with the development of the modern blackpowder sabot. The modern muzzleloader also precipitated many of the modern bullets used by shooters.

Modern Blackpowder Conicals

This is a newly designed bullet, born in recent times and meant to be fired in the muzzleloader. Not too long ago, it would have been appropriate to establish this bullet as all-lead in construction, but that is no longer the case. The Barnes Expander-MZ Muzzleloader Bullet is made of solid copper. This is a deeply hollow pointed pistol-type bullet that comes with its own plastic sabot. Barnes calls it "The first non-lead projectile designed specifically for use in muzzleloading rifles." The goal was "mushrooming," or bullet upset, from an impact speed of 1900 fps down to as low as only 1000 fps. In my bullet box tests (see Chapter 42 on big game hunting with a muzzleloader), the Barnes copper Expander-MZ retained just about all of its original weight. Impact velocity for the 300-grain test bullet was 1500 fps.

Conical Ballistics

Ballistic Coefficient

The concept of "C" is so important that it is addressed more than once in this book. As C goes, so goes downrange retention of velocity and energy. Since sectional density is in reality a preliminary to computing the more important ballistic coefficient value it will be dispensed with here. Maxi and Minie bullets outstrip round balls badly in ballistic coefficient; however, compared with modern jacketed missiles

A good handgun bullet like these Hornady XTPs works well in a muzzleloader because the bullet is actually engineered to open up at handgun velocities, whereas the muzzle-loading rifle drives them faster than handguns for even more bullet upset and energy.

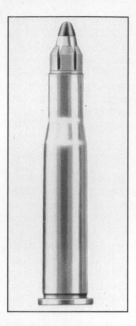

(Right) Remington Core-Lokt bullets and sabots are now a regular item with the famous old company. The large hollow-point promotes quick expansion and delivery of energy.

(Left) The sabot concept is very old. In recent times, sabots have been made of plastic and used with modern cartridges as well as muzzleloaders. Remington's 30-30 Accelerator cartridge is proof of this with a 22-caliber bullet contained in a 30-caliber plastic jacket.

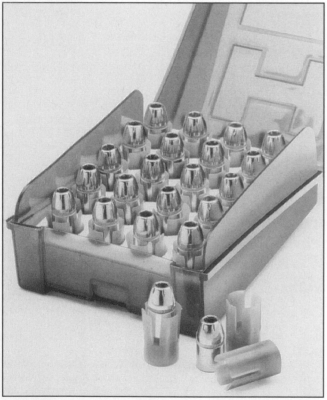

The pistol bullet/sabot concept is so well accepted today that companies like Hornady offer their fine bullets ready to go into the blackpowder field.

for cartridge guns, C figures are still quite unimpressive. That's because blackpowder conicals are, compared with modern spitzer and spire-point bullets, rather squat in profile.

One 45-caliber Minie ball of 265 grains showed a C of .156 (Lyman data). Compared with a round ball for the 45-caliber rifle, with its C of .063, that's good. But it does not match well against even a rather blunt-nosed 45-caliber 500-grain bullet, with a C in the domain of .300. This is important to recognize, for it reveals a ballistic truth about longer-range shooting with blackpowder conicals. The Maxi and Minie are supremely fine missiles, as are the bullets in the other three groups presented here. All are entirely adequate for North American

big game out of the right rifle. But these bullets were never intended to compete with high ballistic coefficient missiles. They just don't have the initial velocity to give them a flat trajectory, nor do they "carry up" as well, which refers to retention of initial bullet speed and energy.

Effective Range

The effective range of the blackpowder conical equates with the round ball, not because of terminal energy, but because of trajectory. Of course, if a skilled shooter can hit way out there with a Minie or Maxi, and especially with one of the long-for-its-caliber standard lead bullets, game of the soft-skinned class can be cleanly harvested. An historical note about the standard lead bullet: When this sort of missile was used by the buffalo runners of the late 19th century, it was effective to long range because it had good sectional density, possessed a pretty good ballistic coefficient, and was fired from an accurate, often scoped, rifle by an expert with a wagonload of experience in range judgement. Shots at 300+ yards were common. Of course, the target was not exactly tiny either, being the chest area of a bison.

With a Sharps or Remington breechloader—scoped—and lots of practice, a couple-hundred-yard shot was no problem. However, with the usual Minie or Maxi humming out at 1500 fps muzzle velocity or thereabouts, a 125-yard limit is not a bad one to observe. (See Chapter 30 on sight-in and trajectory for more information on effective range with blackpowder arms.) Lighter-weight pistol bullets flatten the trajectory a little bit, because they have a higher initial velocity, but they do not have quite the downrange impact of a big lead conical. Even with these flatter-shooting missiles, the 125-yard rule on big game is a good one for blackpowder hunters.

Wind Drift

No bullet escapes the ravages of the wind. The round ball drifts badly in the breeze, but so does the blackpowder conical. One reason for this is time of flight—the actual time it takes for a projectile to go from the muzzle to the target. Obviously, a lower muzzle velocity means a longer time of flight, and the longer a bullet is en route to the target, the more time wind has to work on it. So no bullet, modern or blackpowder, gets away from wind drift.

Surprisingly, depending on the caliber, Maxis and Minies fare only a little better than the lead globe when the wind blows. A 50-caliber (.495-inch) round ball is pushed more than 18 inches off course at 100 yards by a 20-mile-per-hour crosswind, even when it starts at a full 2000 fps MV.

A 50-caliber 370-grain Maxi ball beginning at 1500 fps MV drifts about 16 inches. Since you can't whip the wind, you learn to "dope it

Thompson/Center's Break-O-Way Sabot is designed to do just that: break away from the bullet soon after muzzle exit. The sabot is in two parts—split, as it were, down the center length-wise—with a lubed base wad. When fired, the sabot transfers rifling impetus to the bullet, but only the bullet travels far beyond the muzzle, the other three pieces falling away.

(Below) Just as you can buy pistol bullets as a single component, you can also buy sabots that way. These Hornadys happen to be 54-caliber.

The Thompson/Center Sabot with 275-grain XTP bullet works on the principle of quick expansion, depositing energy into the target.

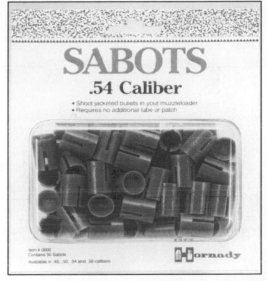

out." Guessing wind velocity is difficult at best, but with practice a shooter can become fairly proficient at this. Along with the wind, range must be judged, too. After all, a bullet drifts much more off course as range increases. Judging wind drift is definitely a challenge, but that's what you got into blackpowder shooting for, isn't it? Furthermore, the target shooter can learn how to dope wind for his purposes, while the blackpowder hunter gets closer to his game to thwart wind drift.

Velocity and Pressure

Forget high bullet speeds with muzzleloaders. The average conical leaves the muzzle at about 1500 fps with a strong hunting load, while lighter-weight pistol bullets go faster. A Knight modern muzzleloader chronographed with a 180-grain Sierra jacketed pistol bullet and sabot reached almost 2000 fps MV with 120 volume RS. But even with heavy powder charges, it's not in the cards for blackpowder guns to push bullets at terrific velocity. A special conical-shooter discussed later achieved an MV close to 1800 fps with a 460-grain bullet, but the rifle was especially built for it with a stout barrel. Looking at a very popular conical-shooting muzzleloader burning a charge of 120 grains volume FFg blackpowder, a 400-grain conical bullet chronographed at 1500 feet per second. That was, as they say, about all she wrote. And not every conical-shooter is granted that large a powder charge with heavy conicals by its manufacturer, either.

FFg and Pyrodex RS work well when conicals are loaded for hunting power. Finer granulations are allowed in some conical-shooters, and they will generally pick up velocity for the same charge of powder. Finer granulations are also acceptable for target loads with conicals. On the other hand, Fg has not produced the best ballistics in calibers 45 through 58 in tests with conicals in muzzleloaders. But Fg does work well in the blackpowder cartridge shooting conicals.

Follow your manufacturer's loading guide. Indications suggest that conicals can raise pressures compared with round balls in the same guns with the same powder charges, but the pressure-test gun does not indicate a problem. Therefore, since the question of conicals and pressure is not answered entirely by the best indicator we have, the pressure barrel, go with tested data. Use the charge recommended by the maker of the firearm for your conical-shooting rifle or pistol.

Rate of Twist

Twist has its own chapter, but do remember that your rifle may or may not be suited for the conical. Conicals, being elongated, require more rps (revolutions per second) to keep them stabilized than round balls do. That is why a 50-caliber round ball rifle can get by with a slow twist, such as 1:66, while a 50-caliber rifle intended for conicals may have a rate of twist twice that fast. One way to be certain is to shoot your rifle at a paper target out to 100 yards. If the holes in the paper are round, the bullets were stabilized when they impacted. If the holes are oblong, then the bullets tipped in flight, or keyholed, striking

123

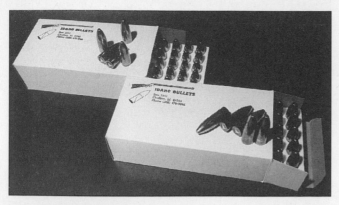

Bullet profile or shape has a lot to do with ballistic coefficient. These Idaho 45-caliber lead projectiles carry a pretty sharp profile and, therefore, pretty good ballistic coefficient.

Another highly innovative muzzleloader conical is from Parker Productions. The bullet has a hollow nose filled with liquid, along with a solid expander ball up front. Upon striking the target, the ball is forced back into the fluid, then hyrdraulic action expands the bullet while weight retention remains very high for long wound channels.

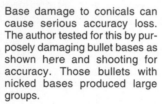

Base damage to conicals can cause serious accuracy loss. The author tested for this by purposely damaging bullet bases as shown here and shooting for accuracy. Those bullets with nicked bases produced large groups.

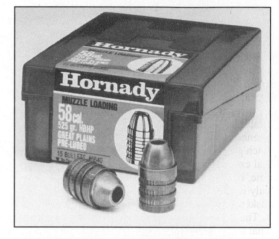

There are dozens of different blackpowder conical designs. These Remington conicals are typical of the trend. They're made to expand readily.

Bullets that are "long for their caliber," so to speak, have high sectional density, while shorter bullets have lower sectional density. These fine Hornady Great Plains conicals have modest sectional density.

the target sideways. If the bullet has keyholed, try a shorter conical. No good? Go back to the round ball.

Special Conical-Shooting Rifles

A brief look at the previously mentioned special custom rifle built to shoot conicals is worth a few lines. The piece turned out to be 11 pounds of dynamite. Although the term "magnum" has been abused to the point of boredom, this rifle was truly a magnum muzzleloader. The barrel was made of quality steel, and every aspect of the rifle was built for strength. Rate of twist was 1:34. The big 460-grain Buffalo Bullet in front of a heavy FFg or Pyrodex RS powder charge gained just shy of 1800 fps at 12 feet to midscreens, which would be a full 1800 fps MV. Its kinetic energy rating: 3310 foot pounds, greater than a standard 30-06 load at the muzzle.

The blackpowder slug gun is mentioned briefly in Chapter 23 on accuracy. It's noted here because it does shoot a special conical bullet with fine accuracy. This super-heavy benchrest rifle is carefully loaded from the muzzle with strict uniformity. It was known as a 40-rod gun because it was often fired at 220 yards, or 40 rods. Ned Roberts noted in *The Muzzle-Loading Cap Lock Rifle*, page 92:

"In times past many remarkably small groups have been made at 40 rods rest shooting with the most accurate muzzle-loading target rifle, equipped with false muzzle and telescopic sight, using various calibers of cylindro-conoidal bullets. Even at 60 to 100 rods [one rod equals 5.5 yards or 5.03 meters], these rifles and bullets when handled by expert rest shots would make groups that very few of our modern high power rifles today can equal."

Sometimes bullets for slug guns were cast in two parts and fitted together with a soft base section and harder nose section. The slug gun proves that cast blackpowder conical bullets can be supremely accurate.

What to Choose

The shooter who wants to fire elongated missiles from his blackpowder rifle has only to choose a firearm with the correct rate of twist. That's all there is to it. The advantages of the conical are clear: more weight per caliber than the round ball with higher retention of velocity and energy downrange. These factors do not give the venerable round ball a black eye. After all, good hunters with round ball shooting guns took game up to and including pachyderms in the past. Good hunters are still harvesting big game with round balls. But for those who want to go conical, the choices are certainly present, and growing all the time.

●

chapter 18

Patches for Blackpowder Guns

"**EVERYBODY KNOWS THAT** the patch was used to seal gas behind the round ball!" I hadn't meant to start a revolution. I just happened to mention in a talk that the patch was not a true gasket and World War III broke out. For many people, blackpowder shooting is far more than a game. It is a way of life. The past, and what was established a very long time ago, is sacred truth, regardless of what science says today. The gentleman who got so upset about my patch/gasket remark probably believes to this day that cloth patches seal expanding gases in the bore behind the round ball. At the same time, he may have no idea how extremely valuable the cloth patch truly is around that lead ball, in spite of the fact that it is not a true gasket.

That's what this chapter is all about—the values of the cloth patch, what it can do, and of course, what it cannot do. There's also a little bit on patching the conical, not with cloth, but with paper. It may seem unlikely at best that the cloth patch, as well as paper patch, could generate more than a fly speck of interest; however, that's just not true. The patch topic is a rather surprising area of study. The smart downwind shooter makes sure he knows quite a bit about that

chunk of cloth surrounding the round ball, and maybe a tad about paper-wrapped conical bullets, too.

Why Patching?

Nobody knows exactly who first came up with the idea of wrapping a piece of cloth or leather around a projectile to take up the windage in the bore, but it was one of man's better shooting ideas. Ramming a bore-size, and sometimes even larger-than-bore-size, lead projectile downbore to ensure a close relationship with the rifling always promoted accuracy, but oversize projectiles were about as practical as pouring a quart of milk into a pint bottle. The use of a go-between was genius-inspired, in spite of the simplicity of the plan. Undersize projectiles work—up to a point—because of obturation. They also function where only fair accuracy is demanded, which is what William Cotton Oswell knew.

That old-time ivory hunter operated in Africa during blackpowder days. He rode out on his horse in pursuit of elephants, and knowing that there was no time to stop and force a well-fitted ball downbore, he didn't even try it. He used a smoothbore for convenient loading ease anyway, and easy cleaning. So Cotton was known to grab a fistful of coarse blackpowder, slap the charge down the barrel, followed by a ball that rolled down like an undersized ball bearing. Accuracy? Forget it. But you don't need a lot of accuracy to hit an elephant at 20 paces, so Oswell's no-patch round ball shooting was a success for him, but not for others.

Shooters needed true accuracy so they could place a projectile right on the mark downrange, whether on paper or game. Over the years, the demand for frontloader precision continued. Rifling was a huge leap forward, even with the lead round ball. In theory, a perfect round ball flies true, without spin, but in reality the ball is not perfect. It took rifling to give it accuracy through rotation. Revolving the ball kept it on course, so to speak. Spinning also averaged discrepancies around a common axis. A lead ball is slightly lopsided, but when spun, this mild lopsidedness no longer steers the round bullet off course, and the lead sphere flies straight on its line of flight.

Spinning a ball was a good idea, but that ball also had to fit closely to the bore in order to gain its best groups. Even with a patch, a badly undersized ball has a hard time achieving accuracy, so there was a compromise. A ball was cast to fit fairly closely to the bore, but undersized enough to ram downbore. The patch acted as a go-between, filling the space (called windage) between the bare ball and the rifling in the bore. More on that in a moment. But first, for the sake of knowledge, what about the true inventor of the patch? Was he an American?

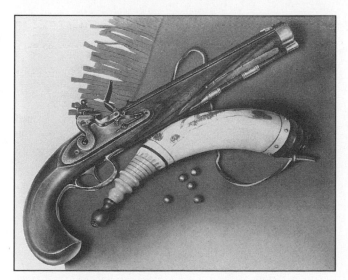

What is a patched ball shooter? A rifle or a pistol that has a slow rate of twist or a smooth bore.

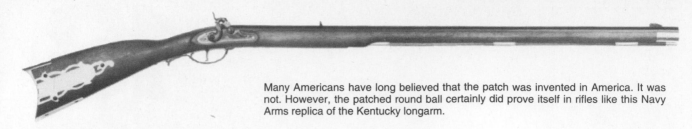

Many Americans have long believed that the patch was invented in America. It was not. However, the patched round ball certainly did prove itself in rifles like this Navy Arms replica of the Kentucky longarm.

Patches in History

Americans have always been accuracy-conscious, and American shooters have long been credited with coming up with the idea of patching the ball. Many shooters still believe that it was a Yankee marksman who wrapped an undersize missile in cloth or leather in order to facilitate loading, while maintaining a close projectile fit to the bore. "Here in America," said John G. Dillin in his fine book, *The Kentucky Rifle*, 1924, "balls were cast smaller than the bore and were enveloped in a 'patch' of leather or cloth to prevent contact with the barrel. This patch enabled Americans to load faster, and to fire longer without cleaning, and to outshoot all others of that time. It was the distinguishing difference between the American rifle and those of Europe." (p. vi)

Well, it is true that Americans took to patching in a big way. Dillin believed the advent of the patch to be a "master stroke," and "the last link forged in the chain of evolution which brought forth a distinctly American rifle," so stated on page 15 of the *Kentucky Rifle*. While Americans took to patching like hogs to rooting, they were not first with the idea. In the text, *Espingarda Perfeyta* (*The Perfect Gun*), a Portugese manuscript dated 1718, the author, J. Joav, wrote concerning early firearms, "Others made barrels with rifling inside, some with more, and others with less rifling, all of them deep and twisted in the form of a spiral. These were loaded by putting the bullet in a little piece of leather of a thin glove, folded only once, dipped in oil, and thus it was pushed down to the bottom in such a manner that the bullet may not lose its roundness..." (p. 341)

The Patch as a Gasket

American blackpowder shooters were forced to credit the invention of the patch to the Europeans. And why not. Shooters of the Old Country also developed jacketed bullets before smokeless powder was invented. So the patched projectile came from across the sea. So be it. But when modern students of muzzleloading were told that the patch was not a true gasket, that's when the oatmeal hit the floor. "Anybody knows," one reader complained, "that the patch is to completely seal the bore off so gases won't get by the ball, which is just what it does, and nobody can prove otherwise." A nerve had been touched. But there were a few facts to consider.

Ed Yard gave the lie to the gasket theory in his *Gun Digest* article, "The Round Patched Ball And Why They Used It," p. 236, 1980 edition. "The inherent inefficiency of the patch as a bore seal was the big factor... of the loss [in pressure]," wrote Mr. Yard. "So we find that no patch really seals," wrote the man who had studied patches with a pressure gun. "Based on the test information presented here and the tabulated data appended, the major function and the practical effect of the cloth patch on a round lead ball in the American rifle is to spin a loose fitting and easily loaded ball to attain real accuracy. It does not really seal the bore," concluded the author. (All quotes are from page 237 of the *Gun Digest* volume.)

I got curious about the patch/gasket story, so I used a chronograph to determine what would happen to muzzle velocity if the patch were left out of the load chain. My work was not conclusive, but it certainly seemed obvious enough that the patch was not acting as a perfect gas seal. Then came another bit of evidence that proved very powerful. After all, anyone could see the results for himself. High speed 16mm movie film was used to capture the muzzleloader going off. And guess what? The movies clearly showed that the first thing to exit the muzzle of the rifle was not the patched ball, followed by smoke. It was smoke first, then the patched ball. Gas was probably jetting down the grooves of the rifling, past the patch.

No big deal, really, but the long-held idea of a ball patch serving as a true gasket was shown to be faulty, if not entirely false. The cloth patch is not a gasket in the sense of a gasoline engine's head gasket. My *Standard College Dictionary*, page 551, says this of a gasket: "A ring, disk, or plate of packing to make a joint or closure watertight or gastight." The last word is the clincher. A cloth patch does not make the bore gastight, but the patch may retard the ball in the breech, promoting inertia, which in turn could help obturate (expand) the projectile to better fit into the lands and grooves of the rifling, which in turn helps retain expanding gases behind the projectile.

While not a gasket in the true sense of the term, the patch is an important part of the load chain. A properly patched round ball can be quite accurate, more than accurate enough for small game hunting, for example.

126

Dark, yes. That's normal and ok. But this patch is not blown, nor is it torn. There is no need to install a protective buffer between this patch and the powder charge with the specific firearm and load used.

This Dale Storey custom pistol delivered excellent accuracy with the round ball and a pillow ticking patch. Three shots cut one ragged hole at 25 yards.

Clearly, this retrieved fired patch shows a big problem. Made of high-grade pure Irish linen, the patch did not fail because of the material. What went wrong? It's up to the shooter to find out.

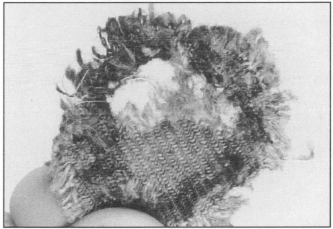

What The Cloth Ball Patch Does

Accuracy and the Ball Patch

Accuracy is a prime and important reason to use the patch. I tested unpatched balls that fit tightly to the bore and they proved fairly accurate. However, they were miserable to load; they did not outshoot the patched ball; and they could have proved dangerous if an unpatched ball migrated upbore, creating a gap between powder charge and projectile. A well-fitted bare lead ball will engrave. That is, it will make contact with the rifling in the bore, thereby gaining spin. But the patched ball does not require engraving for accuracy. In fact, it is hardly marked by its passage through the bore because the patch translates its rotational value to the round ball.

So the patch is a valuable go-between in terms of accuracy, with best accuracy obtained by using a strong, lubricated patch around a ball that fits closely to the bore. Loose balls, even when patched with thick cloth, did not shoot as accurately as balls that were closer to bore size and fit with a strong, but not necessarily thick, cloth patch. In one test session, for example, the best accuracy from a 40-caliber rifle came with a ball close to .400-inch diameter coupled with a .010-inch patch, while the worst accuracy was experienced with by a loose-fitting ball wrapped tightly in a .024-inch thick patch.

Safety and the Patch

Here we go again with the unseated projectile. However, as Ezekial Baker, the renowned ballistician of his time, said in the 1800s, "More accidents happen from a neglect in this precaution than can be imagined, if the ball be not rammed close on the powder, the intervening air will frequently cause the barrel to burst...." (*Remarks on Rifle Guns*, 1835.) The cloth patch deserves full credit in regard to ensuring that the ball remains down upon the powder charge at all times, where I and Ezekial Baker feel it belongs. Therefore, the patch becomes a

safety device when it holds the round bullet down upon the powder charge in the breech.

Pressure on the Powder Charge

As also noted before, blackpowder is supposed to burn better under pressure in the breech than when loosely packed. Certainly there is a vast difference in how blackpowder burns unconfined versus confined, and the data on compressed blackpowder charges seems sound enough. Assuming that it is, and that the slightly compressed charge is best, then the patch serves another important function, because a tight patch retains the ball firmly upon the powder charge, rather than allowing the ball to move away from the charge. I tried testing for this factor. My demonstration was neither elaborate nor foolproof, but here it is. I seated three tightly-patched round balls in three different rifles. The rifles were safely rested in a shooting stand, muzzle up. Then ramrods were dropped down the muzzles, their tips resting on the seated balls. The ramrods were carefully marked. Seventy-two hours later not one ramrod had moved by so much as a hundredth of an inch, indicating that the patched ball held its station on the compressed powder charge.

Uniformity of Pressure and Standard Deviation

When testing for the effects of varying ramrod pressure on standard deviation from the mean chronographed velocity (a test of variance which helps in detecting uniformity among loads), I was at the same time looking at the function of the patch, although I didn't know it. However, the loads with uniform ramrod pressure did best in regard to standard deviation, and it's obvious that the patch had to serve to retain the ball upon the powder charge in a uniform manner or test results would not have been so gratifying.

Translation of Rotational Value

This point is a first cousin to accuracy; however, it's worth noting that tests have shown the patched round ball fired from a rifled barrel does indeed rotate in flight. Obvious? Perhaps, but shooters believed all rifled slugs were merrily twirling through the air, going round and round on their axes, when no such thing was happening. So it is important to note that the supposed go-between value of the patch is a real one, and not a false impression. In short, the patch on the ball really does impart spin from the rifling, something like the sabot translates rotation to a bullet.

Prevention of Bore Leading

This is a gray area. On the one hand, it's as obvious as a toad in your soup that the patch prevents leading. After all, the ball doesn't even touch the bore. On the other hand, very little of a bare ball would touch the bore anyway, since the round ball does not really have a shank, so we must wonder if there would be much leading without a patch. Furthermore, lower velocities don't cause much leading, and the round ball seldom achieves over 2000 fps muzzle velocity.

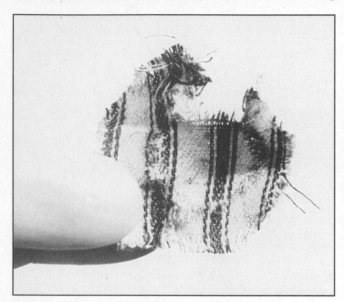

Patches can do some very odd things downbore. This patch has a chunk missing. The smart shooter will collect several more patches downrange using the same load chain, just to see if all patches show the same sign. In this case they did, and a problem with the rifling was investigated.

The patch on the right has only one burned-through spot; however, the patch on the left is a classic example of gas blowing through the grooves. There is a burned out hole where every groove in the bore allowed gas to escape.

Ned Roberts, shown here with one of his muzzleloaders, believed that the cloth patch wore a bore down faster than any other part of the load chain, including the powder charge. However, he also knew how accurate patched balls could be, and he used them.

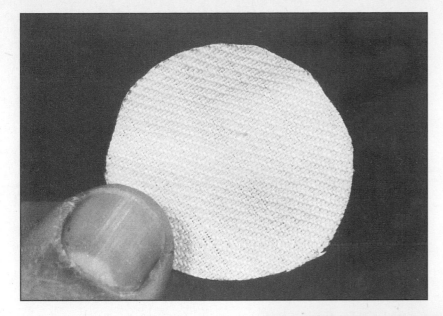

Patching material should be closely woven fabric. Under magnification, this patch shows that trait.

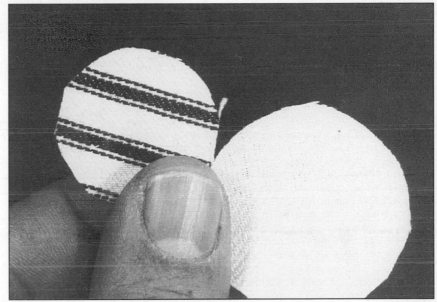

Two good patching materials show their close-weave excellence. Pillow ticking (left) and pure Irish linen (right) make excellent patch fabrics.

Nonetheless, it cannot be denied that the cloth patch does prevent any contact of lead ball with rifling.

The patch also serves to contain lubrication, which in turn helps to keep fouling soft, and to thereby allow more shots in a row before swabbing the bore. Round ball patches do a fine job of containing lube, another important job, and as pointed out earlier, the patch allows a smaller-than-bore-size projectile to be seated, because the patch itself takes up the windage in the bore.

Those are some of the values of the patch. Now let's look at a few more points pertaining to this fascinating bit of cloth.

Reading Patches

In spite of the fact that a cloth patch cannot be expected to seal hot gases in the bore, it is still a vital part of the load chain. The ideal patch is strong enough to withstand ramming home around the ball without breaking through, although any patch, no matter how tight the weave, can be cut by sharp rifling (see cut patches below). How thick should the patch be? Hard to say, but as stated already, best accuracy is achieved by a ball that fills the bore well, rather than an undersized pill. So a patch should be thick enough to make a good solid wall of cloth between the round ball and the rifling, not that it will ever be a true gasket, but it will be better than a thin patch that may fail as it flies up the

bore at high speed. But how can you tell what a patch is doing? You cannot see it in action in the bore, but you can see it afterwards.

What should a spent patch look like? Forget that magical cross so often pictured as ideal. I've never seen one in my life, at least not where the patch is white except for the dark brown or black cross in its center. More often, the fired patch will have a dark circle, not a cross, in the center. Reading patches is valuable. They will be found a short distance downrange and can reveal a lot about the rifle or pistol and its load chain. The patch should remain intact after shooting. If the patch looks like it was soaked in sulfuric acid, something is wrong. It may be *blown*, or it may be *cut*. So be sure to check your round-ball patches to see what they look like. If there's a problem, it may be correctable.

The Cut Patch

One problem encountered with the patch is cutting. This is different from the blown patch, which is discussed next. A cut patch is just that. You can readily see that it was sliced, even if the hole is small and round. What cuts a patch? Generally, it's the lands of the rifling. Do this test: using an unloaded rifle or pistol without any powder in it whatsoever, seat your normally patched round ball into the breech area—normally, but not all the way down into the breech, or it may get stuck there. Try to run it home about where it would usually rest with a

Patches can be any shape, including square, as long as they fully cover the ball in the bore; however, round patches are the rule, and this Patch Cutter from Forster Products cuts a neat round patch every time.

powder charge in the breech. Use your short starter and ramrod or loading stick. Don't fudge on this, even if you are worried that withdrawing the patch might be a problem. This test is worth a little trouble. Now that the patched ball is seated to about its usual depth, place a screw on the end of your loading rod and with muzzle protector in place so that the rifling lands are not damaged, go after the patched ball.

Rotate the screw well into the ball and slowly withdraw the patched ball from the bore. Study the ball first. It should be without blemish. If the ball has been cut or damaged, this could indicate a serious problem with the rifling. Now look at the patch. It should not be damaged. If it is, you've found the culprit, a sharp land edge most likely. Obviously, the patch was not harmed by powder because there was none in the bore. Nor was the patch injured by racing out the bore, because it didn't. A sharp rifling cut it. Two possible remedies: have the bore lapped by a professional gunsmith, or keep on shooting, thereby wearing the sharp rifling smooth.

The Blown Patch

This patch condition reveals itself distinctly. When spent patches have dark holes in them or badly tattered edges, they were most likely damaged by the hot gases of the exploding powder charge. Remedy: use hornet nesting material, one to three layers, on top of the powder charge. Run these downbore with the off-end of your ramrod. Seat them upon the powder charge, then run the patched ball home in the usual manner. Your blown patch troubles will be over.

Will blown or torn patches ruin accuracy? I've seen it happen, but not always. Sometimes a patch is torn or blown, yet accuracy is sustained. Sometimes a patch is torn and blown, and accuracy is ruined. This is proved by using hornet nesting material to safeguard the patch from damage. If the firearm still shoots inaccurately, then the damaged patch was obviously not the cause; but if bullets suddenly begin to group better after safeguarding the patch, chances are the injured patch was the problem. So do read those patches. They can tell a story.

Bore Wear and Patches

Ned Roberts believed that the most severe wear on a bore was delivered by the cloth patch itself. He had a point. Cloth patches can cause bore wear. How can that be? They actually lap the rifling. Imagine rubbing a tight-weave fabric, such as Irish linen or pillow ticking,

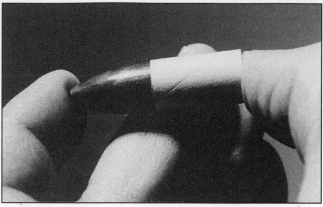

Cloth patches are not the only kind. There are also paper patches that can be installed on conicals like this one. The paper patch is highly admired by some shooters.

through the bore time and again on the end of a ramrod or loading rod. Now think about the same piece of cloth racing through the bore at perhaps 2000 feet per second. So patches may indeed cause bore wear. Do I worry about it? I should say not. Even if there is patch wear, it's not so rapid that a smokepole will go to pot without a lot of shooting. We should be so lucky to have enough leisure time to wear a muzzleloader out by firing it too often.

Patch Materials

What to make patches from? What not to use is even more important. Forget your shirttail, the baby's old diapers, your worn-out boxer shorts and rags from the garage. While denim is all right, there are so many different kinds of denim that I cannot recommend this cloth, although I once did. Today, having sorted and sifted through dozens of different materials, I'm down to only three that please me all the time. These are pillow or mattress ticking, which I classify as about the same fabric, commercial patches, and what is known as "pure" Irish linen. These are all close-weave materials. If you buy pre-cut (even pre-lubed) patches you can rest assured that they are right for shooting. If you buy either linen or ticking, be certain to wash these materials completely to get rid of the sizing.

Sizing is a kind of starch that impregnates the cloth and makes it look nice in the store. It's no good, however, in the bore of the firearm. Rinse the washed cloth and let it dry. That's all you have to do. Your patch material is ready for cutting and lubing. I like to pre-cut my patches and pre-lube them a few at a time. I don't cut patches on the muzzle and can find nothing to recommend that procedure except the romance of it. Cut the patch round, or cut it square if you like. It won't matter. Make it large enough to fully cover the ball, of course.

Warning: Never fire unpatched round balls in your muzzleloader. As stated earlier, the ball could roll away from the powder charge and your barrel might suffer a bulge when fired, or worse. The ball patch is not a gasket by the definition of the term. It is not a ring, disk, or plate of packing to make a joint or closure watertight or gastight. However, the cloth patch is an all-important part of the load chain. It does a lot of important work for so simple a device. And now you know why I promised that the topic of the cloth patch was important and interesting.

The Conical Patch

I became familiar with conical patches not through shooting blackpowder cartridge rifles with lead bullets, but through a friend who could not find exactly the right bullet for a special English rifle he owned. He patched jacketed bullets correctly and solved the problem. However, conical patching is more prominent with lead bullets fired from blackpowder cartridge rifles. The paper patch serves at least three

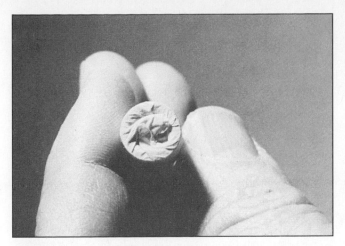

The paper-patched bullet base can look like this.

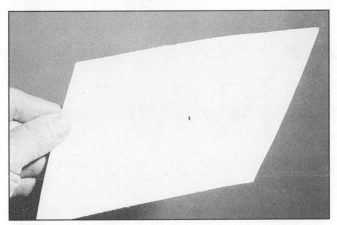

This greatly exaggerated oversized cutout is the proper shape for a paper patch.

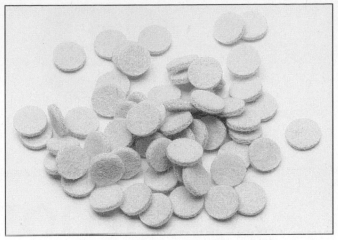

While the author likes hornet nesting material as a buffer downbore between the patch and the powder charge, it is not the only material that works. These CVA Slick Load Wonder Wads can also be used as buffers.

functions. It acts as a bearing surface for the rifling, very much like a cloth ball patch goes between round ball and rifling. Instead of the rifling biting into the shank of the conical lead bullet, it engraves the paper around it. The paper patch is also wrapped under the base of the bullet, which can save the important bullet base from damage. Damaged conical bases have been known to harm accuracy. And the paper patching around a lead bullet serves to prevent any possible leading of the bore. After all, the lead bullet itself never touches the rifling.

Unfortunately, perhaps, the benefits of paper-patching a conical blackpowder bullet are not vital to all projectiles. I have had far too much luck with non-patched projectiles to paper-patch all breechloader bullets as a matter of course. I say this is unfortunate, because it would be nice to recommend paper-patching as a surefire boon to breechloading accuracy and performance. There are many ways to paper-patch a conical bullet. Randolph S. Wright describes an excellent process in his information booklet, *The Paper Patched Bullet*, printed by Montana Armory. Nicely cast bullets can be paper-patched, or cast bullets swaged for uniformity of diameter.

Wright shows how to discover the correct length of a patch for a given projectile by wrapping the bullet with a strip of paper that rolls tightly around it three times, then cutting through all three layers with a sharp tool, such as an X-Acto knife blade. When you unroll the paper, it will have three slits in it. Measure between the two farthest-apart slits; subtract $1/32$-inch from this measurement; and that is the right length of the bullet patch, which can be adjusted later if necessary so that the patch laps around the projectile exactly twice, but no more than twice. The paper patch should be shaped like a rhomboid parallelogram. This particular shape allows the ends to match up at an angle to the bullet.

The ends of the rhomboid parallelogram are cut at about a 30- to 35-degree angle. The bullet is wrapped with two thicknesses of paper patch, since the patch is cut to go around the bullet *almost*, but not quite, two times. This means that the ends of the paper patch will not rest exactly one above the other. Naturally, the ends cannot touch, because one end is wrapped underneath, so the two ends themselves can never meet. The point is, the ends will not match up exactly. There will be a slight space existing between them when the paper patch is cut to the correct length. This also means that the two ends of the paper patch *never overlap*. That is important. Also important: the wrap of the paper patch must be in the opposite direction of the rifling twist. In this manner, the rotation of the projectile in the bore will not tend to unroll the paper patch, as it might if the patch were rolled in the same direction as the rifling twist.

While the process of wrapping a bullet with a paper patch is not involved, it does take practice to make a neat job of it. The parallelogram-shaped paper patch is started with one corner located about where the bullet shank begins to taper toward the nose of the projectile. Of course, having the right paper to begin with doesn't hurt, which is 100-percent rag or cotton content. This is high-grade stationery, and can be purchased at an office supply house, or from Montana Armory. A good thickness is about .0025-inch, which increases bullet diameter by one caliber. A paper thickness of .0025-inch will increase a 45-caliber bullet diameter by .01-inch (one caliber), because the paper goes around the bullet twice, and that increases "both sides" of the bullet shank by .05-inch. As Wright points out, this turns a .448-inch bullet into a .458-inch diameter after paper-patching.

Well-practiced paper patch wrappers use the dry method of wrapping. This is best because the paper falls away after it does its work, whereas wet-patched paper may stick to the bullet downrange, which may harm accuracy, according to the experts. Excess paper extending downward at the base of flat-based bullet can be "tucked inward" to cover most of the projectile's base. Excess paper at the base of a hollow-base bullet can be twisted and pushed into the base cavity. This is a quick look at the fascinating world of paper-patching lead bullets for use in blackpowder cartridge rifles. Those interested in more information should read carefully into Wright's manuscript, which is only nineteen pages long.

We've had a pretty good look at the cloth patch surrounding the lead round ball, and a glance at the paper-patched conical. It's easy to see why blackpowder shooting fascinates so many modern marksmen, not only in America and Canada, but the world over. The materials and the processes are simple, all right, but the work accomplished by cloth and paper patches can be far-reaching and important in many ways, including accuracy and safety enhancement. ●

chapter 19

Blackpowder Chemicals Then and Now

NOT TOO LONG ago, blackpowder magazine articles were filled with interesting "facts" on blackpowder chemicals, from metal-savers to lubes. One fellow wrote that whale oil penetrated so terrifically that if you filled a rifle bore with the stuff at night, it would seep right through the sides of the barrel by morning. Amazing stuff, that whale oil. Another wrote that if your round ball rifle wouldn't shoot, chances were 90 to 1 that it was the lube on the patch causing the problem. It was "not properly balanced." Having run benchtests with every patch lube from saliva to "moose milk," which the oldtimers called water-soluble machinist's oil mixed with H_2O and sometimes a bit of liquid soap, I'll guarantee that while patch lube is important, it is rarely a major criterion for match-winning accuracy in a round-ball rifle.

There has long been a romantic attachment to all things applied to the outside and the inside of muzzleloaders. I'm not sure where it stems from, but if I've heard from one reader, I've heard from a hundred with miracle liquids and pastes that made guns "shoot harder," or with "the best accuracy in the world," or the stuff makes bores last forever. Mainly, these miracle gun care and shooting products come from underneath the kitchen sink or out of the garage. Do any of them work? Of course they do. After all, you can prepare a patch with saliva alone, and it will slide down the bore fairly well, since just about any liquid that is water-like cuts blackpowder fouling to some degree. But don't expect too many repeat shots with "spit patches." I bought into the kitchen sink and garage idea because I like simple solutions to difficult problems, but I don't use these things today.

Modern Blackpowder Chemicals

The fact is, chemical engineers know a lot more about gun care and shooting products than the rest of us, and they have proved it most admirably in relatively recent times. In my years of testing blackpowder guns with many different products from home remedies to the lastest space-age goop, it is only in the past few years that I have run across commercial chemicals that truly do work a lot better than whale oil, cactus juice, almond extract, and so forth. These are, every one of them, products of modern times and trained minds. They work because they are chemically correct, not because they are historically romantic.

Metal Preservers

I don't want to leave the impression that I am against all of yesteryear's gun-care products. Plain old sewing-machine oil still does a good job of keeping oxygen away from metal parts, for example, and give me some rubbing alcohol on a cleaning patch, and I'll shine the bore up bright with it. But all in all, look to the stuff that sells over the counter for best results, and that goes for preserving Old Betsy as well as making her shoot better. Metal preservers are more popular than ever, and they work provably better than ever before, too. There are scads of them available. I got started with Accragard a number of years ago and still use it. It's sold by Jonad Corp.

Venco is another modern company that employs chemists to make gun cleaning and preserving products, rather than counting on what's under the kitchen sink. Their Synthetic All Weather High-Tech Grease comes in a great little tube that puts some right where you want it. Operating temperatures for this metal-saver range from 61 degrees below zero to 350 degrees above zero. The same company has a Rust Prevent metal preservative/lubricant combo. Birchwood Casey offers Sheath Polarized Rust Preventative, another product that works, by test. The rest of this chapter could be filled with similar products, but the point is made—use a modern metal preserver on that frontloader, not only when you put it away for a while, but also to protect the metal when the gun goes into the outdoors.

Lubricants

Lubricants generally serve to reduce friction. Graphite on ball bearings is an example. Graphite reduces the "grab" between each rolling unit. Oil in a car engine prevents galling of working parts. However, blackpowder lubes are a little different. They serve to reduce friction too, but they also perform other important functions, namely breaking down of residue, and metal protection. Putting a tightly fit patched ball down on the powder requires considerable force without lube. So in this application, lubricant serves to reduce friction, creating a slick medium between the cloth patch and the bore of the firearm. But modern blackpowder lube does more. It acts as a cleaning agent.

That service is not entirely unknown among other lubricants, of

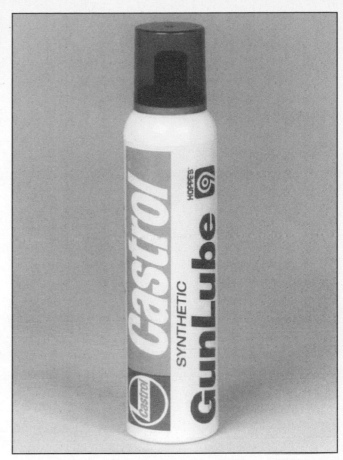

Shooter's Choice Rust Prevent is formulated to thwart rust. It can be sprayed directly on metal surfaces or on a rag, then wiped on.

From the famous Hoppe's company comes Castrol Synthetic Gun-Lube engineered to minimize wear, and it's also a metal protector with the claim its "unique synthetic formula bonds to metal surface."

course. Engine oil helps to pick up grit. But the cleaning process associated with blackpowder lube is especially important, because without any lube whatsoever, very few loads can be fired from a muzzleloader before it takes considerable force to shove the patched ball (or conical for that matter) home. That's because built-up fouling must be attacked. A good lube also protects the bore from attack by moisture and oxygen. For example, when a projectile, patched or not, is seated in the bore with a lube, the lube leaves a coating of itself, and this coating rests in the bore while the gun is carried or left at rest. So modern lubes thwart friction and they protect the bore. They also promote repeated shooting without swabbing between shots. Today, that can mean many, many shots with high-tech products before the bore demands a scrubbing out.

Shoot-All-Day Lubes

My first attempts with lubes advertised as good for dozens of shots in a row were negative. I did not get the promised results. Recently, I've had much better luck with these lubes. I think there are two reasons: First, the products are better than ever. Second, I take care to prepare the bores of test rifles before using these lubes. I have now proved to myself by simple test that they are truly workable products. I have one rifle that fired fifty times in a row without swabbing the bore, using a high-tech commercial paste called Thompson/Center Natural Lube 1000 Plus. This product came my way about five years ago as this is written, and it proved itself worthy of its claims. Today, look for this lube under the name Thompson/Center Natural Lube 1000 Plus Bore Butter. It comes in a tube or a jar. Essentially the same stuff is available as Pine Scented Bore Butter, also in tube or jar.

Success with modern blackpowder shooting chemicals is based on better products, but don't leave proper bore preparation out of the pic-

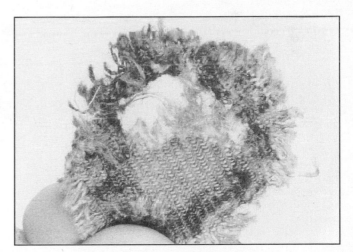

Lube can be an aid against patch burnout, although a severe failure like this one calls for a buffer in between the powder charge and the patched ball. Nonetheless, lubing patches remains a very important process, not only to aid seating the ball, but also for burnout.

ture. I now feel that in order to give any of these modern chemicals a chance, the bore must be prepared beforehand. It is unfair to expect these new lubes to work as advertised when they are used in a bore that already contains some form of lube or metal preserver. So swab the bore out completely before trying any modern lube. Get rid of any traces of previous lube or metal protector, ending up with a bore that is

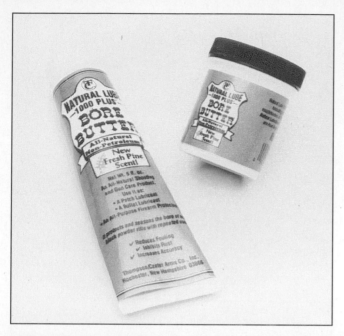

The world of blackpowder shooting has been changed with the advent of the so-called "all-day" lubes. Thompson/Center's Natural Lube 1000 Plus, here in Pine Scented Bore Butter, allows considerable repeat shots without swabbing the bore.

AppleGreen from Michaels of Oregon is another good all-day lube. It smells like green apples, but the new Hot Shot Patch and Bullet Lube is an all-natural, biodegradable product that reduces fouling and protects the bore from corrosion. Michaels of Oregon claims that "Tests have shown that a muzzleloader can be fired hundreds of times without cleaning when this lube has been used."

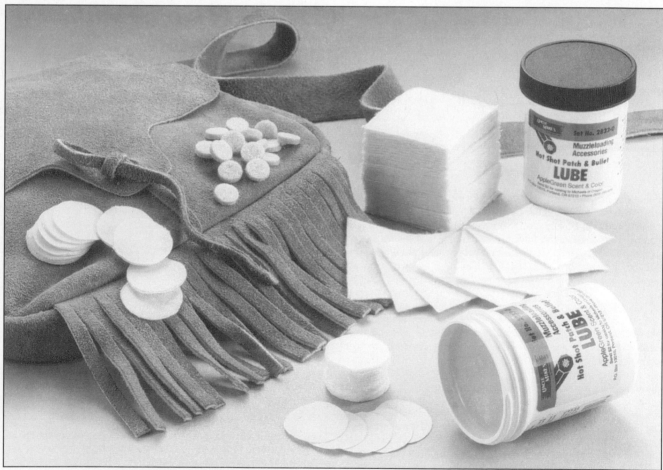

as free of agents as you can get it. Then give the new stuff a try and see how it works in your muzzleloader. As noted above, there are quite a few of these shoot-more lubes available today.

Along with Bore Butter, there is AppleGreen, for example. AppleGreen comes from Michaels of Oregon, which already says a lot for it, considering that company's reputation for testing products. In keeping with the trend today, AppleGreen is biodegradeable, just like Bore Butter. Yes, it does smell like apples, and yes, it does have the color of green

apples. This all-natural lube promises hundreds of shots before bore cleaning is demanded. This statement refers to test conditions, of course, and is not a recommendation for everyday shooting. After a shooting session, clean your frontloader, no matter what lube you're using.

But the new commercial "natural" lubes do break down blackpowder fouling, and they do allow more shots in a row before bore swabbing is necessary. Before going on, let's not forget Ted Bottomley, who started his Ox-Yoke Originals company a number of years ago

This blackpowder bullet clearly shows lubrication in its grease grooves. Obviously, a specific type of lube must be used in this application. If not, the lube won't stick and stay.

FP-10 has a solid reputation as the "Lubricant Elite," but it also works as a patch lube and was used in some of the tests conducted by the author for his book, *The Gun Digest Black Powder Loading Manual.*

with a desire to make blackpowder shooting more enjoyable with less effort. Ted promoted modern lubes, among them his own Ox-Yoke Wonder Lube. All natural, the Ox-Yoke lube proved itself worthy in repeat-shooting tests. Wonder Lube's success prompted further study into blackpowder lubes that neutralized fouling, and Ted deserves special credit for pursuing the shoot-all-day modern blackpowder lube.

There are, as promised earlier, a plentitude of these new multi-shot lubes available, with a trend to produce more of them, and make them even better. They do make blackpowder shooting a more enjoyable pastime, and that's for sure. Before going on, one more modern paste lube deserves special mention, because it is formulated in a special way. This is Pyro Lube. It deserves special mention because it is specially formulated for Pyrodex. After all, Pyrodex is not blackpowder. Pyro Lube is all natural, and it takes into consideration the differences between blackpowder and Pyrodex. It promises the same all-day-shooting-without-swabbing that comes with the other all-natural lubes offered the modern shooter.

Today's miracle lubes allow repeat shooting, as explained above, but they also make later cleanup work much easier because these products perpetually work on fouling during the shooting session. So they keep on working for you even after you shoot your muzzleloader. Meanwhile, don't forget the original purposes of muzzleloader lubricants—to lubricate. Modern lubes function under the ravages of extreme heat and sometimes extreme cold. They work differently from solvents, which is important to remember. Solvents do not have to hold up to the fire of a blackpowder charge, although they may have to remain in liquid form during a very cold day in the hunting field. A good blackpowder lube, then, does many things well: it eases loading, promotes repeated shooting, and aids the cleaning process.

Like the old Snake Oil elixers sold by traveling salesmen, modern blackpowder lubes are sometimes credited with doing just about every-

thing from allowing hundreds of shots between bore-swabbings to making the bullet go faster. Some lubricants can enhance muzzle velocity, but the gains are meaningless on a practical level. A possible velocity increase of a few feet per second may attend certain paste or grease lubes, probably because of their composition more than any chemical reason. Proof of velocity gains across the board with specific lubes is hard to come by. As much as I admire the new biodegradeable natural lubes on the market for what they do, I am not prepared to credit any of them with an appreciable velocity gain.

Types of Lubes

For our purposes, let's consider only four types of blackpowder lubes: liquids, greases, pastes, and creams; with the distinction between the last two tenuous at best. In the past, liquid lubes were considered a good choice for patches because they tended to reduce blackpowder fouling better than greases, creams, and pastes. That is no longer the case—not with our new all-day paste shooting lubes. Remember that blackpowder leaves many salts in the bore after combustion. These salts were broken down very well by liquid solvents, and that is why many lubes of yesteryear for patches were liquids. Blackpowder solvents are still in liquid form, but we're talking lubes here, and today we have cream-like lubes that really do the job, on round ball patches as well as conicals. I still like and use many liquid lubes for ball patches, but these liquids no longer dominate that purpose.

Liquid Lubes for Repeat Shooting

In the old days, the wet ball patch was credited with cleaning the bore every time a patched ball was run home. Was this true? Yes, up to a point. It worked like this: A shot was fired. Then a patched ball, with the patch saturated with liquid lube, was run home. The ball patch, in this instance, performed like a cleaning patch, nullifying part

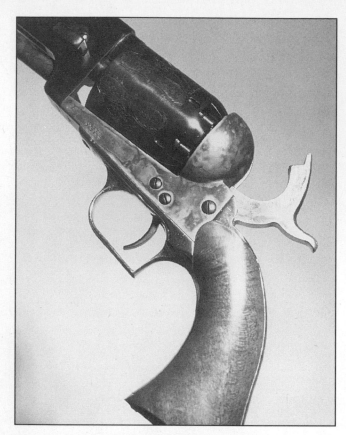

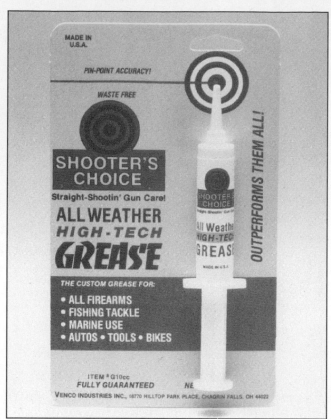

The blackpowder cap 'n' ball revolver is prone to lockup unless lubed properly. Here, a grease-type lube, or at least a gel or cream, is necessary. A liquid lube won't stick.

Sometimes a grease is needed in order to lubricate a particular part, such as the workings of a cap 'n' ball revolver. That's why grease-type lubes like this one are still popular and worthwhile.

of the fouling in the bore. So you shot, loaded up again with a patched ball with the patch well-lubed with a liquid, and you got to shoot again and again, each new patched ball serving to mop up some of the fouling left by the shot before it. Using smallbore rifles, I've gotten off thirty shots in a row without bore swabbing using a liquid lube/solvent such as Old Slickum from the Jonad Corp.

I still use liquid lubes with smallbores. They work fine because of the light powder charges used. Instead of 100, 110, even 120 grains of powder, as with a large bore, only 10 to 30 grains of powder are used in the smallbore. It's a lot easier to neutralize the fouling effects of 10 to 30 grains of powder compared with over 100 grains of powder per shot. In short, the method of using a liquid lube on the patch as a cleaning agent works fine, but with heavy hunting charges in big bores, the modern lube allows more shots before cleaning is demanded.

The Spit Patch (Saliva)

Perhaps the oldest lube of all, saliva is still considered a proper patch lube in certain sectors. The spit patch works. There is no denying that. For plinking, even target work, saliva can get the job done but it does have a few drawbacks. Saliva dries out in the bore quite quickly, especially in a hot barrel. This is easily proved by wetting the side of a warm barrel and watching the moisture evaporate right off the hot steel. Furthermore, if loaded in a cold barrel for hunting, where the shot may not be taken for a while, saliva can promote rust. Big deal? No. Saliva isn't going to turn the breech into a pitted chamber of horrors in a few minutes, but as everyone knows, you don't leave water on steel as a rule. It's a bad policy. Spit patches are fine if tradition beckons. But tradition does not call loudly enough for me to use this method of lubricating the patch.

Sperm Whale Oil

Early America owes a lot to the whale. Whale oil was a true

force in the economy of a baby country. It's a wonderful lube, too. It will not dry out. It will not rust the bore. Unlike certain petroleum products, it won't harm the bore because it does not turn into an asphalt after shooting. Sperm oil is a polar compound that combines with surface oxide films on steel to form a rust combatant. It tends to stay on, too, rather than running off of metal. Sperm oil does not gum up nor turn to sludge. It lasts indefinitely in its original state and it holds up in cold weather. It is not broken down by hot blackpowder gases. It is truly remarkable stuff, and of course as natural as today's blackpowder wonder lubes. It also has many uses other than serving on the blackpowder patch.

Whale oil is a leather preserver and will lubricate many different kinds of working metal parts. But having said all of this, with great admiration, I might add, there is a problem: Whale oil does not attack blackpowder fouling, and therefore does little to promote repeated shooting. Also, it has no power that I can discern in preventing caking of blackpowder residue. Sperm oil is not an animal fat, as I understand it. It's an oily wax. Jojoba oil is chemically similar to sperm oil and shares many of its properties, by the way. I used to use a lot of whale oil but I don't anymore. Whale oil is historically interesting, and worth knowing about. I think every blackpowder shooter should have knowledge of this old-time liquid. But it will not promote all-day shooting.

Synthetic Whale Oil

When romance and tradition bade me to use sperm whale oil, I could find it sometimes, but not all the time. A fellow shooter told me he had gotten away from the real McCoy, however, and was having good luck with a synthetic version of the oil. I tried it, liked it about as well as the real thing, and used the synthetic product for a while. While it did not enjoy every attribute associated with sperm whale oil, the manmade liquid was okay. However, it did not break down the salts left behind by blackpowder combustion any more than whale oil did, nor did it promote repeat shooting, and all in all, I've dropped any interest in this product.

Water Soluble Oil

The generic name for water-soluble oil mixed with water, with a dash of dish detergent if you like, has been moose milk for about a hundred years; however, there is now a trademarked blackpowder lube called Moose Milk and I prefer it far and above the generic liquid. It comes from the Winchester Sutler Co. However, if for historical reasons only, let it be recorded that water-soluble oil can be mixed with water in a 10-percent oil, 90-percent water solution, or up to 25-percent oil with that squirt of liquid dish detergent if you like, and the end result is a perfectly useful patch lube that attacks blackpowder fouling, lubes patches nicely, and serves also as a solvent in cleaning the soot-belcher. It's slippery stuff. If you like it, use it. But I'll take the commercial Moose Milk. It's ready to go and it does a fine job as a patch lube and solvent.

Shortening, Sheep Tallow, and Bear Fat

These fats are slick and therefore they serve as lubricants. They are also easy to use. Melt 'em, apply to the patch, blot the patch on a paper towel, then store the pre-lubed patches in a small metal container, such as the hinged boxes used to hold throat lozenges and cough drops. Handy? You bet! Workable? Sure. But don't expect any breakdown of blackpowder fouling and don't count on these blackpowder lubes where repeated shooting is sought. Historically, these are worth knowing about. In practice, forget them all.

Petroleum Jelly

Petroleum jelly does not dry out, and it provides good lubrication qualities for a shot or two. It's easy to find and cheap, and good for these reasons. It also sticks to a conical fairly well. But it offers no reduction of leftover salts in the bore, and does little to promote repeated shooting. Also, petroleum jelly may leave a trace of itself in the bore after repeated use as a patch or conical lube. Forget it for blackpowder shooting.

Cooking Oils

Peanut oil, olive oil and similar products will serve to lubricate the patch. There's no doubt about it. I tried olive oil as a test because I wanted to be sure I knew what I was talking about concerning this product. I cannot recommend any of the cooking oils as blackpowder lubes, and I don't use them.

Waxes

Beeswax and other waxes can build up in the bore. They may serve

Whale oil has many fine properties, and for the experience, plus the historical value, muzzleloader fans may well want to try some. But there's no doubt that whale oil has been surpassed by modern blackpowder chemicals.

In relatively recent times, cleaning blackpowder guns has changed, due mainly to modern chemicals like Black Powder Cleaning Gel Bore Cleaner. It works on lead fouling, but is most effective on breaking down the residue from blackpowder and Pyrodex.

The original Shooter's Choice solvent was found effective in reducing plastic fouling left in bores from sabots. Plastic fouling is not a huge problem, but it does exist.

Now that blackpowder cartridge shooting is highly popular again, lubes for metallics during reloading are a big consideration. One Shot is from Hornady. It promises to lube cases for resizing in one shot.

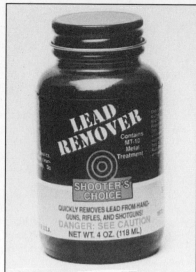

The blackpowder cartridge calls for a special lube, such as this one from the Butler Creek Corporation of Montana. It's especially formulated to lube lead bullets fired in the old-time cartridge guns.

Leading of the bore is not prevalent in blackpowder shooting, mainly due to low-velocity loads. However, it is not unknown. During a test session, leading was found in a blackpowder cartridge revolver, for example, and Shooter's Choice Lead Remover was used according to directions to cure the problem.

as a coating on the Maxi or Minie ball, but all in all they are not the best products to use as blackpowder lubes. I cannot recommend them.

Other Agents

Nobody knows how many different blackpowder lubes have been tried over the years. Certainly saliva and water rank as early lubes, or attempts at lubrication. Beaver fat, pump grease, ammonia, kerosene, peroxide and many other agents have been tried and often applauded by blackpowder shooters. I'm not anti-tradition by any means, and if a shooter truly wants to copy the past in all possible ways, he'll buy some whale oil and work with its weak points as well as its strong points. But there's no denying that today's modern blackpowder chemicals are better than ever, not only for shooting, but also for extending the life of the muzzleloader.

Solvents

Solvents should be considered apart from lubes, even though many solvents work just fine as patch lubes. Solvents are designed to break down blackpowder fouling so that the fouling itself can be removed from the bore of the firearm, as well as from the exterior of the gun. We have a multitude of superior solvents on the market today. A little waltz through the Mountain State Muzzlelaoding Supplies catalog will reveal many good ones. We can't forget the famous Hoppe's name, too, for Hoppe's No. 9 Plus solvent is formulated for blackpowder cleanup. Birchwood Casey's No. 77 Black Powder Solvent is a dandy, too. So is T/C's No. 13 Bore Cleaner, as well as Venco's Black Powder Gel. I apologize for having to leave out so many other good solvents, but I can't forget to mention Hodgdon's Spit Patch solvent. This is the only solvent I used in an extensive test for corrosion. The results of that test are presented at the end of the book in Appendix A. I found Spit Patch a ferocious attacker of blackpowder fouling.

What About Plastic Fouling?

Anyone who thinks that zero plastic wash is left behind when plas-

tic sabots or shotgun wads are fired in the blackpowder shotgun has not studied the situation very closely. One company did, and reported to me that a residue did remain in the bore after firing a number of plastic sabots and wads. I worked on the problem, finding Venco's original Shooter's Choice MC #7 Bore Cleaner and Conditioner useful in removing the plastic fouling.

Leading

Very little leading takes place with muzzleloaders, mainly due to low velocity, even with all-lead projectiles. Nontheless, to believe that there can never be any leading in a breechloading blackpowder cartridge rifle, or for that matter, a muzzleloader firing lead bullets, is to trust too much. There are many modern lead removing chemicals on the market today. Any local gunshop should have at least one of these on hand. An occasional treatment with a lead remover simply cannot hurt the conical-shooting muzzleloader, the blackpowder shotgun, pistol, or revolver, or the breechloading blackpowder cartridge rifle or revolver.

The Blackpowder Cartridge

My own Navy Arms Sheriff's Model 44-40 revolver gets and deserves a thorough treatment with modern blackpowder chemicals, as it should. Blackpowder cartridge rifles earn not only a complete after-shooting solvent cleanup, but also a special lube on the projectile. One of the best lubes for this work is SPG, which is sold through Montana Armory. Rooster Laboratories BP-7 is another dandy lube for the blackpowder cartridge load. Lubes are essential to quality shooting of the blackpowder rifle cartridge.

Sincere apologies, but room prevents inclusion of many fine modern chemicals formulated to make black powder shooting less work and more fun. By all means, every reader is encouraged to study the Dixie Gun Works catalog for its multitude of great blackpowder chemicals, as well as every other catalog that carries products for muzzleloader shooting and maintenance. We are truly lucky to have these agents working for us.

●

chapter 20

Surefire Blackpowder Loading Methods

THERE ARE MANY different trails in the forest of proper black-powder loading methods. The following rules are not chiseled in concrete by Davy Crockett. They just happen to produce surefire loads, which is what we're after.

The Old Ways

Science was in swaddling clothes when blackpowder shooting was the only shooting in the land. That's why so much old-time information came to us by way of a witch's broom. Of course, it wasn't all "bad dope." But naturally the most colorful and fanciful loading theories survived to be preached in modern times. When the embers of blackpowder shooting flared into flame in the 1950s and 1960s, these rules of thumb were disseminated like popcorn at a movie theater. There was a market for blackpowder writing, and everyone who owned a frontstuffer and a typewriter got in on the act, including me.

Unproved myths were told as fact. Shooters put lead balls in the palms of their hands and covered them with blackpowder to come up with a proper load—about like brewing a cure for the common cold with bat wings and wolfbane. Most advice was silly. Some was downright dangerous. Gunstore clerks whose closest association with a muzzleloader was to have one drop on their feet told customers that "you can't overload a blackpowder gun; excess blackpowder just blows away." I heard one representative of a gun company tell a shooter that it was advisable to load more powder behind the heavier conical than the lighter round ball—utter nonsense. He didn't let a little thing like fact get in the way of his blackpowder preaching.

In truth, blackpowder arms are governed by the same laws of shooting that dominate modern firearms. Muzzleloaders do not defy the laws of physics. They are understood through knowledge of interior and exterior ballistics, pyrotechnics, chemistry of propellants, and many other branches of hard science. So when someone tells you that "a perfect load is one grain of powder for every caliber," nod your head and walk away. Sure, you could hunt deer with a 50-caliber rifle loaded with 50 grains of blackpowder, but I wouldn't. Let's build our loads on the scientific facts we have to work with at the moment. It's the best we can do.

Today's Loading Methods

We're sometimes asked to envision a trailblazer cutting off his shirttail for a patch as hostile arrows buzz around his head like angry hornets. He might employ the famous "spit patch" on the spot, or engage in some esoteric voodoo step that will make that frontloader shoot just right. In fact, basic loading techniques center around repeated uniformity. The smart blackpowder shooter has a routine, a step-by-step approach to feeding his muzzleloader the same diet time after time, though of course he can vary the menu from plinker loads to firebreathers right on the spot. The clever blackpowder shooter learns a workable routine and stays with it the way a smart airplane pilot pre-checks his bird before every flight.

Preliminary Considerations

Bore condition is the first consideration in loading any blackpowder gun. The bore must be dry, not damp from lube, and it must be oil-free and grease-free for the most *consistent* results. It's been shown that accuracy may be possible with a damp bore, but since there is no actual control over the degree of film in the bore from one shot to the next, the dry bore is preferred. A dry bore means consistency, while an oily or greasy bore may shoot off-target by a considerable degree. One test rifle placed its group several inches high and to the left when fired with an oily bore. When the bore was consistently dried with a clean patch *after* the rifle was loaded, the group struck center-target again, time after time. Therefore, prior to loading, the bore is swabbed dry. I do this at home before taking the rifle into the field, because this step also gives me a chance to check the function of the half-cock notch and other mechanical details to ensure that the rifle is working properly.

After bore mop-up, the rifle's nipple vent is swabbed free of oil or grease with a pipe cleaner. The nipple may also be removed and the nipple seat and channel to the breech swabbed oil-free. On the flintlock, the touchhole is cleared of any remaining preserving grease or oil. The idea is to maintain a moisture-free route for the source of ignition to travel, whether it's a spark from a percussion cap or flame from FFFFg pan powder.

Loading by volume is the correct way to go. Here, a powder horn is used to dispense powder into a measure. The powder horn is not used to pour powder directly into the muzzle of the firearm.

The correct volume of powder is now poured downbore directly into the breech of the muzzleloader, and then a patched round ball is ready for ramming home on the powder charge.

Loading the Percussion Rifle

Dry-Firing Caps to Ensure a Clear Channel

The oil-free percussion (caplock) rifle is now ready for a pre-load percussion cap or two. Firing caps further dries the nipple vent, also clearing it of any debris which may have invaded the system. The muzzle of the rifle is aimed at a lightweight object on the ground. A bit of paper, small leaf, or twig will do. The percussion cap contains enough force to budge small objects out of the way, provided the nipple vent and bore of the rifle are clear. At the snap of the cap, the leaf, twig, paper—whatever—should jump away from the blast. If it does not, there is probably insufficient expulsion of gas from the muzzle, indicating a clogged bore or channel.

Also, listen for a hollow sound—"Thump!"—when the cap is detonated, not a "Crack!" The sharper sound may indicate that the expanding gases from the percussion cap are trapped before being expelled through the bore of the rifle. If any condition suggests a bore or channel obstruction, the shooter must take care of the problem, running a cleaning patch downbore and a pipe cleaner through the nipple vent and into the nipple seat until the blockage is cleared, so a detonated percussion cap blows small objects out of the way when the rifle muzzle is placed near them.

Oddly enough, the very process of clearing the nipple vent of oil or debris may in fact deposit other debris. The careful shooter runs a pipe cleaner into the vent of the nipple with a twisting motion *after* the firing of the percussion caps, just in case there is any deposit left from the cap itself. This is an important step. Blackpowder hunters work hard for opportunities. It's frustrating to find game and stalk it for a close shot, only to be greeted by a hangfire. So take a few seconds to check for percussion cap fouling. Testing shows that cap debris is more likely deposited when popping percussion caps to dry the nipple than during actual shooting, because the loaded rifle has some blowback which helps to clear the nipple vent, while the unloaded rifle has no blowback.

Loading The Powder Charge and Protecting The Patch

Next, with the hammer eared back on half-cock to allow trapped air to expel itself from the bore through the nipple vent, the powder

No matter the type of muzzleloader—revolver, rifle, shotgun—clearing the nipple by firing a cap or two prior to loading is recommended.

charge is dropped down the bore using a volumetric measuring device.

A patch protector may be installed now if a heavy hunting charge is used. A light powder charge will not normally destroy a patch. I recommend hornet nesting material. The hornet's nest is composed of a

A short-starter is used to force the patched ball down the muzzle. Sometimes a little tap is required to get the ball moving.

After the patched ball is forced partway downbore using the short-starter, the ramrod is inserted into the bore, as shown here.

paper-like substance, and though a match touched to a few leaves of this amazing stuff will annihilate it, downbore the hornet nesting material acts as asbestos when the rifle is fired, totally protecting the patch from burnout.

In my 54-caliber ball-shooting long rifle, I place two to three leaves or layers of hornet nesting material over the muzzle of the rifle after dropping the powder charge. These layers are run down on top of the powder charge with the off end of the ramrod to seat them. I tamp the hornet nesting material on the powder to make sure that it's in position. That takes care of patch protection, because the hornet nesting material becomes a firewall between the powder charge and the patch. I have tested hornet nesting material with heavy hunting charges, and have found the patch unscorched downrange.

The rifle is now loaded with both powder and patched ball. Other patch protectors are possible, including a thin cardboard overpowder wad (lubed), as well as felt disks. But for me, hornet nesting material is the best patch-saver of them all, creating no accuracy problem or gas diversion. I find that a patch protector enhances load uniformity because it acts as a buffer between patch and powder, promising an intact patch every shot. It also provides for a uniformly dry powder charge, rather than one possibly inundated by lube moisture from the patch. In other words, the patch protector itself takes up excess lube from the patch, rather than the powder charge.

Running the Patched Ball Home

The patched ball is loaded by centering the patch over the muzzle with the ball also centered. The short-starter is generally essential to accuracy in most muzzle-loading rifles because tight-fitted balls give best grouping. The stub, or shorter rod on the starter, is used to drive the patched ball into the first few inches of the bore, followed by the longer rod on the short-starter, which pushes the ball farther downbore, followed by the loading rod or ramrod, which fully seats the projectile. Of course, even the veteran blackpowder shooter may sometimes use a looser-fitting ball to facilitate loading, especially when he does not require the best possible accuracy from his rifle. However, never be confused by the difference between a tight patch and a close-fitting

Hornet nesting material will burn if lighted with a match, but downbore its character changes to an asbestos-like substance. A sheet or two will safeguard a patch from burnout.

round ball. My tests show the latter to be more important to accuracy. Of course, blackpowder rifles are individual in nature and thick patches with well-fitted, but not necessarily near bore-size balls, may give best accuracy in certain muzzleloaders.

After-Load Bore Mopping

The bore may be slightly damp now from running the patched ball home because of the lubrication contained by the patch itself. Since we want our bore to remain in a relatively dry state for every shot, this important step is next in line. Run a clean patch downbore a few swipes after the patched ball is fully seated on top of the powder. But wait. Is that safe? Mopping the loaded bore with a cleaning patch is no different from running the patched ball downbore on top of the powder charge, which you just did with a ramrod or loading rod. The same rules of safety apply: keep the muzzle pointed in a safe direction at all times! That means away from you or anyone else. The clean patch will

The patched ball is run home fully upon the powder charge with no air gap in between the projectile and the powder.

A round ball or a conical bullet is simply placed upon the mouth of the cylinder chamber and rammed down over the powder charge.

A flask can also be used to arrive at a proper blackpowder charge and distribute that charge directly downbore, shown here.

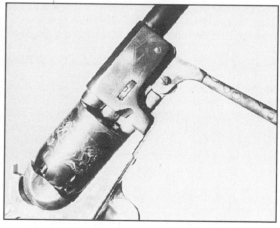

mop up any excess lube in the bore, giving the rifling a uniformly dry bore from shot to shot, and uniformity is, as alluded to earlier, a major goal in surefire loading.

Seating Pressure

Whether seating a conical or a patched ball, try to maintain a reasonably close degree of pressure on the ramrod or loading rod from load to load. Again, this is a matter of uniformity. You don't have to scale-check ramrod pressure. Go by feel, and do not lean your weight on the rod. Remember that haphazard pressure on the ramrod may cause a rise in standard deviation of velocity (more bullet speed deviation from shot to shot), hence a drop in accuracy, so the projectile should be seated firmly on the powder charge with consistent ramrod pressure by feel. Also, remember that blackpowder burns more consistently under pressure, so firmly seat the bullet each time.

What About Conicals?

The conical projectile requires a paste, cream, or grease lube that "sticks" to it. So the elongated missile is simply lubed and rammed home without a patch. The sabot/pistol bullet is treated likewise. A dab

of modern high-tech lube can be placed on the base of the plastic sabot for more continuous shooting without bore swabbing. I've found that a dab of paste lube in the hollow base of the Minie-type projectile works fine. Modern lubes do not damage blackpowder.

Ignition

Ignition for the caplock consists of merely placing a percussion cap on the nipple, and the rifle is ready to shoot. Percussion cap variations from brand to brand exist, but they are not normally a problem. The No. 11 cap is most prominent today. There are many fine caps available. Navy Arms sells the excellent RWS cap. CCI makes a fine cap, too, as does Remington and others. The shooter can use these caps with confidence. Also, a fusil may be employed. This is a device that takes a modern primer in place of a percussion cap. Some modern muzzleloaders have also gone to an ignition system that uses a shotgun primer. But the end result is the same—fire gets to the powder charge in the breech and the rifle goes off.

I prefer a non-corrosive cap. That preference may seem ludicrous. After all, blackpowder fouls the bore anyway, so why worry about the cap adding a little extra smudge to the leftover carbon and salts?

Remember the cap is employed to clear the vent of the nipple prior to loading, after which the rifle may be carried in the field without firing for a day or more. The non-corrosive percussion cap will not damage the nipple or the more important nipple seat during this time period, but firing corrosive caps to dry the nipple vent and seat may attack the metal a little bit.

Loading the Flintlock Rifle

Most of the loading rules above apply to the flintlock, but this system does demand a few tricks in order to gain that shot-after-shot consistency the blackpowder shooter insists on. Clean initial conditions are paramount for reliable flintlock shooting. The touchhole especially must be foul-free for the flame emanating from the pan powder to fly through this channel into the main charge in the breech. The face of the frizzen must be clean (especially grease-free) and of course the bore must be clear of oil or grease for obvious reasons. The flintlock system demands close attention for surefire results.

The first step in reliable flintlock ignition is blocking the touchhole of the rifle. A friend of mine has collected over a hundred original Pennsylvania/Kentucky long rifles. He noticed a number of shooting bags that attended these rifles contained a quill—a feather of one sort or another. The feather proved a mystery for a long time. However, there was one feature about these quills that was universal, and that was a pinched-down section at the tip of the shaft.

It turned out this pinched-down area was a perfect fit in the touchhole. Was the feather used to block the touchhole during the loading process? I like to think so, but of course cannot prove it. But tests have convinced me that blocking the touchhole gives best surefire results.

The idea is to insert a vent pick, pipe cleaner—or quill shaft—into the touchhole prior to loading the rifle. Thus, the touchhole is blocked off and no powder can invade it from the main charge.

Consider this. If the touchhole is filled with powder, the flame from the pan must burn *through* that powder before it can ignite the main powder charge in the breech of the firearm. That's a fuse. Now imagine a touchhole free of powder. The flame from the FFFFg pan powder licks through the channel like the flit of a snake's tongue, darting right into the breech and setting off the powder charge held there. This reasoning, plus trial shooting with blocked touchhole and without, convinced me that the first step in loading the flintlock is to insert some obstacle in the touchhole prior to any other action. I use a pipe cleaner often as not, but I also have several touchhole (vent) picks that serve well for the job.

Dropping the Powder Charge

At this point, the touchhole is blocked. Now the powder charge is dropped downbore followed by the patch protector and patched ball, just as described under caplock loading procedures. After, and only after, the patched ball is firmly seated is the blockade removed from the touchhole. Hopefully, the packed powder in the breech will stay there, rather than trickling into the touchhole. The powder-free touchhole leaves a clear passage for the flame from the pan to ignite the powder charge, instead of a "cookoff," where powder in the touchhole must first burn out of the way.

Dealing with Pan Powder

Experimenting convinced me that a pan full of powder is not the way to go. A pan filled with FFFFg did not work as well as a partially filled pan. I settled on a pan about two-thirds full, with the pan powder

Another important difference in loading the flintlock is blocking the touchhole *before* a powder charge is dropped downbore. In this way, the touchhole with be powder-free, allowing the quick passage of flame from the burning pan powder.

A flintlock requires pan charging before it can be fired. Here, a pan primer is used to drop a small amount of FFFFg blackpowder into the pan.

The pan should not be completely full of FFFFg powder. Two-thirds or so is plenty, keeping the powder to the outside of the pan as shown here. This allows the pan powder to create a flame directed toward the touchhole.

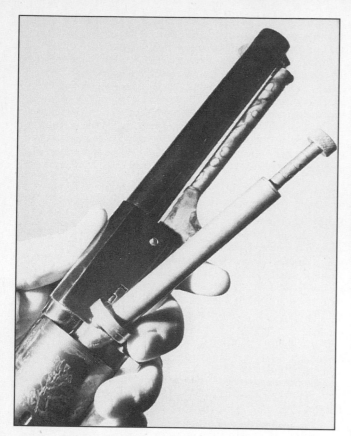

Loading the muzzle-loading pistol is very much like loading the muzzle-loading rifle, but the cap 'n' ball revolver requires entirely different handling. After firing caps on the nipples to dry them, the hammer, as on this Colt revolver, is set on the first notch for loading.

A standard adjustable powder measure, like this one from Uncle Mike's, can be used to measure the proper charge for the revolver and introduce that charge directly into the cylinder chamber, as illustrated.

Surefire flintlock loading includes ensuring the correct position of the flint against the frizzen. Here, the flint makes full contact all across the face of the frizzen, as it should.

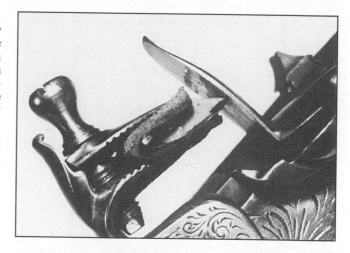

crowded into the *outside* of the pan, not up against the touchhole. My theory may be difficult to prove, but I believe keeping the touchhole free of powder is wise, whether that powder comes from the main charge in the breech or from the pan. Also when powder is ignited on the outside of the pan it seems to provide more certain ignition, possibly because the flame darts across the pan and into the touchhole directly in a more concentrated pattern. I even make it a point to carry the flintlock tilted in my hand when field-walking so that the pan powder remains to the outside of the pan. Nitpicking? Possibly, but in a demonstration for an editor friend who was visiting my home, I fired my Hatfield flintlock a couple dozen times in a row without a single hangfire, let alone a misfire. If the steps noted here promoted that surefire ignition, then they are certainly worth bothering with.

Touchhole Location and Touchhole Liners

Two other factors in flintlock ignition sureness is the location of the touchhole and the touchhole liner. If having a flintlock firearm built, insist that the touchhole be drilled as high on the barrel flat as practical, and not down low on the flat where pan powder can sift into it. The higher touchhole works well, as indicated by several test rifles. Furthermore, if the touchhole is ragged, scored or burned, a liner may be in order. The liner has a smooth interior, and this surface is more conducive to spark travel than is a rough surface. Touchhole liners can be installed by blackpowder gunsmiths at nominal cost.

The Frizzen

Another aspect in certain flintlock ignition is the face of the frizzen. A rough or dirty frizzen may produce feeble sparks. Conversely, a clean frizzen face is best in generating sparks. Remember, it is the frizzen face, not the flint, that produces the sparks. Curled bits of hot metal shower down from the friction of the flint upon the metal frizzen. A dry frizzen face was also noted as conducive to better sparking, hence better ignition.

The Right Flint, and Flint Position

Buy top-quality flints. Cheap flints may not scrape sparks from the frizzen. Also, ensure that the edge of the flint mates squarely with the face of the frizzen. Do this by loosening the flint in the jaws of the cock, allowing the cock (hammer) to fall forward under your control. Line the flint up squarely with the frizzen, and *then* tighten the flint in place while it rests squarely against the face of the frizzen.

Can Flintlocks Really be Counted On?

I think it is useful to address this question here under the topic of surefire ignition. I've read too many times that flintlocks cannot be counted on to go off with any degree of regularity whatsoever. Yet, weren't wars won and lost with flintlock firearms? I'm reminded of the Battle of New Orleans during the War of 1812. Seven thousand trained British troops under General Pakenham were transported to Jamaica, and from there attacked New Orleans. General Jackson had

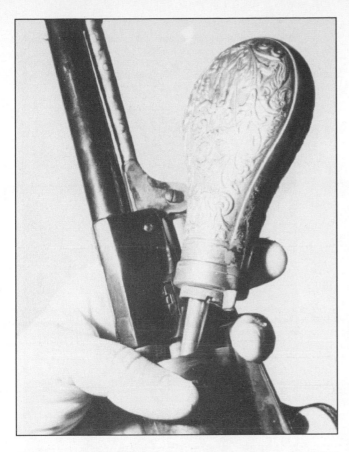

Another way to load the cap 'n' ball revolver with a powder charge is with a flask.

Before capping the cap 'n' ball revolver, a dab of grease or paste is placed on top of the seated projectile.

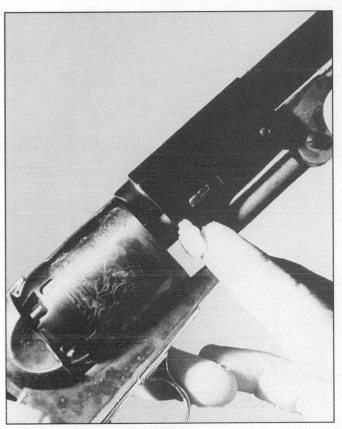

5000 men on hand to meet the attack. Especially, he had a handful of Tennessee sharpshooters at his command. When I first read of this battle, I assumed that the Americans were hiding behind rocks while the British soldiers stood in rank and file. Not so. The Americans were also in rank and file. When the smoke cleared, literally, the British had lost 700 men, with 1400 wounded and 500 taken prisoner. The American loss was eight killed and thirteen wounded. Had the flintlocks of the Tennessee riflemen been as primitive as many modern writers suggest, I doubt that so much havoc could have been wreaked in so short a time. Nor is this story anti-British, for the fine British soldier won many a battle with his flintlock.

Loading the Blackpowder Pistol

The blackpowder pistol is mechanically a muzzle-loading rifle in short form; therefore, it requires no special treatment that has not been covered above. Patch protectors are not normally in order; however, tests with the accurate Thompson/Center Patriot single shot pistol indicated that with some loads a patch buffer did improve accuracy. Flintlock pistols are loaded using the pointers mentioned above for the flintlock rifle, including touchhole blockage, frizzen condition, and using good flints.

Loading the Shotgun

The smokepole scattergun is a caplock these days, with only a few flintlock fowlers available. In either case, follow the general preliminary routines established for caplocks and flinters. Start with a clean gun. Fire caps on the nipples with the same procedure used for the percussion rifle in order to check for a clear bore. Drop the proper powder charge into each bore of the double, being certain to mark the bore that receives the powder to make certain that one bore does not get two charges. Install the wad columns. Choked blackpowder guns, which are finally a reality, provide super patterns with common old-fashioned wads, but may require a little bit of hands-on encouragement in getting certain wads past the choke. This is merely a matter of manipulation, however. Good wads are available from Ballistic Products, Circle Fly, Dixie, Navy Arms, CVA, and many other companies.

The standard wad system uses an over-powder wad. Push this down upon the powder charge and count to five before letting up on ramrod pressure so that any air in the bore will be forced through the powder charge and out the vent of the nipple. The over-powder wad safeguards the rest of the column from scorching due to the burning powder charge. A felt wad may be used next, or another over-powder wad. Blackpowder burning rate and manner allow this simple wad system to provide good results. I have a Navy Arms Full-choke 12-gauge that produces 75 percent patterns with two over-powder wads, the shot charge, and one over-shot wad to hold the column in place, and nothing more. After the shot is dropped, an over-shot wad must be used to contain the load. Never allow the column of one barrel to ride upbore. After firing only one barrel, check to see that the load in the other barrel is firmly ensconced in the breech. Incidentally, there are many other wad columns that are perfectly acceptable in the blackpowder shotgun. The above is only one example.

Loading the Caplock Revolver

The caplock revolver is actually very easy to load. The bore of the barrel must be clean. Also, any form of grease or other element must be removed from each cylinder chamber. Finally, nipple channels should be cleaned with a pipe cleaner. There is nothing new or different about this initial procedure. Clean blackpowder guns have better ignition than dirty ones.

Failure to lube the revolver properly may cause it to lock up. It is easy to see the fouling buildup on this revolver.

After all chambers are loaded and greased, the percussion cap is placed firmly on the cone of the nipple. If the cap is not a tight fit on the nipple, it may back out when the gun is fired, and rotating the cylinder with a loose percussion cap can cause a jammed gun.

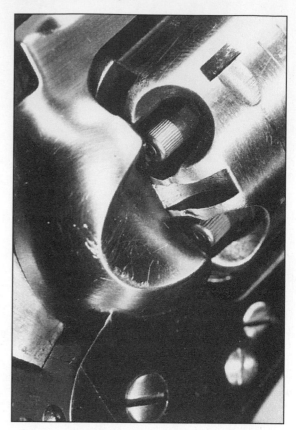

One or two caps must be fired on each nipple in order to dry up any leftover metal preservers, lubes, or other elements not picked up with cleaning patches or pipe cleaners. Then, using pipe cleaners, each nipple is given a quick push-through to guarantee freedom from possible cap debris.

Each cylinder is charged now with the proper amount of blackpowder or Pyrodex. It was common practice in the old days to leave one chamber unloaded, and that chamber lined up with the hammer. This was a safety precaution. Most 1858 Remingtons had safety notches, however, which misaligned the chamber with the hammer, and so all six chambers could be loaded. The Colts had a pin between each chamber and a notch in the hammer which served a similar purpose. A powder measure or flask may be used to deliver the correct powder charge.

The bullet, round or conical, is seated firmly upon the powder charge with the loading rod that rests underneath the barrel. This underlever puts plenty of pressure on the bullet. Do not use excessive force. Option: a lubed buffer wad can be located in between the pow-der charge and the bullet to prevent chainfiring, and to help keep fouling soft.

A dab of grease or lube is then placed on top of each fully seated bullet as it rests in the cylinder chamber, unless an over-powder lubed wad was installed between bullet and powder. This is a very important step. Over-ball grease is generally considered a preventor of the chain-fire (where several or all cylinders fire at once). Grease or lube remains very important, however, in keeping the blackpowder revolver functioning. Shooting without over-bullet grease or lube often means a locked-up gun.

After the chambers are loaded, each nipple receives its percussion cap from a capper, and the revolver is ready to fire.

At the beginning of this chapter, it was admitted that there are several acceptable ways to load blackpowder guns. The bottom line is: if it works for you, and it's safe, and you're happy, continue with your loading method. But don't short-change yourself with shortcuts. Sure-fire ignition is too important for that.

●

chapter 21

The Optimum Load

HERE'S A QUOTE from a magazine: "To start somewhere, lay a ball on the palm of your hand and pour powder over it until it is quite covered. Set your measure for this amount. It should be within ten grains of the proper charge for that ball and rifle, *if you poured the powder over the ball with care* to just cover it." I'm not picking on the author of those words. Heaven knows I've said plenty of foolish things in my life. But it's a perfect example of the sort of advice that flowed like water over a dam not so very long ago. We hear less of blackpowder witchery today, because purveyors of such weighty data have been challenged to prove what they say. Just for fun, go ahead and cover a ball in your hand with powder. Do you hold your hand flat? Do you cup your palm a little? I've tried it. Admittedly, I was prejudiced in favor of failure, and the trick did fail. With flat palm, the powder charge was hefty. With cupped palm, the charge was modest.

Enough of this. Let's get on with building the optimum load. I like that word, optimum. According to my dictionary, it means "The condition or degree producing the best result." That's what I'm after when looking for a good load, especially for a blackpowder muzzle-loading rifle. Of course, it's impossible to come up with one optimum load to suit one given caliber. In the first place, guns vary. A perfect load in one firearm may not be perfect in another of the same make and caliber. Also, optimum for what? Moose, rabbits, targets? In addition, a given rifle may shoot well with several different powder charges. For example, I have a 54-caliber ball-shooter that likes 70 grains volume RS or FFg for target work, but it works just fine with 120 volume RS or FFg, too. However, I'm not about to go big game hunting with the 70-grain charge, even though it is accurate and pleasant to shoot.

It is quite true that just about every muzzleloader will have more than one good load, and while you do not need a big powder charge for paper-punching, for the purposes of this chapter, the optimum load will be *that charge of powder that produces the best ballistic results with allowable and safe pressures.*

What is a maximum load? It varies with the firearm. A heavy barrel like the one shown here has a thick wall, and the manufacturer may allow quite a hefty powder charge because of this strength factor. The last word in maximum loads, however, must come from the gunmaker. He knows his products, and it is his responsibility to say what a maximum load is for them.

Optimum vs. Maximum

The term "optimum" was carefully chosen for this chapter because maximum connotes "all you can get by with safely." Optimum is a little different. A maximum load is safe, of course, but it may not be optimum. For example, a rifle manufacturer may allow 150 grains of FFg blackpowder in a certain 54-caliber round ball-shooting muzzleloader.

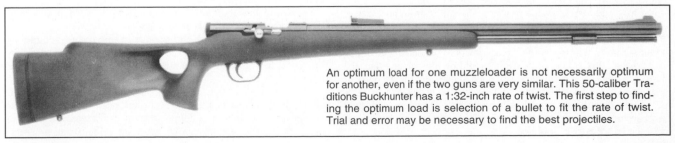

An optimum load for one muzzleloader is not necessarily optimum for another, even if the two guns are very similar. This 50-caliber Traditions Buckhunter has a 1:32-inch rate of twist. The first step to finding the optimum load is selection of a bullet to fit the rate of twist. Trial and error may be necessary to find the best projectiles.

There's a wrong notion on the optimum load, that it should somehow always lean toward the maximum. What about shooting targets? You don't need anywhere near max loads for that.

A cap 'n' ball revolver, like this handsome Colt remake, can hold only so much powder in each cylinder, and therefore it is self-limiting in charge. However, an optimum load for accuracy in this revolver could be far less than chamber maximum. For example, if the cylinder holds 40 grains, only 25 grains could be the accuracy load.

The load chain varies with type of projectile, type of ignition, powder, powder charge, patch, and lube. This load chain is very simple (from left): a fusil with a modern small pistol primer for ignition, followed by a healthy charge of FFg blackpowder and a 600-grain Minie-type conical with a lubed base.

Overloads are possible with blackpowder, but so very easy to avoid. First, do not exceed the limit set by the manufacturer. Second, load by volume using a powder measure. Big bores, such as shotguns, have a very large volume of space within the bore for gases to react upon, and therefore larger bores lean toward lower pressures per powder charge. But factory-set maximums still apply to all muzzleloaders.

That is a *maximum* load. However, the law of diminishing returns may render that load unwise at best, and certainly not *optimum*. In the worst case scenario, the 150 volume charge may actually render *less* velocity than a lighter powder charge, perhaps 120 volume. Or the gain could be nominal for the extra powder burned.

Let's suppose that about 2000 fps muzzle velocity is earned with the 150-grain charge, along with copious smoke and plenty of recoil. Suppose further that 120 grains generates 1900 fps MV, with less smoke, less recoil, and less pressure. Obviously, the 150-grain load, although allowed as a safe maximum charge, is not an optimum load. The optimum big game hunting load for this particular 54-caliber rifle is the 120-grain charge. In short, the extra 30 grains volume isn't worth adding to the charge. It does not pay off in the only thing that counts—bullet speed, which in turn translates into bullet energy.

Just a word about intended purpose: Do you want to hunt a bear with that 54-caliber rifle, or cut a round hole in a target? If it's target shooting you're after, or if the frontloader is serving at rendezvous for shooting games, that big gulp of blackpowder is absolutely uncalled for. Perhaps 50 or 60 grains of FFFg would be more appropriate for target shooting. All the same, as explained above, and for our purposes, the term optimum shall mean that powder charge that delivers peak velocity per pressure from a balanced load chain. The entire load chain is in harmony, not only powder charge, but ball size, patch material and thickness, lube—everything.

Here are a few rules and suggestions pertaining to building the optimum load:

1. Never overload. Do not exceed the manufacturer's maximum recommended charge. Gunmakers are responsible for knowing the safety limitations of their products, and are therefore the right source for maximum safe loads. Optimum loads always fall at, or under, the maximum allowable charge, never over.

2. Begin with a mild powder charge and work up toward the maximum allowable charge in 5- to 10-grain increments, 5 grains for 45-caliber and under, 10 grains for calibers over 45. The reason for using 5- and 10-grain increments is that blackpowder and Pyrodex are not as efficient as smokeless powder, and tiny advances in the charge, such as one grain or two, will not show a provable difference in velocity, trajectory, nor a change in accuracy.

3. Use only blackpowder, Pyrodex, or blackpowder substitute, and no other propellant. We've been through this one before.

4. Load in accord with the proper granulation—remember that kernel size can make a big difference in blackpowder performance, including pressure per velocity, and sometimes accuracy.

5. After you have built an optimum load for one powder brand and granulation, assume that this load is good only for that particular powder and granulation. Blackpowder differs markedly by brand, and different granulations can change both velocity and pressure.

The same goes for Pyrodex. After you have an optimum Pyrodex load, stick with it.

6. Conical projectiles are obviously heavier per caliber than round balls and may call for less powder for an optimum load, but you will often see maximum charges for round balls and conicals that are identical. This is because the firearm is safe to shoot with the maximum listed for the conical, while that same maximum is optimum with the round ball. Adding more powder behind the round ball is to no avail and should not be done. So follow the gun manufacturer's maximum load for both round balls and conicals.

7. Always consider bore size. Small-bore muzzleloaders are more sensitive to pressure than large-bore guns and relatively modest changes in powder charge normally alter velocity more in the small bore compared with the big bore.

Dealing with Old Information

Old data are interesting to read, and students of muzzle-loading are drawn to old-time information like moths to a flame. That's fine, but be cautious! Old-time load information may not work in your firearm for many reasons. Powders have changed. A specific granulation of old could have been very different from a current powder. For example, FFg may have been more like our current Fg for a specific old-time powder. Powder strengths varied, too. So don't trust old-time loading information. It was made for powders we no longer have. Also remember that granulation of blackpowder as we know it was unheard of before the early 1800s.

By the third decade of the 1800s granulation was understood, but definitely not in vogue. Slowly, granulated blackpowder became standard, but when you read about a hunter putting a fistful of powder into his rifle, you don't know what that meant pressure-wise, because you don't know the "power" of his powder, nor its granulation size, and some old-time loading data could pertain to powders of a pre-granulation period. Furthermore, some powders of the middle 1800s were not straight FFg or FFFg, but about a 50/50 mix of the two. Most of these older powders did not have a glaze. Our blackpowder has smoother edges and is glazed. Glazing is supposed to slow the burn rate and render the powder safer and less hygroscopic. Some students of blackpowder say that glazing was useful for sodium nitrate powders, but not for the saltpeter used in our fuel, which is potassium nitrate. Simply put, read old loads for their historical interest but do not rely on them.

Accidents with mistaken propellants often stem from shooters using a powder that is black in color, but which is not blackpowder. There are numerous smokeless powders that are black *in color*, but only in color. They react violently in the breech of the muzzleloader.

A Note on Pyrodex

Pyrodex makes fine optimum blackpowder loads. It always has, right from the beginning, when it was announced in a 1976 letter from Bruce Hodgdon stating that a replica blackpowder would be offered by his company. As promised, Pyrodex was a reality, presented at the 1976 NRA Convention in Indianapolis. More details on Pyrodex appear in Chapter 15.

The Percussion Cap and the Optimum Load

Let's get the percussion cap out of the way right now, as it is the most frequently neglected detail of the blackpowder load chain. First, no special percussion cap is required for optimum loads with Pyrodex. The latest formulations of Pyrodex show very good ignition qualities. Second, yes, there can be slight differences in ignition with various percussion caps. A few muzzleloaders do better with the so-called "hot" cap, such as the fine RWS cap, while still fewer frontloaders need a fusil with modern primers for surefire ignition. So consider trying different percussion caps for that optimum load, but do not expect huge differences in performance from one cap to another.

The Optimum Load and Granulation

We already know that FFg offers a good velocity/pressure relationship for the big-bore muzzleloader. Pyrodex RS substitutes for FFg on a volume for volume basis, so in the big bore, these two powders are usually right for the optimum load because they provide good velocity with reasonable pressures. Meanwhile, FFFg and P are at home in revolvers and small-bore rifles, which was also pointed out earlier.

Assembling the Optimum Load

The following are just a few tips in building an optimum load for your charcoal burner. The steps are sequential. Forgive any repetition on points previously presented. The optimum load is important, and it's worth a careful look.

Step 1: Pick the Proper Projectile

Select the correct projectile based on bullet stability. A round ball or a conical is initially chosen on the basis of twist. Faster rate of twist is for stabilizing conicals, slower rate for the round ball. For example, a Knight MK-85 50-caliber modern muzzleloader of my acquaintance has a 1.28-inch rate of twist. It's a conical-shooter by design. Conversely, a 50-caliber Navy Arms Ithaca Hawken I've been shooting has a 1:66-inch rate of twist, and it's a ball shooter. You will hear marksman after marksman claim their fast-twist rifle groups round

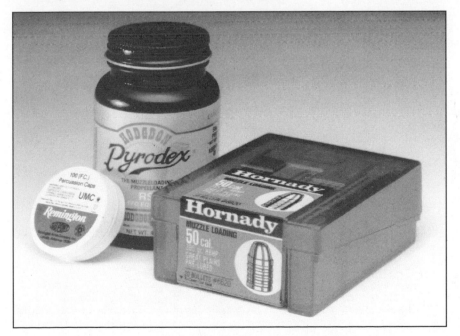

Use only the correct ingredients for that optimum load. Never use smokeless powder, but only a correct muzzle-loading propellant *and granulation,* like Pyrodex RS shown here, with a proper percussion cap and a bullet/powder charge that is within safe and accurate limits.

balls with astounding accuracy, or their slow-twist rifles have great luck with conicals in their guns. That's fine. I always say if it works for you, keep right on doing it.

As for exact bullet match-up for a given frontloader, trial and error remains a good way to find out. But keep records. It's too easy to "remember wrong" when trying to recall which bullet did what in a specific gun. And do keep in mind that RPS, revolutions per second, is the force that stabilizes projectiles. RPS is a product of *rate* of twist (irrespective of barrel length) and exit velocity (see Chapter 27). Therefore, a modestly loaded round ball may stabilize in a faster twist because the

exit velocity is low. Furthermore, as harped on often (and for good reason), muzzle-loading firearms are often laws unto themselves. So don't be too surprised if you find a gun that shoots round balls fairly well from what should be a conical rate of twist, and vice versa.

The right bullet means more than conical or round ball. Bullet diameter also counts. So try to get the right diameter to match your muzzleloader. Although it's true that a round ball closely fitted to the diameter of the bore generally gives best accuracy, an individual rifle may like a looser ball fit. The aforementioned 50-caliber Navy Arms Ithaca Hawken rifle does fine with a .490-inch round ball, for example. The

Preparing to Load the Blackpowder Cartridge

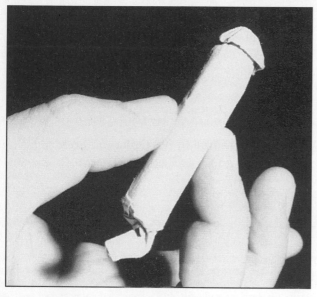

The blackpowder cartridge began life as a paper cartridge. This is an original Civil War paper cartridge containing a 58-caliber Minie.

The old "nutcracker" reloading tool made millions of reloads for shooters well into the 20th century, but of course today has been superseded by modern handloading equipment.

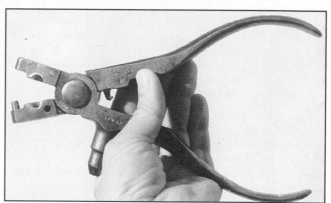

(Left) Before the handloading process begins, the correct shell holder is selected. This one is for the 45-70 Government

(Right) The fired cartridge case is inserted into the full-length resizing die. When the ram bottoms out and the shell is fully into the die, it will have been resized and deprimed.

.495-inch ball did not improve accuracy in that rifle. Bullet configuration is also important. Some rifles prefer one bullet shape over another. Again, it's a matter of shoot and find out. *A balanced load chain is the goal.* All components must be in harmony for best results, and that certainly includes the bullet. Tip: Choose a Minie with a fairly stout skirt if top loads are your interest. A thicker skirt resists flare better than a thinner one.

Step 2: Use the Gunmaker's Suggested Loads

Demand two things from the manufacturer of your frontloader: suggested loads and a maximum load. The optimum powder charge will probably rest between these two. It is up to the maker of the firearm, or its importer, to give the shooter parameters of performance and safety, especially the latter. Most muzzleloader companies do this for the customer. Do not exceed the maximum charge. Even though the maximum load may seem beneath the charge you would like to try, it is your duty to stick to it. Furthermore, the maximum loads I've looked at lately seem right up to snuff. For example, Thompson/Center's suggested round ball load for their 50-caliber Hawken is 110 grains of FFg blackpowder, which in their tests showed an MV of over 2100 fps.

Today, the power behind the great reloads we have lies in the modern press, like this fine RCBS unit.

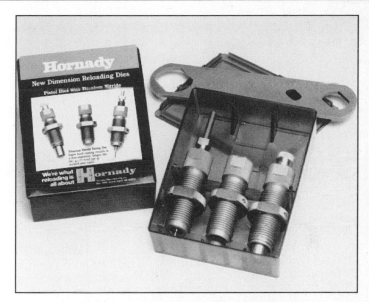

The fired cartridge case must be returned to dimensions that will allow its entry into the chamber of the firearm. The die does that. There is a die for full-length resizing of the case and depriming (expelling the old primer); one for expanding the case neck to the correct diameter and belling it, if desired; and one for seating the bullet to the correct depth and for crimping should the round require crimping of the bullet to help hold it in place.

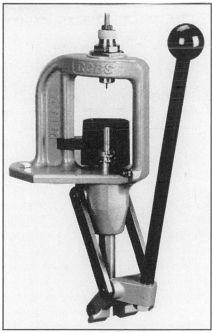

When the round is lowered from the full-length resizing die, it has been returned fairly close to factory specs, and it is deprimed.

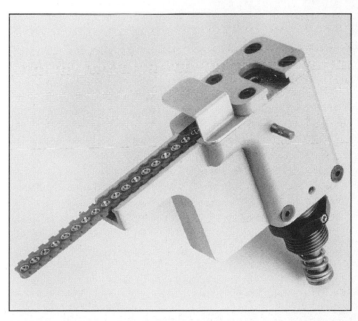

There are various ways to install a primer into the sized case. This RCBS APS Press-Mounted Priming Tool is only one.

That's T/C's maximum for that rifle. The company also lists lighter loads for the same rifle, giving the shooter the parameters just mentioned.

Step 3: Work for Accuracy

Begin working up an optimum load starting at the lower suggested powder charges as your base. For example, the T/C Hawken noted above has a maximum allowance of 110 grains FFg with a round ball, but the booklet that comes with that rifle also shows a low-end load of only 50 grains FFg. While T/C data gives the 110 volume FFg load over 2100 fps MV, the booklet reveals 1357 fps MV for the 50-grain charge. T/C proceeds in 10-grain increments from 50 to 110 grains volume, which makes sense. So do the same, shooting for accuracy, since you already have a handle on velocity from the booklet. Shoot only from a benchrest. Off-hand, sitting, or even prone shooting postures are not stable enough when testing a firearm for accuracy. Again, keep detailed records, and don't rely on memory, nor on only one shooting session. Go to the range several times.

Pretty soon you'll have a good handle on the average accuracy provided by each load. Maybe there won't be enough difference from low

Preparing (cont.)

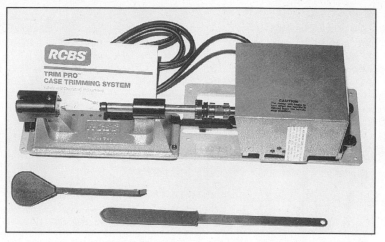

Cases must be trimmed to proper length for safety. An overly long cartridge case can "pinch" in the leade of the chamber, raising pressures. This trimmer is powered.

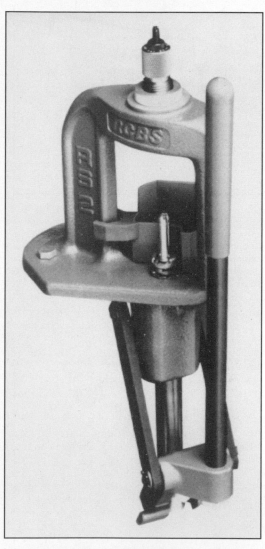

After the proper powder charge has been installed into the sized and reprimed case, a bullet is seated. The round is now ready to fire.

Cases are cleaned not merely for looks, but also to show problems easier, such as a cracked neck or body.

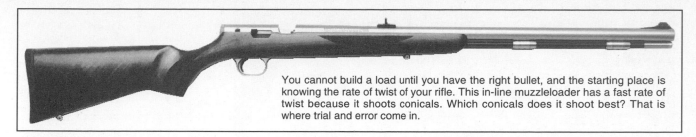
You cannot build a load until you have the right bullet, and the starting place is knowing the rate of twist of your rifle. This in-line muzzleloader has a fast rate of twist because it shoots conicals. Which conicals does it shoot best? That is where trial and error come in.

end to high to matter, which is fine. It just means your rifle is accurate with a wide range of charges, from plinkers and target loads to big game recipes. Seldom will only one charge work out. Normally, a few good loads will surface because most muzzleloaders enjoy a *range* of load acceptability. You have your accuracy load now, and if you're lucky, your muzzleloader will enjoy a broad range of load acceptability, with accuracy occurring at both the high and low bands. On the other hand, you may have to compromise between accuracy and power for big game hunting, sacrificing a little bit of accuracy for more ballistic punch. If so, fine.

Step 4: Finding the Accuracy/Power Compromise

Sometimes a shooter truly has no interest in target, plinking, or small game loads from his charcoal burner. So there is no point in his working with low-end loads at all. He is interested only in the optimum high band load for big game hunting, and he is willing to compromise between top grade accuracy and good hunting accuracy if he has to. In other words, if groups are around 1.5-inches at 100 yards with a light load and 2.5-inches center to center with a heavier hunting load, he chooses the latter for big game. The law of diminishing returns steps in here. There are two ways to work with this important phenomenon. One is to simply check the charts. For example, the T/C 50-caliber Hawken noted above gets 1357 fps MV with 50 grains of FFg blackpowder and over 2100 fps MV with 110 grains of FFg blackpowder.

Using an even 100 grains of FFg, velocity is about 80 fps lower than the 110-grain charge. I'm not suggesting for a moment that 80 fps is meaningless, because it is not. However, suppose our hunter works from a tree stand most of the time, and his shots average about 50 yards. Will 80 fps MV really matter on deer-sized game? Probably not. How about a 90-grain charge? T/C data shows a muzzle velocity about 185 fps below the 110-grain charge. Once again, the hunter decides. Can he get by with a bit less MV when the game is a deer and the range is only 50 yards? Possibly so. Now comes the clincher. If accuracy is markedly better at 90 grains vs. 110 grains, and we're only talking whitetails at 50 paces, that 90-grain charge makes a lot of sense. I said there were two ways to detect the law of diminishing returns. One is checking the charts, which includes information printed in loading manuals.

The other method is to use the chronograph, which is touched on below, so we'll wait for it, but I have to point out that I really like the chronograph. It has no imagination. There is no romance behind its glassy little face. It's all business and it never lies. Well, not unless there's a glint or battery failure or the tester is shooting across the chronograph wires or a bug gets in the way or the shock wave in front of the bullet causes the first screen to start counting prematurely, or the wind is howling or its too cold outside. But all of these things can be controlled. We'll investigate the chronograph and the law of diminishing returns below, but for now, on to Step 5.

Step 5: Juggle Components Safely

This means an intelligent mix and match of different conical and round ball diameters, even various patch thicknesses. It may also mean trying different percussion caps and even different nipples. Lubes can change things, too, although rarely is lube a major factor in accuracy. So mix and match everything in the load chain, including powder type, correct powder charge, ball size, patch, lube, and ignition source. Lots of trouble? Not really. It's mostly a lot of fun at

the range. After all, we got into this game to shoot. Furthermore, once you have established the pet load(s) for your frontloader, you have it—or them—for good.

The Law of Diminishing Returns

I already mentioned the booklet or pamphlet that comes with a muzzleloader. Some offer a lot of data, and it's easy to see where the breakoff point is—where more powder is added for darn little gain. Also useful is a blackpowder manual, such as the *Gun Digest Black Powder Loading Manual*. This book is packed with chronographed data for about 150 different guns.

A chronograph can prove the law of diminishing returns in a hurry. When I was working with the strongly-made Jonathan Browning Mountain Rifle in 54-caliber, the law of diminishing returns jumped right out of the chronograph at me. This fine rifle was allowed a full 150 grains of FFg blackpowder behind a patched round ball. With a 225-grain ball, velocity using a full 150 grains volume FFg was 1970 fps. But 140 grains of powder gave an MV of 1965 fps. If ever the law of diminishing returns—getting little to no gain for additional powder—showed up, here it was. The additional 10 grains of powder showed only 5 fps average gain in muzzle velocity, which, statistically speaking, is essentially meaningless. With 130 grains of powder and the same ball, MV was 1908 fps. The chronograph showed the law of diminishing returns clearly here.

The Chronograph and the Optimum Load

This is quite different from using the chronograph to show the law of diminishing returns. Here, the machine is used to render a whole batch of useful information. My Oehler Model 35P Proof Chronograph gives five pieces of information automatically, and on a printout for every shot string tested. It tells the lowest velocity in the string, the highest velocity, the average velocity for all shots fired in the string, plus the extreme spread between lowest and highest velocity, and it also gives a standard deviation figure. The latter is a sign of variance. The higher it is, the more variance in the load. The lower it is, the more confidence a shooter can have that every aspect of his load chain is balanced and sound. For example, a standard deviation of 10 or 20 fps is very good, while a standard deviation of 100+ fps is pretty bad.

Obviously, the shooter can compute his bullet energy from the average velocity shown by the chronograph, which is important. The chapter on ballistics has more information on energy. The chronograph can also help a shooter pick out a powder and/or granulation that his firearm does well with. Again, this is a simple matter of seeing what velocity and standard deviation results are gotten with different powders and granulations in a given firearm. Perhaps more than anything else, the chronograph takes all of the guesswork out of a load. It tells what's going on and what's not, and no matter how much we may like a particular load, the chronograph will relate its ballistic worth with certainty.

Load Efficiency

Load efficiency is not exactly the same as diminishing returns, although the two are related. Using the same Jonathan Browning Mountain Rifle referred to above, a velocity of 1276 fps was gotten with only 50 grains of FFg blackpowder. Adding a full 100 grains to that charge, velocity gain was only 694 fps. I say "only" because the powder charge was tripled, yet velocity certainly did not increase by anything like three times. As for efficiency, the little 50-grain powder

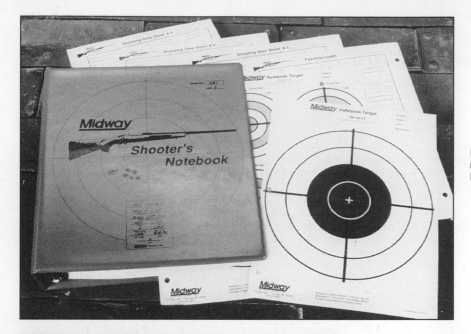

Record-keeping is very important to the optimum load process. This Midway Shooter's Notebook is ideal for keeping such records.

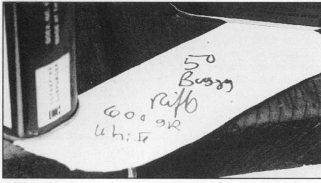

The law of diminishing returns does apply. This chronograph data tape is marked clearly. The rifle is a 50-caliber Storey; the bullet a 600-grain White. On the other side of the tape, the powder charge is marked and, of course, velocity appears. When more powder is added—regardless of the manufacturer's allowance—but little to no velocity is gained, the law of diminishing returns is at work.

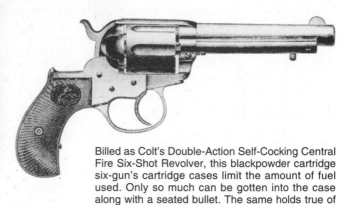

Billed as Colt's Double-Action Self-Cocking Central Fire Six-Shot Revolver, this blackpowder cartridge six-gun's cartridge cases limit the amount of fuel used. Only so much can be gotten into the case along with a seated bullet. The same holds true of the blackpowder rifle cartridge.

charge won hands down, didn't it? Of course, but I think too much can be made of this. While it is true that greater powder charges do not bring equally greater velocity, nonetheless I'm willing to pay the price in extra smoke, noise, and recoil for the added power when power is needed. The 50 FFg charge and 1276 fps provided 814 foot pounds of muzzle energy, while the big powder charge gave nearly a long ton of muzzle energy. So don't let load efficiency rule entirely. Your best big game hunting load will not be as efficient as a lesser load, but it will do more in the game field. Looking at this another way, the most efficient load will seldom be the optimum load.

The Magnum Load

This term has lost some of its charm over the years, but magnum still means "bigger than" and there are "bigger than" blackpowder loads for certain muzzle-loading rifles. The one that immediately comes to mind is the Gonic Magnum modern muzzleloader, which is allowed heavy powder charges with heavy bullets. Another powerhouse is the little Storey Takedown Buggy Rifle, which also uses heavy powder charges with big bullets. So the magnum rifle can use a magnum load, and its optimum rating may indeed be considerably higher than a rifle not rated for so much powder consumption with big bullets. In other words, certain blackpowder rifles are magnums in their own right, and their optimum loads are on the high end of the spectrum.

Handguns, Shotguns, and Breechloaders

Very little attention is paid handguns, shotguns, or breechloaders in terms of optimum loads. This does not mean that optimum loads are impossible in these blackpowder guns. For example, in the *Gun Digest Black Powder Loading Manual, Third Edition*, on page 310, an optimum load for the Ruger Old Army revolver is listed as 40 volume Pyrodex P. That is because 40 volume P with a .457-inch round ball in that revolver proved accurate with top power. Simple as that. In the same book, page 357, the Navy Arms Classic side-by-side 12-gauge shotgun is given an optimum load of 90 FFg with 1 1/4 ounces of shot, because that load earned the optimum rating. The breech-loading rifle is a bit different in that the blackpowder cartridge holds only so much powder, and the goal is normally 100 percent load density, meaning a case full of fuel. So it's a bit pointless to work toward optimum loads when in almost every instance that load will be a case full of blackpowder or Pyrodex.

Putting it All Together

Building the optimum load means putting it all together. Everything is interdependent. If one link in the load chain is weak, the load will be less than optimum. You can't expect a fine projectile to do its best with varying powder charges, for example, or the wrong granulation, any more than you can hope for best results with well-delivered powder charges of the correct granulation if the wrong missile is used. The goal is a balance, and when that balance is achieved, write down the formula because you have your optimum load. ●

154

chapter 22

Blackpowder Ballistics

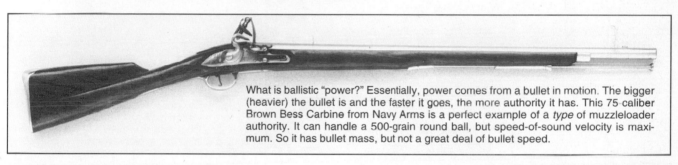

What is ballistic "power?" Essentially, power comes from a bullet in motion. The bigger (heavier) the bullet is and the faster it goes, the more authority it has. This 75-caliber Brown Bess Carbine from Navy Arms is a perfect example of a *type* of muzzleloader authority. It can handle a 500-grain round ball, but speed-of-sound velocity is maximum. So it has bullet mass, but not a great deal of bullet speed.

SPECIAL BLACKPOWDER-only hunts are a major impetus in the continuing popularity of the muzzleloader in modern times. Hunting with a charcoal burner is special, but one fact remains the same no matter the instrument of harvest: Every dedicated outdoorsman who goes forth to procure his own meat wants to do a clean job of it. For muzzleloader hunters, that means understanding the basics of blackpowder ballistics, and that's what this chapter intends to address. Frontloader ballistics are interesting in their own right, too, so knowing what the old-style guns can do goes well beyond the practical. Understanding blackpowder ballistics is a big part of enjoying the sport.

What is Ballistics?

Ballistics is the science of projectiles in motion. There are two general branches of this study: internal ballistics and external ballistics. This chapter is far more concerned with the latter. Internal ballistics is the study of the missile in the bore. When it exits the bore, external ballistics takes over. Ours is a very practical interest, not a scientific one, although no data presented here are guessed at, extrapolated, or wished for. All velocities were taken on an Oehler chronograph, which divulged with the push of a single button the highest velocity in the string, the lowest velocity, average velocity, extreme spread, and standard deviation from the mean velocity. Energies were worked up using the Newtonian formula described below. The energy referred to here is kinetic. We don't do much with potential energy in basic ballistics, which is the inherent "power" of a bullet at rest. Our interest is in the bullet from muzzle to destination.

How Power is Calculated

The only acceptable measurement of power today, as far as modern ballisticans are concerned, is the Newtonian formula. It is expressed in foot pounds of energy, as noted above as kinetic energy, sometimes referred to simply as KE. Every ammo company in our land and across the sea refers to this type of energy. The formula does not necessarily favor muzzleloaders, because the mathematical computation squares velocity, and frontloaders are not known for their high-speed projectiles. At the same time, Newton's energy figure is not such a bad yardstick of blackpowder ballistic authority. For example, the opening of Chapter 32 on big bores speaks of a special William Moore muzzleloading rifle. This monster gun shoots a bullet that weighs a half-pound (3500 grains). At 1500 fps MV, the energy rating is over 17,000 foot pounds. That puts a modern "elephant rifle" like the 458 Winchester at around 5000 foot pounds under the table. If Newton's formula were entirely unkind to blackpowder guns, the William Moore rifle might not show so mightily.

Comparing with Modern Cartridge Power

Comparing the familiar with the less familiar always promotes understanding. For example, most of us are quite familiar with the 30-06 Springfield cartridge—what it can do, and has done, in the big game field and how its energy or power rates. Newton's formula gives the 30-06 more power than my favorite 54-caliber ball-shooting rifle. Is this fair? All in all, I think so, although there is a problem. The problem is that no formula provides a perfect one-to-one correlation between a number and actual effect on game. There are simply too many variables at work. My 54 rifle drops deer-sized game in its tracks with chest strikes, provided I am inside of 100 yards. I prefer getting closer to elk-sized animals. The 30-06s I've used likewise drop deer-sized game instantly with chest hits, but much farther than 100 yards, because streamlined bullets hang onto their velocity/energy better than round balls.

As for "power numbers," my 54 ball-shooting rifle gets 1970 fps MV with a 230-grain round bullet for a KE of 1983 foot pounds. Although my own 30-06 handloads drive a 180-grain bullet at 2800 fps

No guessing! It takes a chronograph to reveal the truth about bullet velocity. This is a look back toward the shooting bench through the three Oehler chronograph screens.

Along with a screen readout, the Oehler Model 35P Proof chronograph also presents its figures on a tape. Five pieces of information are given. Here, the highest velocity was 1282 fps; the lowest was 1236 fps; extreme spread was 46 fps; average or mean velocity was 1259 fps; and the standard deviation was 32 fps.

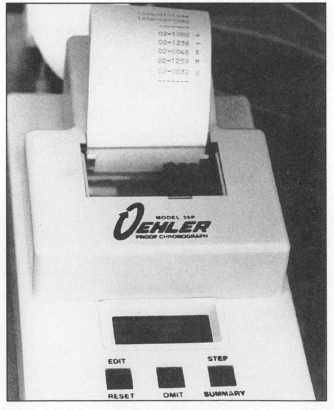

MV, the usual factory load has been standardized for years at 2700 fps MV, so we'll choose it for our energy figure. The '06 shooting a 180-grain bullet at 2700 fps MV gets a KE of 2914 foot pounds, or relatively close to a half-ton more than the 54 muzzleloader in this comparison. That's roughly a third more authority for the '06. Is the 30-06 Springfield cartridge, as loaded with a 180-grain bullet at 2700 fps MV, truly about 33 percent more powerful than the 54-caliber ball-shooter with a 230-grain bullet at 1970 fps? Good question. Hard to answer.

But those are the figures anyway, and perhaps they hold up better than we may think. Nobody has ever come up with a perfect test to prove the true comparative effectiveness between two loads like these. How could you? Gelatin blocks are not wild animals, and wild animals seldom provide the same exact target from one time to the next. Angles differ, as do the weights of the animals, the distances from which they are taken, even the disposition of each beast, from rest to running. So we simply live with the figures. They aren't all bad, after all, even though Newton's formula does not applaud mass as loudly as the big-bore fan would like. Bullet heft is not entirely left out either. By Newton's method a 3000-pound car at only 30 miles per hour, or a mere 44 fps, has an energy of 91,000 foot pounds—so heft of the missile is not ignored by the KE formula.

Big-Bore Fans and Newton's KE

Big-bullet boys have never been thrilled with the scientist's way of measuring ballistic punch. They don't like the fact that velocity is squared. They feel that bullet weight is just as important as bullet speed. Momentum, they say, is just as conducive to bullet penetration and overall harvesting power as kinetic energy. Therefore, these fellows are apt to use another formula to compute ballistic strength. There are several. One is "pounds-feet," a mathematical construct favored by gunwriter Elmer Keith, who was known as a lover of big bullets. Pounds-feet eliminates the squaring of velocity. It hinges on momentum. Momentum is actually mass times velocity, rather than weight times velocity.

Mass is not weight. Mass is a measure of inertia for a given body, to wit: a quotient of the weight of the body divided by the acceleration due to gravity. On the moon, with a gravity only $1/6$ that of earth, a pound weight of anything would weigh $1/6$-pound, but the mass of the object would be the same on the moon as on the earth. Nevertheless, for our purpose we'll use weight times velocity for momentum, not mass times velocity, because weight computes much easier and it works well enough for our purposes. Let's compare a 243 Winchester with a 54-caliber ball-shooter using our weight times velocity momentum formula. The 243, with a 100-grain bullet at 3100 fps MV, earns a momentum figure of 310,000. That's unwieldy, so let's divide the figure by 10,000 to gain a more workable number, which is 31. That's how Keith liked to do it.

Meanwhile, the 54, shooting a 230-grain lead round ball at 1970 fps MV turns up a momentum figure of 45. So we have 31 vs. 45. The muzzleloader wins. But what a can of worms! Remember that no ballistic authority buys the pounds-feet story; not one ammo factory in the world uses it. On the other hand, having hunted big game for many years with many different firearms, modern and otherwise, I'll say I'd rather smack a bull elk in the rib cage with a 230-grain 54-caliber round ball at close range than a 100-grain 24-caliber jacketed bullet at close range. But that's just my personal assessment.

Things get even funnier when you compare fast little bullets with bigger slower bullets. Match a 50-grain bullet from the 220 Swift, for example, with the 54 round ball mentioned above. Energy figures, using KE, are very similar here. But as much as I admire the 220 Swift (one game warden said it was the deadliest deer round he ever used for dispatching animals struck by cars on the highway), I prefer a 230-grain 54-caliber ball from a long rifle over a 50-grain bullet for big game hunting. I cannot agree with the KE figures when these two cartridges are compared. The Swift, with a 50-grain bullet at 4000 fps MV, carries a muzzle energy of 1777 foot pounds, while the 54, with a 230-grain bullet at 1970 fps MV shows a muzzle energy of 1983 footpounds, only a couple hundred foot pounds advantage, which at 100 yards from the muzzle is reversed—the 220 Swift is ahead. It's a sticky mess. I believe in high velocity. But bullet size and weight certainly mean something, too.

Take KE and Live with It

The bottom line is this: Since the only accepted measurement of ballistic power is the Newtonian KE formula, we're married to it, for better or worse. There are many variables in the clean harvesting of big game, so I repeat that no formula offers a one to one correlation between a mathematical number and field performance. Slide Rule Albert, whose closest encounter with a dangerous animal was getting near the lion cage in the zoo, can prove, on paper, that his 223 Zippedy-Do-Da wildcat cartridge has more energy than Black Powder Charley's 50-caliber frontstuffer. But the proof of this pudding is on

How did the 19th-century elephant hunter seeking his fortune in ivory manage to drop the huge pachyderms? Mainly with big bore double guns. Of course, these guns used blackpowder because that's all there was until the late 1800s.

It's the hole in the end of that barrel that really counts when it comes to muzzleloaders. Even blackpowder conicals do best when they're of at least reasonably large diameter. This big bore shoots a matching big bullet.

157

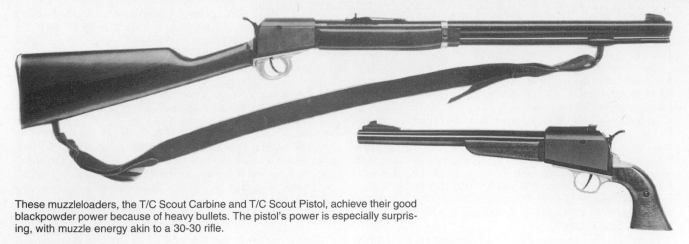

These muzzleloaders, the T/C Scout Carbine and T/C Scout Pistol, achieve their good blackpowder power because of heavy bullets. The pistol's power is especially surprising, with muzzle energy akin to a 30-30 rifle.

the meat pole. A big-bore blackpowder rifle, correctly loaded, has plenty of snort, and it does not have to be a 4-bore firing a chunk of lead the size of a bowling ball to cleanly harvest big game. Normal blackpowder guns, especially in the 50- to 54-caliber range, can be downright effective, shooting decent-sized round balls for deer-sized game, or heavy conicals.

How Effective are Blackpowder Guns?

The Humane Society of the United States engaged in a law suit with the Department of the Interior for allowing blackpowder hunting on public lands (1974). The HSUS claimed that muzzleloaders were inferior, and that they would wound rather than cleanly harvest game. V.K. Goodwin served as friend to the court in this case on behalf of the National Muzzle Loading Rifle Association, and showed that old-style firearms were entirely capable of humanely dispatching big game. The court was entirely satisfied in the matter, and hunting with the frontstuffer continued on public land, as it should have. The case is germain to our look at blackpowder power because it reveals a judgment of law pertaining to muzzleloaders as big game hunting tools. In summary, the law allows them, although specific regulations are often unfounded in ballistic fact.

It's in the Bullet

As the bullet goes, so goes the effectiveness of the muzzleloader, and for that matter, the blackpowder cartridge rifle or revolver as well. Blackpowder guns simply do not register high velocities by modern standards, so bullet size and weight must make up for lack of speed. That's why large calibers (mainly 50 and 54) are so popular for big

game. The round ball gains a lot of mass as caliber increases. We see that by a simple comparison. A 32-caliber .310-inch pure lead ball only weighs 45 grains. If you double the diameter, you won't get 90 grains, however; you'll get a lot more weight. A .620-inch round ball in pure lead goes 359 grains. The conical gains more mass through elongation. For example, you can have a 45-caliber White Superslug at 360 grains, 460 grains, or 520 grains. The Superslug gains weight through added length. Furthermore, don't leave caliber out of the picture. A bullet that starts out $1/2$-inch or larger in diameter is already formidable, even at modest striking velocities.

Ball Placement

I have no idea how many dollars have been paid to outdoor writers for the line, "It's not what you hit 'em with, it's where you hit 'em that counts." That quip makes a great deal of sense, but it also suffers from too much generalization. A lead round ball through neck vertebra number four of a bull elk is worth two shiny jacketed bullets from a 375 H&H Magnum in the hoof. On the other hand, what you him 'em with also counts a great deal. Ask the man who stung a Cape buffalo with a rifle intended for deer-sized game. However, you can't leave bullet placement out of the blackpowder ballistic story, because those of us who have taken up the muzzleloader know that in spite of good ballistics, the average blackpowder rifle is no 7mm Magnum. Blackpowder hunting is a challenge, and the challenge lies in getting close and placing one projectile exactly on target.

Muzzle Energy vs. Downrange Energy

While we dwell mostly on muzzle energies, the fact is, few game

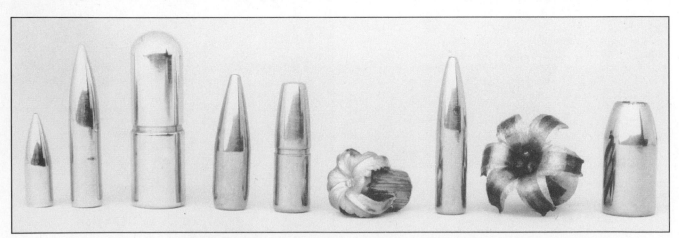

We look for different performance from different bullets. This lineup of excellent Barnes copper bullets includes (far right) the company's Expander-MZ muzzleloader copper bullet. It has a very large cavity in the nose in order to provide expansion even at low striking velocity.

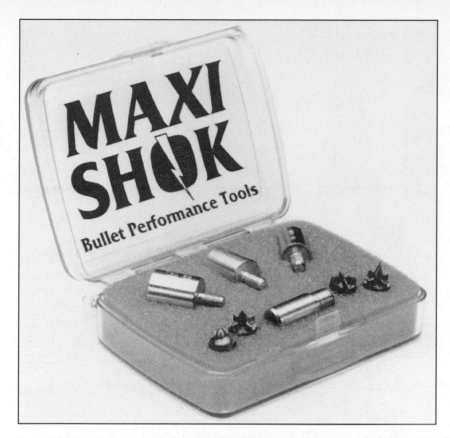

Recognizing the fact that some lead bullets do not expand well on soft-skinned big game, Thompson/Center developed its Maxi Shok tool. It has four different cutting tips: Dot, Star, Cross and Arrow. A tip is attached to the ramrod and it produces an indentation on the nose of the projectile. This indentation does the same type of work associated with a hollowpoint, aiding the bullet in opening up.

(Below) Bullet mushrooming, or upset, transfers energy from projectile to target. That is why Barnes designed its Expander-MZ muzzle-loader bullet to open up dramatically in the target, even at a low 1000 fps striking velocity.

animals are harvested at the muzzle. They're taken downrange. For careful blackpowder hunters, that means about 100 yards or closer, with about 125 yards considered a fairly long shot, since the bullet at this distance drops about six inches with a normal load from a muzzle-loading rifle. So what happens to our blackpowder bullet when it lopes from the muzzle to a hundred yards? If it's a round ball, it loses about half of its initial velocity, and if it's a conical it loses much less speed and power, the exact loss depending upon the ballistic coefficient of the projectile. Some "real life" figures are in order.

A 50-caliber round ball, .490-inch diameter, weighing 177 grains and beginning its journey at about 2000 fps MV, ends up with about 1100 fps remaining velocity at 100 yards. Muzzle energy is around 1500 foot pounds, while 100-yard energy is down to about 500 foot pounds. The figures are rounded off for easy comparisons. The 50-caliber round ball loses about half its original velocity over 100 yards, and due to the importance of velocity in the Newtonian formula, it also parts with about two-thirds of its original horsepower. A 54-caliber round ball, .530-inch diameter, weighing 225 grains and starting at about 2000 fps MV, ends up at 100 yards with a velocity of 1150 fps. Muzzle energy for this 54 round ball is about 1900 foot pounds.

Energy at 100 yards from the muzzle is about 675 foot pounds for the 54 ball. Once again, figures are rounded off for easy comparisons. This projectile gives up about half of its starting velocity, too, with roughly the same energy loss. Very small round balls are devastated power-wise at 100 yards from the muzzle. A 32-caliber round ball that begins with 2000 fps MV and about 450 foot pounds of energy ends up below 1000 fps at 100 yards with fewer than 100 remaining foot pounds of energy.

The elongated bullet fares better, as you would guess. It has a superior ballistic shape and, therefore, it keeps more of what it starts out with. A 600-grain 58-caliber Minie beginning at about 1350 fps is still going over 1000 fps at 100 yards. Its muzzle energy of about 2450 foot pounds has dropped to about 1650 foot pounds, because poor aerodynamics has caused considerable velocity loss. But this conical still does a lot better than a spherical bullet of the same caliber. The facts are clear, but I hope you're not thinking, "Well I always knew that the round ball was no good." That would be a wrong conclusion. All the

same, there's no doubt that conicals hold their velocity/energy better than round balls, with elongated missiles showing fairly well when they have good sectional density.

For example, a White 50-caliber 460-grain Superslug starting at 1460 fps MV is still running at 1200 fps at 100 yards. Its muzzle energy of 2190 foot-pounds is down to 1478, but that "ain't hay." After all, the good old 30-30 Winchester, which has accounted for tons of game meat since it came along in 1895, has only 1827 foot pounds of muzzle energy from a 170-grain flat-point bullet at 2200 fps MV. At 100 yards from the muzzle, this 30-30 load is worth about 1218 foot pounds of remaining energy, or less than the 50-caliber Superslug cited above.

A Glance at Interior Ballistics

The following points about different powders and granulations fall within the domain of interior ballistics—what goes on inside the bore, rather than outside. Yes, muzzle velocities are noted. However, the ballistics we are truly interested in happen between the ignition of the main charge in the breech, and the exit of the bullet.

(Above) Each type of blackpowder projectile offers different ballistic performance based on the bullet's construction, shape, size, and so forth. Shown with a 150-grain 30-caliber jacketed bullet (center) for comparison, these 58-caliber missiles are all different. The round ball will flatten out on a chest strike, while the three Minies may not. Conversely, the Minies can be counted on for very deep penetration.

The old-fashioned round ball remains somewhat a mystery in how effective it can be within its range limitations. Here we have a group of round balls with a 50-caliber Maxi on the left for contrast. The big ball on the right is without doubt capable of more energy delivery than the little ball on the left. And yet, appropriateness is always the key to round-ball choice. The small ball is certainly the choice for small game.

Blackpowder and Ballistics

Blackpowder varies from brand to brand, era of manufacture, components, the quality of those components, origin of charcoal, type of saltpeter, manufacturing methods, not to mention different granulation sizes. These, and other attributes, make a significant difference in delivered energy, hence ballistics. When confined in a firearm bore, blackpowder burns comparatively slowly, although it ignites very rapidly. A time vs. pressure curve read by an oscilloscope gives a clear picture of this pattern, with an extremely sharp upward curve at the beginning of the combustion cycle, then a downturn in the curve, followed by a long, even line to a conclusion. Blackpowder is surface burning. When ignited, the kernel tends to be consumed from the outside in. Therefore, granulation size is very important to burning characteristics. The wise shooter takes note of this factor, choosing his granulation size commensurate with the firearm and the type of load sought, but we've already laid our moccasins on this trail. Now, let's look at different blackpowder in terms of ballistics.

Is More Velocity Always Better?

I like good velocity per charge, and I admit it, but that does not mean that the only factor in choosing a blackpowder by brand or granulation size is bullet speed. There have been some very good powders that burned cleanly with mild pressure increases for a reasonable gain in velocity, but actually less velocity per charge than with some other powders. Currently, only two blackpowder brands are readily found on the shelves, GOEX and Elephant brand. These powders do not burn alike, nor do they render the same velocity per charge. They are—it should be no surprise to the reader by now—quite different in velocity rendered per unit of pressure. But since both produce the desired result, let's move on to ballistics per granulation, since this factor has a more practical application.

No offense to Elephant brand powder, but the concentration is on GOEX here because it is more widely distributed at this time, and therefore more familiar to most shooters. So let's look at a comparison between GOEX FFg and FFFg granulations in velocities rendered.

The round ball does flatten out, but it can also penetrate surprisingly well. This ball was collected from a deer.

While the modern jacketed bullet is a marvel of fine performance, there are instances of the core and jacket going their separate ways, as this illustrates. The all-lead blackpowder bullet fragments very little.

We're dealing with a different kind of energy delivery with blackpowder bullets. The big 54-caliber 460-grain Buffalo lead bullet, shown base-out on the left and upright in the center, can penetrate remarkable well by virtue of its mass. The 180-grain 30-caliber modern bullet on the right depends more on high velocity for its performance.

Recall that FFg comes most-recommended for heavy hunting charges in muzzleloaders 45-caliber and larger. I continue to preach this maxim, but we're talking about ballistics as a science, so how about some proof in terms of tested results? Not many gunmakers recommend FFFg for hunting charges, but a few do. One company is Gonic, for their Magnum model. With a 530-grain bullet and 120 volume FFg, the gun gives about 1455 fps MV. In the same rifle, firing the same bullet but with 120 volume FFFg, velocity is 1498 fps.

That's not a lot of difference, in this powerful rifle with top-end loads. Sad to say, this situation did not exist with all firearms tested. It would be handy if it did. But when testing a Wilderness Mountaineer 50-caliber rifle with round ball, results were quite different. This one is allowed 80 volume FFFg with a round ball. Firing a patched 177-grain ball of .490-inch diameter, 80 grains volume of GOEX FFFg got higher velocity than 110 volume GOEX FFg powder. The lighter FFFg charge registered 1958 fps MV, while the 110 volume FFg charge showed 1905 fps MV.

The Storey Buggy Rifle covered earlier is allowed the use of FFFg blackpowder, but with a big 600-grain White Superslug, FFFg did not surpass FFg by enough to make me choose the finer granulation. In both cases, 120 volume was tested. GOEX FFg produced 1305 fps MV with the 600-grain bullet, while the same volume FFFg got 1410 fps MV. That's 105 fps gain, which is significant, but probably not enough in terms of a balance between velocity and pressure. Conclusion? Tests seem to bear out our original advice that FFg does a great job ballistically in the big bore. But we must also admit that our round-ball Mountaineer rifle really churned and burned with FFFg. Nonetheless, I'm going to stay with my original recommendation of FFg as ballistically fine in big-bore frontloaders.

Blackpowder Ballistics and Bore Sizes

Ballistically, small bores produce considerable muzzle velocity with smaller powder charges because the smaller volume of the bore promotes higher pressures than big bores achieve. A good example of this ballistic fact is a little 36-caliber squirrel rifle burning 70 grains of powder (this charge was for test purposes only and is not a recommended load). Pressure was 15,000 CUP (Copper Units of Pressure, Lyman data). Going toward the other end of the spectrum, a 58-caliber rifle with a test charge of 170 grains of FFg (not FFFg) developed a pressure of about 10,000 LUP (Lead Units of Pressure, Lyman data). Admittedly, we're cheating here, because we're not exactly comparing oranges to oranges. Note that we are also looking at FFg vs. FFFg. No matter, the difference in pressure per bore dimension and consequent volume or space for the powder gases to expand in is the important factor here. Further examples bear this out, but consider that in the above example the 36 was pushing a little 65-grain projectile, and the 58 was driving a big 280-grain missile. By the time we get to the shotgun, bore dimension and powder granulation play an even larger role. The bore is now quite large. Therefore, in spite of what amounts to a heavy projectile, perhaps $1^1/_2$ ounces of shot in a 12-gauge, or over 600 grains weight, pressures are acceptable with correct charges of large-granulation powder, such as FFg or Fg.

Pyrodex Ballistics

Pyrodex yields more energy per weight of charge than blackpowder, but when used volume for volume with blackpowder, as pointed out in Chapter 15, ballistic results are similar. Firing a Thompson/Center 50-caliber rifle with 26-inch barrel, 100 volume Pyrodex RS (70.5 grains weight) was compared with 100 volume (100.0 grains weight) of GOEX FFg blackpowder, both shooting a 385-grain Buffalo bullet. In one test, RS with the 385-grain bullet got 1549 fps MV, while FFg delivered 1439 fps. In another test RS got 1492 fps, while FFg churned up 1451 fps. Ballistically speaking, it's easy to see that RS and FFg gave very similar muzzle velocities volume for volume, which is exactly what we expect.

Pistol Bullet/Sabot Ballistics

Very little discussion is necessary on this topic, because the same

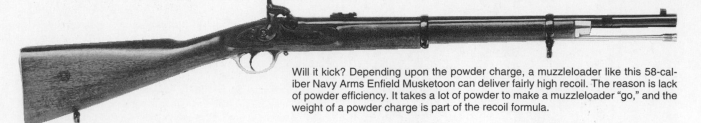

Will it kick? Depending upon the powder charge, a muzzleloader like this 58-caliber Navy Arms Enfield Musketoon can deliver fairly high recoil. The reason is lack of powder efficiency. It takes a lot of powder to make a muzzleloader "go," and the weight of a powder charge is part of the recoil formula.

The Buffalo conical lead bullet, shown on the far left without lube and on the far right with lube, expanded greatly in test media, as indicated by a recovered missile in the center. Expansion like this means energy transfer from bullet to target.

Ballistics does not go out of the window with the shotgun. Each pellet delivers a specific amount of energy to the target. However, since the pellet is very light compared with a bullet, individual pellet energy is low. That is why pattern is so important with the shotgun. Multiple pellet hits make the difference in effectiveness.

rules apply: pistol bullets and sabots are studied ballistically on the basis of Newton's formula, and that's that. For the record, however, here is a typical load from a modern muzzleloader. The rifle is a modern muzzle-loading Knight MK-85, caliber 54, shooting a 45-caliber 250-grain jacketed Hornady hollowpoint XTP bullet with appropriate plastic sabot. Velocity with 120 GOEX FFg blackpowder was 1610 fps at the muzzle for 1439 foot pounds. Using 120 volume Pyrodex RS (82.5 grains by weight) muzzle velocity was 1682 fps, with an energy of 1571 foot pounds.

Recoil

Recoil is considered a function of ballistics here, because the "power" we get from our frontloaders comes with a price, and that price is "kick." As mentioned before, part of the recoil formula is the weight of the powder charge, and since blackpowder is an inefficient propellant in comparison with smokeless, it takes a lot of it to get the job done in a big-bore gun. Not so in the small bore for the very reasons of pressure per bore volume asserted above. A 32-caliber squirrel rifle produces 22 Long Rifle velocity with only 10 grains weight of FFFg blackpowder from a long barrel. But the blackpowder big-bore rifle, as indicated in the comparison below between the 30-06, 54 patched ball rifle, and 54-caliber conical-shooter, can deliver a pretty good kick. Here's how it looks:

Rifle A—30-06 firing 180-grain bullet in front of 55 grains weight of powder. Rifle weight, **9** pounds.

Rifle B—54-caliber firing 230-grain ball in front of 120 grains weight of powder. Rifle weight, **9** pounds.

Rifle C—54-caliber firing 460-grain conical in front of 150 grains of powder. Rifle weight, **10** pounds.

Free Recoil of **Rifle A** = 20 foot pounds

Free Recoil of **Rifle B** = 36 foot pounds
Free Recoil of **Rifle C** = 70 foot pounds

To compute free recoil energy for your rifle use the following formula.

$$RE = \frac{1}{2GW} \left[\frac{bwbv + cwC}{7000} \right]^2$$

RE = recoil energy
G = gravitational constant of 32.2 ft/sec/sec.
W = weight of the gun in pounds
bw = bullet weight in grains
bv = bullet velocity in fps.
cw = weight of the powder charge in grains
C = the constant of 4700 fps, also known sometimes as the "velocity of the charge."

Attention: There are various ways of computing recoil, and the above is only one. Other formulas may render different results. Muzzle velocity for the 30-06 was entered as 2700 fps; muzzle velocity for the 54 ball-shooter was called 1975 fps; and the initial takeoff for the 54 conical-shooter was gauged at 1700 fps, which this special rifle does obtain according to the chronograph. By the way, a 460 Weatherby Magnum's recoil was computed for further comparison. Firing a 500-grain bullet at 2700 fps MV from a 10-pound rifle, recoil energy computed at 116 foot pounds. The big 460 burns over 120 grains of powder in the particular load tested. So recoil can be significant in the muzzleloader, but not so troublesome because heavy loads are not used for target work in the big bores, and the small bores don't generate enough recoil to dislodge a gnat from the barrel.

Ballistics will come up again many times in this book, because performance of our blackpowder firearms is too important to ignore. Whenever we mention muzzleloader effectiveness, we're talking about ballistics. Therefore, several of the following chapters deal either directly or at the oblique with the subject of blackpowder ballistics. For example, what does the rifled musket accomplish energy-wise? (see Chapter 33) How about the blackpowder breechloader? (Chapter 34.) The sidearm? (Chapter 36.) What ballistic impetus can we expect when hunting big game with blackpowder guns (Chapter 42), or for that matter, small game? (Chapter 43.) Or birds? (Chapter 44.) ●

chapter 23

Blackpowder Accuracy

SHOOTING A MUZZLELOADER without reasonable accuracy is just making smoke. Smoke might be enough for the first few outings, but soon enough the rose fades. So what is blackpowder accuracy? Bucket-sized groups are not to be tolerated. Frontloaders are better than that. Certain scoped muzzleloaders produce accuracy in the 1-inch center-to-center category at 100 yards from the bench. I've gotten 2-inch groups with iron sighted frontloaders, but only under prime conditions of no wind, good light, and a solid bench. This chapter is devoted to general accuracy factors pertaining to frontloaders. Specific notes on accuracy for certain firearm groups are included in those chapters. For example, see Chapter 31 for small bore accuracy, Chapter 32 for big bore accuracy, 33 for rifled musket accuracy, 34 for breechloader accuracy, and so forth.

High Quality Loads

Accuracy cannot thrive without quality control, which means getting rid of as many variables as possible in the load chain. These variables destroy the foundation of good bullet clustering. The shooter eliminates them by building a good load chain based on consistency, as pointed out in Chapters 20 and 21. Blackpowder shooters are a romantic lot, which is good; however, this nostalgic link with the past may cause problems—the colorful, but often erroneous, rules of thumb are snatched up as words of wisdom, while some of the more solid tenets are shunned. For example, a downwind shooter may balk at the bulk-loading method, thinking it crude, while clinging to quips such as "load powder until the rifle goes crack! instead of boom! and you'll have the correct charge." He may mistrust the accuracy potential of a round hunk of lead wrapped in a little cloth patch, or scoff at the idea of varied ramrod pressure harming accuracy. There's also an old wives' tale that says, "build a good load by using one grain of powder per each caliber of the ball." This chapter deals with tests instead of myths.

Volumetric Loads and Accuracy

Throwing charges by bulk does not destroy blackpowder accuracy. Having read that accurate blackpowder loads must be scale-weighed, I set out to match the two methods against each other. In the end, the contest was a draw. Both types gave essentially the same results.

In the tests, a battery of rifles was used. Every variable was kept as close to constant as possible, including the use of the same bullet, powder, lube, patches, and caps. The greatest variable at work here was human error at the benchrest. However, instead of a few samples, many targets were fired, which helped control that variable. Further-

Certain combinations promote accuracy. This 300-grain CVA Deerslayer bullet with 90 grains FFg blackpowder produced this surprising three-shot cluster at 50 yards during a test run of the then-new projectile.

more, results showed that the human element was the same for each powder-measuring technique: Group size of each rifle remained the same whether the charge was carefully weighed to a tenth of one grain on a powder/bullet scale, or bulk-loaded with an adjustable powder measure.

Furthermore, standard deviation from the mean velocity did not change with scale-weighed or bulk measured loads, indicating equal reliability for both. Combining target data with standard deviation data, the conclusion that volumetrically derived loads were as accurate as scale weighed loads was inevitable. Below are the velocity and stan-

What are we talking about when we say "center to center" groups? It means the distance between the center of the two farthest holes in the target, and it can be measured as shown here.

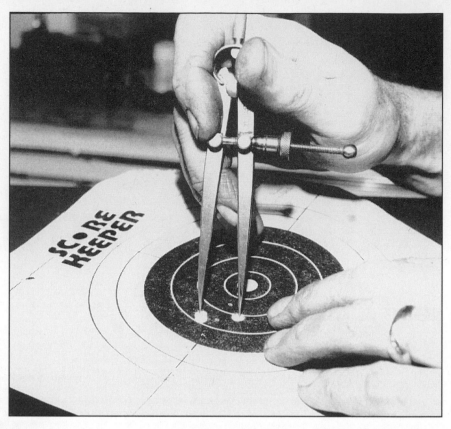

(Below) Of course the "average" muzzleloader is not going to make target-type groups on the range, but understanding the *potential* accuracy of blackpowder guns is essential, because once a shooter knows what these guns are inherently capable of doing, he is no longer satisfied with bucket-sized groups.

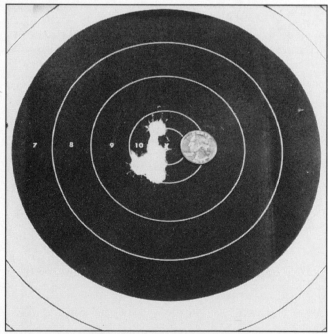

dard deviation factors for one sample load in one sample rifle as an example of this statement.

Rifle: Ozark Mountain Arms Muskrat, 40-caliber
Load: 30 grains by weight and by bulk measure
Powder: GOEX brand, FFFg granulation (fresh)

Volumetric vs. Weighed Loads

Volumetrically Derived Load	Scale-Weighed Load
Velocity: 1584 fps	Velocity: 1576 fps
Sd: 5 fps	Sd: 7 fps

Remember that standard deviation is a number that describes uniformity, and uniformity relates to accuracy. The lower the number the better. If a load has a standard deviation of 10, for example, that means it varies 10 fps from the average velocity. If the number is 100, that means the load has a variation of 100 fps from the average velocity.

The difference between these two loads is statistically insignificant, being only 8 fps at the muzzle for average velocity, and only 2 fps variation in standard deviation from the mean velocity. Test machinery precision alone can account for such minuscule variations. Paradoxically, it is the inefficiency of blackpowder that promotes bulk-loaded accuracy. This factor was discussed in Chapter 14. Lack of efficiency also helps to explain extremely low standard deviation potential, because a slight variation in powder charge cannot bring about a great alteration in velocity.

Blackpowder Measures and Accuracy

A charger was often included with the old-time hand-crafted muzzleloader. A gunsmith probably began with a tube longer than he wanted, cutting it down to reduce capacity until a proper charge was held. William Knight, who makes replica muzzleloader accouterments, sent samples of such chargers. I tried these simple tubes and found them reliable. Modern adjustable powder measures, such as Uncle Mike's, Tresco, Thompson/Center, CVA, and others, also produce accurate charges.

In use, the measures were slightly over-filled; then the barrel of the measure was tapped to settle the charge. As part of the consistency routine, I used a specific number of taps, such as ten, each time. Measures with an integral swinging funnel made it easy to level the powder charge by simply swinging the funnel into place in line with the barrel of the measure. With simple chargers, powder was leveled with a flat (ice cream) stick by swiping excess granules of powder away.

Below, data are presented using the Uncle Mike's adjustable powder measure. The set of three trials is indicative of an overall average and is typical, not special. Although set at 100, the different powders produced different weights by volume. That's normal because powders vary not only in kernal size per granulation, but also in density. Ignore the exact weights per volumetric charge in these samples, but pay attention to their consistency. That's where accuracy is born, in *consistency*.

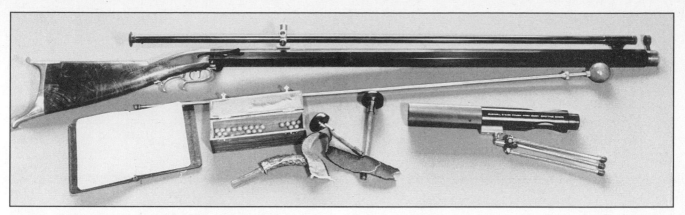

Uncle Mike's Adjustable Powder Measure Demonstration
(powder measure set on 100 volume)

	Vol. Setting (grs.)	Actual Wt. (grs.)
GOEX (GOI) FFFg Granulation		
	100	99.0
	100	98.0
	100	100.0
Average = 99.0 grains weight *		
GOEX (GOI) FFg Granulation		
	100	97.0
	100	97.5
	100	96.8
Average = 97.1 grains weight *		
Pyrodex, RS Granulation		
	100	87.0
	100	88.0
	100	87.5
Average = 87.5 grains weight **		

*Later lots of GOEX blackpowder proved slightly more dense, with FFg producing 99.0 grains average weight with the Uncle Mike's measure set at 100 volume. FFFg produced 102.0 grains weight with the Uncle Mike's measure set at 100 volume. These figures will vary with different powder measures, and different lots of powder.

**Current formulations of Pyrodex RS yield 70.5 to 71.5 weight per 100 volume.

More accurate than your average muzzleloader? Of course. This sidehammer target rifle was built with accuracy in mind from the outset. At the same time, it is not the ideal hunting rifle, due to very long length and excessive weight.

The slight load variations above could not be detected by a sensitive chronograph. Furthermore, standard deviation proved very low for volumetric loads using the above measure, indicating consistency in velocity from charge to charge. Independent testing with a pressure gun also disclosed no difference between powder charge pressures using the volumetric powder measure.

Pressures and Bore Size

As caliber (bore size) decreases, sensitivity to load variation increases. For example, 36-caliber is more sensitive to charge variations than 50-caliber, as indicated by pressure changes. Five grains weight of variation was introduced firing a 36-caliber rifle and a 50-caliber rifle. The additional five grains of GOEX brand FFFg powder to a 36-caliber muzzle-loading rifle loaded initially with 30 grains of powder raised the pressure by 1800 LUP (Lead Units of Pressure). Adding five grains of GOEX brand FFFg to a 50-caliber rifle loaded with a 90-grain charge raised the pressure by only 500 CUP (Copper Units of Pressure) for this one test. Both rifles were loaded with an appropriate patched ball. The natural conclusion is that small bores are more sensitive to powder charge changes than

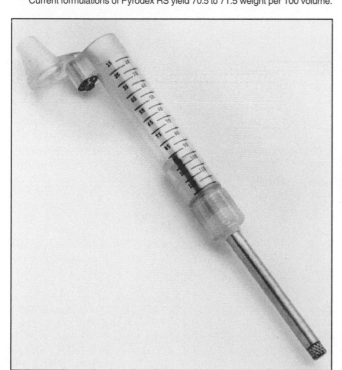

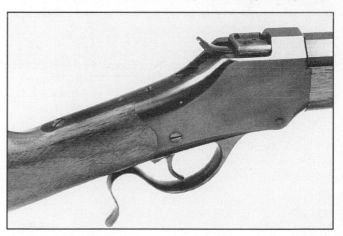

Obviously, there are built-in accuracy differences from gun to gun. For example, the "lock up" on this Winchester 1885 High Wall replica from the Montana Armory helps produce good bullet clustering from blackpowder cartridges.

Accurate blackpowder and Pyrodex loads are built by volume with a measure like this Thompson/Center U-View adjustable powder measure graduated in 5-grain increments up to 120 grains capacity.

Keeping records for an accuracy check is vital. In this case, two matched custom pistols were brought to the bench. They would not shoot the same load and bullet with the same accuracy. Each pistol, although made to be essentially identical, required its own load. Individual firearm variation is an amazing phenomenon.

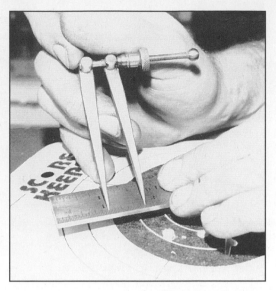

After measuring the center-to-center spread in this group, the calipers are read against a ruler. The group went about 1.5 inches in this case. Shooters interested in gaining the best accuracy from their guns should measure and record groups carefully for future reference.

Why is the Ruger Old Army accurate? Because blackpowder guns shooting appropriate bullets can achieve good accuracy, of course, but also because this cap 'n' ball handgun is precision-made and wears excellent sights.

large bores. This is due to the decreased volume of the smaller bore, offering less room for gas expansion.

Count on volumetric loads. Charges produced with a powder measure provide sufficient consistency for accuracy. Shoot blackpowder and Pyrodex by volume, not weight, preparing loads with a standard commercial measure using a routine designed to promote consistency.

Half-grain Increments

Another fact emerged from powder measure demonstrations. Esoteric pet loads, such as 55.5 grains of FFg or 112.5 grains of FFg (both actual examples) are meaningless. Test machinery could not detect half-grain increments. In practice, a 55.5-grain charge and 55.0-grain charge showed identical results in test rifles, as did 112.5 grains vs. 112.0 grains. In fact, the latter showed no velocity or pressure variation between 112.5 grains and 110.0 grains FFg in a 50-caliber test barrel.

Altering the Powder Charge

Because bore size is effectively chamber size in the muzzle-loading rifle, variations in bore size alter pressures considerably. This point was touched on above, showing smaller calibers as more sensitive to powder charge changes than bigger calibers. The *Lyman Black Powder Handbook* lists a breech pressure of 14,840 CUP in a 54-caliber rifle firing a round ball in front of 150 grains weight of GOEX FFFg powder (p.116). The same study shows a 36-caliber frontloader charged with 65 grains weight of the same powder giving a breech

pressure rating of 15,215 CUP (p.89). Furthermore, a 10-grain increase in the weight of the powder charge, from 140 grains to 150 grains, in the 54 increased pressure by only 300 CUP, while a 10-grain increase of the same powder in the 36 caliber rifle increased pressure by 1465 CUP.

The upshot is that in building the best blackpowder load, consider caliber. Be aware that small bores are more sensitive to minor powder charge variations than big bores. I have a 36-caliber frontloader that will not shoot accurately with 10 grains of FFFg, but accuracy with 15 grains volume is good in that rifle. The reasons for this are probably several, including better obturation (internal bullet upset) with the increased charge, and perhaps more projectile stability due to increased RPS. I cannot say, but I do know that the 15-grain charge is more accurate than the 10-grain charge in that rifle. In a 54-caliber rifle I own, accuracy levels are acceptable with as little as 40 grains of FFFg or as much as 120 grains of FFg behind the same .535-inch missile. A slow 1:79-inch rate of twist probably helps, as well as decreased load sensitivity due to increased bore size, as explained above.

A special test was run for me by a professional ballistics laboratory. Using a pressure barrel, powder charges were steadily increased while a careful accuracy check ensued simultaneously. Beyond the point of diminishing returns, accuracy fell off. With heavy overloads, accuracy was nil. Too much powder seriously increased standard deviation. So don't exceed a gunmakers maximum load, not only for safety, but also for accuracy.

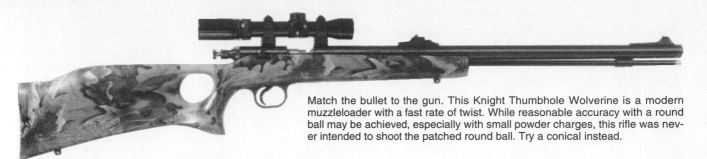

Match the bullet to the gun. This Knight Thumbhole Wolverine is a modern muzzleloader with a fast rate of twist. While reasonable accuracy with a round ball may be achieved, especially with small powder charges, this rifle was never intended to shoot the patched round ball. Try a conical instead.

Projectile Uniformity and Accuracy

Another aspect of blackpowder load quality is the missile itself. Obviously, bad bullets won't make good groups no matter how superior the rest of the load chain is. But how precise are muzzleloader missiles? Accurate bullets can be cast with good lead (pure and uniform), plus proper casting methods. Commercial swaged projectiles are also precise. The results of randomly weighing ten Hornady .535-inch round balls from an off-the-shelf box of 100 resulted in an average of 231.6 grains. The heaviest one was 231.9; the lightest 231.2.

Weight to weight spread was obviously excellent with these balls. Moreover, the actual weight of the 54-caliber (.535-inch) projectiles indicates "pure lead." The perfect weight for a .535-inch round ball would be 230.9 grains. The formula to prove this is: D(3) x .5236 x 2873.5 (where D equals the diameter of the sphere cubed, times the constant .5236 to compute the volume of a sphere, times 2873.5, the weight of 1 cubic inch of pure lead). The average weight of 231.6 proves that the lead purity for the Hornady ball is extremely high. Speer swaged round balls proved equally good.

Home-cast projectiles can also exhibit a high degree of precision. In one example, a batch of .395-inch round balls was prepared using a Lyman mould with lead purified by the methods described in the next chapter. A sample of ten from a population of 100 averaged 93.5

grains with an extreme spread of .8 (eight-tenths of one grain weight). Perfect weight for a .395-inch round ball is 92.73 grains. I miked my products and they were very slightly oversize, which accounted for their heavier weights. Finally, a supply of .520-inch round balls cast from a Lyman mould showed an extreme spread of .5-grain, which is very fine.

A question arose concerning smaller round balls, because percentage differences affect little missiles more than large ones. Ten Hornady swaged round balls in 32-caliber, .310-inch diameter, were sampled. The results were: 44.7 grains, 44.6, 44.7, 44.7, 44.7, 44.7, 44.7, 44.6, 44.9 and 44.7, with an average of 44.70 grains. A perfect .310-inch round ball of pure lead is 44.82 grains, proving how impurity-free these little pills were.

The notes above pertain to cast round balls, but cast conicals can also be precise when "run" with care. A Shiloh 610 mould produced missiles that varied by an extreme spread of only 2 grains, a very small percentage.

Patches and Accuracy

Simply stated, patches are an important part of the load chain. Because it weighs so little, patch weight is a minor factor in load quality. Patch quality with regard to accurate loads means good material (tight weave) and uniform thickness. The latter aspect was tested with a micrometer. Here are the results of ten miked patches, five Irish linen and five commercial patches from Gunther Stifter's Flintknapper Shop in Germany:

Anyone with a micrometer can verify thickness uniformity of his

Patch Thickness Uniformity

.015-inch Irish Linen Patch	.012-inch Flintknapper Shop Patch
.0150-inch thickness	.0120-inch thickness
.0150-inch thickness	.0122-inch thickness
.0145-inch thickness	.0120-inch thickness
.0156-inch thickness	.0117-inch thickness
.0152-inch thickness	.0122-inch thickness

own patches. Patches need not be perfectly precise in thickness from one to the next because they are under compression in the bore, so minor variations tend to cancel out. But significant differences can be a problem. While only a small part of the total accuracy picture, patch thickness variation is a link in the load chain, so patches should be as uniform as possible.

Ramrod Pressure and Accuracy

Another link in the load chain is ramrod pressure. In competition, shooters have been known to use various means of maintaining consistent pressure on the ramrod from one shot to the next. You might observe a shooter placing a forefinger over the ramrod, using it as a pressure gauge. When the finger says *ouch!*, further pressure on the ramrod is released. Slug-shooting muzzle-loading enthusiasts use this or some other other means of maintaining a consistent pressure during projectile seating. But what about the rest of us?

A reader sent an ingenious device for me to test, a spring-loaded metal unit with an insert to accept the topmost part of the loading rod or ramrod. At a specific amount of pressure, the

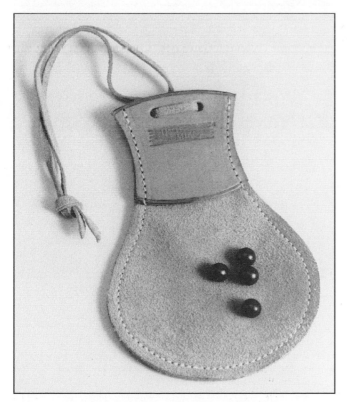

No firearm, blackpowder or modern, can be any more accurate than its bullet. These precise round balls will fly true to the target.

Gonic's modern muzzleloader is accurate. Why? It's a combination of things, as always. The barrel is precision-manufactured with top-grade rifling for starters, but it's more than that. Every aspect of the rifle is built to lend itself to accuracy, right down to the good wood-to-metal fit, as shown here. Firearms accuracy is never a matter of one aspect only.

(Below) Knight's MK-95 Magnum Elite is another accurate muzzleloader. It uses shotshell primers for ignition.

(Below) Accuracy with the round ball normally occurs with a projectile that fits closely to the bore, as opposed to an undersized round bullet. A short-starter is needed to force a well-patched, near-bore-sized ball into the muzzle. This short-starter, with attached muzzle protector, is from Forster Products.

spring collapsed and further seating of the missile ceased, providing a consistent 40 pounds of pressure every time. In an attempt to gather data on ramrod pressure, standard deviation from the mean velocity was measured, first with random pressure from twenty to sixty pounds, and then with consistent a 40 pounds. Here are the results:

Rifle A
Random Pressure = 57 fps standard deviation
Constant 40-pound Pressure = 19 fps standard deviation
Rifle B
Random Pressure = 62 fps standard variation
Constant 40-pound Pressure = 12 fps standard deviation
Rifle C
Random Pressure = 71 fps standard deviation
Constant 40-pound Pressure = 21 fps standard deviation

The conclusion is that consistent ramrod pressure on the bullet/charge promotes low standard deviation. However, I have learned that going "by feel" is good enough. No special device is necessary to regulate ramrod pressure, except for strict testing purposes.

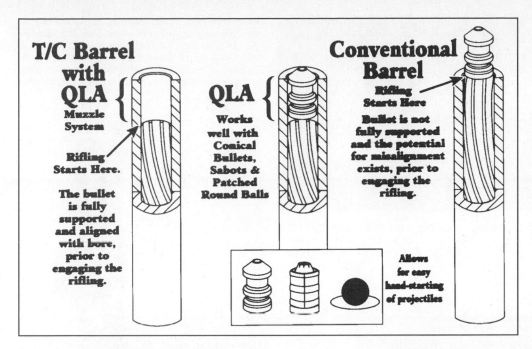

T/C Barrel with QLA
Muzzle System

Rifling Starts Here.

The bullet is fully supported and aligned with bore, prior to engaging the rifling.

QLA
Works well with Conical Bullets, Sabots & Patched Round Balls

Conventional Barrel
Rifling Starts Here

Bullet is not fully supported and the potential for misalignment exists, prior to engaging the rifling.

Allows for easy hand-starting of projectiles

Clearly, how the conical is installed into the muzzle of the firearm can make a big difference in accuracy. Knowing that fact, Thompson/Center developed its T/C QLA (Quick Load Accurizor) Muzzle System, a barrel with a muzzle that helps align bullets to the bore.

Uniformity and Time

What happens to accuracy when Father Time steps into the picture? Bores do change with repeated firing. Depending upon barrel steel, loads used, cleaning faithfulness, and many other factors, rifling wears and accuracy may suffer. As part of load uniformity, it is imperative that the shooter understand maturation changes. The buckskinners of yore did not always have the facility, nor, perhaps the inclination, to scrub Old Betsy after every shooting session. Bores eroded and were "freshed out," which means enlarged to a bigger caliber and re-rifled. Today, we have the time, place, and chemicals to do a good cleaning job, so our blackpowder guns should remain accurate for a very long time through less bore wear.

Modern commercial patch and bullet lubes allow a number of shots before bore swabbing is necessary, in addition to promoting continued accuracy. Dirty bores, on the other hand, did not promote good accuracy, especially in a test of a particularly accurate blackpowder breechloader, which produced excellent groups with a clean bore.

But if group size begins to broaden, something can be done to restore accuracy without freshing out. Sometimes, a new bore "breaks in" after so many shots and accuracy falls off. Generally, an increase in ball size will restore accuracy. For example, if a 50-caliber frontloader was accurate with a .490-inch patched round ball when new, switching to a .495-inch ball may bring accuracy back when the rifle is older. It's worth a try. Consider a larger ball before going to a thicker patch. But do try a thicker patch if the bigger ball is not enough to take up the wear in the bore. Experiment by shooting groups to find out.

Accuracy and Components

Of course components can change in time. The patches you used to buy may no longer be available, for example. Powders definitely change, as proved by volumetric tests with powder measures. But all in all, blackpowder components remain quite constant from year to year. Here are the results of a test that proved this point with data gathered 3 years apart:

Blackpowder Time Test

Charge Wt. (grs.)	Initial MV (fps)	3-Year MV (fps)
80 FFg	1211	1219
100 FFg	1393	1375
120 FFg	1513	1449

The Stabilized Bore and the Fouling Shot

Another factor in blackpowder load quality and accuracy has to do with a stable bore in terms of fouling. Since black powder leaves half to maybe 57 percent of itself in an unchanged state (solids), rather than changing from a solid to a gaseous state more completely as smokeless powder does, there is always a fouling concern. Fouling actually changes bore dimensions, and you don't change bore dimension without altering pressure, which in turn may affect point of impact. The answer to bore consistency rests in two things: first, keep the bore in a good clean condition, and for target work, fire a fouling shot, as explained in Chapter 15 for Pyrodex.

Chapter 19 discussed modern blackpowder chemicals and their ability to reduce bore fouling. What was not amplified in that chapter, however, is the improved consistency of the bore when using the all-day lubes we now have. In other words, lubes are important weapons in the fight to maintain bore stability so that the quality of each load is just as high as the load before it.

Accuracy and Bullets

Tests for accuracy with damaged bullets reveal that the round ball with nose damage may stray off course, while the conical with nose

Aligning a conical properly in the bore is important to accuracy. Forster Products offers their Maxi-Starter for this purpose, and it's a worthwhile tool for the accuracy buff.

The little things can make a big difference in accuracy, right down to ignition. This illustration shows a standard nipple (left) and the Hot Shot nipple (right) in action. Notice the long, clean flame delivery. While ignition in the muzzleloader is generally a small factor in accuracy, it is a factor. For blackpowder cartridges, ignition is even more vital to consistent accuracy.

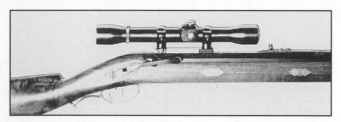

Mounting a scope on a traditional-style muzzleloader like this Storey sidehammer may offend some shooters; however, if a person's eyes are such that iron sights won't work for him, accuracy can only be achieved by changing those sights.

(Below) The fellow who said "You can shoot no better than your sights" was right. This is a special Vernier tang sight from Montana Armory, called the Mid-Range model. The Long Range model is 2 inches taller. It's adjustable for windage and elevation, and it has extra eyecups for different light conditions.

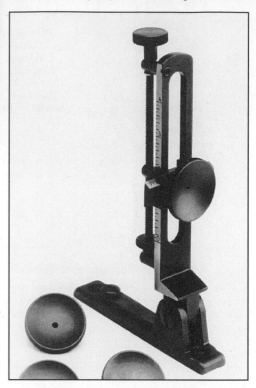

damage tends to remain on course, at least better than the sphere. On the other hand, base damage to conicals definitely steered the elongated bullet off course. The goal is to take care to maintain all bullets in good condition for the best accuracy, when casting, storing or transporting them.

No matter how precise the bullet, if it is not stabilized in flight, good accuracy cannot be achieved. This means using the proper rate of twist for each projectile, as explained in Chapter 27.

A conical bullet forced downbore "cockeyed" tends to stray off course. This problem is addressed in two ways today. There is the recessed bore, as found on Gonic muzzleloaders as well as DGS, Inc. Buggy Rifles. Recessed bores are not new. Also called "recessed muzzles," firearms of the distant past have been found with the bore enlarged at the muzzle. Since the recessed muzzle appears on round ball rifles, we assume they were used to allow easy entry of the patched ball. Short starters are very hard to find with old-time guns and shooting bags. However, the recessed muzzle we speak of on the DGS rifle and Gonic are to align a conical projectile so that the base of the bullet is perpendicular to the bore. A second way to ensure a properly aligned conical is with a device that introduces the bullet squarely to the bore. Such a device is Barnes' bullet aligner, which screws on to the end of a ramrod. It centers the Barnes MZ sabot/conical with the bore.

I've never seen the need to hammer the tip of the loading rod or ramrod on the seated projectile downbore, but this practice continues, and is sometimes recommended as useful. Mishapen bullets have never been known for supreme accuracy, and the banging of the loading rod or ramrod on the seated bullet has no value.

Some bores "like" one bullet over another. The shooter interested in accuracy should try different bullets to see which one his firearm shoots most accurately. Bullets of the same caliber do not necessarily have the same exact diameter.

Accuracy and Caps, Nipples and Sights

Slug guns, the benchresters of the blackpowder world, are given careful consideration where ignition is concerned, leading us to believe that ignition can affect accuracy, if only in a minor way. One theory is that a severe blast from any form of ignition can actually move the load chain upbore before full powder ignition is achieved. I have no way of proving or disproving this theory. Try different percussion caps, however, to see if they make a difference in your charcoal burner's accuracy.

Cousin to ignition, the nipple can make a difference in some firearms. Today, most nipples are well designed, with small base holes for fire exit. Nontheless, be aware that a change in nipple may affect accuracy with some frontloaders.

Sights do not, of themselves, improve the inherent accuracy of a firearm or load, but good sights are required to realize accuracy potential in a charcoal burner and its load chain. Of course, scopes do best. After all, they magnify the target, put both reticle and bullseye on the same plane, and they can be adjusted to suit the individual's specific vision (normally by rotating the ocular lens). If you want to learn what your best blackpowder shooter is capable of accuracy-wise, scope it. See Chapter 29 for more on sights for frontloaders.

You can only shoot as well as you can see, and part of making out a target is not only sights, but the target itself. For more on targets, see Chapter 30 on sighting-in.

That's a look at blackpowder accuracy, albeit not a complete investigation of this vital topic. Read more on the subject in those chapters centering on specific firearms, such as handguns. Meanwhile, keep in mind only realistic expectations for blackpowder accuracy. Sometimes expectations are too high, sometimes too low. Nowadays, there are a number of well-made blackpowder guns that will place three bullets inside an inch circle at 100 yards. Not that long ago, that brand of accuracy was mainly associated with slug guns. I may be wrong, but I think today's best frontstuffers are probably the most accurate charcoal burners we've ever had.

●

chapter 24

Casting Blackpowder Bullets

COMMERCIAL CAST OR swaged round balls are excellent, and good "store-bought" conicals abound. So why would anyone want to cast his own projectiles? There are at least three reasons. First, running ball, as the old-timers called it, is a hobby in its own right, requiring know-how, the right equipment, hands-on experience, a parcel of skill, and a careful attitude. Second, in spite of the wide world of muzzleloader bullets, round and elongated, sometimes a shooter wants or requires something special, so he must make it for himself. When I took possession of my first Storey Buggy Rifle, caliber 45, I wanted a specific bullet for it, something I couldn't buy over the counter. NEI Hand Tools, Inc. came to the rescue with a 390-grain conical, custom mould No. 277A. I also own a flintlock rifle built a long time ago, caliber 42, and for that one a home-cast round ball is necessary. Third, there's money to be saved, especially when lead can be located cheap-

ly or picked up for nothing, which is sometimes possible. For example, I can shoot my 32 squirrel rifle with cast balls for less than a 22 rimfire.

Along with the three reasons cited above, casting bullets is also a part of the complete blackpowder shooter's education and a testament to his self-reliance. The art of casting has gone on for a very long time. Shooters of the distant past had no choice—make your own bullets or don't shoot at all. Only in comparatively recent times have muzzleloader lead projectiles become available at the store. Certainly the American mountain man of the Far West cast his own. He was known to carry a mould among his possibles, and he might even have a bellows for blasting the coals of his fire into red-hot lead-melting nuggets. You can see him now, firelight painting his shadow against a rock bluff as he inspects his newly-formed bullets.

Those who followed the foreworn trails of the mountaineers also

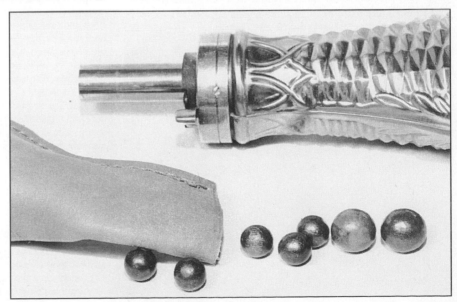

The art of running ball (casting projectiles) is rewarding in many ways, not the least of which is economy. But it's also a hobby in its own right.

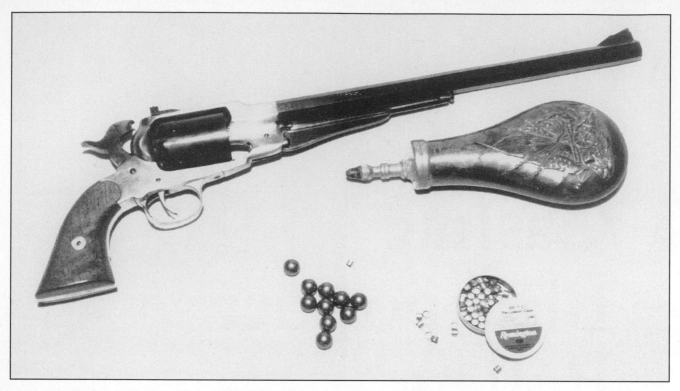

All blackpowder guns shoot properly home-cast projectiles perfectly well, including the cap 'n' ball revolver.

cast bullets for their Sharps and Remington breechloaders. Old-time "buffler hunter" Frank Mayer talked about collecting bullets from defunct bison so he could run 'em again, and Wayne Gard, in his book *The Great Buffalo Hunt*, related the comments of a "buffalo runner," as they preferred to be called, who operated out of Miles City, Montana. Hanna was his name. He said,

> We used the softest lead that we could buy. When a bullet hit a buffalo, it would flatten out like a one-cent piece and tear a big hole in the animal. Generally it would stop on the opposite side against the hide, and the skinner would be able to save the lead and remold it. As lead was high, we reloaded all our shells. After we had killed all the boys could skin in a day, we would go to our shack and reload ammunition for the next day.

The 19th-century explorer could make bullets at his campfire with a mould like this one. While nothing like our modern moulds, this rather primitive unit would turn out a decent-enough round ball.

Casting bullets never went out of vogue. During the Golden Age of American hunting, following the era of the buffalo men, reloading was highly popular, as it remains today, and casting bullets was commonplace. Currently, hundreds of different moulds are available, along with a multitude of excellent casting tools, including special furnaces to melt lead at proper temperatures. Home-cast projectiles are fired by the millions out of modern and blackpowder handguns, even shotguns, not to mention frontloaders as well as cartridge rifles. Do you know how many bullets moulds are available for your 30-06, for instance? I don't, but it's a big number. Instead of bullet casting falling off in popularity, the hobby within a hobby has actually grown in modern times.

Countless pleasurable hours are spent every year by shooters who have a good bullet-making operation in their home workshops or garages. The hobby is so complete that many full-length books have been authored on the subject. The reader is encouraged to locate and read some of these titles to further his knowledge of bullet-casting. See the Library section toward the end of this book.

I have a 40-caliber rifle that shoots bore-sized (.400-inch diameter) bullets best. There are none for sale in my area, so I cast my own. If I didn't know how to run ball, I would have to settle for less accuracy in that rifle. Large round balls are not readily found at the local blackpowder shop either. If you have a 60, 62, or larger caliber rifle, the nearest supply of bullets for it may be your own workshop. A number of shooters have gotten into really big frontloaders, casting bullets for their 4-, 8- and 10-bore rifles. Certain rifled muskets, such as the Whitworth, are worth casting for, too, as are many breechloaders. Making your own lead conicals is the only way you're going to get to shoot some of those great rifles. Finally, running ball at the rendezvous campfire is one more "primitive" skill worth having, if only for its historical value.

Lead, the Shooter's Choice

Lead is a near-perfect element for making round balls and conicals for blackpowder arms. It melts at about 327.5 degrees Centigrade, or 621.5 degrees Fahrenheit, depending upon its state of purity. That's why bullets could be made on campfire coals, especially when the

Tests have shown no accuracy differences between homesmade lead bullets and commercial cast bullets.

The Minie on the left is a perfect example of a well cast lead bullet. It is not wrinkled nor frosty looking, and it will shoot just fine.

The three Minies shown here are all cast projectiles. While they exhibit minor imperfections, they will shoot with accuracy.

coals were encouraged by the quick breath of a bellows. Lead liquefies at temperatures well below some other metals. The Lone Ranger, for example, would have had a tough time making his silver bullets over a campfire, since silver melts at about 960 degrees Centitgrade (1760 degrees Fahrenheit) with a casting temperature around 1030 to 1090 Centigrade. Meanwhile, hardwoods, such as oak, deliver especially hot coals, making solid lead run like water in no time. No wonder our forebears had no problem making their own lead bullets in the field. But campfire casting does not end lead's positive attributes. Along with a low melting temperature, lead also provides excellent mass, which means good heavy bullets for deep penetration on big game, as well as "carry-up," or rention of original velocity and energy.

Furthermore, lead has high molecular cohesion. In other words, it stays together rather than fragments, which is an excellent property where penetration is desired on big game. Also, lead alloys are easy to form, which is important when shooting breechloaders. An alloy is a mixture of two or more metals, such as lead and tin. Alloys of one part tin to thirty parts lead (1:30), or one part tin to twenty parts lead (1:20) work well in blackpowder breechloaders. Such alloys are no problem to attain with lead. So it's easy to see why lead works so well for casting bullets.

Turning Lead into Bullets

Once lead is changed from a solid to a liquid (molten) state, it can be poured into a mould just like the gelatin moulds mother used to make. Liquefied lead runs like water, filling up a cavity. When it hardens, lead retains the form of the cavity—a bullet in this case. Once upon a time, moulds were formed from various materials, even soapstone, but softer old-time brass and iron moulds did not produce perfect bullets. As they grew hotter and hotter, they tended to alter in cavity dimensions so missiles were not entirely uniform, but off-sized bullets obturated to the bore when fired, so all went well enough. You certainly wouldn't miss a deer, elk, or a buffalo because the bullet cast by the fire last night was not perfectly uniform. However, today's iron and aluminum moulds surpass their forerunners considerably, producing much more precise projectiles.

The object of running good ball, round or conical, is to achieve uniformity. The goal is bullets that are not only precise, but alike from one to the next, and one batch to the next. Uniformity is what you get if you know how to cast correctly, beginning with purifying the lead and maintaining proper casting procedures. A good bullet has a smooth

Because lead can be melted at a low temperature, compared with many other metals, it can be worked into useful bullets with simple tools like a melting pot, mould, and a ladle.

Looking at a mould in the open position, we clearly see the shape of the projectile the mould will make. This one is for a Minie, hence the base plug to form the cavity in the base of the bullet.

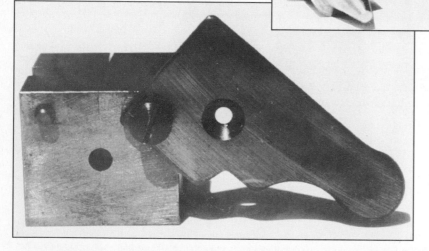

Looking at this mould from the top we see the sprue hole and sprue plate. With the sprue plate closed, molten lead is poured through the sprue hole, filling the cavity and forming the bullet. When the sprue plate is tapped out of the way, it automatically cuts off most of the sprue on the base of the bullet.

"skin," unwrinkled and unblemished. It's not frosty-looking, but is shiny and smooth. A properly-cast bullet does not have air pockets, or streaks of impurities running through it. It is homogeneous. Air pockets and impurities result in a lopsided projectile. Gyroscopic action (spin induced by rifling) tends to equalize the effect of such anomalies, but even rotation cannot entirely overcome lopsidedness.

Visual inspection won't always reveal the bad bullet, either. Traces of "unblended" antimony, tin, or non-lead contaminants are invisible to the naked eye. While bullet spinners and other high-tech tools work well for the benchrest shooter who needs to sort projectiles for a match, they are not appropriate for muzzle-loading. The way for us to make a proper lead bullet is to ensure that it is solid through and through. In order to make this type of bullet you need "pure" lead for bullets intended for muzzleloaders, and a very homogeneous alloy for breechloader missiles.

Why pure lead for frontloaders? In the first place, you won't get 100 percent pure lead because it's not available. The closest I've come across was a guaranteed 99 percent-plus purity, and that's quite good enough. Pure lead is right for muzzleloaders because it upsets nicely due to its softness. Bullet upset within the bore (obturation) is valuable because it is a customizing process, each projectile fitting itself to that particular bore by "fattening out" to take up the windage or to fill the grooves of the rifling. For Minies and Minie-type bullets, the skirt must flare out. Pure lead does that best. Furthermore, even a hard alloy won't prevent a thin skirt from flaring out too much when blasted by heavy powder charges, which turns the skirt into a big dress, which is not good for accuracy. So even for Minies, pure lead is best.

Pure lead helps to effect a seal in the bore, too. We know that the patch, of itself, is not a true gasket by definition. However, the patch holds the ball in place; the ball obturates; and the broadening of the missile itself provides, at least in part, a seal against hot gases behind it. Also, because of its high molecular cohesion, the pure lead bullet

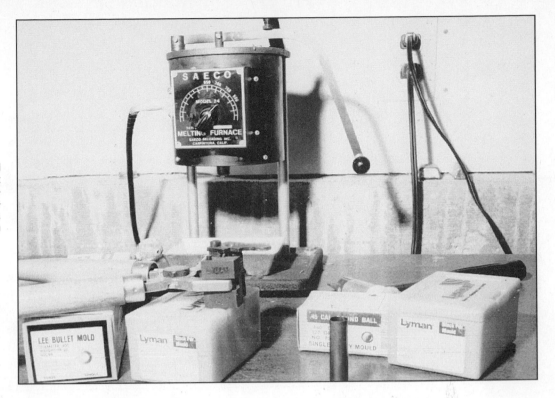

A furnace makes casting lead projectiles easier because it can be set at various temperatures for different size projectiles or for purifying lead. The SAECO brand has been around for a very long time.

hangs together as a deeply penetrating unit, rather than fragmenting. This means long and effective wound channels. Also touched on earlier, pure lead is dense. A lead bullet weighs a lot for its dimensions, and a heavy missile penetrates better than a light one. Consider a bullet made of aluminum. It might fly fast, but how will it retain its energy? How well will it penetrate?

The reader should be aware, for his own shooting knowledge, that bullets have been made of pure copper for a long time. I have samples of a German 30-caliber spitzer of pure copper construction that is at least twenty years old. Barnes now offers pure copper bullets for the muzzleloader and they are excellent, but copper does not fall within the realm of home casting at this time. So our concentration in this chapter remains on lead and lead alloys.

Lead can be expensive to purchase, as from a plumbing supply house or other over-the-counter source. That's why blackpowder shooters are known to scrounge their beloved metal. Lead pipe from old buildings is a good source if you have access to them. Lead telephone cable sheathing is reportedly 98 percent pure. Tire weights are all right, but they are far from pure lead and must be melted down and purified. Lead seals from money bags and railroad cars are, I am told, almost pure lead. But even if you have to buy lead, homemade projectiles still cost less than commercial ones, especially for smallbore round balls like the 32-caliber. A .310-inch ball weighs only 45 grains; there are 7000 grains weight in one pound; so you get about 155 32-caliber pills from a single pound of metal.

The Tools

You don't need a miniature factory to produce fine cast lead bullets, but you do need a few tools. Buy the best. It's cheaper in the long run because good tools, well cared for, last indefinitely. You need something to melt lead with, either a pot or a furnace. The pot is cheap and simple. It may be no more than a metal container, or it may have a built-in heating element, making it a sort of cross between the basic pot and a furnace. A true furnace is another matter. It has heat control and its own electrical heat source. You will need proper clothing, including gloves and eye protection like workshop goggles or glasses. You also need a dipper or a ladle. An old tablespoon is necessary—check the army surplus store for a large military tablespoon. A mould is imperative, of course. And a moulder's hammer is nice, but a hardwood dowel will suffice to tap open the hinged sprue plate.

Safe Casting

There is only one way to cast bullets, and that's the safe way. Casting is just as safe—or as dangerous—as crossing the street. Look both ways when your cross and you won't get hurt. Shut your eyes, and wham! You must pay attention to a few things when casting. Stay away from any source of water. A few drops of water in a pot of molten lead can create a mini-explosion. Cast in a safe place, preferably a spot with a cement floor, and definitely out of the traffic pattern of the home. Kids and molten lead must be kept separated from one another at all times. Wear safety glasses and gloves, and never cast in short pants. Wear long pants only. Wear shoes, not moccasins or slippers, and use a long-sleeved shirt. And you must have ventilation because some of the oxides emitted as products of melting lead can be toxic. Do not inhale fumes coming off of the melting pot or furnace. Stand back a little when dipping molten lead.

Fumes from molten lead can cause trouble. The body will take on lead, giving it up rather slowly, so there can be a cumulative effect. The good news is that such fumes are very limited during the normal course of casting bullets. The NRA ran a test to determine the level of danger associated with bullet casting. A Mine Safety Appliance Company apparatus for detecting lead dust and fumes was placed directly above the melting pot, with only traces of dangerous fumes revealed. Regardless, the wise bullet caster always has good ventilation in his work area, and it's only common sense that he remain well away from the molten lead itself, not only to avoid fumes, but for the obvious reason of getting burned. Finally, there is no reason to super-heat molten lead. The temperatures discussed here are perfectly adequate for cleaning lead, and for making round balls and conicals.

While lead poisoning seems to be an uncommon problem in bullet casting, there is no doubt that the heat source can cause trouble. A number of casters use gasoline stoves and other implements that can produce carbon monoxide. Since carbon monoxide is a colorless, odorless, and tasteless gas, it's difficult to determine its presence. The safety precaution best heeded is to melt lead only in a well-ventilated area, and to always be aware of the heat source's potential for producing carbon monoxide. If a person does get headaches, nausea, or dizziness, he should see his doctor for a checkup. The NRA concluded that, "A maker and user of bullets possibly can be poisoned by lead, under sufficiently extraordinary exposure. This can hardly exist for the individ-

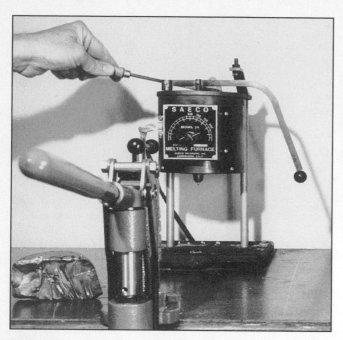

As the lead in the furnace melts, it can be stirred with a long-handled ladle to bring impurities to the top for skimming off.

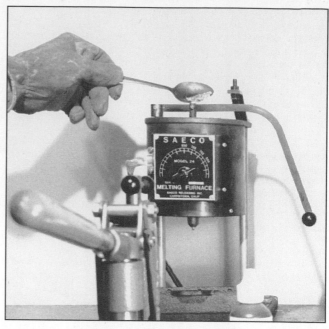

Impurities, called dross, can be skimmed from the surface of the molten lead using an ordinary spoon. Working this close to the molten lead demands hand protection. Dross will be very hot, so it should be discarded on a fire-resistant surface.

ual casting bullets on a scale for his own use, with reasonable ventilation and sanitation."

Purifying or Cleaning Lead

This is not fluxing, but most people call it that anyway. Fluxing is a combination process, a method of promoting the fusing of metals, which is precisely what we are *not* after in the lead-cleaning process. Instead of a fusing of lead with tin, antimony, or any other element or product, we are looking for a separation of lead from any other element or product. In order to clean lead, you have to melt it. Since lead melts at a relatively low temperature, a simple lead pot will suffice, but I

have turned to a furnace for melting lead and casting bullets. Mine is a SAECO. I like the fact that an electric furnace does not, of itself, promote carbon monoxide, and has high capacity plus temperature control. I melt my lead at 800 degrees Fahrenheit. But doesn't lead melt at only 621.5 degrees Fahrenheit? Sure it does, but we are trying to melt everything else that might be in that lead, not just the lead itself. I find 800 degrees suitable for this task. When the lead is liquefied, skim it first. Do not flux at this point. Using an old tablespoon, skim the dross (floating gunk) atop the lead. Carefully sweep this stuff up, part of which may be dirt as well as non-lead elements. Clear this stuff from the surface of the liquid and get rid of it.

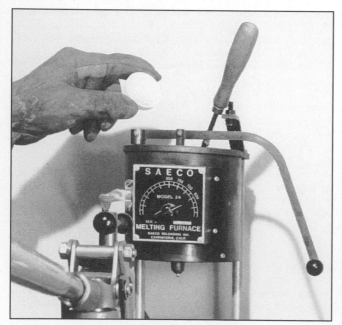

Fluxing, here with a commercial product, is actually a blending process. However, when lead is being purified, fluxing can work to help remove impurities, as well as aid in the formation of a homogeneous product.

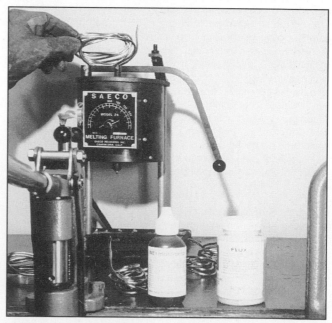

An alloy is a mixture of metals. When making alloy bullets, a specific amount of tin and/or antimony is generally added to the molten lead and fluxed in thoroughly.

Using your ladle or the same large spoon, stir the molten lead. If more dross rises to the surface, skim it away. Now flux the product. Fluxing is not exactly the correct nomenclature here, as already admitted, but it's commonplace terminology for this step. All that fluxing means, for cleaning purposes, is dropping a chunk of paraffin wax about thumbnail-size into the molten lead and stirring it in. If you don't like the resulting smoke, light a match and touch the flame to the fumes above the surface of the lead. Continue to stir. Keep those gloves on, and be certain that your spoon or ladle has a long handle. Those are flames you are dealing with, as well as heat, so keep your head back from the vapors. There's no reason to breathe them in. I know these are common-sense warnings, but you can't be too careful when working with molten metal.

Instead of paraffin wax, you may also use a commercial caster's flux. Dixie Gun Works has a good one. You may accomplish true fluxing (metal combining) from this step, but don't concern yourself. Most of the non-lead products are already skimmed away. Your lead will exhibit traces of tin and/or antimony, but these won't hurt anything now because they are only traces and not a significant volume of your metal. A few heavier impurities may show up during the fluxing process, but they can be skimmed off of the surface with the spoon. The shiny liquid in your furnace or melting pot is now "pure" lead, or at least pure enough. You may now cast bullets.

Casting Temperatures

Only you will be able to determine exactly how hot the lead should be for best casting because the ideal temperature varies with the type of mould used, and with the type of projectile that is cast, as noted below. If you want to go right to work from the pot, do so. You might find that though the lead is a bit hotter than necessary, the heat will warm your mould quickly. So get started. If some bullets are ruined, put them back into the pot and recast them. I generally reduce the temperature of the furnace from 800 down to about 700 degrees before casting round balls. It depends on the size of the projectile. For large conicals, I want a hotter product to fill the mould cavity better.

The proper lead temperature is absolutely vital to a good cast bullet. Lead that is too cool will not fill the mould cavity properly, while lead that is too hot overheats the mould quickly. For long conicals, especially those associated with blackpowder cartridges, I believe a hotter pouring temperature makes sense. In making 500-grain bullets for my

45-70, for example, I use a temperature of 850 degrees. The reason for temperature concern is quality of the final product. Lead that is too hot tends to make a frosty bullet. Lead that is too cool may produce a wrinkled bullet surface, or bullets that have occlusions in them because the mould cavity did not fill properly. If you have problem bullets, just keep on casting. Simply return all bad bullets to the pot for remelting and recasting. As long as you end up with many silver-shiny bullets in the end, you're doing okay. You won't get a perfect bullet every time by a long shot, but with practice you'll learn how to keep the mould at the right temperature to throw the best possible bullets.

Lead Alloy Bullets for Blackpowder Cartridges

While muzzleloader projectiles should be of pure lead, the same cannot be said of bullets for blackpowder cartridges, unless perhaps the rifle bullets are paper-patched. However, in the main, these bullets should be cast of an alloy. Adding tin to the lead makes these bullets a bit harder than pure lead projectiles, although not extremely so. My friend, Dave Scovill, editor of *Rifle* magazine, and an expert caster, believes that the true value of adding tin to lead for blackpowder cartridge bullets is to better fill the mould cavity. It's true, tin does promote lead flow, which in turn aids in filling the mould cavity. Opinions vary on the best possible alloy for this use and one part tin, sixteen lead (1:16) is a well-know mix. Other shooters like 1:20 or 1:30 better.

As long as you're going to cast alloy bullets, you should do it with some semblance of precision, and that means starting out with pure lead and adding a *known* amount of tin, not an estimate. An ideal way to do this is with pure lead ingots and tin bars. Serious shooters bent on winning long-range silhouette matches may wish to purchase pure lead, rather than chancing impurities in home-cleaned lead. At any rate, whether cleaned-up or purchased, the idea is to begin with lead that is as pure as possible. The lead is then melted and cast into ingots. For the sake of simplicity, we'll say our lead ingots weigh 1 pound each. Now it's fairly easy to add the correct amount of tin. For a one- to- twenty mix, ten lead ingots are tossed in the melting pot with a half-pound pure tin bar, which can be purchased. I found my source of tin listed in the yellow pages under "metals," and it came in 1-pound bars to boot, making it very handy to use.

The mixture of tin and lead is blended, as it were, into a homogeneous alloy through fluxing. Now we're talking about true fluxing, not

An aluminum mould can be warmed up a little by touching a corner into the molten lead in the pot for a moment.

After use, one trick that helps keep mould blocks true is to apply a rubber band to the handles to hold the blocks together firmly and in line.

just cleaning the lead, and in my opinion, using a fluxing agent makes sense here. Not that paraffin wax won't do the job, but a true fluxing agent is more certain. The lead/tin product is turned molten at about 800 degrees Fahrenheit, and the fluxing agent is put to work. Precise instructions for use are printed on the label of the fluxing product. The final product will be a true alloy of lead and tin, and an ideal bullet for the blackpowder cartridge.

Remember, of course, that as the metal is used up in the pot, it will have to be replaced with proper ratios of tin and lead. This can be a problem if the alloy has not been maintained. It is possible for the tin to separate from the lead during the casting process. While stopping production just when the finest bullets are coming out is irritating, the caster must be aware that he may have to do this in order to maintain his alloy. I cannot find a problem with quickly refluxing the alloy from time to time.

The Mould

Two major types of moulds are available—iron and aluminum. As with most things in life, both have their good and bad points. Iron moulds are more rugged than aluminum and they tend to last longer. They take longer to heat up, but they also make a relatively long run of good bullets before they get too hot. Aluminum moulds can last indefinitely, provided they are treated with care. They heat up fast, but if they get too hot, they also cool down quickly so you can soon return to work. Aluminum moulds usually cost less than iron moulds. Take your pick. I have both types, and they work fine. Multiple-cavity moulds produce more bullets faster, but the majority of my moulds are single-cavity. I find these fast enough.

Follow the break-in instructions that came with your new mould. They are important to the life of the unit, as well as its continued good service. Warpage of the mould may result in mismatched halves, for example. Opinions concerning mould care vary, but the best bet is to follow break-in and preservation methods as suggested by each individual mould maker. Break-in of an iron mould seems to almost "blue" the metal, somewhat as one might color case-harden steel. When the mould takes on that blued look, it seems to produce its best bullets, and is truly broken in.

Using the The Dipper or Ladle

For our purposes, let's call the dipper a tool with a reservoir and a pouring spout, while the ladle is a more open unit, much like an eating utensil. Both work all right. I happen to have a dipper. You must get the molten metal from the pot into the mould, and this is where the dipper or ladle comes in. First, ensure that your mould is hot. The aluminum mould can be dipped (one corner only) into the molten lead to warm it. Heating the iron mould is best done by running ball until heat is transferred from the liquid metal to the mould blocks. With one continuous motion, introduce the spout of the dipper to the chamfered edge of the sprue hole in the mould. Hold the spout up against the sprue hole. With a turning motion, tip the mould down and the ladle up so molten lead can run into the mould cavity, filling it completely. Hold that position for a couple of seconds then twist or rotate back to the original position and withdraw the spout of the dipper from the sprue entry.

Give the molten lead time to fill the mould cavity. I used to rush this aspect of the job, because I got used to making small-caliber round balls, which can be produced rapidly. When I began making long bullets, especially for breechloaders, I had trouble. Many of my bullets were simply incomplete in form. Allowing the ladle to rest a moment against the sprue hole corrected the problem.

A fin may appear around the base of the bullet. If so, this may mean the sprue plate has warped. Keeping the sprue plate joint lubricated can help to prevent this problem. If the sprue plate does become warped, it may have to be corrected by a machinist who will carefully and precisely grind the plate flat again.

Using the Moulder's Hammer

You don't need a moulder's hammer; a hardwood dowel will work, but I like the hammer better than a dowel. Gently tap the hinged sprue plate with the hammer (or dowel) to move it aside from the sprue hole. Now open the mould handles and drop your new bullet on a soft surface. I use a piece of cardboard set on an incline with a rag of tight-weave material at the base of the cardboard. The bullet rolls gently down the cardboard incline and comes to rest on the non-linty surface of the cloth. It's hot, so don't touch it.

Molten lead is poured into the sprue hole of the mould, using a slight twisting motion of the wrist to bring the mould sprue plate to the horizontal position. The idea is to fill the entire mould cavity with molten lead to form a good bullet.

Usually, pressure with the hammer or a light tap will force the sprue plate open.

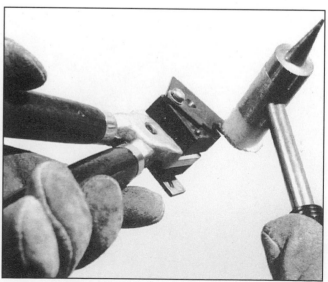

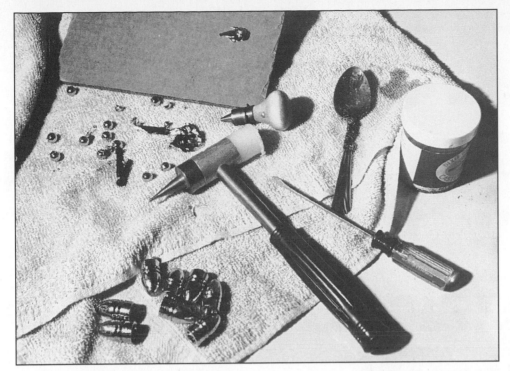

A mould hammer can be used to tap the sprue plate open after the mould has been filled with molten lead.

Product Control

Your home-cast bullets should be just as uniform as commercial products. Check them with a powder/bullet scale. If your bullets vary by more than a percent or so, they are not the best possible missiles. The problem may be that your lead is not pure enough. Or the mould block halves are out of line. Of course, sometimes the trouble lies in the casting technique. You cannot, for example, expect good results unless the lead is carefully poured through the sprue hole so that the liquid evenly and fully fills the cavity within. Timing can cause trouble, too, as noted above. Weighing the bullets does more than verify uniformity. The scale is a detective that can spot air pockets. You know immediately that you have an air pock-

et, or an awful lot of non-lead material, in your bullet when it weighs too light. If your .535-inch round ball mould has been producing bullet after bullet of 230 grains weight and suddenly one shows up that goes 225 grains, that one has an air pocket or a streak of non-lead material in it.

No wonder thousands of shooters cast bullets today for modern as well as old-time firearms. Casting is a hobby unto itself, and well worth doing right. While commercial missiles are superb, and I shoot a lot of them, I admit a touch of special pride when bullets I made myself group well on the target. Hopefully, this introduction to the art of bullet making will prompt the reader to investigate the many full-length books on the topic. As I said, running ball is a hobby within a hobby.

chapter 25

How the Muzzle-Loading Rifle Works

A BASIC UNDERSTANDING of blackpowder firearm function is useful for anyone interested in this branch of the shooting sports. The key to this chapter is the word "basic," for we're interested only in the principle parts that make up these guns and not in a blow-by-blow gunsmith's encyclopedia. Since the non-replica blackpowder rifle is most popular today, we'll dissect one into its major pieces. We'll take a look at safeties, triggers, hammers, plungers, as well as other aspects of the blackpowder gun, with an extremely quick glimpse at the cartridge gun, as well as the revolver. The blackpowder shotgun also receives brief notice, while the muzzle-loading pistol, so much like the front-loading rifle in function, gets only one brush stroke of attention.

The Muzzle-Loading Rifle

This chapter is about the function of the muzzle-loading rifle, with no more than a mention of other blackpowder guns. Our rifle for discussion is a Thompson/Center Cherokee, because it is a straight-out non-replica frontloader, a darn good one to boot, and absolutely typical of the genre. Essentially, a muzzle-loading rifle is a barrel attached to a stock, along with a lock that makes a hammer cock, hold, and fall, and some means of introducing ignition to the powder charge. The breech of the muzzleloader is simply the part of the bore close to the lock that contains the powder charge.

The Barrel

Let's stay with the barrel for a moment. If both ends of the barrel were open, and if a charge were set off within the bore, expanding gases would expel in two directions, fore and aft. That would not make a gun, so the back end of the barrel is closed off, partially sealed, as it were. This job is accomplished by a breech plug. Breech plug fit is so vital to muzzleloader safety and proper function that Thompson/Center individually fits each breech plug to each barrel. If a breech plug were somehow damaged, Thompson/Center would not send another breech plug to just anyone to thread into place. Ideally, a new breech plug would be fitted at the factory, precisely to that individual barrel. On the Cherokee, the breech plug actually performs several functions. It com-prises the closed end of the barrel, or breech, so that expanding gases can only dart forward; it also serves to hold a nipple and a cleanout screw (on my particular model). First, let's dispense with the cleanout screw (or cleanout plug), because only some muzzleloaders have one. The cleanout screw is threaded into a channel that leads to the breech. By removing the cleanout screw, access to the breech is gained. This is somewhat helpful when cleaning the firearm, and it also allows the shooter to introduce a trickle of fine-grain powder should the muzzle-loader malfunction, or should the gun be loaded with ball, but no pow-der, thereby requiring a "booster charge" of fine-grain powder to "blow" the ball out of the muzzle.

As for the nipple, it screws into a threaded hole in the breech plug that leads to the breech. The base of this hole is called the nipple seat.

Continuing with the barrel, there are a few more features worth noting. Not to get too basic, but of course the barrel houses the bore, which in turn carries the rifling, and the frontmost part of this bore is the muzzle, which has a crown on its outermost surface. The barrel also generally holds the sights in place. Sometimes a rear peep sight may be set on a breech plug rather than on the barrel, but on the Cherokee, a typical adjustable open rear sight is screwed onto the top flat of the octagon barrel. Sometimes a rear sight is fitted on the barrel with a dovetail notch.

The front sight of the Cherokee rests in a dovetail notch, and the sight can be moved left or right within the notch. By moving the front sight to the left, the next bullet will strike the target to the right; by moving the front sight to the right, the next bullet will land to the left on the target. In other words, if the group is printing to the right and the shooter (obviously) wants the group to move over more to the left, then he moves the front sight to the right. *So you move the front sight in the opposite direction from where you want the next bullet to place.*

The barrel serves two more functions on the Cherokee. First, it holds a barrel lug, which is also fitted into a dovetail notch. This lug, or tenon, rests in the stock in such a way that a forend wedge or key can be inserted through it. This is what holds the barrel and the stock together (in part) on the Cherokee. Second, on the front underside of the barrel is a rib, and into the rib are fixed two thimbles. Thimbles are

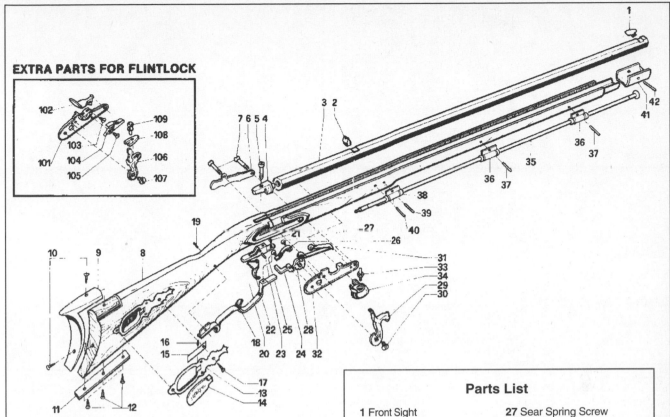

Parts List

1 Front Sight	27 Sear Spring Screw
2 Rear Sight	28 Tumbler
3 Barrel	29 Hammer
4 Breech Plug	30 Hammer Screw
5 Breech Plug Screw	31 Hammer Spring
6 Side Plate	32 Lock Plate, Percussion
7 Side Plate Screw	33 Nipple
8 Stock	34 Drum
9 Butt Plate	35 Ramrod Assembly
10 Butt Plate Screw (2)	36 Upper Ferrule (2)
11 Toe Plate	37 Upper Ferrule Pin (2)
12 Toe Plate Screw (3)	38 Lower Ferrule
13 Patch Box	39 Lower Ferrule Pin
14 Patch Box Cover	40 Tenon Pin
15 Patch Box Cover Spring	41 Nose Cap
16 Patch Box Cover Pin	42 Nose Cap Pin
17 Patch Box Screw (4)	**Extra Parts For Flintlock**
18 Trigger Guard	101 Lock Plate, Flint
19 Trigger Guard Pin	102 Frizzen
20 Trigger Plate	103 Frizzen Screw
21 Trigger Plate Screw	104 Frizzen Spring
22 Trigger	105 Frizzen Spring Screw
23 Trigger Pin	106 Hammer
24 Sear	107 Hammer Screw
25 Sear Screw	108 Top Jaw
26 Sear Spring	109 Top Jaw Screw

the "tubes" that hold the ramrod in place. On the Cherokee, the thimbles are held in place with screws.

The Cherokee has what's known as a hooked breech. This is a rather simple, but ingenious, means of helping to hold the barrel in place, while at the same time allowing the barrel to very easily and quickly lift free of the stock for cleaning. The back of the breech plug has an extension, and this extension locks into a tang. The tang is a metal piece that fits, on the Cherokee, to the upper wrist of the stock via two screws. The back part (extension) of the barrel's breech plug levers into the receptacle provided in the frontmost part of the tang, and when the barrel is settled downward into the stock mortice, the union of male breech plug extension and female tang receptacle join to more or less lock the back of the barrel in place. Of course, the rifle cannot be fired this way, for the barrel would lever up and fly out of the stock mortice. So the all-important forend wedge (key) is installed through the forestock of the rifle through the barrel lug (tenon), strongly linking the barrel and stock of the rifle.

Let's quickly dispense with the ramrod. On the Cherokee, this rod fits through the thimbles and into a hole in the forestock to hold it in place. The ramrod has a bullet seater on the muzzle end and a threaded female jag adapter on the other. The jag adapter accepts many different tools: a jag for hanging onto a cleaning patch, plus worms, screws and other maintenance devices. So much for the ramrod.

The Stock

The stock holds many different parts. The upper forward portion of the stock has the barrel channel or mortice into which the barrel is fitted. The stock may or may not have a forend cap, or metal piece, fitted over the foremost portion of the stock. The Cherokee has a single key through the stock. As previously noted, the key rides through a barrel lug, but on the exterior of the stock, the key goes through two escutcheons that are inlet in place, similar to a keyhole arrangement. These two pieces of metal, shaped like ovals on the Cherokee, rest on either side of the forestock so that the key passes through both of them.

The cutout that holds the lock is called the lock mortice, located on the side of the stock. The lock itself we will skip here, since Chapter 26 deals with various types of locks. The lock is contained with bolts, so there must be holes in the stock to accept these bolts. In most cases, the bolts thread into a lockplate on the opposite side of the stock. While the inlet portion of the stock that accepts the lock is simply a cutout in the wood, it is certainly not a haphazard cutout. It must be neatly and precisely inletted to hold the lock with a good wood-to-metal fit. The lock mortice must also have the proper cutouts to allow the movement of lock parts and insertion of lock bolts. So there's more to the simple lock mortice than meets the eye.

Underneath the lock rests the entire trigger assembly along with the trigger guard. Triggers are discussed later. Looking at the stock, we see a mortice for containing the trigger assembly. Again, the mating of wood to metal must be precise, and as with the lock, the trigger mortice

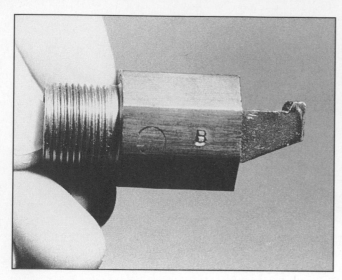

This is a breech plug. It does exactly what its name implies: It plugs the rear of the barrel to form a breech that holds the powder charge. This plug has a hooked breech.

Looking at the breech plug from this angle we see the hook (lower part of picture) and the nipple seat.

must be cut so that trigger parts can properly move. Also inletted into the bottom of the stock below the lock is a mortice to accept the trigger guard. In the case of the Cherokee, the trigger guard serves only the function of safeguarding accidental discharge, by protecting the trigger from making contact with anything. The shooter must extend his forefinger through the trigger guard to reach the trigger in order to fire the rifle. The rearmost part of the Cherokee's trigger guard does serve another function: It supports the hand, not as a pistol grip exactly, but with the same general function. In the next chapter, you'll see a special trigger guard that acts as a hammer mainspring.

The standard blackpowder muzzle-loading rifle with a hammer does not have a separate safety. The hammer acts as a safety when the half-cock notch is engaged. The exception is the modern muzzleloader. These have safeties that function much like contemporary cartridge rifles.

The buttstock holds a buttplate on the Cherokee, which is a rifle-style plate rather than a shotgun-style one. This means it has a curve in

it, while a shotgun-style buttplate is more flat. The buttplate on my particular Cherokee is made of brass. However, as an option, it can hold a recoil pad instead. Basically, that's the stock. Since we're interested in function in this chapter, the style of the stock is not important here.

The Flintlock

No special dissection of a flintlock is included in this chapter, because basic function is similar to the caplock rifle, with the major exception being the lock. There are virtually countless special muzzle-loader designs, and it's not possible, nor practical, to go into them here. For more on the flintlock, see the next chapter, which individually discusses each lock type.

Drum-and-Nipple Conversions

It is worthwhile, however, to talk a moment about the drum-and-nipple setup. Very simple in design, it allows the shooter to convert a flintlock into a percussion firearm. The following is oversimplified,

(Above) This shows the caplock design, including the important nipple seat, where the nipple screws into place. Fire from the percussion cap flies through the nipple (not shown) and into the powder in the breech.

The percussion system is force applied via a hammer to explode a percussion cap. Here, the hammer is drawn to full cock on a Thompson/Center Scout.

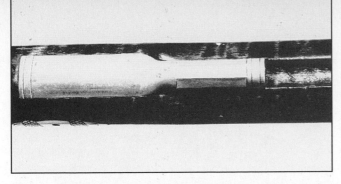

This is called the entry thimble, because it's a ramrod pipe that allows entry of the ramrod into the body of the forestock.

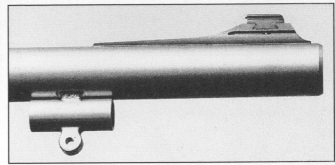

(Above) Shown is a thimble or ramrod pipe, also known as a ramrod guide. This one has an integral sling-swivel eye.

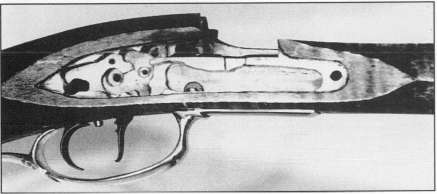

(Left) The lock mortice is the cutout in the stock into which the lock fits.

but it ideally fits our purposes: Drill out the touchhole to a larger diameter and thread this new hole. Then screw in a drum—a cylindrical, threaded piece of metal with a hole all the way through it. Imagine the drum screwed into place where the touchhole of the flintlock used to be.

What's wrong with this picture? Two things. Although the drum threads right into the breech of the rifle barrel and, therefore, lies in association with the powder charge, the other end of the drum is wide open, which means at least some gas would expel through it. Plus, where does ignition come from? It's easy to fix the first problem. Close the open end of the drum with a cleanout screw. Now it is plugged, and it also serves as an avenue to the powder charge in the breech when the cleanout screw is removed—pretty handy. As for igniting the powder charge, drill a hole in the top part of the drum, thread it, and install a nipple. Now a percussion cap can be set upon the cone of the nipple, and when it's touched off, flame will dart through the drum and into the powder charge. Of course, the lock must be changed from a flintlock, with its jaws and flint, to a lock with a hammer that can whack that percussion cap.

The Shotgun

The caplock muzzleloading shotgun is essentially one or two barrels with breech plugs and locks, not entirely different in all respects from the caplock rifle in function. Therefore, we're not going to break one of these guns down into its individual parts. To understand the basic mechanics of the muzzle-loading rifle is to understand the rudimentary operation of the caplock shotgun.

The Pistol

The muzzle-loading pistol is also passed by in this chapter, since it is, in terms of basic function, no more mechanically than a foreshortened muzzle-loading rifle. This goes for both caplock and flintlock pistols.

The Caplock Revolver

The blackpowder cap 'n' ball revolver is a much more complicated firearm than the caplock or flintlock rifle because it has "workings." In fact, later cartridge revolvers were very much a takeoff from this blackpowder handgun. For our purposes, we'll look at the mechanics of the caplock revolver in a very general way. Essentially, it is a frame with a barrel, grips, loading lever, cylinder, and other

The barrel lug or tenon rests on the bottom of the barrel here below the rear sight. Running through this tenon, a key holds the barrel to the stock. The key is also called a wedge.

parts attached. The frame is the body of this handgun. There is, of course, a barrel, which attaches to the forepart of the frame. Onto the barrel a front sight may be dovetailed into place. A frame may or may not have a topstrap, which is the upper piece in between the rearmost of the barrel and the hammer. For example, the famous 1860 Colt blackpowder revolver had no topstrap, while the equally famous Remington Model of 1858 did.

Sometimes revolvers are called "wheelguns." That's because of the revolving cylinder. The object of the cylinder is to hold chambers that align, one at a time, with the bore of the barrel. The word "revolver" says it all. The cylinder revolves on a pin at its center. It moves when the hammer is pulled back, which in turn activates a hand that rotates the cylinder. A cylinder bolt locks the cylinder in place, aligned so that one chamber is "looking" right out of the barrel. There are springs and many other parts that allow the single-action blackpowder revolver to function, but this very basic notation will serve for now. Chapter 20 on loading methods fully explains how the caplock revolver is loaded, which in turn quite clearly describes the function of this handgun, from loading powder into the cylinder chambers to forcing bullets into the same chambers with the use of the loading lever, to putting caps on nipples that are screwed into the back of the cylinder.

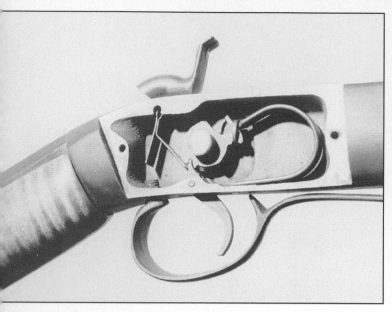

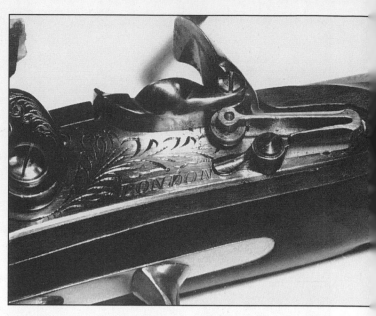

This very basic, and also very good, lock is found on the Mowrey rifle these days. It's a caplock or percussion system with a single trigger, a tumbler, and a mainspring for power.

The flintlock is very familiar in operation, with its pan, pan powder, hammer, flint, frizzen, and flash of flame that darts through a touchhole into the breech. Seen from this angle, the frizzen spring is visible.

The Cartridge Revolver

The function of the blackpowder single-action cartridge revolver is clearly noted in texts on that subject, and is not part of this chapter's goal. The functional principles remain the same as with the blackpowder revolver. The difference is the cylinder, which holds cartridges instead of powder and ball, and, of course, there are no percussion caps, since primers installed into the heads of the cartridges provide ignition. The hammer falls on a firing pin instead of a percussion cap. But all in all, the blackpowder cartridge revolver follows the same principles that make the cap 'n' ball revolver work. Various texts have schematics that clearly show the various parts of the single-action revolver.

The Cartridge Rifle

Two of the most famous blackpowder cartridge rifles of all time are the Remington Rolling Block and the Sharps, both single shots, both employing strong blocks for lockup. We're not going into trigger functions, actions or other aspects of these firearms, which are well-described in other texts. However, a note on the block type of action used by both is in order. On the Sharps, the operation of a lever slides a block down and up. When the block is down, a cartridge can be inserted into the chamber. When the block is in the up position, the chamber is blocked off and the rifle is ready to fire. The Remington has a block, too, but it rolls into position, rather than dropping down and then rising back up. These two actions are both strong in principle and, when made of modern steel, can hold quite a bit of pressure. This is supported by the fact that Remington Rolling Blocks were chambered for smokeless powder cartridges, such as the 7x57mm Mauser, while various falling block actions have been used to chamber many modern rounds, including magnums.

The Modern Muzzleloader

There are various types of modern muzzleloaders which function in different ways. The in-line system gets its name from the fact that the nipple is installed in a bolt of some sort that lies directly in line with the chamber. Therefore, the flame from the percussion cap (or sometime a primer) flies forward directly into the powder charge in the chamber. A bit more on this system in the next chapter.

Muzzleloader Triggers

The topic of triggers for muzzleloaders is at least mildly interesting, and it completes our glance at the function of blackpowder frontload-

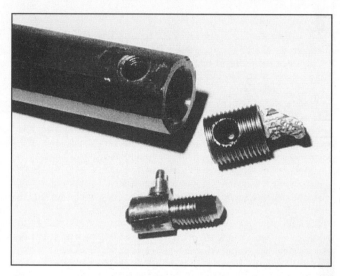

Here is a clear view of the drum-and-nipple system with a hooked breech. The drum is screwed into its threaded channel in the breech plug, and the system is complete.

ers. A basic understanding of triggers can be valuable, because the trigger is vital to the charcoal burner, serving mainly to trip a sear, which in turn activates a hammer, which in turn raps a percussion cap or causes flint and frizzen to produce a shower of sparks. An exception is the modern muzzleloader with modern-type trigger, which is a release-type mechanism instead of a trip type mechanism.

The Basic Single Trigger

Simplest is the single trigger pinned directly into the stock without a trigger plate, as found on old-time fowlers. There is no trigger adjustment in this design, and today it is seldom seen.

Single Trigger with Metal Trigger Plate

Simple, effective and popular, this type is mounted on a metal trigger plate, rather than pinned directly to the stock. The trigger plate acts as an anchor for a tang screw, also serving as a solid base for the trigger to pivot upon. The single trigger does have some travel before engaging the sear, which is bothersome, but this can be corrected by

A Lyman Plains Pistol is shown here in its basic parts. The tang is right above the hammer. The hooked breech, upper right from hammer, fits into the tang for lock-up. The barrel is seen with sights removed. The rib on the underside of the barrel holds the ramrod thimble. Note the dovetail notches for sights. The lock is shown here in the stock, with trigger guard and ramrod below. The barrel wedge or key is bottommost, below the escutcheons.

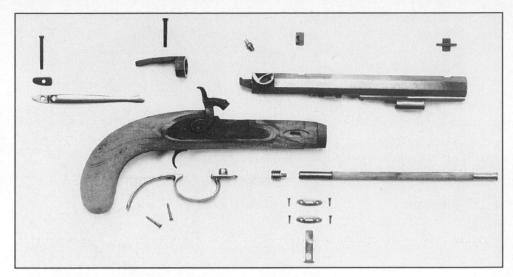

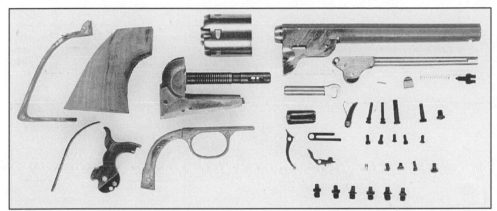

The caplock revolver is obviously more complicated than the cap-lock pistol. This Colt Civil War model, taken down, shows the parts of the gun.

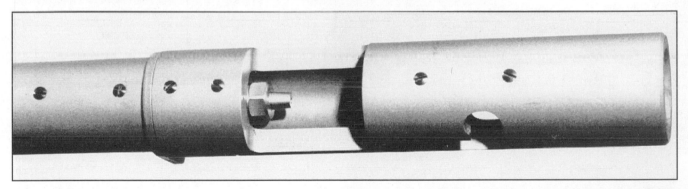

(Above) Here is a clear view of the modern muzzleloader in-line system showing the nipple aligned directly with the bore.

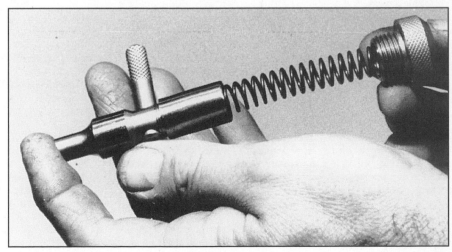

While springs give power to hammers, they also supply the force in the modern muzzle-loader. With this plunger removed, the breech is now clear for cleaning.

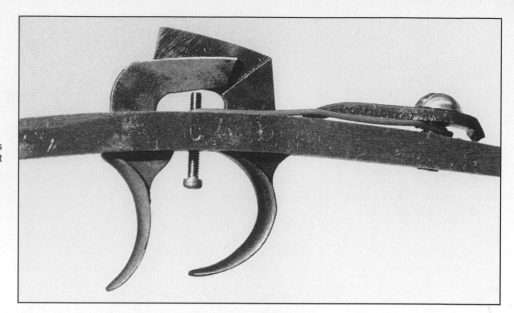

The double-trigger setup allows for a light trigger pull, and a light trigger pull promotes accuracy.

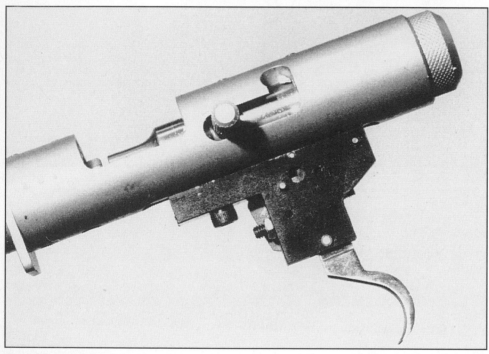

Here we have the in-line ignition system on a modern muzzleloader. Also illustrated is the modern adjustable trigger. The plunger-bolt can be removed for cleaning from the breech end.

installing a weak mousetrap-type spring on the trigger plate to hold the trigger against the sear.

The Single-Set Trigger

The important aspect of the single-set trigger is its ability to offer a very light trigger pull with only one trigger, instead of two. While this type of trigger comes in various sub-styles, the shooter must recognize that it demands precise adjustment for full benefit, and it must be set before the lock can be cocked.

The more advanced single-set multiple-function trigger is unique in that there is but one trigger; however, the gun can be fired in two modes. The trigger can be set by moving it forward until it clicks, or the trigger can be pulled without setting and the gun will fire. Setting the trigger provides a very light let-off, while for fast action, the trigger need not be set.

The Single-Lever Double-Set Trigger

There are two triggers in this system: The rear trigger is the set; the front trigger is the hair. Incidentally, this can be reversed, as on the

Thompson/Center Patriot pistol. Unset, the gun will not fire. But when the trigger is set, the trigger pull is very light. An adjustment screw between the triggers alters let-off. The deeper the screw is threaded upward, the lighter the trigger pull, sometimes measured in mere ounces. *Caution:* This type of trigger can be set too light, so be careful.

The Double-Lever Double-Set Trigger

This trigger is very much like the one previously mentioned, the big difference being that the gun can be fired either in the set or unset position. It is the most common type of trigger found on today's muzzleloaders, as well as on numerous originals. Be careful when using the adjustment screw; do not set triggers too light.

The Modern Muzzleloader Trigger

This trigger is the same as found on modern cartridge guns and will not be discussed further here.

That's a little look at frontloader function, with emphasis on the caplock rifle. The next chapter deals with locks, which are the very heart of blackpowder guns.　●

chapter 26

Basic Types of Blackpowder Locks

LOCKS ARE CLASSIFIED under incidental muzzleloader knowledge; however, there's nothing wrong with that. Many books on firearm history mention locks in passing, but what is a matchlock or wheellock? And, generally speaking, how did these and other lock styles function? Let's take a look.

Firesticks (Hand Cannons)

It's difficult to rank firesticks, or *baston-a-feu* as they were called, as guns with locks. They were truly no more than stoppered tubes made of metal, fairly common during the 15th and 16th centuries in the Netherlands. A shooter—we dare not call him a marksman—poured powder into the tube, followed by some sort of projectile forced downbore on top of the charge. Firesticks, also known as hand cannons, did have one good thing going for them: They could be used as clubs. A firestick might be 2 feet long. Some even had hatchets on one end.

The powder within was ignited through a touchhole at the topmost part of the breech section. For these weapons of war, a soldier carried one of many different implements for ignition. The true "slow match," which was normally a burning cord, may have been used to touch off the charge of powder in the tube. Historians also note burning sticks, heated wires, red hot irons, and even live coals. The military man aimed his firestick in the general direction of the enemy, presented some form of ignition to the touchhole, and boom!

Very little fire escaped from the touchhole due to the Venturi principle, which, oversimplified, means lots more gas escapes from the large hole than the small hole. This principle was seen much later when the German Third Reich aimed "buzz bombs" at Britain. The buzz bomb was a pulse jet that ejected more gas from a larger orifice than a smaller orifice, making it go forward. The firestick employed the human hand with some form of igniter, plus a touchhole, to make up the "lock" of this ancient "gun." One thing is certain: The idea of a touchhole did not fade with firesticks.

Matchlocks

The matchlock was a significant improvement upon the firestick design. We are not exactly certain of the dates, but we do know that a 15th-century German manuscript describes matchlock designs. Now instead of being held by the shooter's hand, a device was mounted upon the firearm itself to hold a slow match or fuse. This time, we're fairly certain that a true slow match was used, sometimes a cord made of hemp fiber, possibly impregnated with saltpeter or some other agent to affect continued but slow smoldering. Along with stocks for holding the gun, the method of retaining the slow match is the major difference between matchlocks and firesticks. We now have our first true lock, a device which mechanically presented ignition to the main charge in the breech.

This was accomplished via a metal holder retaining the slow burning cord. As the cord consumed itself, it could be advanced on the holder. When a trigger device was activated, the holder dropped into place, introducing the tip of the slow match to the touchhole. Apparently, a rather common shape of slow match holder became popular, in the form of a reverse "S," and therefore it was given the name serpentine due to its snake shape. The lower portion of the reverse S protruded below the firearm, representing a trigger on some models. But I have also seen evidence of a true trigger with a spring-activated match holder, not shaped in the S configuration, but simply a curved piece of metal. As always the case in firearm design, there are many variations on a single theme.

On the true serpentine, the upper portion of the reverse S was located above the firearm. When the lower trigger stem of the serpentine was activated, the top stem of the serpentine pivoted forward, the upper nose of the reverse S-shaped piece of metal aimed at a touchhole in the top of the gun. The slow match, held in the upper reverse S portion, then came in contact with the touchhole. This system was altered numerous times. Sometimes the slow match was held atop the barrel, with a type of flash pan used to hold powder and to direct flame from the powder into the touchhole after ignition. The name "matchlock" is therefore very appropriate. It was a true lock device used to deliver a slow-burning match to a touchhole, and it did so mechanically. That last part is important to the lock definition.

The Monk's Guns

The matchlock progressed into many sub-styles. Eventually, it was superseded by what I call pre-wheellock designs. Some histories credit the Germans of the 16th century with one of these guns, which attempted to attach a flint and steel to its side. Certainly, the principle of flint

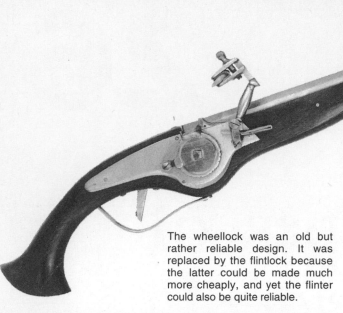

The wheellock was an old but rather reliable design. It was replaced by the flintlock because the latter could be made much more cheaply, and yet the flinter could also be quite reliable.

and steel making sparks was nothing new. Ancient man probably made fire this way. If sparks from a flint and steel could send sparks into a pan, igniting the powder held therein, the flash from the pan could in turn enter a touchhole for ignition of a powder charge in the breech. Matchlocks already functioned on the principle of powder and touchhole anyway, so furthering the basic mechanics was not out of reach.

The pre-wheellock described here was called a Monk's Gun by some. It was supposedly built around 1510-1515, an early example of a spark-fired gun, although it was not a true wheellock. It did have a serpentine, but instead of holding a slow match, it grasped a flint or pyrite. A sort of plunger was used in concert with the serpentine. By withdrawing the plunger quickly, via a thumb or finger ring, a roughened steel bar scraped rapidly against the pyrite held in the jaws of the serpentine, which was a forerunner of the later flintlock cock or hammer. This action produced sparks that hit a touchhole located immediately in front of the serpentine's jaw. While the Monk's Gun was not a true wheellock, one did show up around 1515, history suggests, in Nuremburg. This time it was a real wheellock, for it had a wheel rather than a sliding steel bar. The wheel was an integral part of the lock system. I do not know how this early gun was aimed, or how effective it was, but better wheellocks were on the way.

Snap Locks

Not much is written on the snap lock. It may have predated the Monk's gun, but we really don't know for certain. It appears to be, in design and function, a very late matchlock design, the improvement being a tube containing a bit of smoldering tinder or perhaps a short section of slow-burning match cord. Apparently, the tube concealed the burning end of the match so that it did not give a soldier's position away. Once again, it seems that this specific lock style belonged to the Scandinavian countries of Sweden, Norway and Denmark in the middle to late 1500s.

True Wheellocks

In function, the improved wheellock was not unlike a spring-wound clock mechanism, with a key or spanner used to wind up the wheel. Spring power was harnessed to do the labor, rotating a wheel and mak-

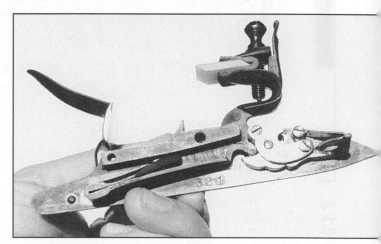

The flintlock is still with us. It's an interesting and colorful lock style depending upon a hammer with a flint, a frizzen for sparks, and a pan with fine powder to produce an ignition flame.

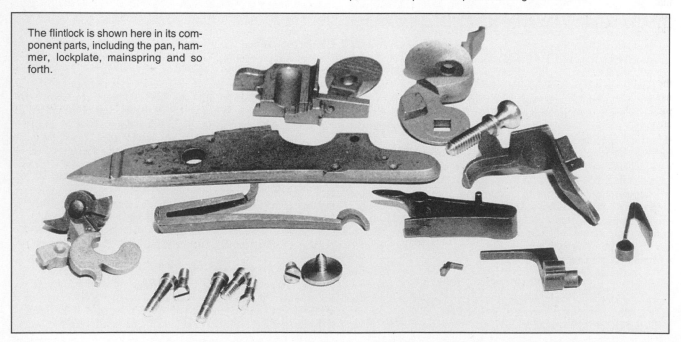

The flintlock is shown here in its component parts, including the pan, hammer, lockplate, mainspring and so forth.

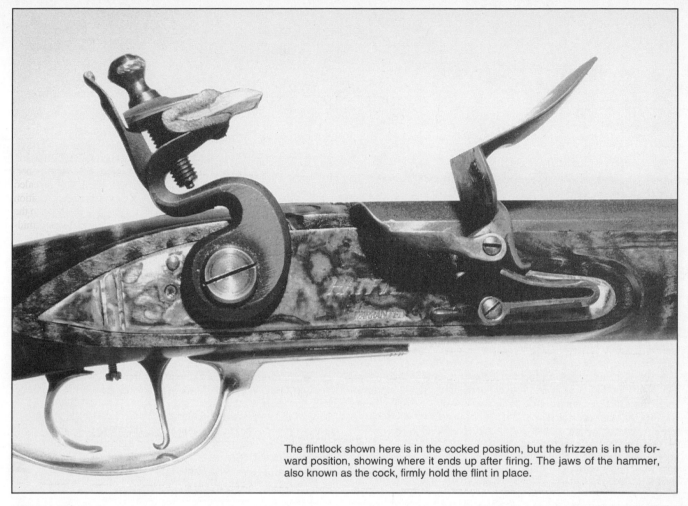

The flintlock shown here is in the cocked position, but the frizzen is in the forward position, showing where it ends up after firing. The jaws of the hammer, also known as the cock, firmly hold the flint in place.

ing sparks. To shoot the wheellock, the flash pan lid was opened, the cock was placed in contact with the wheel, and pressure on the trigger released the wound up spring powered wheel. The speedy rotation of the wheel against pyrites held in the jaw of the serpentine created a shower of sparks. Numerous improvements turned the wheellock into an ignition system of high merit and reliability. Instead of a slow match, or for that matter a single spark, or even a few sparks coming in contact with the pan powder, a shower of sparks fell downward, providing a prime chance for ignition of the powder charge in the breech.

Snaphaunces

Also spelled snaphance, this lock design was clearly a forerunner of the flintlock. It had a true trigger that released a cock (which today we usually call a hammer). More importantly, the snaphaunce cock held a piece of flint in its jaws. The snaphaunce used what was called a steel anvil, but it seems to be a frizzen in practice. This anvil produced sparks, and the sparks were dropped into a waiting charge of priming powder, which sent a flame toward a touchhole. When did the snaphaunce come into being? We don't know. Perhaps as early as 1525, but more likely as late as 1570. The word comes from Dutch or Flemish, meaning "snapping hen," perhaps in reference to a chicken pecking at food?

Flintlocks

The firestick died away, happily. The matchlock fell by the wayside, too, as it should have. The wheellock was actually a good system, but it was expensive to manufacture and not that simple. The flintlock came along to replace it, and we still shoot flinters to this day. Although inferior to the wheellock in operation, the flintlock took over. It was less intricate, less difficult to produce in quantity, and much less expensive, therefore more in tune with the needs of the so-called aver-

age shooter. The flintlock existed in dozens, if not hundreds, of variations, even repeaters, so no single flintlock design can be singled out as definitive. However, the standard flintlock most frequently fired today has a hammer or cock that pivots forward when the trigger of the gun is depressed. This cock has a tang or "comb" section on the backside and a set of jaws. A jaw screw goes through a top jaw, then into a lower jaw, the screw used to cinch together the two parts of the jaw.

So the jaw screw, also known as top jaw screw or cock pin (old-time nomenclature), brings the upper and lower jaws together in a clamping fashion which firmly and securely grasps a flint. A bit of tanned leather, called a flint pad, is usually wrapped around the flint for further security. While tanned leather is common today, as well as on the lock jaws of sporting rifles of the past, thin sheets of lead were often used as flint pads for old-time muskets. The object of the hammer or cock is to present the flint squarely against the face of a frizzen, which was also known as a frizzle, battery, hen or steel—all accepted names for the frizzen. Logically, the frizzen is powered by a frizzen spring, which is mounted on the outside of the lockplate. The frizzen spring also has an old-fashioned name, which is "feather spring."

The object of the frizzen spring is to direct the frizzen out of the way when the hammer-driven flint strikes down against the face of the frizzen. This action diverts the frizzen forward, thereby exposing another important feature of the flintlock, the pan, also known as the flash pan. The pan contains FFFFg (4F) pan powder, the finest granulation of blackpowder currently available to sportsmen, with the exception of Elephant brand FFFFFg (5F). The object of the flintlock is to shower sparks into the pan through the action of the flint scraping curls of hot metal from the face of the frizzen. Eventually, the frizzen can become worn to the point where sparks do not readily generate from contact with the flint. However, with a good flint, well-positioned, and a properly hardened frizzen face, the flintlock works quite well.

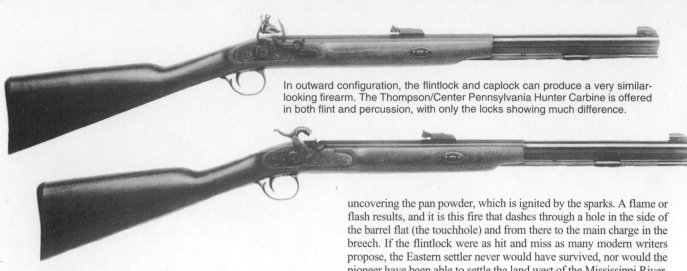

In outward configuration, the flintlock and caplock can produce a very similar-looking firearm. The Thompson/Center Pennsylvania Hunter Carbine is offered in both flint and percussion, with only the locks showing much difference.

Looking at the inside of the flintlock, which faces inward into the stock's lock mortice and cannot be seen when the lock is in place on the firearm, we find a tumbler. The tumbler rotates with the hammer or cock. Half-cock and full-cock notches are integral to the tumbler. These notches are engaged by the sear, which operates via a sear spring. The sear spring must be depressed in order for the nose of the sear to disengage from the notches in the tumbler. Also seen on the inside of the lock is the mainspring, which powers the hammer by forcing the forward side of the tumbler downward. The lock may have a bridle, a piece of metal that spans over both tumbler and sear, providing solid pivot points for both tumbler and sear. The sear releases the tumbler when the trigger is activated, thereby allowing the hammer to fall forward.

There may be a fly in the tumbler, also known as a detent. The fly serves an important function with double-set triggers, preventing the sear from falling into the half-cock notch, and is in effect an override device that does not interfere with the activation of the full-cock notch. The fly allows the firearm to be brought into battery or firing position without the sear hanging up in the half-cock notch. Without a fly in the tumbler, the hammer may not fall fully downward when the trigger is pulled. A stirrup may also be present in some locks. This part rests between the tumbler and spring tip, thereby serving to reduce friction at this location.

The rest of the flintlock story is rather self-evident. The flint scrapes the frizzen, and as the frizzen falls forward, the pan cover is lifted, uncovering the pan powder, which is ignited by the sparks. A flame or flash results, and it is this fire that dashes through a hole in the side of the barrel flat (the touchhole) and from there to the main charge in the breech. If the flintlock were as hit and miss as many modern writers propose, the Eastern settler never would have survived, nor would the pioneer have been able to settle the land west of the Mississippi River. Of course, flintlocks failed to ignite with 100-percent reliability. That's where the term "flash in the pan" comes from, a flash that does not result in ignition. There is also a hangfire, which is a lag between pulling the trigger and the gun going off, and a misfire, which is pulling the trigger without ignition following.

Pill Locks

The pill lock can be thought of (almost) as a cross between a flintlock and a percussion lock. There is a hammer and a cover where normally a frizzen would reside. The pill lock worked with tiny pellets made of detonating powder (impact-sensitive explosive). The falling hammer crushed the detonating pellet; producing a spark that flew into the charge of powder in the breech. The pill lock is often credited to the Scottish clergyman, Rev. Alexander John Forsyth. Apparently, a patent dated 1807 verifies this claim.

Percussion Locks

The percussion lock is also known as a caplock and is currently the most popular ignition system in use. The invention of a workable percussion cap in the 1800s precluded the necessity for pan powder. Ignition was produced by striking the cap, which contained a percussion-sensitive mixture, hence percussion cap. The cap rested upon a nipple, the nipple directing fire from the cap into the charge in the breech—no special granulation of blackpowder necessary. The mix-

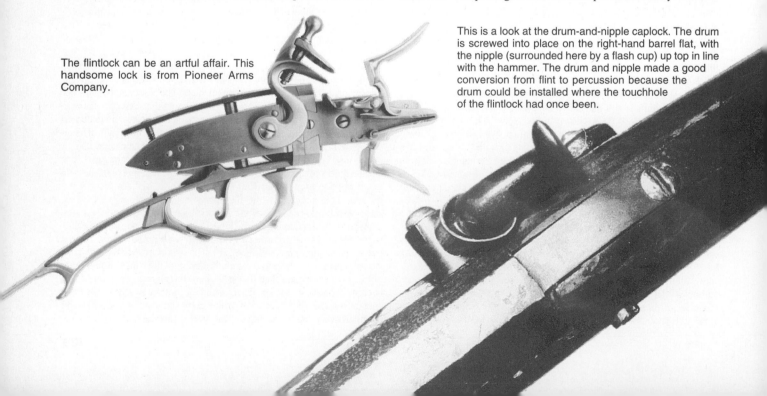

The flintlock can be an artful affair. This handsome lock is from Pioneer Arms Company.

This is a look at the drum-and-nipple caplock. The drum is screwed into place on the right-hand barrel flat, with the nipple (surrounded here by a flash cup) up top in line with the hammer. The drum and nipple made a good conversion from flint to percussion because the drum could be installed where the touchhole of the flintlock had once been.

The back-action lock has the mainspring located behind the hammer. This style of lock can be lean and handsome, and it functions perfectly as well.

Some locks have a detent or fly in the tumbler to allow the nose of the scar to override the half-cock notch for smooth operation. The arrow points to the fly.

This sidelock has an upswing on the hammer. This type is a direct ignition system, since the nipple screws directly into the barrel where the powder charge rests.

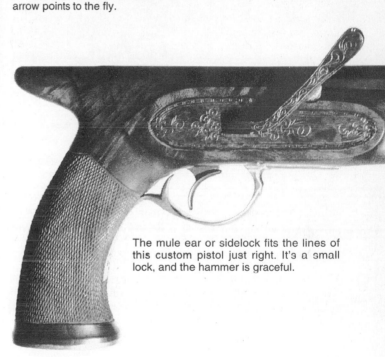

The mule ear or sidelock fits the lines of this custom pistol just right. It's a small lock, and the hammer is graceful.

ture in the cap actually gave off more of a spark than a flame, but that point is academic. Incidentally, there was no single caplock design, as evidenced by the many different percussion locks that exist to this day. As noted in the previous chapter, it was fairly easy to convert flintlocks into caplocks using the drum and nipple system, so even some flinters became caplocks. Plus, the drum and nipple caplock design has been installed on certain new muzzleloaders in recent times. What all the designs share in common is a hammer which hits a cap that is placed upon the cone of the nipple, thereby setting off the firearm.

There are far too many percussion locks to include here. The following are only some of the possibilities. They're interesting from the standpoint of mechanical design.

Front-Action Locks

This is the workhorse lock of the percussion ignition system today, a standard. The mainspring is in front of the hammer as opposed to behind it, hence: front-action lock. This type of percussion lock functions well on single-barrel pistols and rifles. It is also entirely workable for side-by-side double guns, because the locks can be fitted down under the side of the barrel, as on the double-barrel shotgun, for example. The front-action percussion lock is entirely acceptable, and that's why it remains so popular today.

Back-Action Locks

As opposed to the front-action lock, this percussion lock has the mainspring in back of the hammer. This allows the wrist area of a rifle

to be slimmer than that of the front action lock. There is no panel or moulding around the lock in the stock, promoting a more trim firearm design. Also, the back action lock adapts well to swivel-breech firearms and to certain over/under double-barrel configurations. It is excellent for takedown breech systems, too, because the lockplate is located on the wrist of the rifle, rather than up front by the barrel. The back-action percussion lock has been used on slug guns, which are target muzzleloaders with huge barrels.

Front-Action Sidehammer or "Mule Ear" Percussion Locks

A short-lived style because it appeared toward the end of the black-powder era, the sidehammer remains a superb percussion lock design due to its simplicity and reliable function. The main feature of the sidehammer is its *direct ignition* quality. Note that this is not in-line ignition, but it is direct, because the nipple is fitted directly to the barrel. Thus, fire from the nipple darts into the main charge in the breech without the spark routing through a drum and nipple, or any other avenue. Also, the simplicity of design means fewer working parts, although certain side-

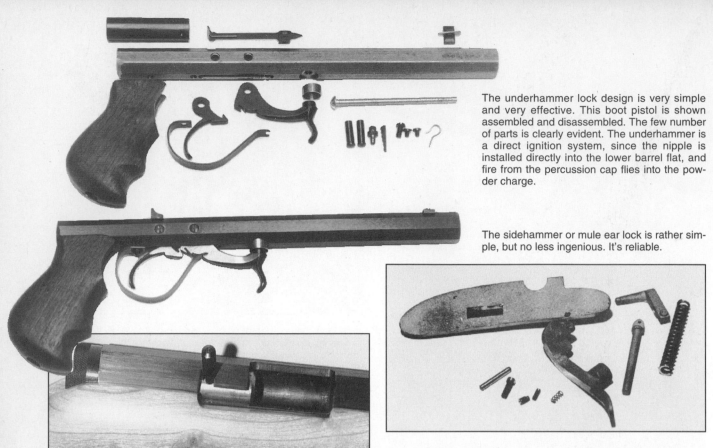

The underhammer lock design is very simple and very effective. This boot pistol is shown assembled and disassembled. The few number of parts is clearly evident. The underhammer is a direct ignition system, since the nipple is installed directly into the lower barrel flat, and fire from the percussion cap flies into the powder charge.

The sidehammer or mule ear lock is rather simple, but no less ingenious. It's reliable.

Is it a lock or an action? This early modern muzzleloader uses in-line ignition, the forward thrust of the plunger setting off the percussion cap.

hammer designs did use complicated trigger systems because of older lock styles with sears activated by lateral instead of vertical movement. Another good feature of the sidehammer is the direction of cap debris away from the shooter. Additionally, many sidehammers have deep hammer noses that tend to contain cap debris. On the other hand, a detriment of older-design sidehammers was a rather long hammer fall. This feature was corrected on the patented Storey sidehammer lock, which employs a strong coil mainspring and a shorter hammer throw. The Storey version of the sidehammer lock also has a sear that moves vertically, which allows the use of a regular trigger.

Back-Action Sidehammer Percussion Locks

Possibly the most versatile percussion lock system, this one works well on single-barrel rifles and pistols, especially for takedown rifles or switch-barrel sets. It functions on a double gun, too, albeit somewhat awkwardly when both hammers are cocked, due to the protrusion of these hammers on either side. This type of lock was built with two hammers on one lockplate and was possibly the best lock for over/under guns. Quick, positive ignition is a high point of this lock type, excellent especially for target guns requiring shorter lock times. On the debit side of the ledger, these locks are somewhat difficult to build. Furthermore, incorporating a half-cock notch is difficult, although it can be done at higher cost in labor.

Underhammer Percussion Locks

This is a simple design. The nipple is screwed directly into the under-barrel of the pistol or rifle. Cap debris is directed downward and away from the shooter. There were many different styles of underhammer locks over the years, most employing a trigger guard as a mainspring, a simple but smart idea. Double-set triggers are not entirely compatible

with the underhammer lock, although they can be used. The underhammer lock is at the core of building an economical muzzleloader. It's also noted as a fine system for the target rifle, due to positive ignition qualities. Underhammer locks have been satisfactorily employed on heavy bench target rifles, as well as on competition shotguns.

In-Line Ignition Percussion Locks

The in-line ignition system is not truly a lock, but neither is it generally associated with actions, so we'll treat it as a blackpowder lock-type system. Not all, but many modern muzzleloaders use in-line ignition, which works on the basic principle of a plunger system. The in-line lock system has undergone several modifications, including a hammer gun with a nipple threaded into a solid breech—no plunger at all, yet true in-line ignition. Popular on modern muzzleloaders, yes, nevertheless the idea is nothing new at all. The Friendship Special over/under shotgun used the in-line system for years. This is an over/under shotgun with two spring-loaded plungers, developed by shooters who frequented the famous Friendship, Indiana, annual blackpowder match. Also, the in-line ignition system was found on converted Enfield and Japanese Arisaka bolt-action rifles turned into muzzleloaders.

This type of ignition is ideal for the modern muzzleloader (see chapter 11). When the trigger is pulled, the cocked spring-loaded plunger darts forward, the nose of the plunger striking a percussion cap seated on a nipple mounted directly into the rear of the barrel. In-line ignition is the result, the fire from the cap flying directly into the powder charge in the breech. This system allows the use of modern-style stocks, modern triggers and safeties, and many other features associated with cartridge firearms.

Admittedly, certain lock chronology discussed in this chapter could be a shade off the mark. That's because no one knows for sure the exact dates for each lock invention. However, the task is accomplished: This chapter's aim is to give the reader a bit of background. To go further and deeper into the subject can become an exercise in "more than you ever wanted to know about locks." Suffice it to say that there were hundreds of different lock designs. Today, we have only a comparative few remaining, the ones that weathered time and trial, as well as shooter opinion and caprice. ●

chapter 27

Understanding Rifling Twist

The Hawken influence led to the 1:48 rate of twist in non-replica muzzleloaders with the Hawken name, such as this well-made Thompson/Center model. The 1:48 twist is known as an "in-between" rate.

THE HAWKEN BROTHERS are often held up as proof positive that the 1:48 twist, a single revolution of the projectile in 48 inches of bore, is precisely correct for all blackpowder firearms. After all, how could the brothers Hawken be wrong? Researchers say that the Hawken boys did rifle everything with a 1:48 rate of twist. Apparently, their equipment was limited to this rate. Furthermore, 1:48 was a standard in those times, as pointed out by Ned Roberts, who said in his book, *The Muzzle-Loading Cap Lock Rifle,* "It appears that the old-time riflesmith's 'standard' twist of rifling for round ball rifles was one turn in 48 inches." Roberts went on to say that some smiths preferred 1:60 and even slower twists.

Captain James Forsyth understood the properties of rifling twist better than many of his predecessors, and certainly some of his peers. Forsyth knew that no single rate of twist could possibly be correct for all calibers. That would be scientifically impossible. Modern gunmakers have known this for a very long time. The 25-35 Winchester, for example, on the market by 1895, had a fast 1:8 rate of twist, because it shot a long 117-grain 25-caliber bullet and Winchester engineers wanted to ensure stabilization. In the same year, the same company introduced the 30-30 round in the same Model 1894 rifle, but you can be sure that the same twist was not used. The 30-30 (called 30 WCF at first) had a 1:12 twist.

I think it's very important to establish the fact early that twist must vary with the projectile. As mentioned often, the round ball, with its low sectional density, requires very little spin to keep it on track, while conicals demand more revolutions per second (rps) for stability. So you cannot talk about rifling twist without talking about caliber and style of projectile. As caliber increases, projectile mass also increases. As projectile mass grows, fewer rps are required to stabilize the missile. In other words, a bullet from a 32-caliber muzzleloader requires more rps to keep it revolving on its axis than a bullet from a 58-caliber rifle. So there it is: No single rate of twist can possibly be correct for all guns.

Projectile style also dictates the rate of twist required for stabilization. For example, certain streamlined projectiles are very long for their

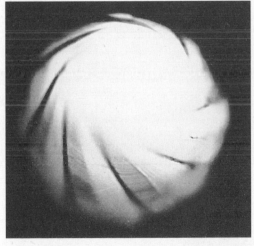

Looking right down the bore of a muzzleloader, we see what twist is all about. The pattern of the rifling forces the bullet to spin.

caliber, and in order to keep them spinning on their axes, rate of twist must be quite fast. A perfect example is Sierra's accurate 69-grain .224-inch diameter MatchKing bullet. This long-for-its-caliber bullet found its way into a friend's 222 Remington rifle with 1:14 rate of twist. The bullet keyholed on the way to the target, hitting sideways instead of point-first. The same bullet was loaded in a Colt AR-15A2 rifle with a fast 1:7 rate of twist, and it flew point-forward to the target, obviously stabilized.

Rifling Twist and Stabilization in Rifles

If anyone, be he writer, veteran blackpowder shooter, or your brother-in-law, insists that the 1:48 rate of twist has some magical property, just nod your head and smile. Don't argue; it won't help. However, at the same time do keep in mind that there is a *range* of acceptable twist for a given projectile. Arguably, there may be one best twist to stabilize a particular missile; however, several different rates of twist will provide proper stabilization for a given bullet. For example, a 32-caliber round ball can be stabilized with a 1:30 or a 1:50 rate of twist, depending in part upon the muzzle velocity imparted to the projectile (more on that later).

Smoothbores, while more efficient and useful in the field than we like to give them credit for, never produced the brand of accuracy sought after by serious marksmen, especially target shooters. The projectile fired from a smoothbore tended to take a line of flight dictated by its distribution of mass. One of the reasons for superb round ball accuracy from a rifled gun, aside from the obvious factor of stabilization which we are now discussing, is the equalization of discrepancies on a common axis. Look at it this way. The lead ball is imperfect. It is not 100-percent homogenous in its molecular distribution. But if you spin the ball on its axis, then the flaws in its character revolve around a common line. By equalizing discrepancies, those differences misguide the missile far less. Therefore, one of the reasons for round ball accuracy—and the lead globe can be surprisingly accurate from a good barrel—is equalization of bullet irregularities through spinning.

Revolutions Per Second

Two and only two major factors dictate projectile rps: rate of twist and exit velocity. The operative word here is *rate* of twist. Barrel length does not matter so much, although it does affect muzzle velocity. Again: The length of a barrel has nothing to do, in and of itself, with bullet stabilization. It's unfortunate that we have come to think of twist as the number of revolutions the missile makes in a given length of bore, because it is misleading and inaccurate. Imagine you have a 50-caliber rifle with a 1:60 rate of twist and a 40-inch barrel, and the ball leaves the muzzle at 2000 feet per second, then you cut the barrel down to 20 inches, but somehow the ball still leaves the muzzle at 2000 fps MV. There would be *no difference* in the stabilization of the projectile out of the two barrels, even though one barrel is twice as long as the other. Think of the rate of twist as a turn in so many inches, but not inches of barrel.

Nor can you look at missile stabilization as a matter of trigonometry. Tangents of angles, graphic triangles, and little line drawings tell nothing worthwhile about rifling twist. This statement will anger readers who have believed in an explanation of rifling twist based on trigonometry. I can only advise these persons to consult a scientist who understands ballistics, or to bring this chapter to that scientist and see what he says about angles, tangents, triangles and such when missile stabilization and rate of twist are under discussion. Captain James Forsyth, mentioned above, worked diligently

Col. Frank Mayer may have been a buffalo hunter with a taste for Sharps cartridge rifles, but here he's shown with something quite different, a long-barreled muzzleloader. Would this long barrel dictate rifling twist? Not directly.

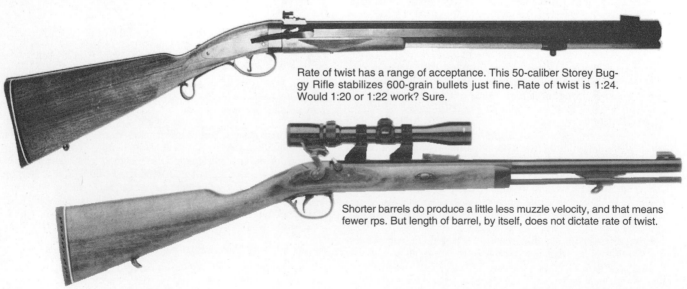

Rate of twist has a range of acceptance. This 50-caliber Storey Buggy Rifle stabilizes 600-grain bullets just fine. Rate of twist is 1:24. Would 1:20 or 1:22 work? Sure.

Shorter barrels do produce a little less muzzle velocity, and that means fewer rps. But length of barrel, by itself, does not dictate rate of twist.

Round balls require very few rps (revolutions per second) for stabilization, but smaller round balls need more than larger ones, due to less mass. These round balls run (from left): .310-inch (45 grains), .350-inch (65 grains), .375-inch (80 grains), .395-inch (93 grains) and .690-inch (454 grains).

on the mathematics of rifling twist back in the 19th century. While his formulas are not of much interest to us today, his conclusions are.

Forsyth concluded, along with Major Shakespear and Sir Samuel Baker, that the larger the missile grows in mass, the less rps is required to stabilize it. Forsyth was right. And he proved his conclusions by having rifles made up with very slow rates of twist. These rifles stabilized large-caliber round balls to long range; they were also quite accurate. Like a train on a track, the bullet can only go where the rifling leads it. Also like a train on a track, the bullet can either stay on its "rails" or it can scoot over them and crash. When the projectile is no longer guided by the rifling, that condition is called, in 19th-century terminology, "tripping over the rifling" or "stripping the bore." No matter what you call it, it spells disaster to accuracy.

Light Loads

Some oldtime gunmakers, including the Hawken brothers, built ball-shooting rifles with comparatively fast rates of twist. These rifles required that the round ball be held to a somewhat modest muzzle velocity because, remember, rps is controlled by rate of twist and exit velocity. When the ball was given too much speed, it stripped the rifling and flew off course. Now you can see where the old idea,

believed by many modern blackpowder gurus as well as their predecessors, concerning accuracy and light loads got started. If the blasted ball had been stabilized to begin with it would have maintained its accuracy in spite of its speed. Not knowing this, the "light-load boys" said that high velocity ruined accuracy, and the notion that "only light loads are accurate" was born and promoted. This falsehood thrived when fast-twist rifles caused the patched round ball to trip over the rifling. Derail a train and it crashes. "Derail" a missile and it cannot fly true.

Round Ball Circumference and Twist

While angles and tangents are relatively meaningless in a discussion of bullet stabilization, it is true that the circumference of a bigger ball is twisting faster than the circumference of a smaller ball. In other words, the speed of a 60-caliber lead sphere on the outside of its surface is greater than the speed of a 50-caliber lead sphere on the outside of its surface, when both missiles are driven at the same velocity from a bore with the same rate of twist. Let's use the 1:66 rate of twist as an example, a single revolution per 66 inches of travel. While the 60-caliber ball's circumference is moving faster than the 50-caliber ball's circumference, there is an equalization factor at work, because the diameter of the rifling changes in exactly the same

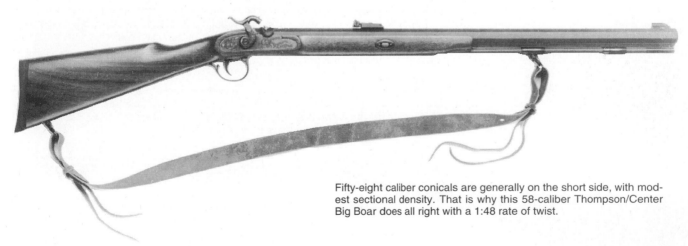

Fifty-eight caliber conicals are generally on the short side, with modest sectional density. That is why this 58-caliber Thompson/Center Big Boar does all right with a 1:48 rate of twist.

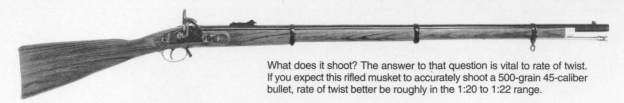

What does it shoot? The answer to that question is vital to rate of twist. If you expect this rifled musket to accurately shoot a 500-grain 45-caliber bullet, rate of twist better be roughly in the 1:20 to 1:22 range.

proportion with the different calibers. So the faster surface movement of the outside of the 60-caliber ball moves over a longer path in the 1:66 rate of twist compared with the 50-caliber ball in the same twist. It all comes out to dead equal. Do not be confused by the preceding: The number of turns in the bore for the 50-caliber ball and the 60-caliber ball are exactly the same in a 1:66 rate of twist barrel.

Here is a mental model that may help produce a picture of the situation. It deals with bolts and nuts. There are two bolts and nuts in this story. One bolt is 1/4-inch in diameter, the other 1/2-inch. Both have, of course, corresponding threaded nuts. Both have 20 threads to the inch, or in rifling talk, 20 turns to the inch. If *either* nut is moved down its respective bolt 1 inch, that nut will make 20 turns. That's the fact we're after in this model. In spite of the diameters of the two bolts being quite different, both are guided identically by their rate of twist. This holds true as long as the nut (or bullet) does not strip out.

This model is important because it reveals several facts. First, no matter if the bolt is 1/4-inch long or 2 feet long, the rate of twist imparted to the nut has to stay the same. Second, the nut can be driven that inch in one minute, one second, or one millisecond, or any other period of time. If the nut is driven at different speeds, the revolutions per time must vary. This is very important to our story because we can see that rps will vary in accord with forward motion of the projectile as it is guided by the rifling.

Rifling Twist in Practice

Now we're getting somewhere. We have some facts to work with. Since rps are related to rate of twist and exit velocity of the missile, we have to consider how many rps are needed to keep an elongated bullet flying point-on and to keep a ball revolving on its axis. A projectile is like a free-moving gyroscope. In order to stabilize a gyroscope, you spin it. Can you spin it too much? There may be a point at which the gyroscope goes crazy when it is revolved too rapidly. However, it is clearly evident that when the gyroscope slows down below a specific spin rate it falls over on its side. When a conical projectile is not given sufficient rps, it may wobble in flight or keyhole. When a round ball is not given sufficient rps, it may depart from its original intended line of flight.

Now let's get practical. I have worked extensively with two custom 54-caliber muzzle-loading rifles. They are very different in their design: One is meant for round ball shooting, the other for conicals. The first carries a rate of twist amounting to a turn in 79 inches (1:79). That is not because I demanded a turn in 79 inches. As noted earlier, a projectile can be stabilized within a range of rps. I simply asked for a slow twist, something around 1:75 to 1:85 and ended up with 1:79, which is fine. This rifle shoots a patched ball remarkably well, as proved when I fixed a 6-24x Bausch & Lomb scope to it and produced five-shot groups of under an inch center to center at 100 yards from the muzzle. The other 54-caliber rifle was built with a 1:34 rate of twist. It will shoot a round ball, but only if the velocity of the missile is held down. Otherwise, accuracy goes to pot. However, it shoots a 460-grain Buffalo Bullet with fine accuracy. Meanwhile, the rifle with the 1:79 twist will not stabilize the conical, as proved by targets with elongated holes in the paper, showing a keyholed bullet.

But isn't a turn in 34 inches still quite slow? Yes, if you compare it with modern streamlined bullets, which have high sectional densities and high ballistic coefficients. However, in the muzzleloader, a 1:34 rate of twist is considered on the fast side, because even the elongated or conical blackpowder bullet is still of comparatively low sectional density and low ballistic coefficient. For example, a 180-grain 30-caliber spitzer bullet may have a ballistic coefficient of over .500, while a 54-caliber 410-grain Minie has a ballistic coefficient of only .137 (Lyman data). Not many rps are needed to stabilize a bullet of such squat proportion. However, a lot of twist is required to stabilize a streamlined, rocket-like projectile.

This fact is born out with the VLD (Very Low Drag) bullet. The 105-grain 6mm VLD bullet keyholed from rifles with 1:10 rates of twist. A special barrel had to be made to stabilize these handsome projectiles, and one turn in 8 inches works well. The 226 Barnes QT (Quick Twist) is another example of a cartridge that demands a fast twist. This 22-caliber wildcat fires a bullet of 125 grains weight, a .226-inch diameter missile that is about 1 3/8 inches long. Experimenters found that a twist of 1:5.5 was needed to stabilize it—a full revolution of the bullet in only 5 1/2 inches of forward motion.

Variables That Affect Twist

There are numerous cause-and-effect gremlins that may bend the laws of linear (Newtonian) physics, seeming to make liars of the most careful ballisticians. Guns can be very different in their performance partly because of "hidden factors" that are often very difficult to assess. For example, depth of groove can make a difference in projectile behavior in a firearm. A deep groove creates a tall land, and a tall land may grip the patched ball quite well, preventing its tripping over the rifling to some degree. So we may run across an accurate ball-shooting rifle with a fairly quick twist. That rifle achieves some degree of accuracy because the tall land grips the patched ball to prevent stripping. Nonetheless, the laws of rps still hold up. They are just bent a little.

Ball fit is also important in this story. If the round ball fits tightly to the bore, and is wrapped in a strong patch to boot, stripping may be somewhat thwarted. When a conical fits tightly to the bore, that can make a difference in the relationship between projectile and rifling.

All along, we've been hinting about bullet shape variations and stabilization from various bores. Consider the conical that has a lot of bearing surface due to its shape. It may, for example, be a round-nose bullet, whereby it's mainly shank with a lot of bearing surface to contact the rifling in the bore. So bullet shape, aside from ballistic coefficient, can have an effect on stabilization.

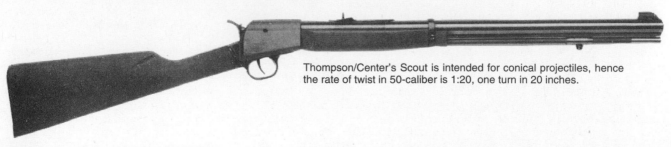

Thompson/Center's Scout is intended for conical projectiles, hence the rate of twist in 50-caliber is 1:20, one turn in 20 inches.

The author's Mulford 54-caliber rifle has a 1:79 rate of twist. This slow twist perfectly stabilizes the 54-caliber round ball.

tions). In working toward the most accurate load possible for your blackpowder firearm, you must consider the rate of twist.

But don't forget that you are dealing with a range of twist. A 54-caliber ball-shooting rifle with a 1:79 rate of twist was mentioned previously. How would you load for this rifle? After much range work, I learned that, although the rifle was built to handle as high as 140 grains of FFg behind a .535-inch patched round ball, velocity peaked with 120 grains—the law of diminishing returns was at work. Therefore, my optimum charge in that 54-caliber rifle became 120 grains of FFg with a .535-inch round ball and .013-inch thick Irish linen patch for a muzzle velocity over 1900 feet per second. However, the same rifle was also accurate with a 70-grain charge of FFg (same ball and patch), as well as with only 40 grains of FFFg—the latter a decent plinker that got 175 shots out of a pound of blackpowder.

Too Much Twist?

Overstabilization can occur. At least, we think so. Dr. Lou Palmisano, a vascular surgeon, invented what is at the moment the world's most accurate cartridge, the 6mm PPC. He tried various rates of twist, ending up with 1:14 because his goal was stabilization of a 70-grain target bullet, and that was just enough twist. Overstabilization with round balls and lead conicals may not be part of the same story. However, it is possible for a bullet to ride over the lands in the bore when twist is simply too sharp. Also, spin costs in bullet speed because it takes energy to rotate a missile. The energy from a powder charge is used up in various ways, including propelling the bullet forward, heating the barrel and spinning the bullet (rotational energy).

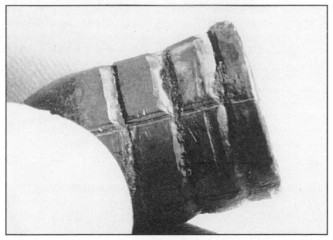

It does not take much "bite" on the bullet for rifling to do the job. The very minimal engraving shown on this Minie was sufficient to rotate it for stability in a 1:48 rifle.

However, none of these shooting variables alters the fact that slow-twist is best for the round ball, while a quicker twist is suited to the conical. So you know more about rifling twist, but what good is this information? The knowledge is useful for selecting a rifle/handgun and for loading the appropriate missile with an appropriate powder charge commensurate with rate of twist (as well as other rifling considerations, such as groove depth, land height and projectile shape varia-

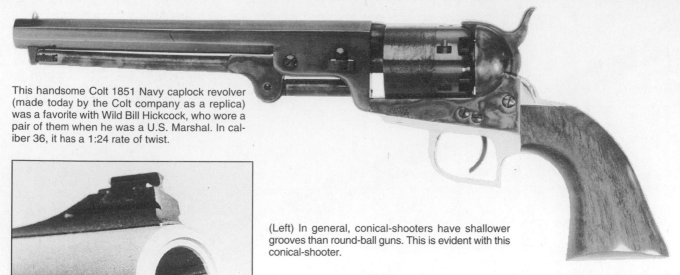

This handsome Colt 1851 Navy caplock revolver (made today by the Colt company as a replica) was a favorite with Wild Bill Hickcock, who wore a pair of them when he was a U.S. Marshal. In caliber 36, it has a 1:24 rate of twist.

(Left) In general, conical-shooters have shallower grooves than round-ball guns. This is evident with this conical-shooter.

Rotational velocity is easily retained, while forward velocity is not. Therefore, once the ball is spinning merrily away on its axis, it tends to remain on track because rotational velocity is not easily lost. The ball or conical will spin all the way to its final destination. Forward velocity loss can be roughly in the 50-percent realm for a round ball over the course of only 100 yards, but the original rate of spin is reduced by only a few percent over the same distance.

Handguns and Twist

Don't handguns have faster rates of twist per caliber than rifles? Usually they do, and this may lead us to believe that barrel length is, after all, an important factor in achieving rps. However, the reason sidearms carry faster rates of twist per caliber is because of their reduced exit velocity compared with rifles in the same calibers. If you had two 30-caliber barrels—one 30 inches long and the other 3, and both with 1:10 rates of twist—both would stabilize a 180-grain bullet if the velocity imparted by both barrels were sufficient to realize enough rps to spin the bullets and keep them flying point-on. The problem is that a 3-inch 30-caliber barrel is not long enough to provide sufficient fuel combustion to deliver the required energy and full velocity potential to drive a 180-grain bullet fast enough for stabilization. So, if you want to kick rps up, you must increase the rate of twist in that short barrel.

Generally speaking, and this statement does not include the modern high-velocity single shot pistol, sidearms fire bullets of lower section-

al density per caliber as compared with rifles. A 38 Special, for example, might shoot a 158-grain bullet, while a rifle of about the same caliber (around 35) might shoot a 200- or 250-grain bullet. Therefore, the rate of twist for the one-hand gun, while generally higher than a rifle of the same caliber, need not be immensely higher. For example, the Browning 45-70 single shot rifle enjoys a 1:20 rate of twist, while the 45 Colt pistol has a 1:16 rate of twist. The rifle shoots a 45-caliber bullet in the 400- to 500-grain class, while the pistol shoots a much shorter bullet in the 200- to 250-grain domain.

That's a look at rifling twist and consequent projectile stabilization. If you want a rifle to shoot a round ball, choose a slow rate of spin, but remember that the twist must also vary in accord with caliber. A smallbore ball-shooting rifle will have a faster rate of twist than a big bore ball-shooting rifle. For example, my Hatfield 36-caliber Squirrel Rifle has a proper 1:40 rate of twist because its .350-inch ball only weighs 65 grains, and it requires higher rps to stabilize it than a larger missile demands. The Navy Arms Ithaca Hawken 50-caliber ball-shooting rifle carries a 1:66 rate of twist, and it shoots with fine accuracy. On the other hand, my 50-caliber Storey Takedown Buggy Rifle has a 1:24 conical-shooting rate of twist.

And if you don't think blackpowder arms manufacturers have been listening, look around. Their new modern muzzleloaders designed for conical shooting carry comparatively fast rates of twist. CVA's 50-caliber Apollo Shadow goes 1:32, as does Dixie's In-Line Carbine. Knight's MK-85 has a 1:28 twist in 50-caliber, and the Navy Arms 45-caliber Whitworth has a 1:20 rate of twist. Meanwhile, the Wilderness Rifle Works Summit in 50-caliber has a 1:66 twist for ball-shooting. The 1:48 twist, by the way, is far from dead. It appears in many firearms today, such as the Navy Arms Parker-Hale 58-caliber Musketoon, and does very well in rifles like that. But there is no such thing as one magical rate of twist for all muzzleloaders. We know that for sure. ●

The Knight Hawkeye Pistol, 50-caliber, is a modern muzzle-loading pistol and, as such, is designed for conical shooting with a 1:20 rate of twist. But this rate of twist in a short barrel is not oriented specifically for high sectional density bullets. The Hawkeye does best with sabots and relatively short pistol bullets.

chapter 28

Ignition

Blackpowder ignites readily. That's why the flintlock system, as found on this fine Dixie Gun Works replica of the U.S. Model 1816 flintlock musket, works so well. If blackpowder were difficult to ignite, flame from pan powder would have a hard time making a gun go off.

Ignition for this Navy Arms Kentucky pistol is by the flintlock design, with FFFFg pan powder creating a flame that is directed through a touchhole and into the powder charge in the breech.

THIS IMPORTANT CHAPTER is on a small but vital topic—ignition. Without it, there would be no muzzleloaders, nor any blackpowder cartridge guns, for that matter. The last chapter dealt mainly with ways of delivering some type of fire or spark to a charge of powder in a breech. Locks and ignition intertwine, to be sure, but they are not the same topic. After all, when we touch on primers for blackpowder cartridges, locks are left behind altogether.

Fire

There is little doubt that fire was the first igniter of blackpowder in guns. Whether by heated iron, smoldering coal, slow match, hot wire or any other source, fire (heat) was used to set off the firestick of old, as well as the matchlock. In a sense, fire remained an igniter throughout snaphaunce and flintlock history—that is, fire from burning powder.

Flintlocks continue to rely on fire for ignition; however, this fire is caused by sparks igniting a special source of fuel. These sparks are produced by a flint striking a frizzen, or in the case of the true wheellock, pyrites or other sparkgivers that light an initial charge of powder, which in turn sets off the load of powder in the gun's breech. Fine-grain powder was developed early in gun history for this very reason: ignition. Until recently, it was safe to say that FFFFg (4F) was the finest-grain "pan powder" available over the counter. Today, Elephant brand has a finer grind yet—FFFFFg (5F).

I'm not sure if FFFFFg is truly necessary in a flintlock; however, it certainly has worked extremely well in a 42-caliber flintlock rifle that came my way from my mentor, Professor Charles Keim, the Alaskan master guide who got me started on the charcoal-burning trail. There isn't much more to tell about powder as an igniter. We know that it rests in the pan of the flintlock, the idea being to create a quick flame that darts through the flashhole and into the main charge of powder in the breech. And that's about it.

The Tube Lock

No, this is not a return to locks. The tube lock is of interest here *not* because of any special mechanical function. It's worth noting because of its ignition. The tube lock itself was essentially a hammergun. It was what the hammer hit that made it interesting—a small tube filled with fulminate of mercury, an impact-sensitive explosive. The falling hammer crushed the tube, and the resulting explosion created a flash for ignition in the breech. The tube lock was patented by famous English gunmaker Joseph Manton in 1816. I present it before talking about the percussion cap, which predated the tube by a couple years, because the percussion cap is so much more important.

The Pill Lock

Once again, it's not the lock we're interested in, but the source of ignition. Actually, the pill lock goes back in time even farther than the tube or percussion cap. It's noted as the invention of Scottish clergyman Rev. Alexander John Forsyth, who patented the device, my sources show, on April 11 of 1807. This is noteworthy because Forsyth is credited by other historians as the inventor of the percussion cap, which may or may not be so. The pill itself was a tiny pellet filled with what my literature calls simply "detonating powder." A hammer fell upon the pill, setting it off, which in turn exploded fire into the powder charge in the breech of the gun. So much for the pill lock of Rev. Forsyth.

The Tape Primer

The tape primer has been credited to dentist Edward Maynard as an invention of 1845. Again, we are well out of sequence with the percussion cap of 1814, but it's better to get some of the minor, but interesting and important, ignitions out of the way before talking about caps. Imagine a roll of toy caps. The tape primer was essentially two strips of tape with interspaced fulminate pods lodged in between the

Muzzleloader ignition may be caused by a standard percussion cap, such as the No. 11 (far left) in front of a nipple made for that size cap. Next is the much larger top hat or English musket cap, with its appropriate nipple. On the right is a type of fusil that is made to hold a modern primer. This fusil replaces the standard nipple, threading right into the nipple seat.

The Thompson/Center Scout Pistol mounts the percussion cap directly into the breech and in line with it.

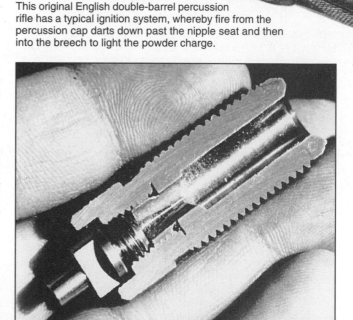

This original English double-barrel percussion rifle has a typical ignition system, whereby fire from the percussion cap darts down past the nipple seat and then into the breech to light the powder charge.

strips. The tape was fed over the top of the nipple mechanically by a ratchet arrangement. Each time the hammer was eared back, the tape advanced to offer a new priming pod under and in line with the falling hammer nose.

Disc Primers

Even more out of sequence is the disk primer of 1852, an invention of Christian Sharps, whose name is hardly unfamiliar to blackpowder cartridge fans. These were tiny copper discs launched forward one at a time by the action of the hammer. The discs contained fulminate of mercury, making them explode when struck by the hammer nose. The sparks emanating from the exploded disk flew into breech to ignite the powder therein.

The Percussion Cap

Gun history is involved and convoluted. It's pretty difficult to pin down all invention dates. Fortunately, patents help, but not all great inventions were noted by patents. As previously mentioned, Rev. Forsyth is very often credited with the percussion cap. Other sources disagree. Captain Joshua Shaw of Philadelphia is noted in at least some arms histories as the father of the percussion cap, dating back to 1814. It's probably so, but be that as it may, the concept of installing fulminate within a small metal container (cup) was ingenious. It changed shooting forever, because the modern primer is no more than another style of percussion cap.

This interesting cutaway shows how the Knight MK-85 modern muzzleloader delivers the flame of the percussion cap directly into the powder charge in the breech. The jet of flame is actually concentrated right at the base of the powder charge.

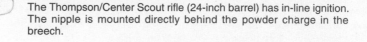

The Thompson/Center Scout rifle (24-inch barrel) has in-line ignition. The nipple is mounted directly behind the powder charge in the breech.

This Navy Arms Le Mat replica revolver is typical in its ignition pattern. Nipples are mounted directly into the base of the cylinder, which in effect makes for in-line ignition. The flame from the percussion cap strikes the rearmost portion of the powder charge in the cylinder chamber.

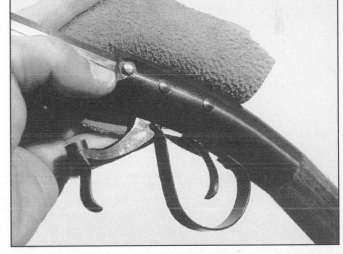

The Thompson/Center Hot Shot nipple has a vented cone to aid in bleeding off gas, rather than having trapped gas force the percussion cap away from the cone of the nipple.

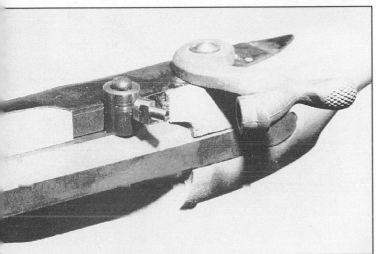

The drum-and-nipple system is clearly seen on this rifle. The drum is the portion that screws directly into the barrel flat where a touchhole would appear on a flintlock rifle. Into the drum two things are fitted: first, there is a cleanout screw, which has a regular slot; and second, there is the nipple itself.

There is direct ignition and in-line ignition. The two are not the same. This underhammer represents direct ignition. A percussion cap is screwed into the lower barrel flat so that flame is ushered directly into the powder charge, but not in line with it.

Fulminates are percussion-sensitive; that's why the term "percussion" cap was used to describe the device. Give a fulminate a good whack and it explodes, producing flame. Once harnessed within a container, its powers could be controlled and channeled. In this scenario, the idea was simple: When struck by the hammer of a firearm, the cap explodes, sending a flame of ignition into the breech of the firearm via a nipple.

Of course, the cap must provide a flame capable of traveling through the nipple vent (a channel or passage) and detonating the main charge in the breech of the firearm. In order to do this, the spark has to be large enough to get the job done, but that's not all. A good cap must also be consistent. It should offer similar duration of flame from one cap to the next from the same box. In other words, it should emit a spark which lasts long enough to reach the powder charge every time. So a proper percussion cap is more than a little item that goes "Pop!" It has to be a precise detonator with specific repeatable and reliable properties.

The big push for years was the hot cap. "Hot" meant that the cap threw a heck of a flame. The logic of the hot cap did not hold up. Those who tested the situation decided that a "cool" cap could be just as effective as a hot one. The ideal cap delivered just enough spark to bring about ignition, but not an excessive explosion. A cap that was more powerful than necessary caused pieces of its own body to fly in all directions. The too-hot cap could also all but "blow itself out." It went Crack! on the nipple, but the flame seemed to spatter, for lack of a better term, rather than channeling under control through the vent and into the powder charge. Excessive cap debris also clogged the nipple, impeding the progress of flame when the next cap was fired.

This situation could cause a misfire, or at least a hangfire. Nowadays, the problem is mostly settled. Once there were many different cap styles and brands in many degrees of power. Today, percussion caps have standardized—up to a point. So the hot/cold cap theory is mostly, but not entirely, outdated. The best cap offers a sustained spark, instead of an instant but short-lived blast.

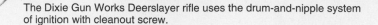

The Dixie Gun Works Deerslayer rifle uses the drum-and-nipple system of ignition with cleanout screw.

There is variation in how the body of a cap reacts to detonation. Some of these caps are almost intact; others have missing parts that have broken away; still others are split, but intact. Opinion varies as to which body style is best. The fact is, they all work well.

The English top hat or musket cap is much larger than the standard percussion cap.

There are still cases where a cooler cap would suffice to ignite the powder charge, and it's up to the shooter to test various caps to see which one works best on his particular muzzleloader. *Brands do vary.*

Furthermore, one cannot consider the cap without considering the nipple. The Hot Shot nipple, for example, is vented so that excess explosiveness can be bled off instead of channeled toward the breech of the firearm. Therefore, while some of the old-time shooting greats, such as A.C. Gould, called for the cool cap, we don't have to pause long to consider the problem today.

Today's Hot Cap

Currently, there are caps being sold that are called hot. Do not avoid them. The term has stuck, but these caps are not *excessively* forceful, with only a couple minor exceptions that I've seen in the past decade. For example, the fine RWS cap is noted as a hot cap. It's strong but also well-controlled, with terrific reliability, dependability and consistency. The overly hot cap is easy to spot. It darn near blows itself off the nipple in bits of shrapnel.

The Non-Corrosive Cap

In a practical sense, the use of a corrosive or non-corrosive cap should be moot. Blackpowder and Pyrodex leave fouling in the bore, and the firearm has to be cleaned anyway, whether the cap is corrosive or not. But in reality, non-corrosive versus corrosive does matter: Before loading and in the process of readying a caplock muzzleloader, one or more percussion caps may be fired on the nipple to clear oil away and to ensure a clear channel from the cone of the nipple into the breech. Corrosive particles or "smudge" may end up in the nipple, nipple seat, breech, and actually downbore all the way to the muzzle. Meanwhile, the firearm may not be fired for some time, and these corrosive elements are at work. No big deal, but I still like the non-corro-

sive cap because it won't leave problem residue in the gun when used for clearing before loading. Of course, for target shooting and plinking, the corrosive cap is fine. The muzzleloader is shot successively, then cleaned soon after.

The Waterproof Cap

What about waterproof caps? Some caps withstand moisture better than others, right down to running water over them. This is an easy-enough test in itself. Dampen a cap at the range. Fit it on the nipple. Fire the cap. Waterproof caps tend to go off a lot more reliably than non-waterproofed caps under such test conditions and in the often-damp hunting field. Let common sense prevail. If your cap is an economy brand, and its use is centered on target shooting and plinking, waterproofing is of no consequence. But on a hunting trip, the waterproof or "lacquered" cap is not a bad investment. By the way, caps that were not waterproofed a decade ago now are. So be sure to test yours.

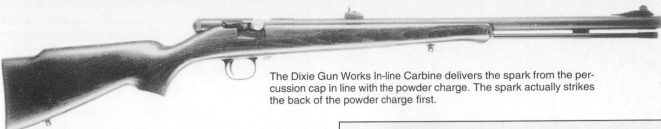

The Dixie Gun Works In-line Carbine delivers the spark from the percussion cap in line with the powder charge. The spark actually strikes the back of the powder charge first.

Cap Sizes and Construction

Just as all caps are not created equal in terms of waterproofing and degree of potency, neither are they the same in size. The No. 11 percussion cap is a standard today that fits most nipples. A No. 10 is often better on some revolvers. There are variations in cap sizes from brand to brand. For example, a No. 11 cap in Brand A may be identical in size to a No. 12 cap in Brand B. Check for yourself. Buy a can of caps. See how they fit the nipple(s) on your favorite guns. Apart from the usual Nos. 10, 11 and 12 caps, there is the English musket cap, also known as the "top hat cap." It fits most rifled muskets, such as the Navy Arms Whitworth rifle. Top hat or English musket caps are much larger than standard caps, and they throw a big spark. If top hat caps are difficult to find, investigate the possibility of a conversion nipple. A conversion nipple has a large threaded base section that fits the seat of the musket, but the cone of this nipple is sized for a standard cap instead of a top hat cap.

Another percussion cap criterion is construction of the body, basically ribbed or smooth. Both work equally well. A more important cap body difference is malleability. Some caps are brittle and fragment upon firing; others are more malleable. They tend to smash against the cone of the nipple where they either fall off of their own accord or can be flicked away. The latter are preferable, but not essential.

Working with the Percussion Cap

This information was broached in our chapter on proper loading of the charcoal burner, but a brief reminder is valuable. An accepted practice of clearing the nipple of the firearm prior to loading is popping percussion caps beforehand. Remember that bits of metal may end up within the channel or vent of the nipple. This scenario may cause an opposite and undesirable result—cap fouling within the nipple vent that could prevent the clear passage of flame, causing a misfire or hangfire. So pop caps before loading up, but run a pipe cleaner into the cone of the nipple afterward to ensure that no cap debris remains behind. This is so important that mentioning it again seemed vital.

Caps can be installed neatly on nipples with either spring-loaded in-line or gravity-fed magazine cappers. There are even special models available for individual guns, such as Ted Cash's superb magazine cappers for cap 'n' ball revolvers.

Percussion caps can be tested and evaluated by the shooter using a screwbarrel pistol, such as the Dixie Gun Works Derringer Liegi, 44-caliber. **Keep all parts of your body clear of the pistol in this test**. Do not allow anyone else in the area when you are testing caps. The use of the pistol is entirely safe unless the shooter does something outrageously foolish. For the No. 11 size, unscrew the barrel from the pistol, expose the breech and place a percussion cap on the nipple. Fire in a darkened place, such as the garage at night. A flame/spark will jet from the pistol. The shooter can visually determine for himself the duration of the spark and its size, as well as the shape of the flame—long and slender, short and wide, or in between. Cap brands can be readily compared with this useful visual inspection method.

Nipple Designs

The number of different nipple designs would astound the unsuspecting blackpowder shooter. There have been hundreds, and there is no value in listing them now. However, be aware that we still have various nipple styles, so it may pay to check out a few for your own edification. Dixie Gun Works' and Mountain State's catalogs show several different models. Incidentally, be sure to obtain the correct thread when you select a nipple for your blackpowder guns. There are several,

This nipple shows the flat base with pinhole, a proper design meant to help control expulsion of gas back through the vent of the nipple when the gun is fired.

including the 1/4-28 thread, 6-.75 and 6-1mm metrics, 12-28s and other thread dimensions.

Avoid the "straight-through" type nipple. They have a large orifice from base to cone, as opposed to nipples with flat bases and pinholes. The flat-based nipple with pinhole is my preference. I discard the other type. Even though the Venturi principle (the basis of the ramjet) allows very little gas to escape through the small port in a nipple when that gas has an escape route through a much larger exit, be wise—buy nipples that have a flat base with a pinhole. They better control the direction of gas in the system.

Use the correct cap for your firearm's nipple, never a cap that is too loose or too tight. Also, never use a nipple that does not fit perfectly because it could blow out of the nipple seat. And never prime a nipple by putting powder down into the cone. That is not the proper function of a nipple. Such priming of the nipple could cause flying cap debris. Also, go with a flash cup around the nipple to contain debris and to prevent damage to stock wood. Tedd Cash makes a good one.

The percussion cap may be a simple device, but it's the spark plug of the caplock firearm—rifle, pistol, shotgun or revolver. Without that little flame, there is no ignition. Know your percussion caps and the nipples that they rest on.

Cap Safety

Let common sense prevail. Caps are obviously sensitive to blows—that's what makes them detonate. So don't hit 'em with anything. They can also be set off by heat. So store them away from heat sources. Also, carry caps safely, in special cap boxes (as made by Tedd Cash) in their original containers, in cappers (in-line or magazine), or in some other sturdy holder where a blow won't set 'em off. And by all means, keep percussion caps safely stored where children cannot find them. Also, as noted earlier, make sure your caps fit the nipples of your guns. Forcing a percussion cap on a nipple is unwise for obvious reasons. Snug fit, yes, so the cap does not fall off the cone of the nipple when you're hunting. But an overly tight cap fit is asking for trouble. Use a capper, too. It keeps caps away from fingers.

The fusil uses a modern primer that is detonated by the tiny protrusion inside of the cover (left), which acts like a firing pin.

The Modern Primer

Today, modern pistol, rifle and shotgun primers see action on muzzleloaders. This, in itself, is not new. A device known as a fusil also has been around for years. A fusil takes the place of a regular nipple, holding a modern primer instead of a percussion cap. For example, there's Hubbard's Mag-Spark, which takes shotgun primers. Also, modern primers are the ignition source for blackpowder cartridges. Chapter 34 on breechloaders discusses primers for blackpowder rifle cartridges. Chapter 36 on handguns mentions primers for blackpowder cartridge revolvers.

Ignition, Accuracy and Failures

On the one hand, ignition can be wrongly blamed for ill accuracy in a muzzleloader. I look for many other factors before deciding that I have, for example, the wrong percussion cap or the wrong nipple (see Chapter 23). However, I am equally certain that non-uniform ignition hurts accuracy, and that notion comes from experience, not careful testing. Firearms that "go off" erratically do not impress me; I want reliable and repeatable ignition. This is no more than common sense at work. Flintlocks that don't discharge on cue really cause trouble. The shooter may actually alter his point of aim between trigger pull and ignition. Conversely, flinters that fire with good lock time are easier to control and to shoot accurately.

There are many different reasons for ignition failure, including bad caps, cap debris, poor firearm design, damp powder, the wrong nipple, poor touchhole location, a badly made lock (either percussion or flint), improper loading practices, burned out touchhole and so forth. See Chapter 39 on troubleshooting the blackpowder gun.

The Hangfire

Also written as hang-fire, this means ignition delay. The gun still goes off, but not instantly. Pssttt! - Boom! That's the sound of a hangfire. **Warning:** Continue to safely aim the firearm downrange if it does not go off right away. The longest hangfire I am aware of is 1$^1/_2$ minutes between trigger pull and the gun going off.

The Misfire

The misfire sounds like this: Click! The gun does not fire at all after the trigger is pulled. Once again, continue aiming in a safe direction following a click. It could be a hangfire instead of misfire.

The Flash in the Pan

This term, as with so many others from the world of shooting, entered our everyday language many years ago. A flash in the pan is a person who starts something he does not finish. With a flintlock rifle, it is a flash from the priming powder with no ignition of the powder in the breech.

That's a look at the all-important subject of ignition. Blackpowder guns that don't fire with reliability are absolutely frustrating. They can cause missed targets and lost opportunities on game. Don't let either happen to you. ●

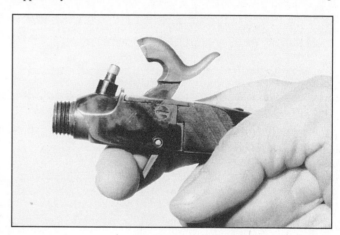

The Dixie screwbarrel pistol, with barrel removed, makes a safe and useful percussion cap test device.

The blackpowder cartridge rifle is now popularly loaded with a magnum-type modern primer in the cartridge case. Even though blackpowder is easy to ignite, shooters report that the magnum primer gives best consistency.

chapter 29

Sights for Blackpowder Guns

Although some shooters may feel that they do not even look at their sights when they "snapshoot," the fact is, the bullet can only go where the sights are aiming.

"ALL I DO is snapshoot. I don't even aim. I don't know where the sights are looking when I pull the trigger. But I hit the target." The old-timer who told me that was a gold and silver miner who lived a little south of Patagonia, Arizona. Cradled in his arm as he spoke was his favorite Model 94 Winchester 30-30 carbine. It sure sounded good to me. "Snapshooting," he called it. Forget aiming. Just up and fire and let instinct do the rest. I considered the old man's words for years, and actually came to be a fair snapshooter myself. One day, the reality of it struck me. Aim without sights? Not hardly. Allowing that a rifle is sighted-in to begin with, where can the bullet go but where the sights are "looking"? Reversing the thought, haphazardly pointing the muzzle in the general direction of the target won't work because the bullet has to go where the sights are aiming. The oldtimer shot fast, you bet. He also aimed unconsciously. After all, if his rifle's sights were truly off target, his bullet would be, too.

The concept of sights was born early in gun history. We don't know for certain, but it's a good guess that shooters quickly learned how futile pointing a firearm was. True, many armies from the Mongols to the Britishers shot bows with good accuracy. The Japanese archer was known as a remarkably good shot. These bows had no sights. Aim was perfected in various ways. The term "instinctive shooting" came along to describe the process of aiming effectively without sights. I call it reflexive shooting. Coordination between hand, eye, and brain seems to make it work. I have no doubt that guns were fired in a similar manner for some time. The British Brown Bess musket was fired effectively on the battlefield for years. However, precision of aim was never possible without refined sights. And so it was that sights of many different types came along.

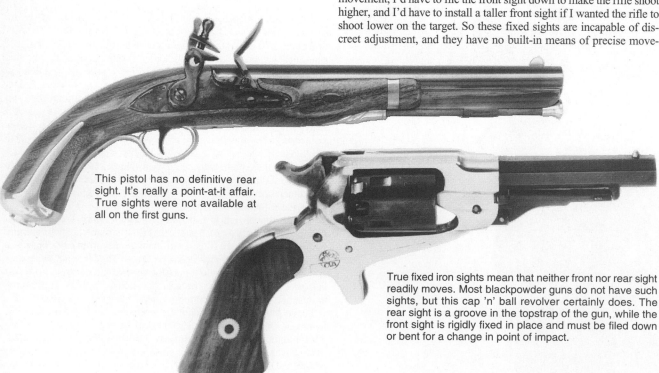

The globe front sight is a "ball" of metal. It served very well on 19th-century muzzleloaders and still functions well today. This one is on a Navy Arms Volunteer rifled musket.

Fixed Iron Sights

One of the earliest forms of sights was the true fixed set. These appeared on some early firearms. I've seen them on single shot black-powder pistols. The front sight was an immobile blob. The back sight likewise, with a slit in it. The front sight fit optically into the notch created by the slit in the rear sight, if a lump of metal with a cleft in it can be called a sight. And that was that. Truly precise aim was impossible with these sights because the shooter could not generate a clean, repeatable sight picture. But they were no doubt much better than pointing the muzzle of the gun in the general direction of the target and firing.

Updated Fixed Iron Sights

Today, our so-called fixed iron sights are only partially that way. For example, my 42-caliber flintlock, patterned after an ordinary long rifle of Early America, wears what must be termed fixed sights, for they have no built-in device to allow adjustment, such as a ladder underneath a movable rear sight. The front sight is dovetailed directly into the top flat of the barrel about an inch behind the muzzle. The rear sight is a plain metal fixture with a V-slit, also attached in a dovetail notch, but about 8 inches in front of the pan. These sights, which lie very low on the barrel, provide a clean sight picture. They're a far cry from the primitive fixed sight noted above.

When R. Southgate originally built this rifle, he very carefully matched sights to performance. The round ball hits right on the money at about 75 yards, and a touch low at 100 yards. For a 42-caliber round-ball rifle, this sort of sight-in is ideal. The little pill shouldn't be fired on deer-sized game at anything that resembles "far off." On the one hand, this Southgate flinter does wear fixed sights, and yet on the other hand, both front and rear sight can be moved. If I wanted the point of impact to change on the target, I'd slide the front sight horizontally in its dovetail notch or the rear sight horizontally in its dovetail notch. If I wanted the next bullet to hit to the right on the target, I'd nudge the rear sight a bit to the right, or I would slide the front sight a little to the left.

Remember that you move the front sight in the opposite direction of where you want the next bullet to go, and the rear sight in the direction you want the next bullet to land. Since the rear sight has no up or down movement, I'd have to file the front sight down to make the rifle shoot higher, and I'd have to install a taller front sight if I wanted the rifle to shoot lower on the target. So these fixed sights are incapable of discreet adjustment, and they have no built-in means of precise move-

This pistol has no definitive rear sight. It's really a point-at-it affair. True sights were not available at all on the first guns.

True fixed iron sights mean that neither front nor rear sight readily moves. Most blackpowder guns do not have such sights, but this cap 'n' ball revolver certainly does. The rear sight is a groove in the topstrap of the gun, while the front sight is rigidly fixed in place and must be filed down or bent for a change in point of impact.

This is a fixed rear sight in that it has no built-in adjustment capability. However, it can be drifted in its dovetail notch for windage.

This full buckhorn rear sight may not be mechanically ideal, but neither is it a poor sight. It has a very fine notch, and a clear sight picture is possible. Less "horn" is found on the semi-buckhorn rear

ment. That's the fixed iron sight as we see it today on various muzzleloaders. On replica pistols, as well as certain longarms, the true old-fashioned fixed sight can still be found, correct historically, but it's not much of an aiming device.

Fixed Iron Sights on Cap 'n' Ball Revolvers

Original-style cap 'n' ball revolvers may have fixed sights that are rigidly in place. For example, the Remington Model 1858 blackpowder revolver that I own has a barleycorn front sight integral to the front of the barrel. There is no dovetail notch to slide it in. About the only way to make the revolver hit to the right or left by manipulating the front sight is to bend it, hopefully without breaking it off in the process. If I want this revolver to shoot higher, the front sight must be filed lower. If I want this revolver to shoot lower, a gunsmith will have to build the front sight up with additional metal. As for a rear sight, it is merely a groove in the topstrap of the frame and absolutely nothing can be done to change point of impact with it. My Colt Dragoon blackpowder revolver uses the same type of front sight in shape, but the rear "sight" is a notch cut in the hammer nose. When the gun is cocked the hammer nose is elevated and the notch is easily aligned with the front sight. The

In contrast to the fixed rear sight, this Lyman rear sight is adjustable for elevation.

handgun sights described above are not target shooting affairs, but at close range on large targets, they more than suffice.

Buckhorn and Semi-Buckhorn Fixed Sights

There are countless other fixed sights found on blackpowder guns, far too many to detail each one here, even if we could find them all. But the buckhorns and semi-buckhorns deserve mention. These rear sights can be adjustable, but on many frontloaders they are not. In theory, the buckhorn rear sight is about useless. It supposedly covers the whole target, making precise aiming impossible. In real life, however, these sights work well enough. I don't favor them, but I have used semi-buckhorns and buckhorns, finding that the target never got blotted out for me. The buckhorn sight is simply a rear sight with "horns." When they curl around almost to touch, that's the full buckhorn. When they do not project so far up (less curl), they're semi-buckhorns.

That's a little bit about the fixed iron sight found on some muzzleloaders. When set in dovetail notches, they allow windage (horizontal) adjustment. Elevation (vertical) adjustment is generally accomplished by filing the front sight down, installing a taller front sight, or having a gunsmith add metal to an existing fixed front sight.

Adjustable Iron Sights

The Ladder-Adjustable Open Sight

There are numerous ways to accomplish iron sight adjustment. One is the ladder, which rests beneath the back sight. The ladder has graduated notches cut into it. On the shortest notch, the rear sight rests lowest on the barrel, and the firearm hits its lowest. On the tallest notch of the ladder, the rear sight is elevated to its highest position and the rifle

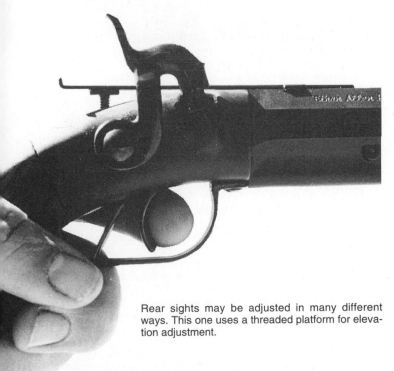

Rear sights may be adjusted in many different ways. This one uses a threaded platform for elevation adjustment.

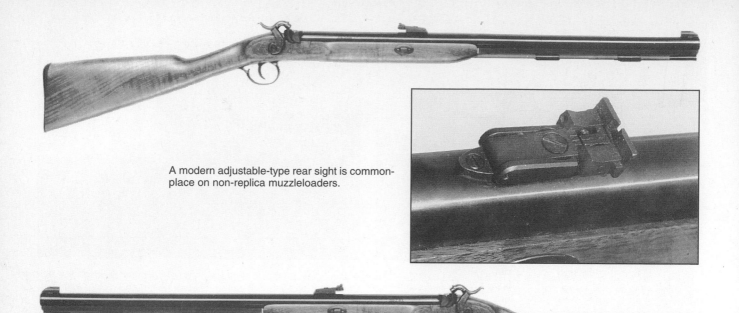

A modern adjustable-type rear sight is common-place on non-replica muzzleloaders.

The ladder-type adjustment is very old and remains with us to this day.

hits its highest. The ladder-type rear sight is normally coupled with an ordinary non-adjustable front sight, but the front sight may rest in a dovetail notch, giving it some lateral movement. The ladder rear sight often rests in a dovetail notch also. While elevation adjustment is accomplished with the previously-described notch-in-the-ladder arrangement, windage is accomplished by drifting the rear sight in its dovetail notch, or the front sight in its dovetail notch.

The Modern Adjustable Open Sight

Today there are many different types of adjustable open sights. The rear sight on my T/C Hawken, for example, is fully adjustable. That is, it carries both elevation and windage screws. When I turn the elevation screw downward, the rear sight rises up. Remember what we said about rear sights: Move them in the direction you want the next shot to hit, so the rear sight rising means the next bullet also rises on the target. A screw adjustment on this T/C sight also moves the rear sight left or right for windage changes. There are too many variations of the adjustable iron sight to treat them all here. However, the type is easily recognized: adjustment is in the rear sight, which can be moved up or down, left or right, normally with a screw.

Receiver Sights

On blackpowder guns, the term "receiver" seems unfitting, since the majority of sootburners have no true receiver. However, the name is accurate because it refers to a rear sight that mounts around the breech area. There are quite a few blackpowder guns that wear receiver sights these days. My Storey Buggy Rifle is one. The gunmaker has fixed a

Williams adjustable receiver sight to the top flat of the barrel, as far back on the barrel as it will go. If this were a modern rifle, the sight would be mounted on the receiver, for it certainly rests in the same domain. It's important to mention two other names for receiver sights: aperture sight and peep sight. Whether you call it a receiver sight, aperture sight or peep sight, its function is the same. The object with these sights is to look through the hole in the rear sight, focus on the front sight, and put the front sight on the target.

The most outstanding problem I've encountered when trying to teach the use of the peep sight is convincing the shooter that once he sees the front sight through the aperture, the work of the aperture is ended. Do *not* try to center the front sight in the middle of the peep. Keep your face firmly on the buttstock. Continue looking *through* the peep. Place the front sight on what you want to hit and control the trigger normally with proper squeeze. Bang! It's a hit. The human eye automatically centers the front sight in the aperture because that is where the greatest light is. One more time: Don't try to consciously center the front sight in the peep. Look through the peep and put the front sight on target. That's it.

Peep sight adjustments occur in various ways. Staying with the Williams receiver sight on my Buggy Rifle, it adjusts for both windage and elevation. The aperture rides in a dovetail notch integral to the sight itself. A screw holds the aperture in place. Want the rifle to hit to the left? Loosen the screw, slide the aperture to the left one increment, then retighten the screw to hold the aperture in place. Notice I said move the sight only one increment. There is no click adjustment on this Williams sight. However, tiny white lines on the top of it in front of the aperture serve as markers, so the aperture can be moved a precise number of increments.

Elevation on this Williams peep sight is accomplished with a sliding platform. The aperture is dovetailed into that sliding platform. The platform is clearly marked in white increments on the right-hand side. Slide it forward and the platform rises. The rifle shoots higher. Slide it rearward and the platform descends. The rifle will shoot lower. This is how the Williams receiver sight adjusts, but as noted, there are many other ways to change point of impact with peep sights. Some of the more sophisticated models have built-in click adjustments, for example, with $1/4$-minute-of-angle moves. One click equals about $1/4$-inch at 100 yards.

A quick and simple way to change elevation is seen on this Colt Model 1861 musket. The taller rear sight flips up for farther ranges.

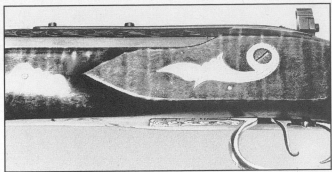

This handsome peep sight (which is adjustable) is the Storey model, as found on this Storey custom muzzleloader.

The Tube Sight

Some feel that the tube sight was forerunner to the scope. It certainly did look like a scope as it rested above the rifle barrel, but it had no glass in it. These were simply long metal tubes mounted full-length atop long rifle barrels. The tube was adjustable, and it carried some sort of aiming device. The tube sight isolated the view of the shooter for easier concentration on the target, and it effectively blocked out superfluous light, and therefore any glint from the sun's rays. In further describing one tube sight I saw, it had an eye cup and in the cup was a tiny hole. A peep sight, if you will. Mounted in the other end of the tube was a globe front sight. This was simply a round chunk of metal, hence "globe." This peep and globe sight resided in the shade all the time and produced a pretty effective sighting device. Remember, the tube sight had no optical lenses, so it must be classified as an iron sight.

The Scope Sight

Two distinct camps exist concerning scopes on frontloaders. The first camp is occupied by purists who feel that blackpowder guns represent a challenge which is best met with old-time sights. These individuals prefer plain open sights. They don't even care for aperture sights, in spite of the fact that peep sights were found hundreds

The tube sight had no glass in it at all. It was, as its name implies, a long tube. This one has external adjustments just like some target scopes. It's on a Storey custom sidehammer muzzleloader.

of years ago on certain crossbows. So originality of sights does cause a problem for this camp. If a peep sight, for example, was found on an original Hawken, or a scope on a breechloader, then what? But their point is understandable, too. Want the challenge of blackpowder shooting? Then accept all of it, including the more basic sights.

The other camp has an entirely different point of view. The people who reside here believe that blackpowder guns have plenty of challenge, even when scope-sighted. After all, range is still limited, and the game remains a one-shot affair. Those who occupy this camp believe that scopes are right on muzzleloaders. They help a shooter place his bullet better, which means more perfect hits. Some see the scope as an ethics question. They believe it is a hunter's responsibility to use sights that allow ideal bullet placement for the quickest and surest harvest. "Just get closer, the way you're supposed to," say the lads from the other camp.

Not allowed on muzzleloaders for special "blackpowder-only" hunts in some states, the scope is legal for hunting with frontloaders in the majority of states today. To scope or not to scope—that is the question. As stated above, some believe that mounting scopes on blackpowder guns is cheating, because the glass sight reduces the challenge of hitting the target, while others feel that scopes on muzzleloaders can actually conserve game, since the scope allows for better bullet place-

This Vernier tang rear sight has a disc that moves up and down for changes in elevation. It is, of course, a peep sight with a large target-type disc having a very small aperture (peep hole).

209

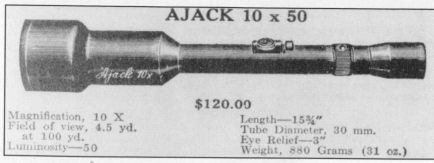

AJACK 10 x 50

$120.00

Magnification, 10 X	Length—15¾"
Field of view, 4.5 yd. at 100 yd.	Tube Diameter, 30 mm.
	Eye Relief—3"
Luminosity—50	Weight, 880 Grams (31 oz.)

The scope is hardly new. Riflescopes were available in the 19th century, picking up momentum in the early 20th century, but not reaching full acceptance until the 1950s, approximately.

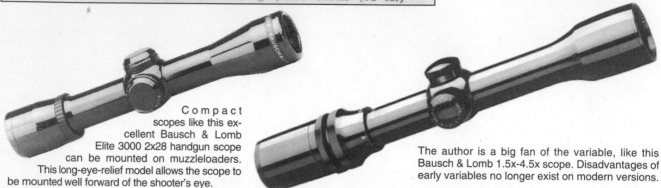

Compact scopes like this excellent Bausch & Lomb Elite 3000 2x28 handgun scope can be mounted on muzzleloaders. This long-eye-relief model allows the scope to be mounted well forward of the shooter's eye.

The author is a big fan of the variable, like this Bausch & Lomb 1.5x-4.5x scope. Disadvantages of early variables no longer exist on modern versions.

ment and a more certain harvest. What won't wash is: "Scopes are wrong because they're inventions of recent times." That notion is provably incorrect. History says so. Scopes have been around for a very long time.

There's good evidence that some of the mountain men attached scopes to their muzzleloaders. Scopes were certainly available in the 19th century for those who wanted to use them. Riflescopes were born of terrestrial telescopes. Johann Lipperhey of Middleburg, Holland, a spectacle maker, came up with the first working model of a terrestrial telescope in 1608. It was crude and of little practical value until Galileo, the noted Italian astronomer, took the idea to new heights, creating a viable instrument. Because of his efforts, Galileo, not Lipperhey, is credited with the first true telescope. It was carried far and wide by explorers, finding its way into Jim Bridger's possibles during the fur trade era. Bridger was known to climb vantage points for a study of the area via his telescope.

As far as riflescopes go, William Ellis Metford, an Englishman, had a practical model by 1824. A Metford scope sight was attached to an experimental rifle by Colonel George Gibbs, a rifle shooter of the day. Gibbs' scope was said to be 8x, which was an approximation. Regardless of the exact magnification, the Colonel's scope did work. Metford's claim was simple. He claimed that mounting one of his scopes on a rifle would improve chances of putting bullets right on target, and he was willing to prove it to anyone. Oddly, scopes did not catch on. It took years for hunters to gain trust in glass sights. Younger shooters may find it hard to believe, but even as late as the 1950s, some of our country's more proficient outdoorsmen claimed scopes were a gimmick.

It was after World War II that riflescopes gained a toehold in America. Even then, it was more a matter of tiptoeing into the shooting world, not racing headlong into broad use. Leaders of the hunting community began not only using scope sights, but writing about them. Jack O'Connor, for example, extolled the virtues of the scope sight, stating that it was ideal for hunters who wanted to put bullets right where they belonged on big game. Elmer Keith, another famous gunwriter, wrote the same, as did Colonel Townsend Whelen, Paul Curtis, Stewart Edward White, and other respected shooters of the day. From distrust, the scope sight moved into the ranks of faithful use. It's no wonder that modern American shooters look to the glass sight even for their charcoal burners.

Which Scope?

The riflescope built for today's modern big game cartridge rifles are the same scopes used on muzzleloaders. These scopes offer high optical resolution; a good clear sight picture, in other words. It is just as important to see well through a scope mounted on a frontloader as it is to make the target out clearly when shooting a cartridge gun. Optical resolution pays off when the target is in the brush, or any other hideaway terrain, because the shooter can better make out where to aim. The blackpowder scope must be ruggedly constructed so that lenses remain mounted within the tube following recoil. Reticles must be proper. The duplex is the most popular, with its heavy, easy-to-see crosswires coupled with thinner wires in the center for precise aiming. And, of course, the scope must be capable of precise adjustment. In short, the muzzleloader scope is the same scope you might buy for your 30-06 rifle. The same goes for scopes that see duty on blackpowder breechloaders.

What Power?

The brush hunter is well equipped with a 2.5x scope, or even a scope of less magnification. Conversely, the target shooter whose goal

While some blackpowder shooters may disagree with scopes on muzzleloaders, there are many blackpowder guns now wearing the glass sight.

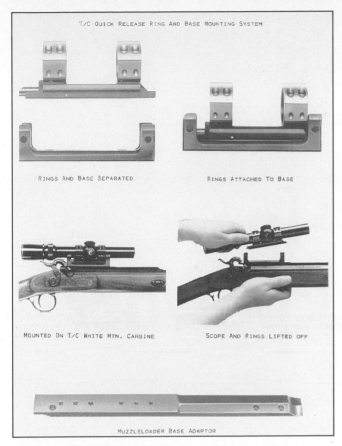

Scope mounts for muzzleloaders are no longer a problem. Thompson/Center, for example, offers many different types of mounts that serve on frontloaders.

is tight clusters will find high scope magnification ideal. You will never see a modern benchrest match won with a low-power scope. There's a reason for that. Benchrest shooters know that a super-magnified target allows for greatest precision of aim. So appropriateness is the byword. If all hunting takes place in the thickets for close-range whitetail deer, a low power scope is perfectly fine. For greater precision of bullet grouping, more magnification is appropriate.

The Variable Power Scope

I've long been a fan of the variable. Today's variable is so good that it is probably the best all-around choice for modern cartridge rifle, muzzleloader, or blackpowder breechloader for hunting purposes. The once-upon-a-time negative features of variables are long over with. Today's variable is optically excellent. It has superb reticles. It is capable of precise adjustment. I'm open-minded on most matters of guns and shooting, but find arguments in favor of fixed-power scopes weak and flimsy. "Oh, I don't like big scopes." Well, variables may have been oversize by a little bit at one time, but not now. They are nicely compact today. "But I really prefer 6x over any other magnification." Very well. Leave your variable on 6x if you insist, but think it over: is 6x ideal for brush or any close-range fast-action shooting? And can you prove that 6x is better than a higher magnification when you want to place a bullet just right?

The variable can do it all. While I seldom use a scope sight on a muzzleloader because I remain confident with open sights, and because I often hunt where scopes on muzzleloaders are not even allowed, my big game bolt-action cartridge rifles all wear variables. The variable scope changes the very nature of your rifle. Suppose a 2.5-8x variable is used. The rifle is carried into the hunting field with the scope set at 2.5x for a wide field of view, all the easier to get on target if the game is very close or on the move. Now I'm on a ridge, glassing. There's a buck, bedded down. I can close the gap

on him, and I do. But I don't shoot with the scope on 2.5x. The scope is adjusted up to its highest magnification so that I can best see exactly where to place the crosswires on that bedded animal. As for the in-between settings, they can be useful. Hunting antelope, for example, there's no reason to leave the scope on 2.5x. I'll never need that wide a field of view. I might put the scope on 5x or 6x as I hike along. No sir, if I'm going to scope at all, variable it will be most of the time.

Scope Mounts

These days, many muzzleloaders come drilled and tapped for scope mounts. The factory does it for you. Now it's a simple matter of buying the correct mounts to match the firearm, and that's it. For example, my Knight modern muzzle-loading single shot pistol is tapped for scope mounts. It comes from the factory with two holes drilled and tapped over the breech section, and two more behind the port. The latter two are used to attach a Williams open adjustable sight from the factory, but with the sight removed, these two holes are for scope mounting. For those blackpowder guns that do not come pre-drilled and tapped for scope mounts, a gunsmith can do the job. Choose the appropriate mounts beforehand, and bring them, as well as your scope, to the gunsmith so that he can actually attach the scope for you, precisely, correctly and neatly.

Older Eyes and Scopes

Young eyes have the ability to accommodate, or focus fore and aft, nicely. In the next chapter, this factor is more fully explained. For now, let it suffice that when our eyes do not focus as rapidly or as well as they once did, the scope sight can be a real boon to us. Not only is the target magnified with a scope, you see, but visual accommodation is also put aside. While iron sights require the shooter to focus on the rear sight, front sight, and target, and with the peep sight demanding focus on the front sight and target, the scope's picture is on one flat two-dimensional plane, demanding the eye to focus only on that plane, so to speak.

Summary

Simple iron sights can be fine. In some cases, they are truly the only correct sight. For example, I'd not want to alter the fixed open iron sights on my Lancaster-period long rifle. On the other hand, my Storey Buggy Rifle's aperture sight is perfectly at home on that specific rifle, and I do not feel any sense of inappropriateness when I fire that rifle with its modern Williams peep sight. As for scopes, the choice is purely personal. Old eyes may demand them, and thousands of blackpowder shooters simply prefer them. The choice, in the end, is personal. ●

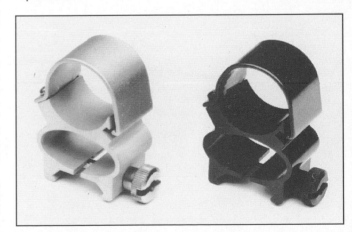

Once a problem, today there are many different scope mounts available for muzzleloaders. These Thompson/Center Weaver-style rings are of the see-through detachable type, allowing the use of scope or iron sights on the same rifle.

chapter 30

Sighting–In The Blackpowder Firearm

SLUG GUNS, WHICH are heavy benchrest-type blackpowder rifles, have produced groups that rival modern arms. Just as an example, consider the .734-inch group, center to center, that was recorded not at 100 yards, but 200 yards from the muzzle, and not for three shots, or even five, but for ten shots. Another slug gun printed a group that measured .505-inches center to center, this for five shots, but still at 200, not 100 yards from the bench. Establishing faith in blackpowder accuracy is important, but the potential of any firearm can only be realized with good sights, distinguishable targets, careful gun management, and of course proper sight-in. This chapter is about sighting-in a blackpowder firearm to best advantage.

All bullets from all guns travel in a curve known as a parabola. The 220 Swift, for example, is a hotrock if ever there was one, with its little missile speeding away at 4000 fps MV. And yet, the rocket-like Swift does not "shoot flat," not even to a mere 100 yards, and blackpowder guns don't shoot nearly as close to the baseline as modern high-velocity cartridges. But that's okay. We got into muzzle-loading for the challenge, right? Well, part of the challenge is getting closer to game, as well as understanding the arc taken by the bullet, be it from a muzzle-loading pistol or rifle, a cap 'n' ball revolver, blackpowder cartridge six-shooter, or blackpowder cartridge rifle.

You cannot sight-in a muzzleloader without understanding its tra-

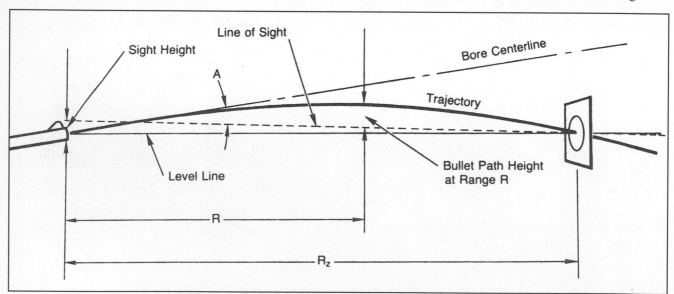

This schematic of a parabola shows the potential path of a bullet in flight. The shooter must know the trajectory of his firearm in order to sight-in to best advantage.

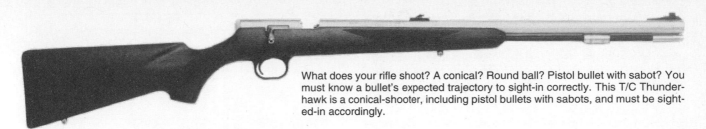

What does your rifle shoot? A conical? Round ball? Pistol bullet with sabot? You must know a bullet's expected trajectory to sight-in correctly. This T/C Thunderhawk is a conical-shooter, including pistol bullets with sabots, and must be sighted-in accordingly.

jectory. Line of sight simply means the line produced with our vision. The eye makes an absolutely straight "line" to the target when we look at it, be the target a flat piece of paper with a bull's-eye, a gong, a metal silhouette, or a deer. This line of sight is straight, but not necessarily horizontal. You can see why. Suppose you're standing on a hill, looking down at your target. The line of your vision from eye to target is straight, but not horizontal. In making a model of this, a term called "baseline" is used to represent a horizontally flat line from bench to target. The line the bullet takes, of course, is anything but flat. It runs along a curve, which is a parabola.

The bullet takes off "nose up," aiming above the line of sight. This is called the line of departure, where the projectile leaves the baseline to rise on its journey to the target. It crosses the line of sight twice, once fairly close to the shooter, and again at the zero point. This path is known as the bullet's *trajectory*, with zero being the exact location of sight-in. If we were shooting a 50-caliber muzzleloader—just as an example—with a round ball, the ball would cross the line of sight twice, once up close, and again "out yonder" where the gun is sighted-in. Remember, the base line is flat, but the path of the bullet, round or conical, is curved.

The round ball from a well-loaded 50-caliber blackpowder big-bore rifle crosses the line of sight the first time at about 13 yards, and then for the second time at around 75 to 80 yards, falling *below* the line of sight after that. It's important to understand line of sight and parabola, because when we sight-in our gun, we're using both, whether we know it or not. By understanding the path the bullet takes along the line of sight, both above it and below it, we can hold appropriately for both close-range targets and targets within the limit of our firearm. Take a look at the suggested sight-ins for various blackpowder guns coming up, and you'll see what I mean.

How Far Can They Shoot?

Long-range target blackpowder shooting is definitely in the bargain. I emphasize *target* shooting. You can't wound a target, and so if you wish to shoot at one at a thousand yards, go ahead. In fact, thousand-yard matches are nothing new in the smokepole business. Such matches were held in the 19th century, and a few continue to this day. What we call rifled muskets were often employed to put bullets downrange all the way to a thousand yards, and with acceptable accuracy. The Whitworth rifled musket is a perfect example of a muzzleloader capable of such far-away accuracy. Naturally, those muzzles were aimed skyward to the proper degree in order to land a bullet in the bull's-eye at a thousand yards. However, as noted earlier, we have no rifle of any vintage that shoots "flat," so even the fast Swift and current magnums require a severe muzzle-up attitude to put a bullet on target at a thousand yards.

Another long-range shooting game practiced today by dedicated blackpowder shooters is the *silueta*, to use the Mexican name for this sport. It means firing at metallic figures, offhand, at long range, using single shot blackpowder cartridge rifles. The sport is touched on in Chapter 50. As for long-distance shooting at game, I'm against it. Sure, a practiced rifleman with an accurate muzzleloader or blackpowder cartridge rifle can put his bullets on the money "way out there," as the 19th century buffalo runner proved over and over again. But I prefer closing in for that one good shot when the target is a deer instead of a metallic cutout or paper bull's-eye. I think it's the sporting thing to do.

Getting the Range

In target shooting, the firearm can be sighted-in specifically for a given range, even a thousand yards. Midrange trajectory (how high a bullet rises between the muzzle and the target) is of no consequence here. Suppose a target shooter does sight his musket in for a thousand yards, and suppose this sighting puts his bullet a few feet high in between the muzzle and that distance? No harm done. But for game shooting, we'd not want to be forced into a guessing situation with a sight-in that puts the bullet very high at midrange. Naturally, it all depends upon the firearm and the load. But all in all, and as a crude rule of thumb, I consider 125 yards far enough for big game muzzleloaders, whether loaded with round balls or conicals.

Because they were good at judging range, the buffalo hunters of the 19th century were able to make hits at very long range with their blackpowder cartridge rifles.

Conicals and round balls describe very similar parabolas from muzzle to 125 yards for normal guns with normal loads. Faster-shooting pistol bullet/sabot combos shoot a bit flatter, as pointed out below. As for the round ball, it starts out faster than the conical, somewhere in the 2000 fps domain, but loses its initial velocity rapidly, while the conical begins its journey slower, generally in the 1500 fps realm, but retains its initial velocity better. The end result is that either missile lands roughly a half-foot low at 125 yards, and 6 inches, I think, is as much variation as I want to work with on a big game animal, where chest strikes are called for. Again, see the sight-in suggestions for various guns below.

Sight Adjustment

The two terms used most often when discussing sight-in are windage and elevation. Windage is horizontal movement. When we adjust our sights to move bullet impact to the left or to the right, we're effecting a windage change. Elevation is a vertical movement. When sights are adjusted to move bullet impact either higher or lower on the target, we're effecting an elevation change.

Before any sighting can be accomplished, a shooter must know how to adjust the sights on his firearm. We've touched on this a bit earlier in our chapter on sights. Fixed sights are manipulated by drifting them to the left or right in their dovetail notches, when these sights are mounted in dovetail notches. That takes care of windage. Fixed sights are adjusted for elevation by filing the front sight down, installing a new, taller, front sight, or having a gunsmith make a front sight taller by welding metal on it. To review: move the rear sight in the direction you want the next bullet to hit, left for left, right for right. Move the front sight in the opposite direction you want the next bullet to hit. A taller front sight makes the gun shoot lower. A shorter front sight makes the gun shoot higher.

Sights readily identify themselves in terms of adjustment. The ladder/dovetail rear sight, for example, is clearly altered in one of two ways. For elevation, move the ladder to either raise or lower the rear sight. For windage, move the rear sight to the right or to the left by drifting the sight in its dovetail notch. Other iron sights have different means of adjustment, sometimes a set screw that holds part of the sight firmly in place, for example. Loosen the screw, move the sight appropriately, retighten the screw, and that's how adjustment is accomplished. There are far too many variations to broach here, plus, as stated, most sights have obvious adjustment features.

Scopes are very simply adjusted. Remove the turret caps. Under these caps you'll sees arrows. For windage, the arrow may have an "R" for right. For elevation, the arrow may show an "Up." Some scopes vary this feature, but all clearly show windage and elevation directions. If the shooter wants the next bullet to hit to the right, he rotates the knob to the right in the direction of the arrow. If he wants the next shot to hit to the left, he rotates the knob in the opposite direction of the arrow. For up/down (elevation) adjustment, the knob is rotated appropriately. Scopes offer calibrated adjustments. Sometimes the adjustments are marked off with lines, each line signifying a value at 100 yards. Directions with the scope give the values. Or the values may be written underneath the scope turret cap. Sometimes adjustments are arranged in "clicks." Rotating the knob results in actual clicks that you can feel. A single click may be worth one-quarter minute-of-angle.

Actually, a minute of angle, also written as MOA, means $1/60$th of one degree. For our purposes, which is practical shooting, a minute of angle is considered 1 inch at 100 yards, 2 inches at 200 yards, 3 inches at 300 yards, and so forth. So in shooting, a quarter minute of angle is only $1/4$-inch of movement at 100 yards, a very fine adjustment that is found on quite a number of riflescopes these days.

Sight Radius

A good concept to know about in sighting-in a rifle is sight radius, because it tells us why a sight picture may be so much clearer for us with one iron-sighted rifle over another. This applies only to iron sights. Scopes have one flat field, totally eliminating sight radius as a

After learning the trajectory of his blackpowder cartridge rifle, this shooter knows where to set his rear sight for a given range. Note how high the disc is set here. It should be. The buffalo cutout at which he is aiming is 900 yards away!

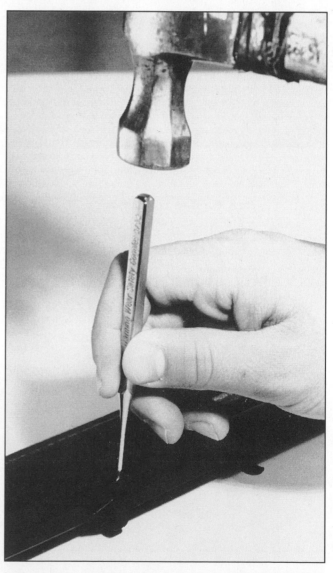

Using a punch, this "fixed" rear sight is drifted in its dovetail notch for windage.

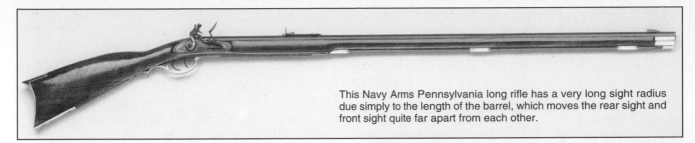

This Navy Arms Pennsylvania long rifle has a very long sight radius due simply to the length of the barrel, which moves the rear sight and front sight quite far apart from each other.

A modern telescopic sight like this one makes it very easy to sight-in the rifle. And once sighted, it remains so. Scopes are more reliable than ever.

The scope is exceedingly easy to adjust. This elevation turret has an arrow with the word "up" to indicate how to adjust to raise—or lower—the aimpoint. To make the gun shoot lower, make the adjustment in the opposite direction of the arrow.

criterion for a good sight picture. But with iron sights, sight radius can make a difference, sometimes a big one, depending upon the eyes of the shooter. As noted, aging eyes are less adept at focusing from rear sight to front sight to target.

Sight radius is simply the distance between the front sight and the back sight, but what a difference this can make in clarity of sight picture. Dedicated iron-sight shooters have been known to ask their gun-makers to leave the back sight completely off of their longrifles at delivery. Then these fellows go to the range, carrying a back sight that has a flat base. They rest this sight on the top barrel flat with the rifle firmly on the bench. By sliding the rear sight slowly back and forth along the barrel, while at the same time carefully checking for sight picture, the shooter finds that perfect "sweet spot" where the rear sight is placed just far enough from the front sight to afford the owner of the rifle the clearest possible picture. Then the sight is marked for position and the rifle is returned to the gunmaker, who fits the rear sight into a dovetail notch right where it belongs. I'm a firm believer in a long sight radius as eyes "mature." On my long-barreled 42-caliber flint-lock, for example, the distance between front and rear sight is exactly 28 inches.

Furthermore, where the rear sight is placed in relation to the eye is vital for that clear picture. While sight radius is what we're talking about here, it's just as important to recognize that as eyes grow older, moving the back sight farther up the barrel can really pay off. When the open rear sight is too close to the shooter's eye, he cannot focus on it. The rear sight becomes a blur, and a blur is not what we want in gaining a clear sight picture. The rear sight on my aforementioned 42-caliber flintlock is a full 9 inches forward of the base of the breech, and about 14 inches away from my eye. With this arrangement, I can see the back sight clearly in relation to the front sight, affording me a good sight picture.

Sight Picture

Before going on, let's stop briefly to talk about sight picture. Without a good one, sighting-in can be an exercise in futility. For a scope, a clear sight picture is created by adjusting the ocular lens, which simply means screwing it inward or outward until the picture is clear. For the

receiver sight, the main criterion for a good sight picture is the size of the aperture or peep hole. For hunting, a slightly larger aperture is preferred. It is easy to see through, plenty accurate, fast to use, and it affords a nice bright sight picture. With open iron sights, frame of reference is important. If the front sight optically fills, fully, the rear sight notch, you cannot achieve a precise aimpoint. There should be a little bit of light showing on both sides of the front sight as it rests in the notch of the rear sight. In this way, you have a frame of reference. You can tell when your sights are truly lined up.

Some shooters prefer the so-called Patridge open sight style, where the front sight presents parallel sides, appearing as a rectangle that fits into a rectangular or square notch in the rear sight. Shooters can line this sort of sight up readily, because light clearly shows on both sides of the front sight as it rests in the rear sight notch. Plus, it is very easy to align the top of the front sight right across the top of the rear sight notch. This affords a very clear sight picture. The six-o'clock hold is often used with the Patridge sight, which means resting the target optically right on top of the post-like front sight, like a pumpkin sitting on a square-topped board. This way the target is not covered up by the sights, which further clarifies the aimpoint. Of course, the gun is sighted-in to print its bullet upward from the topmost of the front sight, so the bullet will strike the center of the bull's-eye.

As with so many shooting terms, this one has multiple meanings. Today, the globe front sight is usually hooded, with interchangeable inserts of various styles. In the old days, a globe front sight could be a simple small ball of metal. This is mentioned because the globe sight could be placed directly on the bull's-eye, unlike the Patridge sight noted above. So there are definitely different ways to achieve a sight picture. Personal preference has led to dozens of different sight styles to please different shooters.

Projectiles in Flight

Bullet Drift

Bullet drift is not vital to blackpowder sight-in, but before moving on to wind drift, which is very important in shooting blackpowder guns accurately, this term must be dispensed with. Bullet drift is the normal

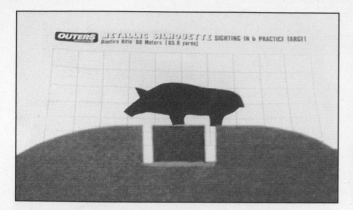

The six o'clock hold places the front sight underneath the target's aimpoint.

The hunting or center hold means placing the front sight directly upon the target.

horizontal departure of a bullet from the line of sight in the direction of the rifling twist. Right-hand twist rifling makes bullets rotate right, and bullets that rotate to the right tend to drift a bit off course to the right, in the direction of their spin. Left-hand rifling twist encourages bullets to drift left. This is a minor factor in practical shooting.

Wind Drift

Wind drift, on the other hand, is a very prominent factor in sighting guns, and in shooting them accurately under field conditions. Also known as wind deflection, this is the condition of the bullet drifting left or right of the line of sight due to the power of the wind. And it sure doesn't take much wind to move the usual blackpowder bullet well off course. Time of flight is an important factor in wind drift. It's easy to see why. The longer it takes a bullet to go from muzzle to target, the more time wind has to act on it. Super-fast bullets have a very short time of flight, which means the wind does not have as long to push on them. But time of flight is not the only criterion in wind drift. The mass of a bullet helps it to maintain its path, as clearly shown by the 220 Swift versus 30-06 Springfield comparison below. So let's look at these modern examples before going forward with this important subject. Comparing the 220 Swift with a 30-06 Springfield reveals a lot about wind drift. The following figures are rounded off for easy reference.

220 Swift - 50-grain bullet - 4000 fps MV
Wind velocity: 20 mph (miles per hour)
Bullet drift at 100 yards = 1.75 inches
Bullet drift at 200 yards = 7.00 inches

30-06 Springfield - 180-grain boattail bullet - 2700 fps MV
Wind velocity: 20 mph
Bullet drift at 100 yards = 1.25 inches
Bullet drift at 200 yards = 5.25 inches

The 220 Swift bullet covers the ground fast, but it's light. The 30-06 bullet starts out much slower than the Swift, but the bullet has better windbucking properties (higher ballistic coefficient), and so it wins the duel in bullet drift at both 100 yards and 200 yards in the same 20 mph wind. Now check out a couple of round balls, 36-caliber versus 50-caliber.

36-caliber round ball - 65-grain bullet- 2000 fps MV
Wind velocity: 20 mph
Bullet drift at 100 yards = 28 inches
Bullet drift at 200 yards = 113 inches

50-caliber round ball - 182 grains - 2000 fps MV
Wind velocity: 20 mph
Bullet drift at 100 yards = 18.5 inches
Bullet drift at 200 yards = 80 inches

Clearly, the larger, heavier round ball, with its better ballistic properties, outshines the smaller round ball when it comes to wind deflection. At the same time, both round balls fly well off course in a 20 mph wind. Imagine what a powerful wind would do to these projectiles. The importance of wind drift cannot be overstated. Good marksmen know what it is, and they know what to do about it, at least to the best of their judgment in applying Kentucky windage, as described below. Hefty blackpowder conicals do better in the breeze, but are still thrown off course considerably. A 50-caliber Maxi-Ball, 370 grains weight, at 1500 fps MV, drifts 16 inches at 100 yards in a 20 mph wind and 60 inches at 200 yards in the same wind, while a 54-caliber 400-grain conical starting at 1500 fps MV drifts 11 inches off course at 100 yards in a 20 mph wind and 43 inches off the beam at 200 yards in the same wind. These conicals do better than round balls, but still suffer badly from wind deflection.

Bullets begin to drop as soon as they leave the muzzle. Of course, initial drop is minute, while drop at long range is pronounced. Obviously, bullet drop is vital to sighting-in, and must be considered as we look at Arkansas elevation, coming up.

Kentucky Windage and Arkansas Elevation

Even the relatively fast-starting pistol bullet/sabot from the modern muzzleloader suffers pronounced wind deflection and bullet drop. Now that both of these have been examined above, it's time to talk about handling them. The "good shot" applies both Kentucky windage and Arkansas elevation to overcome these problems, at least to some degree. Nobody can guess the exact velocity of the wind under field

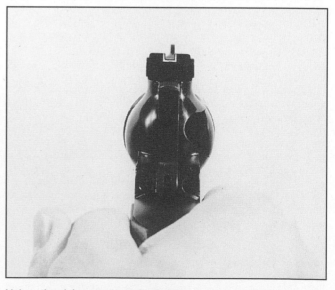

Unless the sights are properly aligned, you're going to miss, no matter how well your firearm is sighted-in.

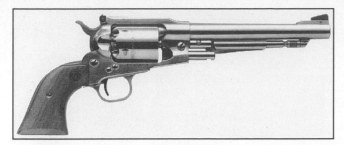

Good sights are a must for hitting the target consistently, as well as sighting-in confidently. The sights on this Ruger Old Army are not only adjustable, but they also allow a clear sight picture.

Sighting the Lyman 57 SML Micrometer Sight is a matter of simple adjustment with clearly marked indicators.

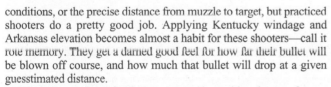

You cannot hit it if you can't see it, and see it well. That's why sighting-in requires a target with a clear aimpoint.

conditions, or the precise distance from muzzle to target, but practiced shooters do a pretty good job. Applying Kentucky windage and Arkansas elevation becomes almost a habit for these shooters—call it rote memory. They get a darned good feel for how far their bullet will be blown off course, and how much that bullet will drop at a given guesstimated distance.

Kentucky windage simply means guessing, with a degree of accuracy, how far to hold off a target so that the bullet will be blown by the wind back onto the target by the time it arrives there. One time I was hunting mule deer in a powerful wind. My chance came from only 80 yards or so. But even at that distance, I knew the round ball of my 54-caliber long rifle would drift over as much as three feet. I applied a dose of Kentucky windage to the sight picture by aiming at an imaginary target suspended in midair two feet to the left of the buck's chest cavity. Even with that hold, the round bullet hit a bit farther back than I wanted, but it was still in the vitals and I had my deer, cleanly. Without Kentucky windage applied, it could have been a disaster—a round ball striking a ham instead of the vitals.

Arkansas elevation means guesstimating the range or distance from muzzle to target and allowing for bullet drop. I have darned little trouble with Arkansas elevation because I do my best to get within close range of game, and on the target range I know how far away the object of my shot is, so I can hold over to make a hit. Nonetheless, it's well to have a feel for bullet drop elevation, even when the shot is not all that far, and certainly when you don't want to resight your rifle to aim at a gong or other target. Knowing how to apply Arkansas elevation is the mark of a practiced shooter, and it comes from understanding bullet drop with a given firearm for a given load, plus practice.

Good Targets

If you can't see it, you can't hit it, and if you cannot delineate the aimpoint on your target clearly, sighting-in will be a problem. Fortu-

nately, targetmakers caught on in the last decade or so. They realized that black bull's-eyes are just fine, and still entirely viable, but that black bull's-eyes on off-white paper were not the only worthwhile aimpoints. Many new targets have come along, including the revolutionary Shoot-N-C display target from Birchwood Casey Co. This target comes in various styles, but the main feature on all is the same. You see the bullet hole instantly. Wherever the projectile passes through the target, a black coating flakes off approximately two times the size of the hole, leaving a bright halo around each bullet hole. This makes sighting-in fast and even fun.

However, of most importance in selecting a target is a clear and defined aiming point. The marksman must be able to line his sight up on the target so there is no question about aim. This sounds basic and obvious, but I've watched countless attempts at sight-in fail because there was no way to take a precise aim at a target. Iron sights can be a problem, especially open irons, because they optically cover a large area downrange, making a precise sighting point absolutely essential to any hope of getting the firearm on target. I've turned to bright orange dots, for example, using the six-o'clock hold noted earlier. With this hold, the sight picture is better defined because the bead or post of the front sight does not cover up the bull's-eye. It rests just underneath the bull's-eye. I hold so that the very top of the bead touches the very bottom of the bull's-eye. For a hunting gun, I want the group to print where the bead and target touch. This way, I know when I put the bead right on the game, that's where the bullet will go. But for target work, I prefer the group to print into the center of the bull with the six-o'clock hold.

Practical Sight-In Guide

Every big-bore muzzleloader can be sighted-in for a flat 100 yards. Every smallbore muzzleloader can be sighted-in for 50 yards. Such sight-in will suffice, but here is a simple practical sight-in guide for

Using a bench is the best way to sight-in, with the forearm of the rifle firmly rested on a solid yet padded support.

various blackpowder guns that might help to develop a more refined sight-in.

Big Bore Rifle—Hunting Load

Rifle calibers 45 through 58, whether shooting a round ball or a conical, can be pre-sighted at 13 yards for starters. This normally puts the bullet back into the line of sight at about 75 to 80 yards, considering a round ball at about 2000 fps, and a conical at about 1500 fps. Another good way to sight-in the big-bore hunting muzzleloader with hunting loads is to center your group an inch high at 50 yards. An accurate muzzleloader should shoot a fairly close group at 50 yards, so determining where the center of that group is located is not impossible. Sighted to hit an inch high at 50 yards, the round ball or conical will hit an inch or so low at 100 yards and about a half-foot low at 125 yards.

Medium Round Ball Rifle Calibers

Calibers in the 38 to 40 range shooting round balls at around 2000 fps can be sighted dead-on at 50 yards. This puts the round ball an inch low at 75 yards, and two or three inches low at 100 yards.

Smallbore Rifle Calibers

Calibers 32 and 36, shooting round balls loaded to about 1500 fps for small game, should be sighted dead-on at 50 yards, which is about as far as most hunters will shoot at small game. Start by sighting dead-on at 25 yards, which gives a 50-yard zero. So sighted, these light missiles strike about an inch low at 75 yards, which is certainly not enough to miss even a small game animal. Loaded to shoot at 2000 fps, the little 32 or 36 can be sighted dead-on at 75 yards, giving these smallbores a practical range of about 100 yards.

Modern Muzzleloaders with Pistol Bullets and Sabots

A 50- or 54-caliber modern muzzleloader firing a 44-caliber 240-grain pistol bullet in a sabot with a muzzle velocity of 1600 fps, can be sighted-in to print the group 2 inches high at 50 yards. This puts the group about 1.5 inches high at 75 yards and on the money at 100 yards. From there, the bullet drops about 3 inches at 125 yards. While it's possible to hit out to 150 yards on big game with the 44-caliber bullet so loaded, I'm not too excited about shooting past 125 yards. The bullet drops 7 inches at 150 yards when the rifle is sighted-in as suggested here. Bullet energy is down to around 775 foot-pounds.

Sighting-In Tips

1. Be sure your blackpowder gun has good iron sights, and if it wears a scope, make certain that the mounts are correct, and secure. You cannot sight-in with lousy open or peep sights, or with an insecure scope.

2. Make certain that your firearm is sound. You cannot sight-in with a broken firearm. Tighten all screws, including sight screws.

3. Be certain of your load. You cannot expect an inaccurate load to print consistently in the same group on the target.

4. Use a benchrest for sight-in, and use it right. Make certain that both the forend of a rifle, plus the toe of the stock, are well padded and secure on the benchtop. A rifle should be set up so that it all but aims itself when properly resting on the benchtop. The shooter should be comfortable with both feet flat on the ground and spread apart a bit for stability. The left hand for a right-handed shooter may grip the forestock if recoil is a problem, but if recoil is not a problem, it's better to rest the left hand flat on the benchtop. The right hand for a right-handed shooter controls aim, as well as the trigger.

5. As noted above, use a target you can see, which means a target with a well-defined aimpoint.

6. Start by just getting on the paper. Do not frustrate yourself with 100-yard shooting when the gun has never been sighted-in before. Get close. Start out at only 10 to 15 yards. Adjust sights to hit dead center at that close range, then move out to 100 yards.

7. Know your trajectory before trying to sight-in. There is no point in sighting-in a big game rifle with its top load for only 50 yards, for example, when even with round ball that rifle deserves to be sighted dead-on at 75 to 100 yards.

If using a bench is not possible, at least use a rest, like this box with a cutout padded with tanned leather, and shoot from the prone position to promote steadiness.

The right elbow of a right-handed shooter should be well-planted on the bench for sighting-in, not floating.

The Breechloader

The blackpowder cartridge rifle can usually be sighted-in for 100 yards. Considering the 45-70 Government round, for example, with a 500-grain round-nose bullet at 1200 fps, sighting to print the group three inches high at 50 yards will give a 100 yard zero, with the bullet falling about 6 inches below the line of sight at 125 yards, and 9 inches below line of sight at 150 yards.

The Blackpowder Handgun

The cap 'n' ball revolver, as well as most muzzle-loading pistols, do well with a starting group at 25 yards that is about 1 inch high. This puts the bullet back on target at 50 yards. The same sight-in applies to the blackpowder cartridge handgun.

Don't Forget Protection

When sighting-in, don't forget eye and ear protection. This goes for any firearm. Also, when shooting the big bores with full-power loads, consider a pad between your shoulder and the buttstock. There's nothing sissy about being comfortable at the bench. Do not use a sandbag between your shoulder and the buttplate, however, as this may cause a stock to break. There are special shooting pads available at gunshops that have give, which treats both shoulder and rifle stock well.

Sighting-in is a relatively simple matter on the face of it, but it does entail a number of important issues, as indicated in this chapter. Sighting-in is worth all the study and bother. After all, an unsighted firearm is less than effective. ●

chapter 31

The Great Little Muzzle-Loading Smallbore

SQUIRREL RIFLES. THAT'S what they called 'em in the old days. By virtue of modest caliber, these longarms were designed for small game—squirrels of course, but also rabbits, 'possum, even wild turkeys—all the littler stuff that can be turned into wholesome meals as common as fried rabbit or as regal as Thanksgiving gobbler dinner. Smallbore blackpowder rifles are overlooked these days because the number one reason for "going blackpowder" is to get in on those special "muzzleloader" only big game hunts, where sub-bores don't fit in. Too bad they aren't more popular, because smallbore blackpowder rifles are fun to shoot, darn near recoilless, easy on the pocketbook, great for plinking and practice, better on the target range than most shooters think, and downright interesting to mess with.

What is a smallbore muzzleloader? According to Captain Dillin, in his fine book, *The Kentucky Rifle*, the most popular smallbores of the golden era of shooting in Early America carried either 110 or 150 round balls to the pound. The first translates into 36 caliber (.350-inch lead sphere), while the second comes up 32 caliber (.310-inch pill). Ned Roberts, in his book *The Muzzle-Loading Cap Lock Rifle*, said that "The 100 to 220 to the pound gauge—36 [about] to 28 calibers—were called 'squirrel rifles.'" Nothing has changed very much. The two most popular smallbore calibers today remain 32 and 36, and many of us continue to refer to muzzleloaders in these sizes as squirrel rifles. Arguably, other calibers qualify. Smaller-than-32-caliber rifles can be custom ordered. Calibers 38 up to 40 also offer decent economy, light recoil, and small-game-hunting ability. But the truth is, smaller-than-32s are darned hard to find, and anything over 36-caliber is entirely unnecessary for small game and plinking. So 32 and 36 it shall be for our purposes, while those who prefer slightly smaller or larger squirrel rifle calibers are welcome to their personal choices.

Versatility

Smallbores are overlooked in part, I think, because they're not considered versatile. "What would I do with one?" a shooter might ask. "Aren't they only good for small game?" Actually, smallbores do quite a bit. Hinted at above were several smallbore uses, including target work and general practice. Getting right down to it, there's good argument that squirrel rifles are the most versatile of the frontloader clan. They're wrong for big game, and less than perfect for rendezvous shooting sports, but let's not concentrate on their shortcomings. Let's look at their good points.

Small Game

Obvious, yes, but smallbores are ideal for small game, and I wish more modern shooters recognized the fact for their own sake. They're missing one of the great joys of smokemaking—huntin' littler critters with a smokepole. Small game hunting is popular, and should be. Some sort of small game generally resides darn close to home: it's easy to get to; you don't need a guide; it doesn't require "safari" dollars to hunt them. Furthermore, small game seasons are usually long. When big game season is finished, many small game hunts are still going. Cottontail rabbits, for example, abide just about everywhere in North America and in other countries, such as Australia. In some areas, cottontails are in open season from half a year to all year long. And squirrels are a hunting joy in their own right. It's a privilege to stalk bushytails amongst the trees. There are some hunters who prefer Mr. Chatterbox to all other game.

Wild turkeys are the big game of upland birds, and there is no reason in the world that they cannot be hunted with the smallbore blackpowder rifle in either caliber 32 or 36. I've taken several with both. The challenge rises a notch or two over other firearms for these great prizes, but I've never lost an opportunity on a gobbler because I had a squirrel rifle in hand. A 32- or 36-caliber long rifle will also pot "Ben's bird" every time with solid ball placement. The largest game I'd personally go after with a smallbore is the javelina of the Southwest and Mexico, and at that I'd only try for these animals if I could be sure of a perfect shot. They're small, as big game goes, but a lot bigger than wild turkeys. Wild pigs, as they're fondly called, field dress in the 30- to 40-pound range, and they can be pretty tough little guys to bring down. But a cool shot can drop one of these musk hogs with a single small-caliber ball, when it's placed right.

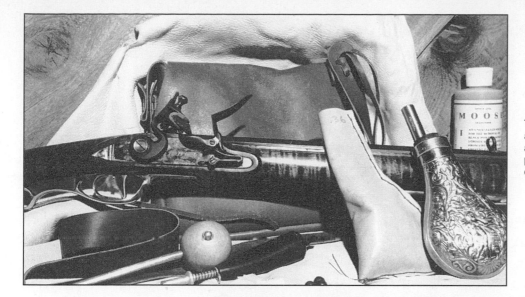

The smallbore "squirrel rifle" does a lot more than harvest squirrels and other small game. It's also a plinker, practice piece, turkey-taker, and much more.

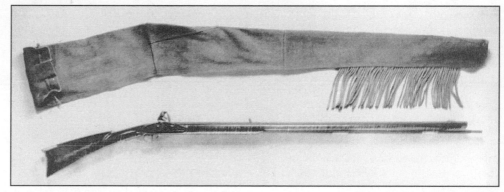

This Cumberland flintlock rifle comes in 36-caliber. It's from the Wilderness Rifle Works in Waldron, Indiana, and it's a smallbore capable of a lot of work and play.

Economy Shooting

A 32-caliber squirrel rifle can be fired for fewer pennies per pop than a 22 rimfire, if the shooter casts his own ball (from scrap lead) and makes his own percussion caps (using the Forster Tap-O-Cap tool). The little 32-caliber patched ball achieves 22 Long Rifle muzzle velocity with only 10 grains of FFFg blackpowder—that's 700 shots for one pound of propellant. The larger 36-caliber smallbore costs a bit more to shoot, but it's not expensive either. The 36 does demand a bit more lead: 65 grains per round ball instead of 45 grains, but it costs no more in percussion caps, and it only takes 15 grains of FFFg blackpowder or Pyrodex P to achieve 22 Long Rifle muzzle velocity.

Target Shooting

Smallbores are accurate. My most-used sub-caliber smokepoles in 32- and 36-caliber produce five-shot groups under an inch center to center at 50 yards—and that's with iron sights. They are not ideal for serious longer-range competition for the simple fact that a larger ball

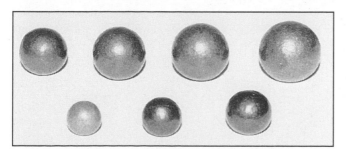

There is no doubt that the larger round ball has the potency, while the smallbore's round balls, 32 lower left and 36 next to it, are for the smaller jobs. But what good jobs those are.

has a better chance of cutting the black (bull's-eye) when it hits in that area than a smaller ball. That's just the way things work out, size-wise. In other words, if you just miss the black with a 32-caliber ball, a 50-caliber ball would have counted as a bull. Also, the larger projectile drifts a bit less in the wind, which can help on the target range. Plus, the little pill is not as good as the big one for certain games, such as splitting the round ball on the axe (see Chapter 49). Again, it's a simple matter of size: the bigger ball provides a better opportunity than the smaller one. Nonetheless, watch out for the guy with a pleasant-shooting smallbore. He just might win the target match.

Plinking and Practice

These are both enjoyable and useful, enjoyable because there's no pressure involved in either one—rolling a tin can is accomplished just to see the tin can roll. And there's a lot of transfer value from shooting the smallbore to mastering the big bore. If a marksman can manage to put little bullets on target with a smallbore, chances are he'll be pretty good at putting larger bullets on target with his big bore.

Because smallbore bullets out of squirrel rifles demand careful attention to the wind and range—wind blows 'em off course pretty readily, and little pills tend to develop a fairly pronounced trajectory curve—the squirrel rifle is interesting to shoot. And, as noted above, once a marksman gets good with his smallbore, proficiency with the big bore is not very far off.

Recoil and Noise

Blackpowder big bores are known to develop pretty fair recoil with heavy hunting charges because they demand a lot of powder to gain reasonable muzzle velocity, but smallbore blackpowder rifles don't develop enough kick to bruise a ripe tomato. Don't worry too much about noise pollution either with the well-mannered squirrel rifle. In either caliber 32 or 36, shooting regular small game loads, the small-

221

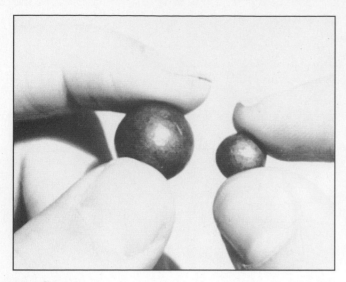

Bigger isn't always better. It depends upon the task at hand. If power is needed, the larger ball wins hands down. But when small game is the goal, the little ball is the right choice.

The smallbore is perfect for brush-hunting cottontails, taking them cleanly and without undue waste of edibles.

bore muzzleloader is a comparatively quiet rifle. This trait is also very positive for the landowner, who doesn't hear the roar of a cannon on his back forty when you're hunting his land.

Smallbores are fun to shoot, and they carry darned good ballistics. After all, bigger isn't always better. You don't need a baseball bat to down a gnat. Smallbore ballistics are just right for many applications where big bores just don't fit in. The 32-caliber squirrel rifle, for example, can be used on small game up to gobblers, with perfect head shots on javelina tossed in. Furthermore, power varies greatly with the charge, making the little 32 just as versatile as promised earlier in this chapter. Ten grains of powder behind the 32-caliber round ball produces about 1200 fps in a 24-inch barrel. In my studies of the 22 rimfire, I found that most brands of Long Rifle ammo develop a muzzle velocity roughly in that domain.

So that's enough power to bag a bunny or drop Mr. Bushytail from a tree limb. My chronograph shows about 1650 fps with 20 grains of FFFg. Going up to 30 grains of FFFg produces a strong wild turkey load with the 45-grain ball departing the muzzle at 1871 fps for an energy of 350 foot pounds, again in the 24-inch barrel. At close range, the 30-grain charge equals the authority of the 22 Winchester Magnum Rimfire cartridge. Longer barrels produce even more muzzle velocity with this charge. The 41.5-inch barrel of the Dixie 32-caliber rifle delivers close to 2100 fps with 30 grains of FFFg for about 440 foot pounds. A 22 LR firing a 40-grain bullet at 1250 fps earns 139 foot pounds of muzzle energy. So you can see that the above are pretty good ballistics for close-range shooting.

The 36 is also a dandy caliber for pure enjoyment and great ballistics. Its .350-inch round ball weighs 65 grains. In my Hatfield flintlock 36-caliber rifle with 39.5-inch barrel, 20 grains of FFFg delivers 1471 fps. Thirty grains gives it 1799 fps, and 40 grains weight of FFFg drives the patched ball at 2023 fps with 591 foot pounds of energy. The 32's super economy of 700 shots per pound of powder isn't duplicated by the 36, and loads under 20 grains may not always deliver top accuracy, nor the flatness of trajectory desired. However, 350 shots per pound is still economical shooting, and some 36s do shoot accurately with only 15 grains of fuel. The 36 ball "carries up" better than the 32, with greater terminal ballistics, and it bucks more breeze. But all smallbore muzzleloaders are prey to wind and none is a powerhouse. These points were discussed earlier.

The smallbore squirrel rifle is associated with small game, and rightfully so, but a 32- or 36-caliber round ball properly loaded is fine wild turkey medicine.

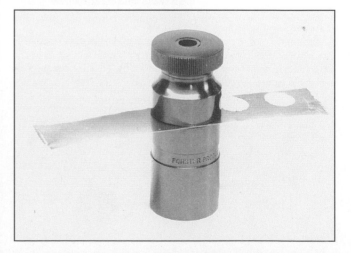

Forster Products' Tap-O-Cap is used to form percussion cap bodies from light gauge metal. The result is an extremely cheap cap, which promotes smallbore economy.

So Which is Best?

Either caliber means a lightweight rifle, because a heavy barrel is not necessary for a smallbore. The squirrel rifle can be long-barreled, but it doesn't have to weigh much. My long-tom Hatfield hefts under seven pounds. Either the 32- or 36-caliber rifle with muzzle velocities in the 1500 fps range can be sighted to print the group's center of impact an inch high at 25 yards. So sighted, the lead balls chew the X-ring out at 50 yards, clustering about two inches low at 75 yards, which is farther than most of us need to shoot at small game with a charcoal-burner (or any other rifle). Most of my rabbit and squirrel shooting takes place no farther than 25 yards from the tip of my boot. The squirrel rifle in either caliber is also simple to manage and easy to take care of.

Because modest powder charges are burned, the heavy-duty caking and fouling common to the big-bore hunting rifle is not present in the squirrel rifle. Modern shoot-all-day lubes work very well with smallbores. (See Chapter 19). Many shots can be fired simultaneously in the field without serious bore swabbing. They're both great, but the 32 wins for economy. It's a case of less being more. The 32-caliber 45-grain round ball is large enough to get the job done with less shooting cost. Loaded up to snuff, the 32 is fine for game up to the size of wild turkeys, and while smallbores can take javelina with perfect ball placement, larger calibers are more suited to the task. So I give the nod to the 32-caliber squirrel rifle over its slightly bigger 36-caliber brother.

Good Fuel for Squirrel Rifles

FFFg blackpowder and Pyrodex P are proper squirrel-rifle propellants. Pyrodex P is especially clean-burning. Hodgdon Powder Co. has tested P in smallbores, finding it ideal for either the 32 or 36. The small powder charges required in the squirrel rifle do not cause excessive pressure with the finer-granulated Pyrodex P powder. Swabbing between shots is unnecessary with Pyrodex, as clearly shown in Chapter 15, and this powder can be used *in the same volume, not weight,* as FFFg blackpowder. Since Pyrodex is less dense than blackpowder, it offers more shots per pound. Meanwhile, FFg granulation is wrong in the smallbore. It yields low velocity. Pyrodex RS (Rifle/Shotgun) is also unwarranted in the squirrel rifle. FFFFg is equally out of place in smallbores. It is unnecessarily fine-grained for the main charge in these muzzleloaders. That leaves FFFg and Pyrodex P as best in the squirrel rifle.

Smallbore Bullets

The least expensive way to shoot the smallbore is to cast your own round balls, as described in Chapter 24. This is especially true when scrap lead can be found free or at small cost. Also, as explained in Chapter 24, casting is a hobby in its own right, and the shooter may wish to run his own smallbore pills for the pure fun of it. However, the smallbore fan is also well-provided for with commercial round balls in both 32 and 36 calibers from several major companies. I've found

The smallbore rifle makes a great starter for the young shooter. There is no objectionable recoil; noise is comparatively mild; and the rifles themselves are generally modest in size and weight.

Smallbores, like this 32-caliber Mowrey, are ideal for lady shooters who want to get started in blackpowder shooting.

(Left) Smallbore muzzleloaders make fine plinkers. If round balls are cast, and Tap-O-Cap percussion caps are used, the smallbore costs less to shoot than a 22 rimfire.

(Right) You never outgrow a smallbore rifle. The little shooter is always ready to put supper in the pot, a neat round hole in a paper target, or to make a tin can leap into the air.

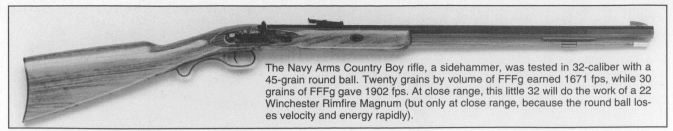

The Navy Arms Country Boy rifle, a sidehammer, was tested in 32-caliber with a 45-grain round ball. Twenty grains by volume of FFFg earned 1671 fps, while 30 grains of FFFg gave 1902 fps. At close range, this little 32 will do the work of a 22 Winchester Rimfire Magnum (but only at close range, because the round ball loses velocity and energy rapidly).

Smallbore doesn't mean only rifles. Black-powder pistols and revolvers have long been offered in sub-size calibers. This Navy Arms replica of an 1862 Police revolver is caliber 36. There are also 31-caliber sidearms.

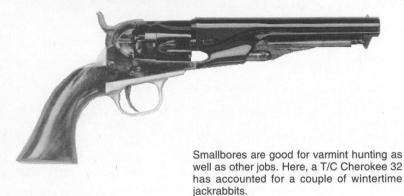

Smallbores are good for varmint hunting as well as other jobs. Here, a T/C Cherokee 32 has accounted for a couple of wintertime jackrabbits.

commercial pills excellent. For the shooter who plinks and practices a great deal, casting is probably the best way to go, while the person who mainly hunts small game with his squirrel rifle may wish to buy a couple boxes of round balls at the beginning of each season and call it good.

Incidentally, matching smallbore bullets and patches is no problem. Try first to match the ball to the bore. That is, select a ball which is not terribly undersized. Then choose the proper patch thickness to ensure a tight ball/patch fit. However, if you must use a thicker patch to take up the windage in the bore, don't worry about it. My Hatfield 36-caliber flintlock fires a cast or commercial .350-inch ball with a .017-inch-thick Ox-Yoke patch. Maybe a slightly larger ball would be better, but I don't have a .355-inch mould, and I'm well satisfied with the 50-yard under-an-inch clusters that rifle provides.

Getting in on the Fun

So make your squirrel rifle a 32 or a 36—a 32 if you want more economy. Remember that neither ball carries as far as its larger cousins, which is good in small game country. You don't need, nor do you want, a ball to travel far under such conditions. Sure, with head shots, all muzzleloading rifles are small-game worthy, but take note of the high points for 32s and 36s, and get in on the fun. While the smallbores are not nearly as popular as the big bores, gun companies have seen fit to offer quite a few squirrel rifles. Smallbore rifle style is of no consequence. Historically correct longarms are nice, and there are, of course, original squirrel rifles available in good condition, but if these oldtime rifles are of collectible quality, don't shoot them. Their value may be destroyed. The modern-made muzzle-loading rifle is just as much fun to shoot, and it has no collector value. Flint or percussion? Correctly designed and manufactured flintlocks, properly loaded, go off like clockwork, and if a hunter or target shooter wants to add a touch more nostalgia to his sport, a good flinter will do it. Smallbore flintlock rifles are, however, difficult to locate, and a custom rifle may be called for.

But there are plenty of good smallbores to choose from today. How about the long-barreled (42-inch) J.P. McCoy Squirrel Rifle, caliber 32, percussion or flint? Mountain State Muzzleloader Supply company

offers this rifle with curly maple stock. Or there's Dixie Gun Works' Tennessee Squirrel Rifle in 32-caliber with its 41.5-inch barrel, 13/16-inch across the flats. In kit form or ready to shoot, this smallbore strains the ballistic potential from the 32. With only 10 grains of FFFg, the 45-grain round ball scoots out at close to 1300 fps. Twenty grains delivers almost 1800 fps. The Navy Arms Country Boy rifle is offered in 32- or 36-caliber with 26-inch octagonal barrel. This caplock is of the "mule ear" or sidehammer design for quick ignition. It weighs only 6 pounds and is available as a kit or finished. The little Cherokee from Thompson/Center, with adjustable sights, also weighs 6 pounds with its 24-inch barrel (25 inches to the end of the breech plug). It shoots the round ball nicely with light powder charges, and will also handle a conical bullet due to its 1:30-inch rate of twist.

CVA offers its Squirrel Rifle in 32-caliber also. This is another 6-pounder, kit or finished form. It has a 25-inch barrel. It's a caplock, and triggers are double set (multiple lever). If you like old-time looks and handling characteristics, check out the Wilderness Rifle Works fullstock Cumberland Rifle in 36-caliber. It comes in either percussion or flint, and it has a 39-inch long barrel with ball-shooting rifling twist. It also has double-set triggers, iron or brass furniture, curly maple wood, and comes as a kit or factory-finished. The long barrel really takes advantage of the fuel supply, and the Cumberland will get around 2000 fps MV with 40 grains of FFFg with the 65-grain round ball. And don't forget the Mowrey Squirrel Rifle, sold through Mountain State Muzzleloading Supplies. This little caplock rifle comes in 32-caliber. The list goes on, including the Traditions Pioneer in 32-caliber, as well as the same company's Frontier Scout in 36-caliber. A couple of underhammer squirrel rifles are back, too, as this is written. These are the Hopkins & Allen Heritage, and the Buggy Rifle, either in 36-caliber from Mountain State.

That's a look at the smallbore—small in caliber, big in performance and enjoyment. Chapter 43 on small-game hunting speaks further on these wonderful muzzle-loading rifles, including a few more ballistic figures. Easy to shoot, easy to clean, efficient, versatile, and just plain fun, a blackpowder squirrel rifle should be on everyone's "must have" list.

●

chapter 32

The Blackpowder Big Bore

FAST BULLETS ARE deadly. There's no doubt about that. Where allowed by law, good hunters who are also good marksmen have taken all manner of big game with small, speedy bullets. An Arizona hunter harvested a dozen deer in a row with a dozen shots from a 22-250 firing a 50-grain bullet at 3800 fps. That's all well and good, but the other side of the ballistic coin is pretty shiny, too—the heavy projectile at modest velocity. That's what blackpowder big-bore guns are capable of—firing big bullets a lot faster than most things travel on this planet. Sometimes we forget just how fast blackpowder bullets do fly. A 45-70 stuffed with blackpowder won't rocket a bullet downrange like a modern high-velocity cartridge. On the other hand, a 500-grain lead missile at 1200 fps isn't a snail's pace either.

The William Moore rifle mentioned briefly in Chapter 22 really intrigues me, not that I'd want one, although I wouldn't mind touching a few rounds off just for the thrill of it. This 19th-century rifle, by Moore of London, is a 2-bore. Remember that means two round balls to the pound. There's 7000 grains weight in one pound, so the Moore cannon launched a round lead missile that weighed half a pound, or 3500 grains. Velocity was chronographed at 1500 fps. Data shows that it took 800 grains of blackpowder to do that. Muzzle energy of this 2-bore ran 17,491 foot pounds. Any way you want to compare that, it's an astonishing figure. A 30-06 Springfield with 180-grain bullet at 2700 fps produces 2914 foot-pounds, while the 458 Winchester, which is capable of dropping an elephant, fires a 500-grain bullet at 2100 fps for just under 5000 foot pounds of ballistic authority.

"Knockdown" Power

Chapter 22 on ballistics dealt with some of the ways power is computed and discussed. However, I wouldn't want to begin a big-bore discussion without touching on a term that's been buzzing around for the better part of a hundred years: Knockdown.

"The fast little bullet is good, but it doesn't have knockdown power, like a bigger slower bullet." Are shooters who talk that way right? Do big bullets really knock animals down? The laws of physics say no, and they've said no for a very long time. Way back in 1903, *Mechanics for Engineers*, by Maurer, Roark, and Washa, refuted the knockdown argument. Here is how they told it:

> The forward momentum of any bullet can be no greater than the rearward momentum of the recoiling gun from which it was fired—actually it is less due to loss of forward momentum in powder gas—[so] the

bullet exerts on the object it hits no greater impulse than the gun exerts on the shooter. (page 270)

Clearly, the authors of *Mechanics for Engineers* vote "not possible" when it comes to rolling game animals head over heels with a bullet. Or if you could, it appears that recoil of the gun would knock the shooter over, too. The term "impulse" was used in the explanation above. It has a special meaning. If you lift a five-pound bucket of sand one foot high, you've done five foot pounds of work. Now you can set the bucket back down slowly, or you can drop it to the floor. If you drop the bucket, it picks up speed (and momentum), hitting harder than if you just set it down. That harder hit from dropping the bucket is called impulse. All of this is important to us before we go forward with big-bore calibers, because we need to know what they cannot do, as well as what they can do. What the big bullet does *not* do is knock game animals off their feet.

Stopping Power Versus Knockdown Power

Big bullets don't knock game down, but they can stop ponderous animals. That's evident from the fact that hunters in Africa who must thwart the charge of a Cape buffalo or become peanut butter beneath the beast's hooves never select the 220 Swift to do the job. They choose big bullets instead. Kinetic energy simply isn't the whole story in stopping power, in other words, although it certainly counts. Energy is energy, all right, but how it is delivered makes a big difference in performance. Consider that a blow from a two-pound hammer is capable of delivering a thousand pounds of force, yet that blow only drives a nail a short distance into soft pine wood. At the same time, a 38 Special with a mere 250 pounds of force may send a bullet through several inches of the same wood.

Part of the reason for this is work versus time, or how long energy is applied. When an object is moved any distance (any object, including a bullet), time is consumed. It does not matter what the item is, or how it is moved. The hammer blow consumes only a tiny fragment of time. The 38 Special bullet applies itself longer to the wood. This is all-important to energy delivery. Big bullets are capable of doing several things, including achieving deep penetration, which creates long wound channels as described in Chapter 42 on big game hunting. Keep this in mind as you touch on the various calibers discussed in this chapter, from sub-40s to 2-bores. Super speed belongs to none of them. But bullet mass and diameter does.

Big Bore Calibers

In Africa, a big bore is generally considered to be one that is over 375-caliber. A 375 is a medium caliber. African hunters don't consider the 30-06 a big bore, while in America, Canada, Australia, and other lands, just about anything over 22-caliber is often classified as a big bore. For our blackpowder purposes, this chapter deals with calibers sub-40 through 2-bore. Recall that the round ball is undersized so it will fit downbore when wrapped in a patch. When we speak of a 40-caliber sphere, for example, it may actually be .390- or .395-inches in diameter. Before looking at calibers, let's get a mindset for one thing: caliber means something in itself. A projectile starting at 50-caliber, for example, is already "bigger around" than some smaller bullets become after mushrooming (bullet upset). Also, in Chapter 24 on running ball, lead was credited with the property of cohesiveness. Lead molecules like each other. They tend to stick together. So a large-caliber lead bullet at modest striking velocity doesn't normally come apart in game. It proceeds as a unit, retaining most of its original weight. Round lead balls recovered from big game generally show a loss in original weight of only a couple percent. Now let's look at some specifics.

Less Than 40-Caliber

Under 40-caliber is not big bore in the context of this work. However, we cannot leave out a brief mention of blackpowder cartridges for single shot breechloaders and lever-action repeaters that fall below this caliber size, yet perform well on deer-sized game, and sometimes larger animals. The 32-40 Winchester, for example, is a decent deer cartridge for those who hunt carefully, get close, and place bullets where they belong. The 38-55 Winchester is even better. Many a moose has fallen to that old blackpowder cartridge with one well-centered bullet. So there are sub-40 blackpowder rounds, and quite a few of them that, while not big bores in the blackpowder world, do perform well on big game.

Knockdown is mainly a myth. Even big bores don't bowl game over. This mule deer buck will not be knocked over by a bullet, even one of very big caliber.

Blackpowder firearms gain power from bullet mass, and even with conicals, mass comes with large caliber. The 50-caliber Buffalo Bullet (right) weighs close to 400 grains, more than double the heft of the 30-caliber modern bullet (left).

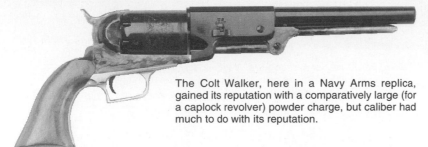

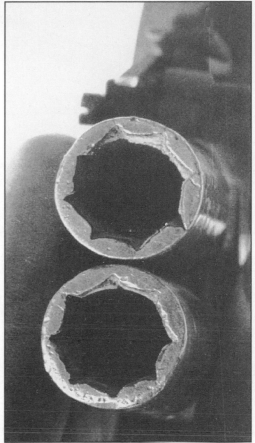

The Colt Walker, here in a Navy Arms replica, gained its reputation with a comparatively large (for a caplock revolver) powder charge, but caliber had much to do with its reputation.

Bullet upset, or mushrooming, in effect enlarges the caliber of the projectile. The unfired 50-caliber bullet on the right is, of course, 50-caliber. The same bullet extracted from test media (left) is 75-caliber.

This 19th-century screwbarrel pistol got its effectiveness from bore size, as did all of the early pistols of any power. The rifling-like configuration at the muzzles were probably for tool fit.

Caliber 40

I'm including 40-caliber with big bores, but not because of the round ball. Even at 2000 fps, the 40-caliber round ball, weighing under 100 grains, is no powerhouse. It generates less than 250 foot pounds of energy at 100 yards. However, 40-caliber does make the big-bore grade when it's a conical. The White Superslug, for example, mikes at .409-inch diameter. It weighs 280 grains, of if you prefer more heft, there's a 40-caliber Superslug at 320 grains weight. You don't have to get the long 40-caliber Superslug screaming from the muzzle to achieve good performance on big game. Even at velocities not too far above the speed of sound, these high-sectional-density bullets penetrate well. For the big game hunter who wants a 40-caliber muzzleloader, long conicals fill the bill.

Also viable are 40-caliber blackpowder cartridges, of which there were many. Some old-time hunters preferred the 40-caliber single shot breechloader above all others. Shoot one and you'll see why. They're pleasant to fire, and yet they get the job done in the hands of a cool shot. Bullet placement is always vital with all calibers, but when blackpowder projectiles are not terribly large, caliber-wise, bullet placement becomes even more crucial to clean harvesting. The 40-50 Sharps Bottleneck cartridge didn't hold a heap of blackpowder, but it pushed bullets in the 300-plus-grain class at speed-of-sound muzzle velocities with powder charges ranging from 45 to 50 grains. Give me a 300-grain 40-caliber bullet at around 1100 fps in a decent rifle and I'll promise you a venison dinner.

I said there were plenty of 40-caliber blackpowder cartridges around in the 19th century. The 40-50 Sharps is but one. There was also a 40-70 Sharps Bottleneck round. It held around 70 grains of fuel. The 40-50 pushed a 277-grain bullet at a little under 1200 fps. The 40-70 fired the same projectile at about 1400 fps. We could spend an awful lot of time looking at 40-caliber blackpowder cartridges, so let's taper off. But before leaving those interesting 40s altogether, consider one more Sharps round, the 40-90. Also a bottleneck, the 40-90 could shove a 300-plus-grain bullet at about 1600 fps and a 400-grain bullet at about 1400 fps. Just a mention of the 38-40 Winchester cartridge and we'll push on. It worked well in sidearm or rifle, and was actually a 40-caliber round despite its name. It fired 180-grain bullets in the 1200 fps domain.

Caliber 44

The 44-caliber round ball is no heavyweight. The particular mould I had in this caliber threw a ball that went .435-inch in diameter for a weight of 124 grains. Even at a muzzle velocity of 2000 fps from my long-barreled flinter, muzzle energy was only 1102 foot pounds. That's not a lot of power. At 100 yards, velocity was down to about 1150 fps for a remaining energy of only 364 foot pounds, sufficient to stop the charge of an angry mouse, but not enough to lend confidence for big game. Because of this poor showing, energy-wise, I'd call the 44-caliber ball-shooting rifle OK for deer hunters who get close and put the ball perfectly on the money, and that's about it.

Blackpowder lead conicals for 44-caliber muzzleloaders are not prevalent. However, the caliber does OK as a blackpowder cartridge in the cap 'n' ball revolver, the blackpowder cartridge revolver, and as a 44-caliber pistol bullet with sabot. We'll save pistol bullets and sabots for down the line. Let's talk about the 44-caliber blackpowder cartridge first. The 44-40 Winchester was never a powerhouse, but it was chambered in sidearms as well as in rifles, and good shots did all right with the short-cased round on deer-sized game. It's still factory-loaded and chambered in newly manufactured guns as well, which says something for it. Colt, Navy Arms and many others, for example, offer revolvers

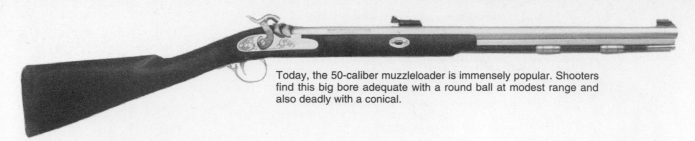

Today, the 50-caliber muzzleloader is immensely popular. Shooters find this big bore adequate with a round ball at modest range and also deadly with a conical.

in caliber 44-40. Other 44-caliber blackpowder cartridges were more powerful. The 44-60 Winchester was available in Winchester's single-shot rifle. It, along with the 44-60 Sharps Bottleneck round, pushed a bullet close to 400 grains weight at about 1250 fps. There were many other 44s, including the long 44-100 Remington Creedmoor and 44-100 Ballard, but that's enough on 44s for our purposes.

Caliber 44 blackpowder pistols exist, but they're of little ballistic interest. But the 44-caliber cap 'n' ball revolver remains highly popular. In the cap 'n' ball, caliber 44 is a big bore. While I do not consider any of these 44s, including the Colt Walker, suitable for large game, these sidearms were powerful weapons in gunfights, and they saw plenty of military service. A peek at cartridge history reveals a heap more 44s, especially blackpowder cartridges for breech-loading rifles, but interest turns now to the 45s.

Caliber 45

While the literature may not agree, I know that 45-caliber was highly popular in the Kentucky/Pennsylvania long rifle. I say so having studied these rifles in large private collections as well as museums. I've always wondered about it, too. A 45-caliber round ball is not exactly a Halloween pumpkin. In .445-inch diameter, it weighs only 133 grains and delivers under 350 foot pounds of energy at 100 yards, about like a 9mm Luger pistol at the muzzle. The 45 makes a decent rendezvous caliber, and will harvest deer-size game at modest range with that oft-repeated ideal bullet placement. But that's about it.

On the other hand, 45-caliber conicals can get downright hefty, producing a heap of power. In the Navy Arms Parker-Hale Volunteer rifled musket, a bullet in the 490- to 500-grain weight range leaves the muzzle at over 1500 fps for a muzzle energy darned close to 2500 foot pounds. This long 45-caliber lead bullet carries the better part of a long ton in energy at 100 yards. As a 45-caliber pistol bullet with sabot, this caliber is also good, as touched on below. And, of course, there were many fine 45s among the blackpowder breechloaders of the 19th century, including the still-factory-loaded 45-70 Government cartridge, king of the modern blackpowder silhouette game.

So many super 45s could be named that the rest of this chapter would be no more than a listing of them. Consider the whole line of Sharps rounds in that size, for example, such as the 45-90-2.4 Sharps, the 45-100-2.6 Sharps, the 45-110-2$7/8$ Sharps, and of course the his-

torically interesting 45-120-3$1/4$ Sharps, which propels a 500-grain missile downrange at over 1600 fps. So, as big bores go, the 45-caliber round ball is nothing to write home about, but the long-bullet 45s—be they conicals for muzzleloaders or bullets for blackpowder breechloaders—are to be reckoned with. They continue to drop big game with one well-placed shot, and always will.

Caliber 50

In my opinion, the round ball has just reached true big game potential in caliber 50, with a .490-inch ball weighing 177 grains. At 2000 fps, or close to it, the 50-caliber lead sphere will harvest deer-sized animals in their tracks up close and at the 100-yard mark, with 125 yards an outside limit. While many an elk has dropped to 50-caliber round balls in recent times as well as the past, I think getting close to these big animals is essential for good clean harvesting. Up close, placing the round ball directly into the boiler room is more likely, and the 50-caliber ball retains sufficient velocity and energy at the 75-yard mark and closer to offer a good long wound channel. My son Bill Fadala dropped a bull elk with his Ithaca 50-caliber one afternoon in Colorado. The bull was close. The ball passed through the breadth of the chest region and the big animal fell on the spot.

Fifty-caliber blackpowder conicals are also excellent. My handy 50-caliber Storey Buggy Rifle goes with me often these days because it carries so readily, but in the meanwhile it also has prime big game potency. Not long ago, as this is written, I managed to harvest a buck mule deer with the rifle as well as a buck antelope. Each took no more than one step after the projectile landed. Firing the big 600-grain 50-caliber White Superslug, the Buggy Rifle achieved close to 1400 fps for more than 2500 foot pounds of muzzle energy. Although short-barreled, the little 50 packs the mail out to 100 yards with remaining energies that go over a long ton. In short, experience has taught me that 50-caliber round balls can be deadly, while 50-caliber blackpowder conicals carry even more impressive force.

Caliber 50 can turn a well-made muzzle-loading pistol into a powerhouse, too. The aforementioned Thompson/Center Scout is capable of driving bullets in the 350-grain class at muzzle velocities close to 1300 fps. Muzzle energy is well over a half-ton. Even with heavy handloads, the powerful 44 Magnum has a hard time besting this blackpowder sidearm energy-wise. Hornady shows a 300-grain bul-

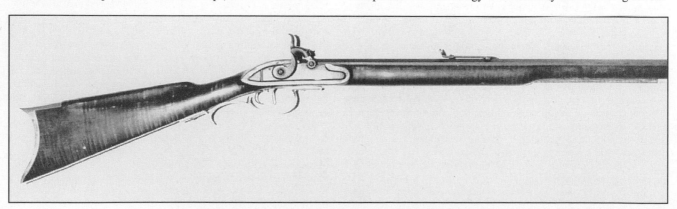

This Mountain State Mountaineer shoots a 50-caliber round ball. The round ball is dependent on bore size for its effectiveness on game, and as the ball grows in caliber, its mass increases out of proportion.

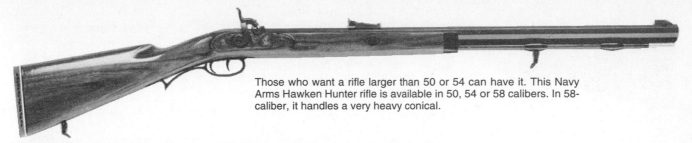

Those who want a rifle larger than 50 or 54 can have it. This Navy Arms Hawken Hunter rifle is available in 50, 54 or 58 calibers. In 58-caliber, it handles a very heavy conical.

let traveling at 1250 fps in the powerful 44 Remington Magnum for a muzzle energy of 1041 foot pounds. Of course, special Garrett ammo for the 44 Magnum achieves even more energy, but you get the point—a 50-caliber blackpowder pistol firing conicals can show very well for itself.

As far as 50-caliber breechloaders go, there were a couple dandies. Records show, for example, that the 50-70-1³/₄ Sharps was well liked by the buffalo runners of the late 19th century. The 50-70 can still be loaded today, and when it is, expect a bullet of over 500 grains weight to head downrange at about 1050 fps. The 50-90-2¹/₂ Sharps is more potent yet. It pushes a 605-grain bullet out of the muzzle at close to 1250 fps. Then there is the bigger-yet 50-140 3¹/₄ Sharps. Historians say this big boy was a bit late for buffalo hunting, but it can still wow modern shooters with a 605-grain bullet at about 1500 fps.

The 50s are darn good blackpowder guns in various styles and types. A half-inch diameter bullet makes it so, even in the round ball configuration. Give me a careful, practiced, and experienced hunter who is also a good shot, and I'll lay odds that he'll "make meat" with his 50-caliber long rifle with no trouble at all. Now let's look at calibers a little bit bigger than 50.

Calibers 52 and 53

Calibers 52 and 53 have shown up, even in modern times, and both as round ball rifles, in my experience. There's not a thing in the world wrong with either one, as proved by the Allen Santa Fe Hawken in 53-caliber firing a round ball of .520-inch diameter weighing 211 grains. But these in-between calibers simply are not popular today. The 54, however, is very popular.

Caliber 54

By the time caliber 54 is reached, the round ball is a pretty good-sized missile. Not only more than a half-inch in diameter, it also takes advantage of more bullet weight. A 54 ball, .530-inch diameter, runs 224 grains weight, while one in .535-inch diameter goes 230 grains weight. In good muzzleloaders the 54-caliber lead sphere can be driven at around 2000 fps for a muzzle energy not too far away from a ton. Of course, round balls are round balls, and this good muzzle energy looks more like 675 foot pounds or so at 100 yards. That's why I suggest an outside limit of 125 yards on deer-

The author counted on a 50-caliber rifle with a heavy bullet to down this bear. Blackpowder big bores can take any game in the world, if the bore is big enough.

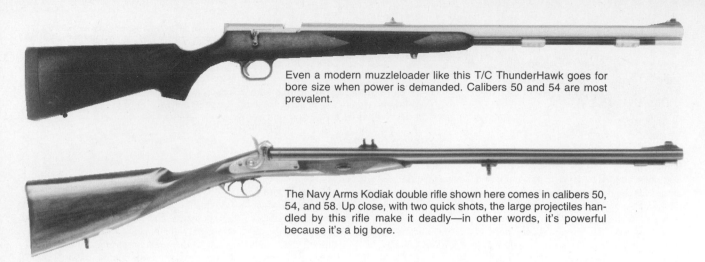

Even a modern muzzleloader like this T/C ThunderHawk goes for bore size when power is demanded. Calibers 50 and 54 are most prevalent.

The Navy Arms Kodiak double rifle shown here comes in calibers 50, 54, and 58. Up close, with two quick shots, the large projectiles handled by this rifle make it deadly—in other words, it's powerful because it's a big bore.

size game even with the 54-caliber round ball, and 75 yards and under for elk-size animals.

Conicals in 54-caliber are capable of big punch. For example, in testing Knight MK-85 modern muzzleloaders in 54-caliber, driving Hornady's 425-grain Great Plains Bullet well over 1500 fps was no big trick, with muzzle energies over 2200 foot pounds. Another powerhouse load was with the excellent Buffalo Bullet, 54-caliber, weighing 460 grains. Speeds of over 1400 fps at the muzzle with this deep-penetrating bullet achieve over a ton of muzzle energy. In a bullet box (the testing device mentioned in Chapter 42), the Buffalo Bullet 54 conical proved itself by running through a lot of tough media.

Caliber 56

Bullets of 56-caliber have been offered over the years, but they're of too little concern for our discussion. So let's move on to a caliber that has seen a lot of use over the years, the big 58.

Caliber 58

I'm no fan of the 58 for round ball, although it's certainly not a bad size. It's just that once the round ball goes past 54-caliber, it takes a good deal of powder to achieve reasonable muzzle velocity and good trajectory. There are many examples to prove this. My own Navy Arms Hawken Hunter, a very well-built rifle with a 1:60 rate of twist, shoots a 58-caliber round ball accurately, but even with a 140-grain FFg powder charge, velocity is only 1550 fps at the muzzle. Muzzle energy is under 1500 foot pounds. As sounded early in this chapter, caliber means something in itself, so the 58 ball does harvest big game impressively, but all in all, 54-caliber round balls scooting away faster from the muzzle are as effective. The Lyman company ran tests with 58-caliber round balls. Charges running 190 grains of FFg rendered velocities just over 1800 fps at the muzzle for a muzzle energy under a ton, and a 100-yard energy of roughly 750 foot pounds.

The 58-caliber conical is quite another matter. This size was popular very early in history, with many "army guns" going caliber 58, including muskets carried in our own Civil War. Rifled muskets in this caliber continue to abound. The Navy Arms Parker-Hale Musketoon in 58-caliber is allowed 100 grains by volume of Pyrodex RS, which provides a muzzle velocity over 1200 fps and a muzzle energy just short of 1700 foot pounds, this with a 505-grain Buffalo Bullet. Navy Arms' longer-barreled 1853 3-Band Enfield musket did better with the same 100-grain RS charge, turning up closer to 1300 fps. Thompson/Center's Big Boar rifle in 58-caliber fires a 560-grain Maxi-Hunter bullet at 1350 fps with 120 grains of FFg for a muzzle energy of 2267 foot pounds. One more time—caliber is important in its own right, and a 58-caliber missile is pret-

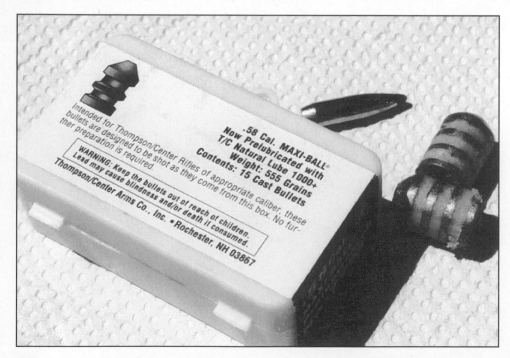

Caliber 58 is the largest size that is popular at this time, although far from the largest blackpowder bore size available. The big 555-grain 58-caliber T/C Maxi-Balls on the right are capable of taking any big game in North America within the reasonable range of the firearm.

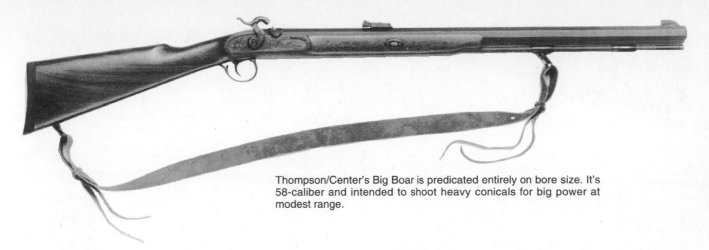

Thompson/Center's Big Boar is predicated entirely on bore size. It's 58-caliber and intended to shoot heavy conicals for big power at modest range.

ty darned "wide."

The 58-caliber pistol must not be forgotten. On paper, this big-bore single shot one-hander simply isn't exciting. It fires its big bullets too slowly to achieve high muzzle energies. But let's take a look anyway. How about a 58-caliber 500-grain Minie at 800 fps? Energy-wise, we have only 750 foot pounds, but 500 grains is good weight, and 58 is a big diameter. Or a 570-grain Minie at just under 700 fps for about 650 foot pounds of muzzle energy, even less dramatic on paper, but still an awful lot of bullet.

Caliber 60

I have tested one round ball rifle in 60-caliber. It fired a round ball that weighed over 300 grains, but the best muzzle velocity I got with the powder charges I was willing to burn didn't much top 1600 fps. Yes, a 60-caliber lead ball of 300 grains is formidable, even at 1600 fps, and for those who desire a 60, Godspeed. But get ready to burn some powder.

Caliber 62

Likewise for the 62-caliber round ball. It's a big boy, all right, and there's no doubt of its potential, but 62s demand lots of powder to achieve reasonable muzzle velocity.

Calibers 69, 75, and 12 Gauge

I'm lumping these together. In my own 12-gauge shotguns I've had good luck with patched .690-inch diameter round balls running 494 grains weight. I'm not sure what could be achieved from a 12-bore gun firing big round balls in front of heavy powder charges. I followed accepted procedure in my tests, burning rather light charges of Fg blackpowder, with 80 grains earning around a thousand fps and about half a ton of muzzle energy. Not to play a broken record, but of course such a large round bullet is deadly at close range by virtue of mass and caliber alone. As for 75-caliber, which falls roughly, too, in that 12-bore realm, the British Brown Bess Musket was a 75, and with its comparatively modest powder charge behind a ponderous round ball, it was effective for its purpose.

The 10-Bore Rifle

A red tinge on the horizon of true big-bore rifles indicates a modest new dawning of interest in these truly large-caliber blackpowder firearms. Turning to old-time data, we find one 10-bore rifle shooting a 698-grain bullet in front of 5 drams of blackpowder. Since a dram is 27.34 grains weight, that's a powder charge of 136.7 grains weight. Because this weapon is noted as a "jungle gun," I'm assuming close-range work. This particular 10-bore produced only 1316 fps muzzle velocity. The big bullet made up for unimpressive velocity to some extent, and this not-heavy load did develop 2685 foot pounds of energy at the muzzle. But that's not much for a 10-gauge rifle, which was capable of far more potency.

The 8-Bore Rifle

I did locate more impressive data for an 8-bore rifle, being a 10-dram powder charge (273.4 grains of blackpowder) behind a heavy 1257-grain bullet for a muzzle velocity noted as a flat 1500 fps. The chart showed a muzzle energy of 6273 foot pounds. My calculations say 6282 foot pounds. Close enough. Obviously, the 8-bore is some kind of powerful.

The 4-Bore Rifle

The 4-bore reveals itself with that single number, four. Four round balls to the pound. That means a round ball would weigh something like 1750 grains (or a little less, considering sub-sizing for a patch). An original 4-bore load I ran across showed a lighter round ball, sized down for a patch, or perhaps this 4-bore was not quite 4-bore. Anyway, the spherical bullet was listed at only 1250 grains weight with a muzzle velocity of 1460 fps and a muzzle energy listed as 5912 foot pounds. My figures show 5918 foot pounds. Another 4-bore load from the old days is more impressive, a conical weighing 1882 grains driven by 12 drams of blackpowder (328 grains weight) for 1330 fps and a listed muzzle energy of 7387 foot pounds. I get 7394 foot pounds.

The 2-Bore Rifle

Enough was said of this caliber with early remarks about William Moore's 2-bore playtoy firing half-pound sphericals with tank-stopping muzzle energies. Bore size really comes to the fore in these truly big guns, proving that while Newton's formula for energy definitely

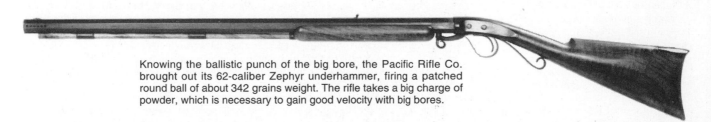

Knowing the ballistic punch of the big bore, the Pacific Rifle Co. brought out its 62-caliber Zephyr underhammer, firing a patched round ball of about 342 grains weight. The rifle takes a big charge of powder, which is necessary to gain good velocity with big bores.

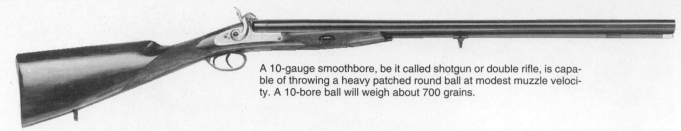

A 10-gauge smoothbore, be it called shotgun or double rifle, is capable of throwing a heavy patched round ball at modest muzzle velocity. A 10-bore ball will weigh about 700 grains.

Yes, big-bore blackpowder guns, like this short-barreled shotgun that can shoot 12-gauge patched round balls, will have a bit of recoil. Part of the reason is the large powder charge required to gain reasonable velocity in a large-bore muzzleloader.

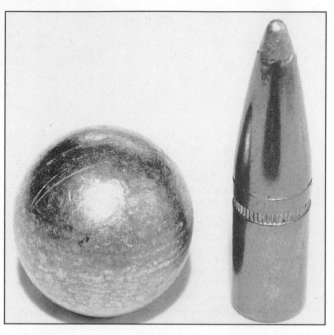

Twelve-bore is about 75-caliber, an awful lot larger in diameter than the 30-caliber modern bullet compared to it here. Experience shows that a 12-bore round ball like this from close range, given a reasonable velocity with a stout powder charge, is capable of taking any big game in North America.

sides with velocity, when bullets get really heavy, energies rise in spite of modest speeds.

The Modern Muzzleloader

I've singled out the modern muzzleloader for its own niche because it handles pistol bullets with sabots as well as lead blackpowder conicals. You cannot argue with success. The largest elk recorded in the blackpowder records was dropped with one pistol bullet fired from a modern muzzleloader using a sabot. Furthermore, certain rifles shoot pistol bullet/sabot combinations admirably. And for pure power, well, there is plenty of snort in those relatively fast-moving pistol bullets. After all, these 44- and 45-caliber projectiles are supposed to be big game worthy at regular handgun velocities. How could they be anything but deadly when fired even faster from muzzleloader rifles?

And yet, the same modern muzzleloaders firing big conicals do pack a heavier punch, right down the line. For example, a 45-caliber 250-grain bullet with sabot earned just shy of 1700 fps in front of 120 grains of Pyrodex RS in a modern muzzleloader with 24-inch barrel. That's worth darned close to 1600 foot pounds of muzzle energy. However, the same rifle whipped out a 425-grain conical right at

1550 fps for 2265 foot pounds of muzzle energy. The story repeats often. My Storey 50-caliber Buggy Rifle with a 44-caliber 240-grain pistol bullet achieved a muzzle velocity of 1750 fps for a muzzle energy of more than 1600 foot pounds, but the same rifle firing a 530-grain Gonic conical achieved a muzzle energy of 2152 foot pounds for 1352 fps and 2575 foot pounds with a 600-grain Super-slug at 1390 fps.

Many different muzzleloaders will handle pistol bullet/sabots these days due to the switchover to faster rifling twist that is prevalent. There is nothing wrong with the power these guns deliver with pistol bullets, and it is often more punch than necessary for the task at hand. If they work, use them. That's my advice. But don't think of them as more powerful than blackpowder conicals, for they are not.

Recoil and Big Bores

We know that blackpowder guns require heavy doses of powder to achieve good velocity, and that more powder means more recoil. So it's no surprise that big bores, which can burn very heavy charges in hunting loads, do kick when loaded heavily. Also part of the recoil formula is bullet weight, which means that heavy conicals, combined with big powder charges, up recoil another notch. How do we deal with this? In the hunting field, we ignore recoil. At the bench, we prepare for it with proper cushioning of the shoulder, as suggested in Chapter 30.

This has been a look at the blackpowder big bore in rifle, pistol, and revolver, cartridge as well as muzzleloader. It's a mini-education in what blackpowder guns can do, power-wise. They can really deliver the cargo, as long as the bullet is big enough.

●

chapter 33

Rifled Muskets— Rugged Hunting/ Target Guns

IN A PREVIOUS roundup of gun groups, there were originals, replicas, non-replicas, customs, modern muzzleloaders, plus another equally viable long arm, the rifled musket. You may recall that the contradictory name of this firearm was mentioned. We think of muskets as military smoothbores. The rifled musket does present the configuration of the old-time musket; however, it's rifled, hence: "rifled musket"—musket by design, with a rifled bore. Muskets were originally rugged military long arms, but some of them metamorphosed into target pieces. Naturally, those had rifling. Today, we must classify most currently-manufactured rifled muskets as replicas. Some aren't (like the Volunteer introduced below), but most models are close enough to originals to deserve the title "replica." These long arms are sturdy, field-worthy, and mostly historically correct. They also make one heck of a hunting/target rifle, although the modern blackpowder shooter hasn't entirely figured this out yet, as proved by the modest numbers of rifled musket fans.

The rifled musket is not a bulky, unwieldy, hard-to-carry, difficult-to-load, inaccurate chunk of metal and wood. Ok, a few aren't that handy to pack around. Heavy? Some are. A fine rifled musket offered today is the Volunteer from Navy Arms. To be fair, the Volunteer does not truly replicate a rifle of the past, but it is a cousin to the Whitworth by design. It's a rifle of quality manufacture, accuracy, and strength. Navy Arms allows a full 130 volume FFg or 130 volume RS for a muzzle velocity of about 1500 fps with a 490- to 500-grain 45-caliber projectile. That's more power than the 45-70 Government cartridge, which drives similar bullets at 1200 fps mv with blackpowder loads.

The Volunteer rifle I have weighs 10½ pounds. That's not light, but I loathe flyweight rifles anyway, so I get along fine with the heft of the Volunteer. In the field, it's well balanced and most of all, stable, which is worth more to me than packing a wristwatch-light rifle that you cannot truly control. If you think this is false, test it for yourself. Shoot a heavy rifle from the sitting position at a gallon-size plastic milk container at 100 yards. Now do likewise with a flyweight. I'm betting on more hits with the heavyweight. That doesn't mean we need to tote a benchrest rifle into the hunting arena. It simply suggests distrusting the flyweight until you truly try one out, not off the bench where you can hold it down, but in the field using offhand, sitting, kneeling, and other standard shooting positions.

Granted, the rifled musket's military appearance can be a turn-off, but it is quite a rifle all the same. I don't say it's for everyone, but the rifled musket may be just what some big game blackpowder hunters need for rough, rugged conditions. Furthermore, these long arms appeal to a sense of history on one hand, and a desire for the practical on the other. As for history, the rifled musket goes way back in time. My dictionary defines this shoulder arm as an "archaic smoothbore," a muzzleloader of course, historically assigned to the period from about 1622 to 1786 when the Musketeers, or French Royal Body Guard, carried them. Our own Springfield Rifled Musket, as it was called, served the American military once upon a time; musket it was by design, but smoothbore it was not.

As for practical, the rifled musket, especially those like the Volunteer and Whitworth with adjustable sights and fast-twist conical-shoot-

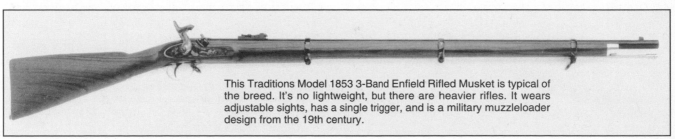

This Traditions Model 1853 3-Band Enfield Rifled Musket is typical of the breed. It's no lightweight, but there are heavier rifles. It wears adjustable sights, has a single trigger, and is a military muzzleloader design from the 19th century.

233

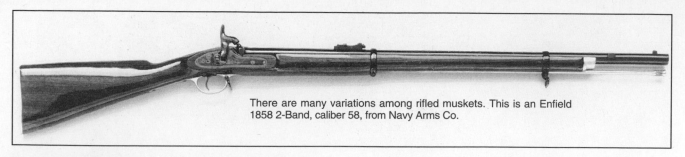
There are many variations among rifled muskets. This is an Enfield 1858 2-Band, caliber 58, from Navy Arms Co.

Val Forgett, president of the Navy Arms Co., with an antelope taken with a Navy Arms Volunteer rifled musket.

ing bores, offer power with accuracy, which is not a bad combination in the big game field. Today, a number of importers bring in muskets of various designs, most of them in calibers 45 and 58. The 58s shoot projectiles classified earlier as standard lead conicals, very much like those associated with cast bullets for cartridges. They generally shoot Minie-type missiles, although they will certainly handle other projectiles as well. Bullets for the 45-caliber rifled muskets are long for their caliber, while 58-caliber bullets are less streamlined, with a lower ballistic coefficient. Both types carry plenty of weight.

The Volunteer 45-caliber rifled musket mentioned above handles 600-grain—and heavier—bullets. An example of a 58-caliber rifled musket, the Dixie U.S. Model 1861 Springfield (with 40-inch barrel, but only 8 pounds weight) normally shoots a Minie-type bullet that runs between 500 and 600 grains. It's important to note that not all muskets are long of barrel. For example, the Navy Arms J.P. Murray 1862-1864 Cavalry Carbine wears a 23-inch barrel. It's also 58-caliber. It weighs in at 7 pounds, 9 pounces. It has a blade front sight, and the rear sight is adjustable for windage only, accomplished by drifting the sight in its dovetail notch. A rifled musket like the J.P. Murray is terrific in close quarters, such as wild hog hunting or for whitetails in the thicket.

The Zouave

There is a rugged 58-caliber hunting rifle known as the Zouave. It's imported by several companies, including CVA, whose gun I tested. The CVA Zouave is bored 58-caliber, which is typical. It has a fixed front sight and adjustable rear sight. The barrel is 32½ inches long, and overall weight is 9½ pounds. Sorry to be a broken record, but this Zouave was just what I promised earlier: rugged and ready for action. In tests, it fired a 530-grain Lyman Minie (mould No. 577611) at 1100 fps MV with 80 volume FFg blackpowder. Before taking a brief look at the Zouave in history, another Zouave deserves mention. It's the Navy Arms Zouave Carbine with 26-inch barrel. For those who like the Zouave, but want a shorter barrel for brush hunting, this is it. Allowed 100 volume Pyrodex RS, the shorter gun gave 1064 fps MV with a 460-grain Minie.

The Zouave in History

The Zouave were fighting men noted for their special uniforms as well as their military ability. They wore what began as North African-style apparel, which became altered considerably over time. French fighters made up companies of Zouaves. Captain George B. McClellan, a U.S. Army observer in the Crimea, held the Zouave in great esteem. He influenced others to similar thinking, and a Zouave-like soldier emerged in Eastern America. Eventually, there were several Zouave-type fighting groups in middle 19th century America, and both the North and South had Zouave units during the Civil War. To this day, the colorful outfits remain before the public, because teams of Civil War reenactors still wear Zouave-like uniforms.

Historian Joe Bilby noted one Zouave uniform as "red fezzes with blue tassels, blue jackets trimmed in red, red shirts and sashes, blue and white pillow ticking baggy pants, blue and white horizontally striped hose and white gaiters." He also said, in his article "Zou-Zou-Zou-T I G R R R" for the Dixie Gun Works *Blackpowder Annual 1995*, that "guns carried by Zouaves did not differ from those of other volunteer regiments." Therefore, the Zouave name, generally attached to the Remington rifled musket of 1863, goes somewhat unfounded and mysterious. A 69-caliber smoothbore was more likely found among the Zouaves. "There is no evidence," Bilby points out, "that any of the Zouaves were issued the colorful Remington Model 1863 or 'Zouave' rifle." So when the modern blackpowder shooter speaks of the Zouave, he means the rifle style described above, and not the soldier.

Caliber 45 or 58?

Choosing between the 58- or 45-caliber musket isn't difficult. The 58 is at home in timber and brush where the shooting is close for wild boar, deer, even elk and moose with the right load. Sheer caliber promotes its game-taking ability. A 58-caliber missile makes an impres-

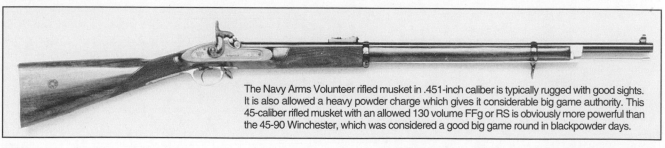
The Navy Arms Volunteer rifled musket in .451-inch caliber is typically rugged with good sights. It is also allowed a heavy powder charge which gives it considerable big game authority. This 45-caliber rifled musket with an allowed 130 volume FFg or RS is obviously more powerful than the 45-90 Winchester, which was considered a good big game round in blackpowder days.

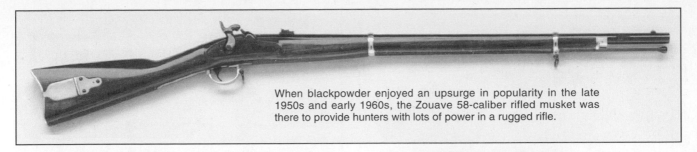

When blackpowder enjoyed an upsurge in popularity in the late 1950s and early 1960s, the Zouave 58-caliber rifled musket was there to provide hunters with lots of power in a rugged rifle.

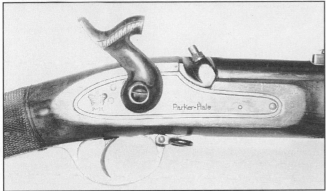

This close-up of the Navy Arms Parker-Hale Whitworth shows the fine big lock and also the English top hat-type nipple size, traits of rifled muskets.

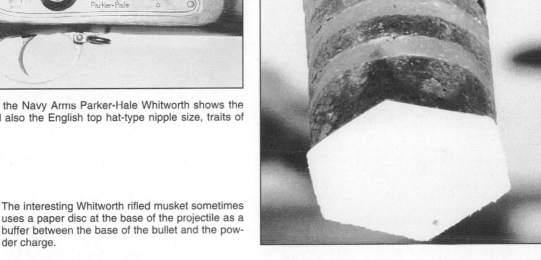

The interesting Whitworth rifled musket sometimes uses a paper disc at the base of the projectile as a buffer between the base of the bullet and the powder charge.

sive wound channel. The 45 caliber rifled musket marches to a different drum beat. These rifled muskets are generally a little more refined than their larger-bore counterparts and can be more accurate for longer-range shooting. The oft-mentioned Parker-Hale Volunteer Rifle and Parker-Hale Whitworth Military Target Rifle must be noted again as prime examples of 45-caliber rifled muskets, the former a "magnum" among the breed with conical bullets in the 500-grain class at around 1500 fps MV. We already admitted that this gun outpowers the 45-70 Government when the latter is loaded with blackpowder. It beats the larger 45-90 Winchester, too, which is a good blackpowder cartridge. That speaks highly of the Volunteer's power.

The Whitworth Rifled Musket

The Volunteer is interesting ballistically, but the Whitworth is more intriguing historically. It's called a target rifle because it was devised as a long-range instrument. Whitworths were used in our own Civil War by both the Blue and the Gray for picking each other off at far distances. The Whitworth also achieved fame as a 1000-yard match rifle in its time and the replica version can still slap bullets through the bullseye at great range. Sir Joseph Whitworth was a master toolmaker. He was born in 1803, died in 1887, and in between created his special shoot-far rifle. Whitworth's claim to fame was a bore of hexagonal configuration, supposedly the voodoo that made bullets group close together. Bullets were cast to match the esoteric bore style. I say Sir Joseph's oddball rifling system had precious little to do with its penchant for shooting straight. The Whitworth rifle clustered its bullets so well because the firearm was carefully made and its missiles were properly stabilized. Rate of twist for the Whitworth was 1:20 inches, by the way, which is maintained in the replica we have today from the Navy Arms Co.

The long bullet flew point-on all the way to a thousand yards because it had enough initial spin to keep it rotating on its axis. While forward velocity is lost fairly rapidly, rotational velocity holds up quite well over long range. This is an important point for this discussion, because once the Whitworth bullet was stabilized, there was sufficient spin to keep it rotating on its axis at long range. Naturally, if the bullet is barely stabilized at the muzzle, it may yaw or keyhole before it reaches a long-range target, because even a little loss in rps (revolutions per second) destabilizes it. There was something else that made the Whitworth shoot true: good bullets. This factor is not often mentioned in the Whitworth story, but if bullets are not well-cast, accuracy is wishful thinking only. Rotation or spin overcomes some bullet eccentricity, but when a missile is lopsided, or its base is at the oblique, chances of good groups fall into the "slim to none" category.

Ned Roberts, one of the principals behind the 257 Roberts cartridge, was a fan of the Whitworth. He tried various bullets, including a 530-grain design that provided excellent results at the target range. My replica Whitworth test rifle was caliber 451, had a 34-inch barrel, and weighed 9$^1/_2$ pounds. It produced good accuracy with a number of well-cast bullets ranging from 480 to 550 grains weight. I also tested the Volunteer, which gave excellent results, too. Both rifles used musket caps for ignition. If you recall from Chapter 28, these are the English top hat caps, which are much larger than the standard percussion cap widely used today. Mine were RWS brand. These caps produced good fire and positive ignition for all replica rifled muskets I tried. Away from the target range the Whitworth and Volunteer rifles proved their mettle at long range in the field. Good balance made for easy carrying. The long sight radius and the refined iron sights encouraged solid bullet placement on deer-sized targets out to about 200 yards *in my target tests*. For game shooting I prefer getting closer than 200 yards, even with an accurate and powerful conical-shooting rifled musket.

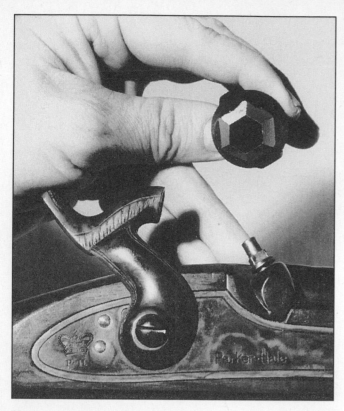

The great Whitworth rifled musket had a punch for cutting paper discs that are then placed in between the base of the projectile and the powder charge as buffers.

I wanted to try the Whitworth and Volunteer muskets to satisfy my own curiosity, as well as for reporting. I didn't shoot past 200 yards because in spite of these rifles' ability to group bullets well beyond that distance, blackpowder is blackpowder, so trajectory is looping, and it's wrong to shoot at game so far away that hold-over becomes guess-work. I kept game-shooting ranges to around 150 yards, about 25 yards greater than I allow myself to shoot with other muzzleloaders. Naturally, these ranges are subject to the hunter's ability to judge distances, and are not exact. As I carried these two muskets in the field, I was impressed with how good they could be as big game hunting tools.

Big Game Effectiveness

I hunted only antelope with my test muskets, so I have no personal knowledge to pass on concerning larger game. But I did learn something important—a comparatively slow-moving conical placed in the lung region does not always signal a solid hit. By that I mean that the shooter cannot always tell that he scored. In some cases, the bullet sailed right through those prairie goats without telling effect. I always

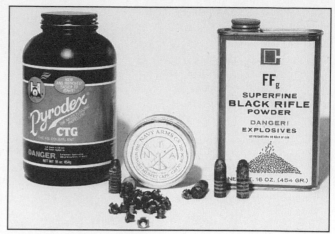

Two good powders for the rifled musket are Pyrodex CTG and GOEX FFg. Pyrodex RS (not shown) is also a fine choice in that it gives higher velocity per volume than CTG. Note large top hat caps in the center and long 45-caliber lead bullets.

follow up on every shot to make certain that it was a miss if it looks like a miss. Not that the antelope went anywhere. But typically they did dash off forty or fifty yards before falling. Incidentally, these were doe antelope allowed on Additional Tag licenses, and therefore lighter in weight than pronghorn bucks, which means under 100 pounds on the hoof. I don't like wasting meat, but I had to "go for bone," on one 'lope in the interest of possibly saving a lot of game down the road.

A hit in the shoulder as opposed to behind the shoulder was, to apply the badly overused term of today, "awesome." Shoulder-hit pronghorns went down immediately and stayed down. A general rule of thumb is to go for bone with lead conicals and try for the boiler room with round balls. I feel confident that this is a good rule, in spite of a little lost meat with the shoulder strike. The round ball flattens out fairly well on lung tissue, and I've had great luck with behind-the-shoulder hits, while heavy conicals did not always produce instant results. On the other hand, the long conical bullet penetrated remarkably well, and is no doubt terrific on really large game, such as elk, again with strikes in the scapula (shoulder blade) region.

Are rifled muskets legal for blackpowder-only hunts? This is a good question. It's easy to see how someone could be confused concerning the legality of a rifled musket on a muzzleloader-only big game hunt, because these rifles do not look like other muzzleloaders. However, the rifled musket is a frontloader in every regard. It shoots only blackpowder, Pyrodex or blackpowder substitute, and it must be allowed on primitive weapons hunts. In short, the rifled musket is a muzzleloader, single shot, loaded one bullet at a time with one charge of powder.

While a good multiple-lever (double set) trigger is more crisp than either the Whitworth or Volunteer triggers, both of these single-stage (non-set) triggers were better than I expected, breaking at around 3 1/2

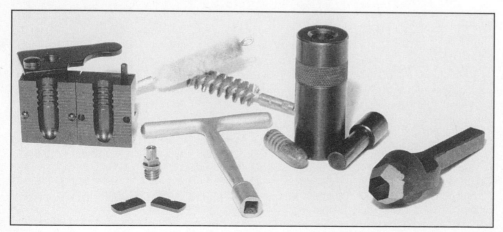

Shooting the rifled musket is a game unto itself. These tools belong with the Whitworth (from left): a bullet mould, extra nipple, extra sights, nipple wrench, bore mop, bristle brush, bullet swager with punch (bullet resting to the left of swager), and a punch to make paper discs for buffers.

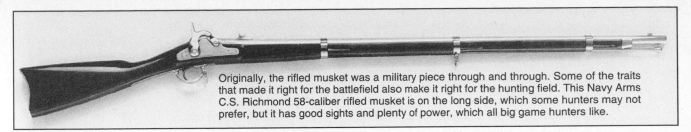

Originally, the rifled musket was a military piece through and through. Some of the traits that made it right for the battlefield also make it right for the hunting field. This Navy Arms C.S. Richmond 58-caliber rifled musket is on the long side, which some hunters may not prefer, but it has good sights and plenty of power, which all big game hunters like.

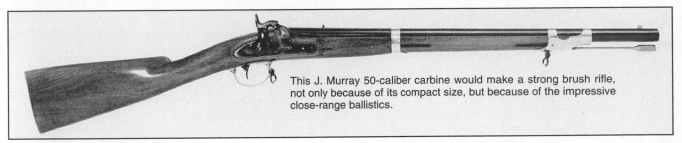

This J. Murray 50-caliber carbine would make a strong brush rifle, not only because of its compact size, but because of the impressive close-range ballistics.

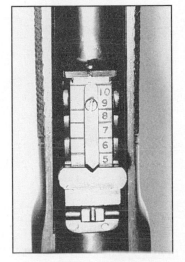

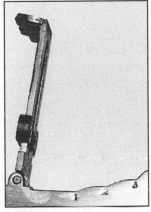

(Above) A trait of many rifled muskets is good sights. This rear sight allows a clean sight picture, and it is adjustable.

(Above) Because there were so many variations in rifled muskets, the modern blackpowder shooter has a wide variety of choices in remakes of this fine old rifle style. These are Enfields from Navy Arms in various configurations.

In the raised position, the sights on this rifled musket are set for very long range. Rifled muskets have provided accuracy out to 1000 yards.

pounds with minimal creep. However, in all fairness to the reader, not all rifled musket triggers were all that good. If I owned some of the muskets I tested, a blackpowder gunsmith would be doing a job on the trigger. It is impossible to make some of these triggers truly crisp and light in letoff, but they can be improved safely by an expert.

Loads for Rifled Muskets

Powder charges were created volumetrically with an adjustable powder measure for all rifled muskets tested, just as they are for any frontloader. In my tests, only blackpowder and Pyrodex were used. Accuracy differences between 70-, 80-, and 90-grain powder charges were impossible to detect with either the Whitworth or Volunteer rifles, as well as other rifled muskets tested. Because of a high degree of accuracy and good ballistics, I originally determined that I would shoot thin-skinned game at up to 200 yards with the Whitworth or Volunteer, but as stated above, I changed my mind, even though deer-sized paper targets were easy to hit at 200 yards with either rifle. I backed off to 150 yards as my maximum because of bullet drop, and the fact is that in field-shooting my groups were not good enough beyond 150 yards. Even though off-the-bench groups averaged four to six inches center to center at 200 yards for both three-shot and five-shot strings, away from the rest I could not keep all bullets inside a six-inch bull at 200 yards, meaning I'd need a rest in the field to ensure long-range hits. So I decided on a personal 150-yard limit, even with the accurate Whitworth and Volunteer.

Accuracy and Bullet Selection/Volunteer and Whitworth

Accuracy with standard conical bullets was good in both rifles. No hexagonal projectiles were fired in the Whitworth. Good luck with these guns prompted shooting with several rifled muskets. For the sake of comparison and curiosity, I bench-tested the Brown Bess smoothbore in contrast to the rifled musket, which taught me a good deal about shooting a smoothbore. In this particular test, I was able to hit a six-inch bullseye at 50 yards with the big 75-caliber Brown Bess and its round ball, but that was it. I never got good enough with the gun to fill me with big game hunting confidence. Up close, however, there is no doubt that it could do the job on big game with proper ball placement.

Light vs. Heavy Powder Charges

Light powder charges were OK for fooling around with rifled muskets, and they did prove accurate, but definitely not powerful. For example, the 1861 Springfield rifle was not given a maximum powder charge recommendation by the manufacturer, and the only sanctioned load I found was 60 grains of FFg, which reduced the rifle's ballistic effectiveness to popgun status, even with a 58-caliber bullet. If the Springfield can truly handle no greater powder charge, I'd eliminate it from the field as a big game rifle. The Zouave Carbine cited above, with its 26-inch barrel, digested an allowed 100 grains of FFg for far more valid hunting authority.

Rifled Musket Big Game Hunting Loads

Rifle:	Parker-Hale Whitworth Rifled Musket
Caliber:	45 (.451-inch)
Barrel:	36 inches
Twist:	1:20
Bullet:	490-grain Lyman cast, #457121
Load:	90 grains volume FFg (manufacturer's maximum recommendation)
Muzzle Velocity:	1306 fps
Muzzle Energy:	1856 foot pounds
100-yard Velocity:	1112 fps
100-yard Energy:	1346 foot pounds

Specifications

Rifle:	Parker-Hale Volunteer Rifled Musket
Caliber:	45 (.451-inch)
Barrel:	32 inches
Twist:	1:20
Bullet:	490-grain Lyman cast, #457121
Load:	130 grains volume FFg (manufacturer's maximum recommendation)
Muzzle Velocity:	1474 fps
Muzzle Energy:	2365 foot pounds
100-yard Velocity:	1254 fps
100-yard Energy:	1711 foot pounds
Load:	130 grains volume Pyrodex RS (110 grains by weight, manufacturer's maximum recommendation)
Muzzle Velocity:	1514 fps
Muzzle Energy:	2495 foot pounds
100-yard Velocity:	1290 fps
100-yard Energy:	1811 foot pounds

Specifications

Rifle:	Navy Arms Zouave Rifled Musket
Caliber:	58
Barrel:	32 1/2 inches
Twist:	1:48
Bullet:	530-grain .577-inch Minie, Lyman #577611
Load:	80 grains volume FFg (manufacturer's maximum recommendation)
Muzzle Velocity:	1095 fps
Muzzle Energy:	1411 foot pounds
100-yard Velocity:	953 fps
100-yard Energy:	1069 foot pounds

Specifications

Rifle:	Navy Arms Model 1841 Mississippi Rifled Musket
Caliber:	58
Barrel:	33 inches
Twist:	1:48
Bullet:	530-grain .577-inch Minie, #577611
Load:	80 grains volume FFg (manufacturer's maximum recommendation)
Muzzle Velocity:	1102 fps
Muzzle Energy:	1430 foot pounds
100-yard Velocity:	959 fps
100-yard Energy:	1083 foot pounds

Specifications

Rifle:	Navy Arms JP Murray Artillery Carbine Rifled Musket
Caliber:	58
Barrel:	23 1/2 inches
Twist:	1:48
Bullet:	530-grain .577-inch Minie, Lyman #577611
Load:	70 grains volume FFg (manufacturer's maximum recommendation)
Muzzle Velocity:	980 fps
Muzzle Energy:	1131 foot pounds
100-yard Velocity:	882 fps
100-yard Energy:	916 foot pounds

Specifications

Rifle:	Navy Arms 1853 3-Band Enfield Rifled Musket
Caliber:	58
Barrel:	39 inches
Twist:	1:48
Bullet:	505-grain Buffalo Bullet Co.
Load:	100 grains volume Pyrodex RS (71.0 grains weight, manufacturer's maximum recommendation)
Muzzle Velocity:	1274 fps
Muzzle Energy:	1820 foot pounds
100-yard Velocity:	1101 fps
100-yard Energy:	1360 foot pounds

Rifled Musket Big Game Hunting Loads

Specifications

Rifle:	Navy Arms Parker-Hale Musketoon Rifled Musket
Caliber:	58
Barrel:	24 inches
Twist:	1:48
Bullet:	505-grain Buffalo Bullet Co.
Load:	100 grains volume Pyrodex RS (71.0 grains weight, manufacturer's maximum recommendation)
Muzzle Velocity:	1226 fps
Muzzle Energy:	1686 foot pounds
100-yard Velocity:	1043 fps
100-yard Energy:	1220 foot pounds

Specifications

Rifle:	Dixie 1861 Springfield Musket
Caliber:	58
Barrel:	40 inches
Twist:	1:48
Bullet:	505-grain .577-inch cast bullet, Lyman #575213
	(Note: although mould read 505 grains, actual bullet weight was 517 grains)
Load:	60 grains volume FFg (from literature; Dixie offered no maximum recommendation)
Muzzle Velocity:	753 fps
Muzzle Energy:	651 foot pounds
100-yard Velocity:	676 fps
100-yard Energy:	525 foot pounds

Specifications

Rifle:	Dixie Zouave Carbine
Caliber:	58
Barrel:	26 inches
Twist:	1:56
Bullet:	460-grain Minie, Lyman #575213-OS (Old Style)
Load:	100 grains volume Pyrodex RS (72.5 grains weight, manufacturer's maximum recommendation)
Muzzle Velocity:	1064 fps
Muzzle Energy:	1157 foot pounds
100-yard Velocity:	883 fps
100-yard Energy:	797 foot pounds

Specifications (for comparison purposes only)

Rifle:	Dixie Brown Bess Musket
Caliber:	75 (actually 74-caliber bore)
Barrel:	41³/₄ inches
Twist:	Smoothbore
Bullet:	494-grain .690-inch round ball
Load:	00 grains volume FFg (from literature; Dixie offered no manufacturer's maximum recommendation)
Muzzle Velocity:	809 fps
Muzzle Energy:	718 foot pounds
100-yard Velocity:	680 fps
100-yard Energy:	507 foot pounds

The Murray Carbine was granted 70 grains of FFg for reasonable close-range punch on deer-sized game with a 58-caliber missile, but I was not impressed with its power as compared with the Zouave. The Parker-Hale Musketoon, with its handy 24-inch barrel, was more like it for hunting: it was loaded with 100 grains volume Pyrodex RS (71.0 grains weight) and a 505-grain 58-caliber Buffalo Bullet at 1226 fps MV for a muzzle energy rated at 1686 foot pounds. While the energy figure isn't too exciting, that thumb-sized missile at over 1200 fps is.

A factory-recommended charge of 80 grains of FFg gave the Mississippi Rifles enough power for game up to moose and elk size *at close range with proper bullet placement*. The test bullet was a 530-grain Lyman cast from mould #577611. I found the long-barreled Zouave on par with its shorter brother because the former was given a 100-grain FFg maximum load while the latter was allowed 80 grains of the same fuel. The 32¹/₂-inch Zouave with 80 FFg was a ballistic twin to the Zouave Carbine with 100 FFg volume—for all practical field value.

The 1853 3-Band Enfield was another rifled musket that provided big game hunting power. I ran into a little problem in initial testing, not with the rifle, but with its test bullet. Accuracy fell off at 100 volume Pyrodex RS. The otherwise excellent 505-grain Buffalo Bullet suffered skirt failure at this limit by flaring too much. That was corrected by the Buffalo Bullet company, which is not unusual, since the principals behind this operation are always seeking the very best results from their products. The current Buffalo Bullets with hollow bases all have proper skirt thickness, according to my test. This makes the Enfield rifled musket a prime hunting rifle with the fine Buffalo Bullet and full-throttle 120 volume charge of powder.

Final Choices

The rifled musket is one more viable choice for the blackpowder shooter, not only for big game hunting, but for target work. While the blackpowder silhouette game is officially restricted to breechloaders, the Whitworth would certainly hold its own in a match. And the shorter rifled muskets are workable in any setting where shots are close. When I think about hunting wild boar in thick cover, these big-bullet, compact rifles seem right, and of course they'd perform equally or better on less truculent game at close range.

The Whitworth rifle is an overall dandy piece, although it is not granted the same powder charge allowed in the Volunteer, and for that reason only I'd probably choose the latter for hunting big game where a maximum range of 150 yards was prudent. The rifled musket should be considered by a hunter who wants power from an it-can-take-it rifle. These military-minded pieces were built to withstand field duty. They're blue collar all the way, honest and hard-working. But in brush, bramble, briar, black timber, jackpine thicket and catclaw jungle, these blackpowder hunting rifles deserve more than honorable mention. Some hunters will find them absolute winners. ●

chapter 34

The Single Shot Breechloader

A REBIRTH OF the blackpowder cartridge is credited to the interesting and challenging blackpowder silhouette match so popular the past few years. There is no doubt that this fascinating shooting game, which has become a national phenomenon in recent times, has influenced shooters to look at the blackpowder cartridge. But there is more to it than that. The blackpowder cartridge gun itself has created its own impact. Three types of blackpowder firearms have emerged victorious, not just rifles. The blackpowder cartridge revolver is definitely popu-

lar. Navy Arms Co. alone offers several, not only in the famous Colt style, but Remingtons, Schofields, and others as well. Then there's the blackpowder lever-action repeater, also available again widely in replica form. But in this chapter, it's the single shot breechloader that we're interested in, the only firearm allowed in official blackpowder silhouette matches.

There were literally dozens of different blackpowder single shot cartridge rifles around in the 19th century. This is not the

The original Colt six-shooter holstered here never went out of vogue. Now they're hotter than ever due to the new (as this is written) sport of cowboy action shooting. Today, there are many replica revolvers available in the old tradition, as sold by the Navy Arms and others.

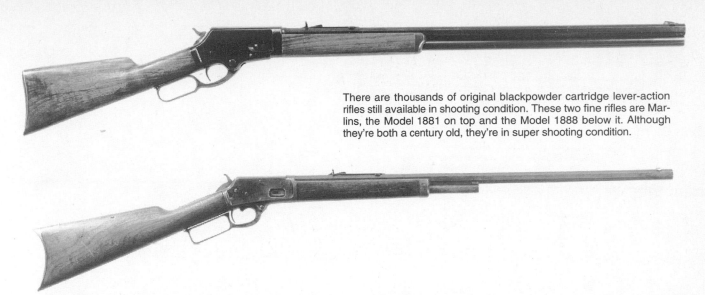

There are thousands of original blackpowder cartridge lever-action rifles still available in shooting condition. These two fine rifles are Marlins, the Model 1881 on top and the Model 1888 below it. Although they're both a century old, they're in super shooting condition.

forum for discussing each one, because that would take a book of its own. The two that stand out most are the Sharps and the Remington rolling block. Not to say that these are the only 19th century single shot breechloaders offered today. They are not. We'll touch briefly on the fine Winchester High Wall, too, just to prove that point. But the Sharps and Remington get the most play because, according to record, they were most popular during the latter part of the 19th century when buffalo runners, as they preferred to be called, roamed the vast plains in search of shaggies. A little bit of history about the sad bison slaughter is in order also.

The Sharps and the Bison

The Gun that Shaped American Destiny. That's what Martin Rywell called the Sharps rifle in his 1979 Pioneer Press book. Rywell's title is on the money. The Sharps rifle did help to shape American destiny. It filled the hands of countless pioneers trekking westward. According to Rywell, an Indian gentleman by the name of American Horse said, "Emigrants passing up the South Platte River to Colorado between 1858 and 1865 were largely armed with Sharps military rifles." The Sharps found its way all over the West, and of course played a role in the American Civil War, where almost 100,000 rifles and carbines served Union troops, according to Rywell. Colonel Hiram Berdan's Sharpshooters finally got their Sharps rifles in 1862, using them during the Seven Days Battles of June and July of that year.

The Sharps rifle is also credited with wiping out up to sixty million bison over an area that covers half of the United States, a place so vast that even today with vehicle travel it would take months to see only a particle of it. Of course, the few hundred to as many as 10,000 hunters (nobody knows the figure) could not slay even twelve million bison (low figure) in only a few years, let alone sixty million (high figure), which some say roamed the plains. Picture the professional buffalo hunter traveling in mule-drawn wagons, and on foot, killing that many animals. But many factions did want the bison out of the way. The Army did, because the Plains Indian warrior depended on the shaggy for meat, and even shelter. The railroad did, because running a train through a buffalo herd was no way to cross America. Settlers did, because bison were all over the land they wanted to build on.

So the shoot was on, and it was a lousy thing to do. But think about it. If you turned a pack of full-time modern hunters loose today with four-wheel drive trucks, let alone mules and wagons, they wouldn't stand a chance of killing up to sixty million breeding bison over countless acres of U.S. landscape, not to mention British Columbia, Alberta, and Saskatchewan—the bison's range in the 1800s. The United States Congress of 1870 debated giving up on the Far West because many of the members felt a war with the Plains Indian could not be won. But if the bison were wiped out, then what? No bison, no Plains Indian. So shooting buffs was encouraged.

But buffalo runners, "sportsmen," settlers, soldiers and all the rest

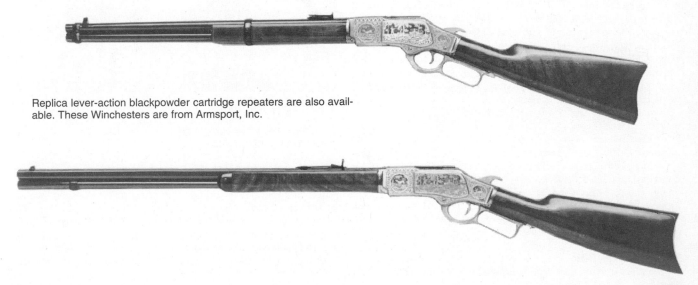

Replica lever-action blackpowder cartridge repeaters are also available. These Winchesters are from Armsport, Inc.

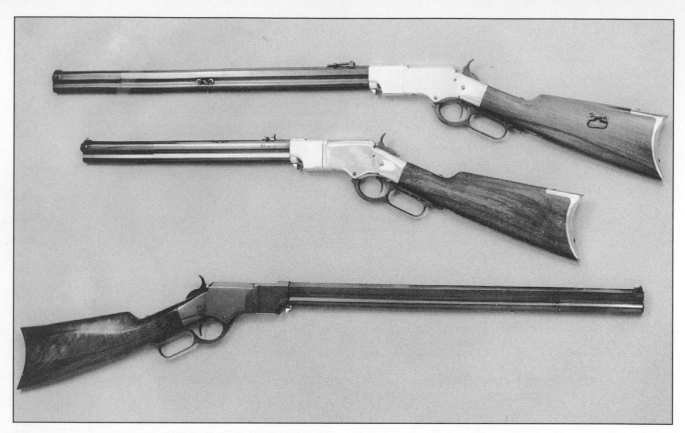

Navy Arms offers several blackpowder cartridge lever-action repeaters, including the Military Henry rifle (top), the Henry Trapper (middle) and the Henry Iron Frame model (bottom).

didn't come close to killing up to sixty million bison in the few years the animal was hunted without mercy on his home ground. Something a lot smaller than a bullet finally did the trick, however. It happened when cattle were introduced into the West. Anyone who underestimates the power of a tiny microbe or virus should consider that the black plague wiped out one-third of the entire population of Europe in the 1500s, and Spanish influenza dropped more people shortly after World War I than all the bombs and bullets spent in that conflict. But all of this is beside the point. The Sharps was still one of the two rifles that buff runners carried on their mission impossible—to rid the West of the shaggy.

The Greatness of the Sharps Rifle

The Sharps cartridge rifle was challenging, of course. It allowed one shot at a time with blackpowder, which required regular cleanup after shooting, but the rifle was still fired much faster than a muzzleloader, and the Sharps wasn't really that hard to take care of. It cleaned from the breech end, after all. Col. Frank Mayer, a buff runner, wrote of flushing the bore with cold water, followed by urine, supposedly to introduce a touch of acidity, with a hot water rinse afterwards. Also to its credit, the Sharps rifle was offered in a huge array of worthy cartridges, far too many to list here. And it was truly a long-range rifle, capable of winning thousand-yard shooting matches, as well as dropping "bufflers" at several hundred yards, especially when topped off with German-made target scopes.

It came in many styles, including a special Creedmoor version for long-range target work (and hunting). A clue to some of the Sharps' great blackpowder cartridges exists in a catalog that shows 40-100 through 50-100 "brass, centre fire re-loading shells" running from $1^{11}/_{16}$ inches through $2^{1}/_{2}$ inches in length, costing from $22 to $30 per thousand. Low blackpowder pressure was easy on brass, too, and the Sharps company promised that "with proper care in re-loading" a cartridge case could be reloaded up to 500 times, provided that each case was cleaned after use to prevent deterioration. The Sharps was great, all right, so good that it can still get the job done today.

Replica Sharps Rifles

There are a number of replica Sharps rifles on the market today.

The blackpowder cartridge shooters helped to boost the popularity of the single shot blackpowder cartridge rifle. Firing on targets at great distances is the name of the game. This shooter was aiming at a buffalo silhouette at 900 yards.

Montana Armory deserves a big mention, or at least its headman does. John Schofftstall, president of Montana Armory, is a blackpowder single shot rifle expert, and far more than a mere fan of the firearm. He's dedicated to making superb single shot breechloaders, mostly Sharps models, but also a New Model 1885 High Wall. John has done a great deal of worthwhile testing of blackpowder cartridges, too, and his data pamphlet is a fine source of information on the subject. Other Sharps imports include an Italian-made model from Armsport and Dixie Gun Works.

The Remington Rolling Block

Many buffalo hunters preferred the Sharps, but others leaned toward the Remington rolling block breechloader. Joseph Rider is credited with the rolling block design built by Remington. In fact, some models were called Remington-Riders. The rolling block principle itself was hardly new, of course. Flobert's rolling block design came very early in gun history with its pivoting breechblock. Rather than a breech sliding up and down (the falling block), the rolling block's breech did just that—it rotated. The action was quite strong for its time, as proved by the fact that it later held smokeless powder rounds such as the 7mm Mauser. W.W. Greener wrote about the Remington-Rider in his revised 1910 edition of *The Gun and its Development*. He said:

> The rifle was tried at Wimbledon as long ago as 1866, and attracted considerable attention at that time, in consequence of the extraordinary rapidity with which it was loaded and fired; as many as fifty-one shots were discharged within three minutes. . . . The mechanism consists essentially of two pieces, being the breech-piece and extractor, and the other the hammer breech-bolt. This breech-piece and hammer-bolt each work upon a strong centre pin.

In function, the breechblock rolls back; a cartridge is inserted directly into the breech, and then the breechblock is rolled forward again to lock the action. The significant factor here is the hammer, which falls down into a slot within the breechblock itself. The Remington rolling block remains available to this day, just like the Sharps. Navy Army has long championed the continuance of the design with several well-constructed replicas. Also highly worthy of note is a special Remington rolling block rifle from the Schuetzen Gun Co. This rifle is custom built by R.P. (Richard) McKinney, who was once top gunsmith for U.S. Army Marksmanship Unit Custom Shop, Montana Armory, etc.

The model I tested had wood that goes beyond handsome to another realm, and workmanship was impeccable. In caliber 45-70 Government, the McKinney Remington Rolling Block, with 28-inch, straight octagon, 1-inch Douglas barrel, filled me with confidence. Considering iron sights, plus trajectory of the 45-70 Government round, I'd still get within 150 yards of big game, but I'd have no trouble "making meat" with this fine rifle. I managed to bang the dickens out of a two-foot square gong at 300 meters with the rifle. McKinney's custom Rolling Block is the J.P. Gemmer model, incidentally. McKinney's Rolling Block is proof of the design's continued worth, for no one would build so fine a piece on a poor action.

The New Model 1885 High Wall

This is another fine single shot blackpowder cartridge breechloading rifle in replica form from Montana Armory, with all the old-time nostalgia left intact. It deserves mention because it's a handy-toting rifle worthy of the big game field as well as the target range. It's actually chambered for a smokeless round, the 30-40 Krag, as well as these blackpowder cartridges: the 32-40 Winchester, 38-55 Winchester, 40-65 Winchester, and 45-70 Government. The receiver is the early octagon-top thick High Wall variation with the later coil spring action. There are other refinements, such as a smaller-diameter firing pin, but standard High Wall sights prevail, and the receiver, lever, hammer, and breechblock are all case-colored. A 28-inch, octagonal, Douglas 4140 premium barrel is machined to the High Wall No. 3 taper. A Marble's standard ivory bead sight rests up front, with a No. 66 long blade flat-top rear with reversible notch and elevator. The stock is premium American black walnut with satin finish.

Modern Ballards

Also worthy of mention are the Ballard rifles of current manufacture. These nice rifles are fully machined and carbon case-colored like

The American bison, popularly known as the buffalo, made the single shot blackpowder cartridge rifle famous in the 1800s. These huge animals, the largest four-footed beasts on the continent, roamed the entire West. Some experts claim there were sixty million of them at one time.

The Sharps action with the falling block in position for shooting (above) and the view directly into the open chamber (right).

the originals. Furthermore, they're legal for official blackpowder cartridge silhouette matches, due to the time period of the Ballard and its single shot blackpowder cartridge design. The company, in fact, builds rifles for specific matches, not only blackpowder silhouette, but others such as the Montana Long Range match. Modern-made Ballards are available from Rifle Works and Armory in Cody, Wyoming. The company also deals in obsolete brass cartridge cases manufactured to original specifications for Sharps, Ballards, Remingtons, Maynards, and other blackpowder cartridge firearms.

A Few Blackpowder Cartridges

There are plenty of single shot blackpowder cartridge rifles offered today, but it is also important to take a look at a few of the cartridges still available today for the single shot breechloader fan. When I say a few, I mean it. There were hundreds of different cartridges available in the 19th century. The following are noted only because they can be loaded today without too much fuss, and they deserve continued use. Cartridge history is not included in this brief roundup because there are many books devoted to this subject, including loading manuals. Consider *Cartridges of the World* by Frank Barnes, available from DBI Books, as one good source.

The 32-40 Winchester

Actually, this was more a Bullard, Ballard, Remington round in the 19th century, as I understand it, but the Winchester 32-40 configuration survived and is still fired. Some marksmen consider this a match/target cartridge. The round was well liked by the old-time Schuetzen shooter, especially with 190- to 200-grain bullets. Although not a powerhouse, the 32-40 is definitely capable of cleanly harvesting deer-sized game, especially in the hands of an expert rifleman/hunter. A 165-grain bullet achieves 1360 fps MV with 35 grains of FFFg blackpowder, and about 100 fps less with a 190-grain bullet and 35 grains of FFg. Fine accuracy is the byword for the 32-40, along with mild recoil. Its shortcoming is lack of bullet weight compared with the other cartridges that follow.

38-55 Winchester

This round fires bullets in the 300-grain class at about 1250 to 1300 fps using 45 grains of FFg or FFFg blackpowder. The 38-55 Winchester stayed "on line" for many years. It was considered by some a better

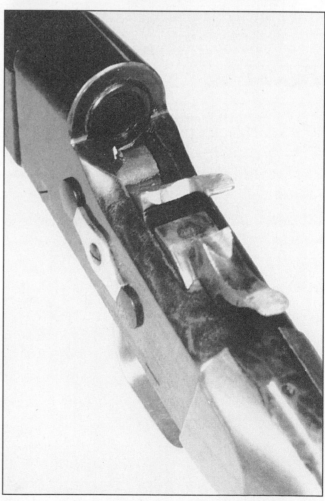

This is the Remington rolling block rifle with the action in the open position, ready to take a round, which is simply fed directly into the breech.

This original Remington rolling block rifle is fitted with a long-range Vernier tang sight. The rifle was photographed at a blackpowder cartridge shoot where it was used against a buffalo silhouette at 900 yards.

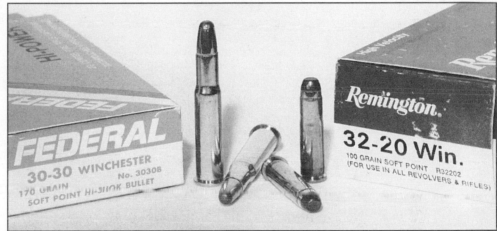

Remarkably, the little 32-20 Winchester cartridge, here compared with a 30-30 round, remains factory loaded and is chambered in modern-made guns. It is legal for cowboy action shooting games today.

round than the 30-30 for deer and even larger game. Today, the 38-55, loaded with blackpowder, is applied as a fine target round, as well as hunting number. I'd not hesitate to go into the brush with it for deer, especially when loaded with a good lead projectile, such as the 305-grain bullet from a SAECO mould. Hunting yes, but the 38-55's claim to fame today remains its penchant for accuracy. It is also mild when compared with larger-bore blackpowder cartridges burning more powder behind heavier projectiles.

40-65 Winchester

This is a dandy little cartridge that I became acquainted with when testing a prototype of a new Sharps rifle that has yet to hit the market as this is written. My test bullet was a 390-grain lead conical. It took off at 1300 fps MV in front of 60 volume GOEX Cartridge-grade blackpow-

der. The 40-65 Winchester is pleasant to shoot, and could be a possible candidate for silhouette games. The near-400-grain weight bullet should knock the ram over, but I do not profess personal knowledge of that occurring. I do know that within reasonable range, I'd be confident on big game with the rather mild-mannered 40-65. Cases can be formed from 45-70 brass, by the way, or purchased from The Complete Handloader, Inc. The 40-65 is seeing a little more play lately, and is being chambered in several single shot rifles.

40-70 Sharps BN (Bottleneck)

Simply another nice blackpowder cartridge. My tests revealed no clear advantage over the 40-65 Winchester in terms of muzzle velocity, bullet for bullet, but obviously a few grains more powder will show. My test rifle put a 330-grain bullet downrange at 1350 fps MV with a full 70 volume FFg.

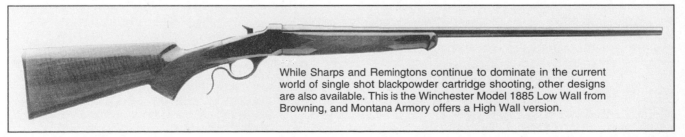

While Sharps and Remingtons continue to dominate in the current world of single shot blackpowder cartridge shooting, other designs are also available. This is the Winchester Model 1885 Low Wall from Browning, and Montana Armory offers a High Wall version.

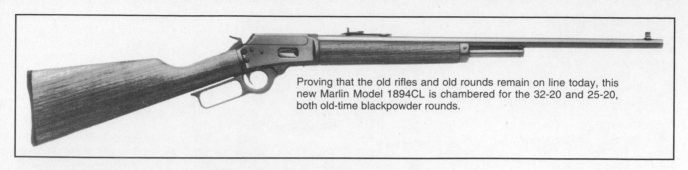

Proving that the old rifles and old rounds remain on line today, this new Marlin Model 1894CL is chambered for the 32-20 and 25-20, both old-time blackpowder rounds.

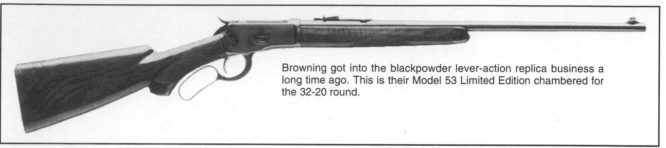

Browning got into the blackpowder lever-action replica business a long time ago. This is their Model 53 Limited Edition chambered for the 32-20 round.

The 40-90s

The 40-90s are not as popular as they might be, but I like them all the same. So did Frank Mayer, who claimed that his 40-90 was more than sufficient for bison hunting. They make a lot of sense to me, although I'd be hard pressed to prove their value over other blackpowder cartridges. They intrigue me, although they should not, because they're just more of the same—more powder behind the same bullets fired by the 40-65 and 40-70 rounds just listed. The 19th century was packed with blackpowder rounds, and there were many different 40-90s. *Cartridges of the World* shows several, but the list is not complete. Here are just a few to consider: 40-90 Bullard, 40-90 Sharps Straight, 40-90 Sharps Bottleneck, 40-90 Ideal, and the 40-90 Ballard, not to be confused with the Bullard.

Bullet weights vary quite a bit in the 40-caliber blackpowder cartridge category, with lightweights running around 260 grains to heavier missiles in the 500-grain class. Incidentally, the 40-90 Bullard and 40-90 Ballard are about opposites in design. The Ballard is long and lean, while the Bullard round is bottlenecked and squat. A velocity of 1427 fps with a 370-grain bullet is noted for the Ballard, while the Bullard shows a 300-grain bullet at 1569 fps MV. I suspect that with a 400-grain bullet and a case full of Fg blackpowder, the 40-90-class cartridge would have to be suitable for silhouette, and certainly big game without undue recoil. My testing of a 40-90 Sharps Bottleneck provided over 1300 fps MV with 85 grains Fg blackpowder and a 425-grain bullet.

44-Caliber Blackpowder Rifle Cartridges

Currently, there's not much play on the 44-caliber blackpowder rifle cartridge, but there were many available in the 19th century, and they were well liked. The 44-60 Sharps fired a 396-grain bullet at 1250 fps MV according to Barnes, while the 44-70 Maynard pushed a 430-grain projectile at 1310 fps MV. There was a 44-75 Ballard Everlasting, too, a 44-77 Sharps & Remington, a 44-90 Remington, as well as a 44-90 Sharps Bottleneck, the latter driving a 520-grain bullet at 1270 fps MV (Barnes). Other 44s include the 44-95 Peabody "What Cheer" and the 44-100 by Remington and Ballard. The former is historically interesting because it was chambered in the Remington-Hepburn or No. 3 Long-Range Creedmoor rifle (Remington rolling block); the 44-100 Remington was designed as a match round from the start.

The 45-70 Government

There were many 45-caliber single shot blackpowder cartridge rifles around in the 19th century. These included, among others, the 45-75 Winchester, 45-75 Sharps Straight, 45-100 Ballard, 45-100

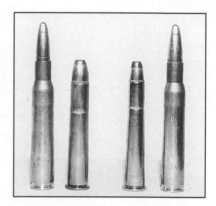

Two of the most popular blackpowder cartridges of their era were the 38-55 and 32-40 Winchester rounds. Flanked by two 30-06 cartridges for comparison, the 38-55 (second left) remained popular in a smokeless version well into the 20th century, but the 32-40 (second right) fell to the widely used 30-30 round.

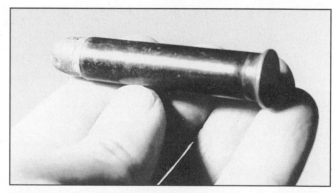

There were scores of blackpowder cartridges in the 19th century, in numerous variations. This 45-caliber blackpowder round was a large rimfire cartridge.

Remington, and others. But of the 45s, only the 45-70 Government remains on the factory list to this very hour. The round is so important to blackpowder silhouette shooting, as well as hunting, that I've given it its own story. See Chapter 45 for more on this amazing cartridge.

The 45-90 Winchester

This is, for all practical purposes, a longer 45-70. It was introduced in 1886. Claims were made that the extra 200 fps did great things for the 45-90 over the 45-70. More powerful, yes, but in my own tests it did not blow the 45-70 away. Nonetheless, it's a good cartridge, able to push 500-grain bullets at about the same MV associated with 400-grain bullets in the 45-70.

The 45-90 2⁴/₁₀ Sharps

This 45-caliber blackpowder cartridge is commonly called a 45-90 today. Montana Armory shows date of introduction as June 8, 1877, by the Sharps Rifle Company. It was touted as a fine long-range target round, credited with winning 1000-yard matches as late as 1900. A target load is listed as 85 grains of Fg blackpowder with a 500-grain bullet at 1226 fps MV.

The 45-110 Sharps

This cartridge, also known as the 45-110-2⁷/₈ Sharps, is credited as being the largest 45-caliber cartridge originally designed for the Sharps rifle. It was chambered in the Sharps 1874 model for the buffalo hunter who wanted plenty of long-range power. Montana Armory shows the cartridge developing 1360 fps MV with a 500-grain bullet and plenty of accuracy. A lighter 430-grain bullet is listed at 1430 fps MV, with a lighter-yet 325-grain missile at 1596 fps MV, all burning 110 grains of Fg blackpowder.

The 45-120 Sharps

This dandy cartridge is well received today, although it is not the

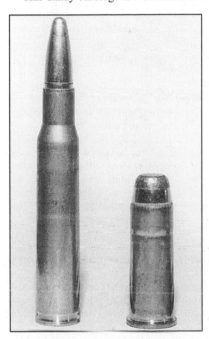

The squat 44-40 Winchester (right) is shown with a 30-06 for comparison and is currently chambered in many guns, both revolver and rifle. It is still factory-loaded today.

round of choice among serious blackpowder cartridge silhouette shooters. The 45-70 Government holds that honor. Nonetheless, the 45-120 remains a fine cartridge. There is a problem, however, not with the round itself, but with its history. Students of the Sharps, who have studied the company's rounds as well as its rifles, insist that Mayer and others who touted the 45-120 as an important cartridge among bison slayers were incorrect. The cartridge, these individuals say, and feel they can prove, was not widely in use until after the era of the bison hunt in America. Barnes notes that the 45-120 was introduced in 1878-79 for the Sharps-Borchardt rifle, "though there is no documentary evidence that the Sharps factory offered rifles in this caliber." John Schoffstall states that, "Although never offered in the original Sharps rifles, the 45-120-3¹/₄ Sharps is the big 45-caliber of the New Sharps line." He confirms his own statement with the clear remark, "The 45-120-3¹/₄ was never chambered in an original Sharps rifle."

In spite of lineage problems and historical pitfalls, the 45-120 is a heck of a round. It's also known as the 45-120-550; it's a powerful number. The latter nomenclature, from blackpowder days, means 45-caliber with 120 grains of blackpowder and a 550-grain bullet. My own test of the 45-120 in a modern-made Sharps rifle was entirely satisfactory. I tested the cartridge under the name 45-120-3¹/₄, which was another blackpowder designation, this one meaning 45-caliber, 120 grains of blackpowder, with a case length of 3¹/₄ inches. A 480-grain Lyman bullet was tested. This was from Lyman mould number 457121 listed as 490 grains, but it produced bullets of 480 grains weight with No. 2 alloy.

A charge of 110 grains GOEX Fg blackpowder propelled the 480-grain bullet at 1482 fps MV for a muzzle energy of 2341 foot pounds. Remaining velocity at 100 yards was 1230 fps for an energy of 1613 foot pounds. This sort of ballistic authority is obviously strong enough for any big game on the continent, considering the heft of the bullet. Furthermore, the fine Sharps replica rifle was accurate. On calm days, putting five shots into a 6-inch bullseye at 200 yards from the bench posed no problem. Mayer claimed that, "At distances above 500 and up to 1,000 yards, the 45-120-550 Sharps with patched bullets is absolutely unsurpassed by any weapon known to man." Overstatement, yes, but not an entirely outlandish claim. Its only problem was that if the Sharps was never chambered for the 45-120, which rifle was Mayer speaking of concerning the use of the round on buffalo?

The 50-70-1¹/₃ Sharps

This round was also known as the 50-70 Government, and it apparently saw much use for hunting in the 1800s. Tested in a New Model 1875 Sharps Sporting Rifle, the cartridge delivered 1071 fps MV with a 500-grain bullet and 70 grains GOEX Fg blackpowder. While this is

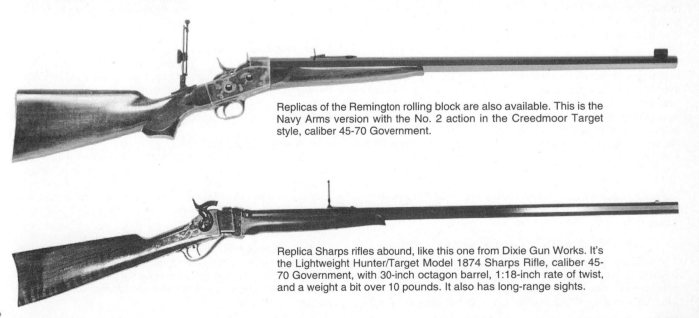

Replicas of the Remington rolling block are also available. This is the Navy Arms version with the No. 2 action in the Creedmoor Target style, caliber 45-70 Government.

Replica Sharps rifles abound, like this one from Dixie Gun Works. It's the Lightweight Hunter/Target Model 1874 Sharps Rifle, caliber 45-70 Government, with 30-inch octagon barrel, 1:18-inch rate of twist, and a weight a bit over 10 pounds. It also has long-range sights.

While handloading for the blackpowder cartridge differs from making smokeless powder rounds, it also has many similarities, including the basic handloading equipment, such as this outfit from RCBS.

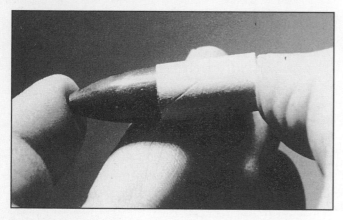

The paper-patched bullet is the choice of some blackpowder cartridge shooters. The entire blackpowder cartridge game is specific, right down to bullet styles.

not a lot of bullet speed, it does amount to almost 1300 foot pounds of energy. Another test was with a 528-grain lead bullet at 1051 fps MV for 1320 foot pounds of muzzle energy.

The 50-140 Sharps

There were many 50s in use in the 19th century. We'll pass over the rest, but a word on the 50-140 is in order. This was a truly big round, and it was not an original Sharps chambering. In fact, the cartridge appeared circa 1884, three years after the Sharps company no longer manufactured rifles. This was also the last year of widespread buffalo hunting. The cartridge was known as the 50-140-3$^1/_4$, as well as the 50-140-700, the latter for a 700-grain bullet. Tested with various 50-caliber bullets, the Big Fifty drove a 638-grain NEI-cast lead projectile at 1413 fps MV with 140 volume Pyrodex CTG grade powder for a muzzle energy of 2829 foot pounds. Today, the 50-140 Sharps is chambered in a currently-manufactured Sharps rifle, the Model 1874 Long Range Express from Montana Armory.

Handloading the Blackpowder Cartridge

This in-depth topic is best handled in books written for the sole purpose of loading this ammunition. For our purposes, however, a few notes are in order, some of them very important. See Chapter 21 for some loading information, but for more in-depth data, check for books

Loading for the black-powder cartridge is very specific, including the choice of primer. Today, magnum primers like these CCI No. 250s in Large Rifle size are popular.

in print on the subject, as well as the library listing in the Reference Section of this text. Here are a few points to consider:

1. A Case Full of Powder

It is common practice to load large blackpowder cartridges such as the 45-70 Government with very small charges of smokeless powder. Personally, I do not care for the practice because of the safety factor. While it is just as difficult to prove as the short-started bullet warning sounded often in this book, the fact is, some rifles have been destroyed with light smokeless powder charges in large old-time cartridges. Recently, a shooter experienced such a catastrophic failure with a Sharps replica in 45-70, the rifle blowing up, and the shooter suffering a badly injured left hand. I shoot the blackpowder cartridge with a case full of blackpowder or Pyrodex, and that is my recommendation.

2. How to Get More Blackpowder in the Case

The use of a drop tube is well-accepted as a means of filling a case with blackpowder, a very old and widespread practice. The reason we cannot readily get into a case what the old-timers did is the change of case design through time. The old balloon-head case actually had a greater interior powder capacity because the head was thinner. These weaker cases were replaced with those that have more metal in the head, and therefore are much stronger and safer. That's why it's difficult, for example, to get 70 grains of Fg powder into the 45-70, which was certainly intended to hold 70 grains in the old days.

3. Use Magnum Primers

Only in relatively recent times did I learn from those who live and breath blackpowder cartridge shooting. Writers such as Mike Venturino taught me that the magnum primer did a better job in the larger blackpowder cartridge than milder ones. I have since enjoyed success with the Federal 215 large rifle primer and similar magnum primers.

4. Paper-Patched Bullets

Paper-patched bullets have shown good accuracy for me and I now understand why others have recommended them in the past. However, I have to say that tests are in order rifle for rifle, for I've also gotten fine accuracy with plain lead bullets. The blackpowder cartridge fan, however, should consider paper-patched bullets as viable and worth trying.

Final Note

Getting started in blackpowder cartridge shooting is not difficult. However, I strongly recommend that you study the subject thoroughly, and there are many books on the subject. See the Reference Section for these, as well as for periodicals that deal with the old-time cartridge and its guns.

chapter 35

The Blackpowder Shotgun

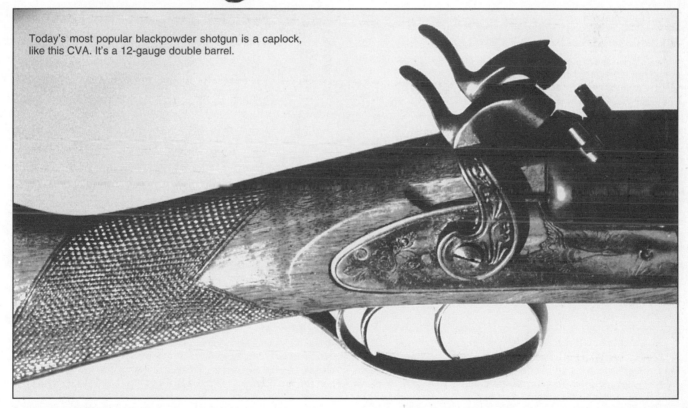

Today's most popular blackpowder shotgun is a caplock, like this CVA. It's a 12-gauge double barrel.

I CALL THE blackpowder shotgun the most obedient firearm in the muzzleloader battery—because it is. If I were turned loose in my favorite outdoor stomp and had to make a living off the land with a frontloader, I'd have to choose the shotgun, double barrel, please, percussion; 12-gauge will do. Why a smokethrowin' scattergun? Because this shotgun can do it all, that's why, from small game and waterfowl to large fauna. Harvest rabbits, wild turkeys and grouse with #5 shot, #8s on quail and dove, #4 Buck for campside safety should a rabid coyote come calling, a good dose of steel shot on ducks and geese. With patience, a deer or other big game can be taken with a 494-grain patched round ball (up-close shooting only).

Although we don't talk or write about it that much, don't you think that the blackpowder shotgun did a lot for the pioneer? I have to guess that on the old homestead it was the shotgun that brought in a lot of meat, and for protection when danger struck home, multiple projectiles filling the air like African killer bees had to be a pretty good way of saying, "Please leave me alone." Versatility is the word I'm looking for. The charcoal burning shotgun is versatile. It can do many things, and you don't need a lot of gear to run one well: just a little powder, shot or lead to make shot, patched round ball, percussion caps, and some form of wads (home-cut will work). In addition, because of its large smooth bores, the muzzleloading shotgun is easy to clean and maintain. Do you wonder why I call it the most obedient firearm in the blackpowder gun line?

(Above) The lure of the smokethrowin' shotgun is not so much in the challenge itself, for these shotguns can be made to produce very fine patterns, but in the thrill of shooting the old way.

A 10- or 12-gauge blackpowder shotgun is capable of any type of shotgun hunting. Here on the plains of Wyoming, Fadala took sage grouse with his Navy Arms double gun while hunting with Dean and Cindy Zollinger.

Modern Use

All the nice things I've said so far would go for naught if performance were lacking, but it is not. In no way will the old-time sootburner whip the great shotgun ammo coming from today's plants, because Winchester, Remington, and Federal are creating the finest shotshells ever. At the same time, believe it or not, the well-loaded blackpowder shotgun is not too far behind it's shell-shucking brother. This makes the muzzleloader great for trap, Skeet, upland gunning, waterfowl, even deer. The only problem I've encountered in the field is hunting with partners who don't realize that I have to reload after touching off both barrels. Sometimes they forge ahead, jumping birds, while I tend to business. But that's just a matter of education and communication.

The percussion shotgun is by far more popular than the flinter today. A flint fowler is a beautiful thing to behold and handle, but it has a longer lock time (the time elapsed from trigger pull to ignition) than the caplock. With practice and proper follow-through, the fowler is entirely effective and I have nothing against using one, such as the

Navy Arms Mortimer 12-gauge; however, most shooters are better off with a caplock. It's a matter of practicality. I've also noticed over time that hunters turning to sootburners are more than willing to accept the challenge, even a lighter game bag, but they do demand some success, and the great fowler is best suited to a truly dedicated downwind shooter.

Single barrel or double barrel? I have nothing against the single barrel, and there are many good ones to choose from nowadays. In some instances, the single is actually the better choice, for it is light and fast, and where no more than one shot is normally needed, as in rabbit hunting under certain conditions, the one-shot shotgun is all right. Furthermore, there are some single shots that play two roles. Thompson/Center's New Englander is one. It's a 12-gauge with an interchangeable 50-caliber rifle barrel.

The side-by-side blackpowder percussion shotgun in 12-gauge is the most popular choice. The 10-gauge is also a beauty, especially for waterfowling, or any shotgun work demanding a heavy shot charge. Several good double guns are offered in both gauges. Most of my blackpowder scattergunning today centers around the Navy Arms

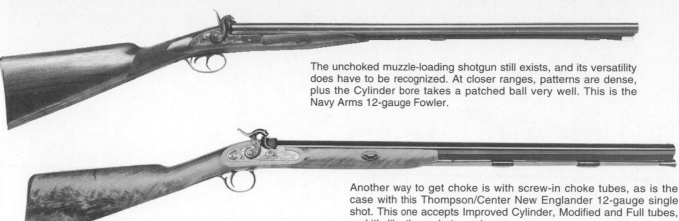

The unchoked muzzle-loading shotgun still exists, and its versatility does have to be recognized. At closer ranges, patterns are dense, plus the Cylinder bore takes a patched ball very well. This is the Navy Arms 12-gauge Fowler.

Another way to get choke is with screw-in choke tubes, as is the case with this Thompson/Center New Englander 12-gauge single shot. This one accepts Improved Cylinder, Modified and Full tubes, and it's like three shotguns in one.

T&T, a 12-gauge with Full and Full chokes, because the shooting I do requires close patterns. We'll cover choke later, but first, what does bore size mean in a scattergun? This is an important question, and the answer may surprise you. A 12-gauge is not always a 12-gauge, for example. A look at the Actual Gauge further on will be educational.

Bore Size

Shooting on the Wing, an 1873 publication by "An Old Gamekeeper," contains a rundown of shotgun bore sizes. Gauge/caliber relationships, according to the Old Gamekeeper, were as follows:

Bore Sizes

Gauge	Caliber in Inches
1-gauge	1.669
4-gauge	1.052
8-gauge	.835
10-gauge	.775
11-gauge	.751
12-gauge	.729
13-gauge	.710
16-gauge	.662
20-gauge	.615
24-gauge	.579
28-gauge	.550
32-gauge	.526
36-gauge	.506

Looking at this another way, the 10-gauge is a 78-caliber firearm; the 12-gauge is 73-caliber; the 16-gauge is 67-caliber; the 20-gauge is 62-caliber; and the 28-gauge is 55-caliber. This information is useful and instructional, but remember that gauges were not always the same size over the years. Today, that's especially true of muzzle-loading shotguns.

Actual Gauge

The actual gauge of the blackpowder shotgun may vary from its stated gauge. Many, if not most, modern 12-gauge blackpowder shotguns are 13-gauge; many 10-gauges made today are actually 11-gauge. The reason for this is that a 13-gauge shotgun handles loading components normally intended for a 12-gauge shotgun shell, and an 11-gauge frontloader handles components for a 10-gauge shotshell. Remember that the wad column goes *inside* of the shotshell, so, the wads are smaller than the inside diameter of the modern shotgun bore. The last two shotguns I bought were exactly this way. Be certain when buying or ordering wads that you get the right ones. If looking for standard wads (not the one-piece plastic wad, but the older fiber cushion and nitro card types) you may have to order 13-gauge size for your 12-gauge and 11-gauge for your 10-gauge, as I must do for mine.

Choke

In the past, importers and manufacturers of blackpowder shotguns insisted that these old-time scatterguns did just fine without choke. Shooters like me, who asked for choke, were mostly ignored. It took a long time, but at last we have what should have always been: choked bores. For example, my Navy Arms T&T (Turkey & Trap) shotgun, noted above, has Full and Full bores. Other blackpowder shotguns now have screw-in choke tubes offering immediate pattern alteration by inserting the choke tube of choice. Jug chokes are being cut into Cylin-

The shotgun also helped tame the West. No one wanted to come up against a load of shot at close range, and many a homestead was guarded with a blackpowder scattergun.

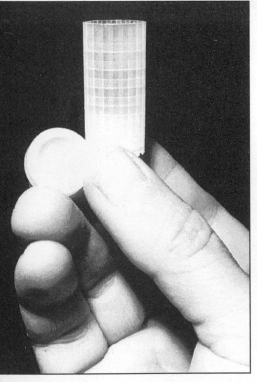

A powder measure may be used in preparing a correct charge for the shotgun.

Blackpowder shotguns will accept many different types of wads, including one-piece modern types like this one from Ballistic Products, Inc.

der-bore blackpowder shotguns, too, closing the usual 45 percent Cylinder pattern to Full choke and even Extra-Full. Myron Olson specializes in jug choking, and can be reached at 605-886-9787. A 10-gauge shotgun that Myron jug-choked for me produced 80 percent patterns after he was done, a far cry from the original 45 percent pattern. Other professional gunsmiths will install choke tubes in blackpowder shotguns that will accept them. Not all shotguns can be fitted with choke tubes, however—modern as well as blackpowder guns—because there must be sufficient barrel wall thickness to accept the screw-in tube. Check the smiths in your area for this work.

A 10-gauge flintlock shotgun with Full choke was measured with precision calipers. Outside diameter at the muzzle was .965-inch, but the end of the barrel was belled because outside diameter six inches behind the muzzle was down to .900-inch. This belling at the muzzle provided a thick barrel wall at that region. The inside diameter of this old shotgun was .778-inch at the muzzle for 1 inch into the bore. At the 1-inch mark, the bore sharply opened in a cone shape to .831-inch. Choke tubes were mentioned above. How thick a barrel wall is needed to allow the installation of these tubes? Here are a set of figures offered by one choke tube company:

- The 10-gauge should have an outside barrel diameter of .900-inch, with an inside bore diameter of .781-inch, leaving a wall thickness in the muzzle region of .0595-inch.
- The 12-gauge requires a minimum outside diameter at the muzzle of .825-inch with a maximum bore diameter of .736-inch. To put it another way, the barrel wall must be .089-inch divided by two which leaves .0445-inch before the tube can be installed.
- The 20-gauge outside diameter must be a minimum of .700-inch

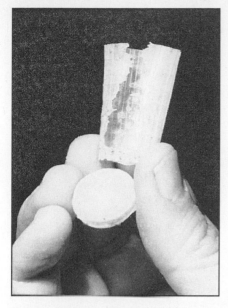

Just as a wise shooter checks for spent patches downrange, he must also look for his wads to see how they fared. Even the finest wads may not work in specific guns with certain loads.

The blackpowder shotgun can be loaded with one-piece wads. Powder and shot can be carried in film containers.

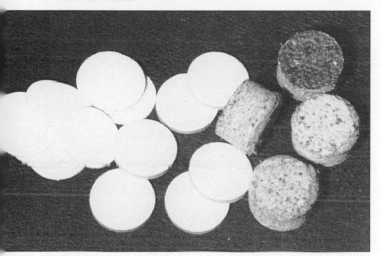

Today, the blackpowder shotgunner is well provided for with standard wads like these over-shot wads (left), over-powder wads (center) and cushion fiber wads (right).

with a maximum bore diameter in the muzzle region of .626-inch, leaving a barrel wall thickness of .037-inch.

When the great blackpowder gunsmith, V. M. Starr, installed his recessed choke he used a constriction of .021-inch for Full choke, .015-inch for Modified and .010 for Skeet choke, these predicated upon the use of cardboard wads only, he said. As for standard constrictions per choke, here are some figures:

- The 10-gauge shotgun with a .775-inch bore diameter will be choked .035-inch for Full choke, .017-inch for Modified and .007-inch for Improved Cylinder.
- The 12-gauge with a .729-inch bore diameter will be choked .035-inch for Full, .019-inch for Modified and .009-inch for Improved Cylinder.
- The 20-gauge with a bore diameter of .617-inch (yes, there are variations in the standardization of bore sizes; the Old Gamekeeper's data gave the 20-gauge a bore size of .615-inch) calls for a constriction of .025-inch for Full, .014-inch for Modified and .006-inch for Improved Cylinder.

The Jug Choke

Art Belding, a blackpowder shotgun fan, reported that jug-choking his 12-gauge shotgun did wonders for it. It's also known as a recessed choke. The work was done by Joe Ehlinger in Addision, Michigan. It was patterning about 45 percent before the job. Afterward, using #5 copper-plated shot with a pellet count of 185, and 90 grains volume GOEX FFg blackpowder with the V.M. Star wad method described earlier, here is what happened: For eight shots on the pattern board, the shotgun delivered 88, 92, 82, 91, 87, 81, 77, and 83 percent patterns. That's an average of 85 percent, or Extra-Full choke.

Wad Columns and Patterns

I hope I have not left the impression that a Cylinder-bore shotgun is worthless. For quail shooting at close range, the no-choke gun may be ideal, for example, but if you have a Cylinder bore and want a denser pattern, don't despair. Even if it cannot be fitted with choke tubes, it can be jug-choked. Also, one-piece plastic wads can tighten patterns. Any plastic fouling left behind must be removed, as with Venco's Shooter's Choice solvent. So the use of plastic wads need not cause a problem. I use certain plastic wads even for close-range shooting, because I demand good patterns, and chokeless shotguns don't always deliver pretty patterns, even at close range. Choke does more than concentrate shot. It can also improve patterns by eliminating holes and giving more even dispersion of pellets. The use of a one-piece plastic wad system acts, in a way, like a choke by helping to create a denser pattern.

Different Wad Columns

Numerous wad column configurations can be used with the blackpowder shotgun. The simple cardboard wad, associated with the late V.M. Starr, has worked well since the beginning of shotgunning, and it still does. Heavy-grade cardboard wads $3/_{32}$-inch thick may be cut from display signs with a punch. These heavy wads are used over the powder charge—two of them, one atop the other. Period. A thin cardboard wad is placed over the shot. And that's all there is to the cardboard wad column. As Starr pointed out, "You can put in more wads on the powder if you wish, or if you enjoy cutting them, but my experience tells me that you are just wasting your time." Starr did not use felt wads. He did, by the way, win several shotgun matches against good shooters using modern shotguns. The blackpowder charge, com-

Sometimes patterning at 40 yards is impractical. This shotgun was patterned at only 20 yards to see how it would do on close-up bird shooting.

Patterning is essential. If you don't know how your shotgun patterns, you don't know its effectiveness, or lack of it.

posed of Fg or FFg, does not tend to destroy wads, which is part of the reason for these simple cardboard cutouts holding up so well in the old-style scattergun.

The long-time standard wad column consisting of a thick cushion fiber wad over the powder and a thin cardboard wad over the shot—as

Choke Sizes

As always, standardization is difficult to find. However, the following choke designations with pattern density for each are a sound assessment and a useful guide. The only possible exception is Cylinder, which is noted as under 45 percent. I have tested Cylinder-bore guns that were 45 percent, putting them in the Improved Cylinder II category, but some lapover is expected, and this does not negate the value of the choke listing.

Choke and Pattern Density

Extra-Full = 80 percent or tighter
Full Choke = 70 to 79 percent
Improved Modified = 65 to 69 percent
Modified = 55 to 64 percent
Improved Cylinder I = 50 to 54 percent
Improved Cylinder II = 45 to 49 percent
Cylinder = under 45 percent

in the days prior to the star crimp—also works fine in the blackpowder scattergun. The modern plastic wad offers another workable method of loading the blackpowder shotgun. Remember that the one-piece plastic wad functions well because many muzzle-loading shotguns are a gauge undersize. The wad made to fit inside the hull of the 12-gauge modern shotshell is sized properly for a 13-gauge muzzleloader bore. If your gun is a true 10- or 12-gauge, you can still buy 10- or 12-gauge size blackpowder wads from Dixie Gun Works or Ballistic Products, Inc. Never use any type of wad which may become lodged in the bore. A critical time in powder charge combustion is the escape of the missile, no matter what the projectile is, round ball, conical, or for that matter a shot charge. Pressures may rise if the wad does not start upbore when expanding gases strike it, so be certain that the wads fit properly. The wad column must be free to escape upbore.

Patterning

The pattern is the key to performance. Of course, pellet velocity and energy are important, too. But it's no trouble getting good shotgun velocities with the old-time shotgun. However, obtaining the best possible pattern sometimes requires a little research. In other words, *pattern* that shotgun. Find out what it is doing so that you can improve upon it if improvement is called for. The blackpowder shotgun requires patterning just as surely as its modern counterpart does if the shooter is to know for certain what his shotgun is providing in shot distribution. The process is not difficult. A large sheet of paper is pinned up at 40 yards. The shotgun load is fired dead center into that large sheet of paper. Using a cardboard cutout 30 inches in diameter, a circle is drawn around the *concentration* of shot on the paper. The holes within that 30-inch circle are counted and the number is divided by the number of pellets in the load. If your shotgun fired 100 pellets and fifty of them were in that 30-inch circle, the pattern would be 50 percent, or Improved Cylinder choke.

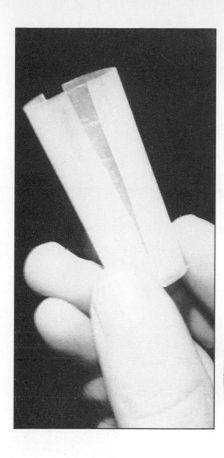

Sometimes slitting a one-piece plastic wad is necessary in order to achieve a good pattern. It's a matter of experimenting on the pattern board.

An original 19th-century shot flask. It could be carried via a strap around the neck of the hunter. That's what the metal hook is for on the bottom of the flask.

This 19th-century shot flask is adjustable for a 1½- or 1¼-ounce shot charge. Here, it is oot on 1¼ ounces.

Check *the way* the shot is distributed on the paper, too. Your shotgun may put half of its charge within the 30-inch circle, but are there large holes in the pattern? Naturally, one try won't tell the story. The serious shooter should pattern his shotgun several times, until he has a very good picture of the average pattern that gun develops. I have seen shotguns which patterned 75 percent, but the patterns had big holes in them. If this happens, it's wise to juggle load components. Sometimes a simple change in shot size will do the trick. If #8 shot is not working, try #6 shot, for example. Or you may wish to alter the balance in shot/powder *by reducing the powder charge* and not by adding more shot. This does not always make for a denser pattern, but I have seen holes in patterns disappear by simply reducing the powder charge by 10 volume or so. Do not juggle components at will. Follow the basic prescription of the shotgun manufacturer.

The Powder Charge

Forget FFFg in the blackpowder shotgun. Though this granulation is not a problem unto itself, with many types of wad columns the faster-burning FFFg may flame through the over-powder wad. When the wad which supports the shot charge is damaged, the pattern will suffer. FFg, Fg and Pyrodex RS are the preferred powders for the muzzle-loading shotgun. Pressures are lower with the larger granulations, and the burning curve is smoother. As for Pyrodex, I've had good luck with RS and also with CTG granulations. Cardboard burnout with the larger granulation powder is uncommon. Initially, I believed that FFFg powder was blowing patterns because it was blasting the shot charge and wad column from the bore with an excessively high muzzle velocity. I should have known better. Wad burnout was the problem. So leave FFFg out of the blackpowder shotgun picture. Incidentally, rumor has it that CTG will be discontinued. If so, rely on RS because it works fine.

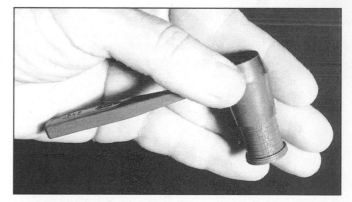

A simple adjustable measure like this one is all the blackpowder shotgunner really needs to measure his loads when he uses the volume-for-volume approach to loading.

The weight of the powder charge—maximum as well as recommended load—is in the hands of the shotgun manufacturer. If there is no maximum load information with your new shotgun, write the gunmaker. After you learn the maximum charge allowed in your particular gun, do not exceed it. Of course, the blackpowder shotgun is generally allowed fairly good charges of Fg and FFg because the volume of the bore is so large that resulting pressures are low. Also, a single scoop may be used for both shot and powder charge, making loading a cinch. This means that the blackpowder shotgun will han-

255

There are many ways to carry shot and powder. The 35mm film canister in the foreground is one way—shot in one canister, powder in another. Horns also work. The rearmost horn is for powder, while the other is for shot.

dle the same *volume* of powder and shot. This has nothing whatsoever to do with the weights of either; it is strictly a volumetric proposition. If your shotgun is allowed 90 grains of FFg, set the powder measure at 90 and use that setting for both the shot charge and the powder charge—*by bulk*.

You don't have to load this way, and you may not want to. Safe juggling was mentioned earlier. A reduction of powder while leaving the shot charge as is often results in marked improvement in pattern density, but you must pattern your shotgun to find out. I've seen so many individual differences in loads from various muzzle-loading scatterguns that it's almost impossible to recommend one set of loads for all shotguns. One 12-gauge Full & Full shotgun gave best results with the most powerful loads allowed in that gun—a heavy shot charge and a heavy powder charge. Another gun gave best patterning when the powder charge was dropped by about 20 percent below the recommended weight. My 12-gauge is allowed a full 1½ ounces of shot, but I often load only 1 ounce with only 70 grains of FFg blackpowder for upland birds at close range. What versatility from a single firearm!

Overloads

Blackpowder can be overloaded, as established in Chapter 14. Likewise for any propellant. The blackpowder shotgun, due to large bore size, develops low pressures per powder charge, which may be deceiving. True, a 12-gauge loaded with 100 volume FFg behind a full 1½ ounces of #4 steel shot develops under 6500 LUP, as compared with a 50-caliber muzzleloader, for example, shooting a 370-grain Maxi with 100 volume FFg for about double that pressure. However, you still never overload a blackpowder shotgun. Bad things can happen if you do, especially if there is a "kink" in the load chain. A bore obstruction, for example, can cause a burst barrel, as can short-starting the wad system. Also, a too-tight wad column may cause big trouble, because it

The blackpowder shotgun load is actually quite basic. From the right: a percussion cap to set off the powder charge, a wad system (one-piece or standard), some shot, and an over-shot wad to hold the shot in place.

A capper can be used to install the percussion cap on the nipple of the shotgun. It's handy and safer this way.

Loading the blackpowder shotgun is basic, but every step must be handled with care. After popping caps to ensure a clear passageway, a correct powder charge is dropped downbore.

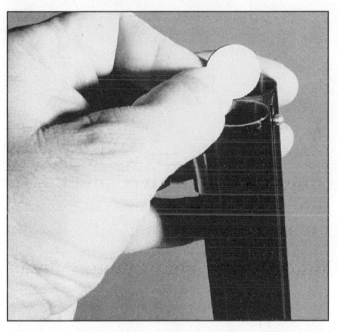

After putting a one-piece plastic wad, or standard wads, downbore and loading the shot, an over-shot wad is installed to keep the shot in place, and then the shotgun is ready to cap for firing.

cannot start upbore readily, meaning very high breech pressure. An overload of powder, plus any of these problems, can be dangerous. Don't overload.

Lead Shot

Cubed shot was used once upon a time. A sheet of lead was literally chopped into small cubes, sized as desired by the shooter. Talk about fliers! However, while these little cubes of lead were anything but aerodynamic, they apparently worked to some degree, as reported by old-timers who used them. At close range on small edibles such as quail and cottontails, the cubes buzzed their way through with considerable effect. Since these and similar game are usually harvested at under twenty paces, cubed lead sufficed. Another lead shot from long ago was called swan shot. Before telling about this, please heed this

warning: don't try to make any. It's not that great to begin with and it is hazardous to work with. A person could be badly burned in the process. Swan shot was made by pouring molten lead through a screen and from there into a cooling medium, such as water. The shot hardened when it hit the liquid, often with a tiny tail on each pellet, remindful of the south end of a swan flying north. Historically interesting, yes, but swan shot has been outstripped considerably by modern shot and there is no need to make any of it.

Steel Shot

Steel shot is created from soft steel wire, formed into pellets, annealed (heating followed by slow cooling), and finally coated. The process is not unlike the manufacture of steel ball bearings. It has taken us a very long time to learn how to live with steel shotgun pellets. Only recently have most of us stopped complaining, putting our energies into learning how to manage steel, as well as understanding its strong points. Not that steel shot is perfect, because it is not. But it's far better than we first thought, and quite effective when applied correctly, from choosing the right pellet size, to understanding pattern shape.

The Bad Points

Steel shot is lighter than lead shot. Pellet for pellet, you can count on steel to weigh about thirty percent less than lead. This is not good. If you recall the discussion on the good properties of lead, you'll remember that density is one of them. Being heavy, it forms into a weighty projectile that retains its initial velocity better than a lightweight missile. Steel pellets lose velocity faster than lead pellets, and that means they carry less energy down range to the game. To offset this factor, we've gone to larger pellets for steel shot, carrying this too far, in fact, so that many hunters fire loads that contain too few pellets for a dense pattern.

The Good Points

Is the glass half empty or is it half full? That's a question asked about attitude. A person with a good attitude sees the glass half full, while others see it half empty. So it is with steel shot. We can sit around complaining about its lack of weight per pellet, or we can look on the good side and learn how to use this stuff to full benefit. Steel shot has some wonderful properties. Here are a few:

1. No Fliers

Soft lead shot deforms in the bore, creating pellets that look like miniature flying saucers. These are called fliers, and they soon depart the pattern. Of course, up close some of these fliers might actually hit the target, working in effect to broaden a pattern. But all in all, fliers are not good. Steel shot is hard, and does not deform into fliers. Each pellet retains its shape and stays in the pattern.

2. Steel Shot is Round

We have long worked toward a truly round pellet for shotgunning. Lead can be made into pellets that are quite round, but steel pellets are rounder, from start and certainly to finish (bore to target).

3. Steel Shot is Uniform

Steel pellets are uniform, one to the next. You don't have much difference in size within a given pellet number. If you buy a bag of steel No. 5s, you'll have a whole bunch of pellets that run .120-inch diameter.

4. Steel Shot Creates a Short Shot String

For years, all we heard was how great it would be if we could only get a short shot string. Well, steel provides it. While lead pellets tend to string out long, and due to fliers, also somewhat wide, steel sends out a shorter, narrow shot charge. I call that good, because the pattern is dense and lots of pellets land on target. But it is, admittedly, bad, too, because the shorter, narrow shot string is harder to hit with.

Non-toxic shot is the law now, and we must all obey. Steel shot is to be used only on waterfowl, and in certain areas on other game. Steel is not the only legal shot, by the way. The law truly calls for non-toxic shot, which steel is, but so is bismuth. Bismuth is legal, and it is for

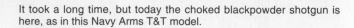

It took a long time, but today the choked blackpowder shotgun is here, as in this Navy Arms T&T model.

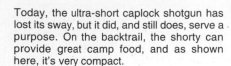

Today, the ultra-short caplock shotgun has lost its sway, but it did, and still does, serve a purpose. On the backtrail, the shorty can provide great camp food, and as shown here, it's very compact.

sale through Ballistic Products, Inc. Bismuth is an element, Bi, and it makes a very fine shotgun pellet, non-toxic and effective. The problem with bismuth pellets at the moment, aside from widespread availability, is cost. An 8-pound container of BB, #2, #4, or #6 shot destroys a hundred dollar bill.

Steel Shot in Muzzleloaders

Sad to say, steel shot is not entirely workable in the blackpowder shotgun, but it could be. Scoring of bores has been experienced, and shooters have complained about sub-par killing power on game. This does not have to be, and two things will correct the problem. The first is shouting for special steel-shooting muzzle-loading shotguns, just as we had to holler about choked guns. We need more steel-shooting blackpowder shotguns, period. Noted in the 1996 *Gun Digest* is one such gun, the Navy Arms Steel Shot Magnum in 10-gauge. It's choked Cylinder & Cylinder, and that's not so bad. As explained below, steel shot patterns differently than lead, and with plastic wads, this open-bore gun delivers closer to old-time Modified choke results *with larger steel shot sizes*.

The other half of the problem is learning to shoot steel shot patterns. First, pick the right shot size. The use of super-large pellets means fewer per charge, resulting in skimpy patterns. There is no need to use overly-large shot on game, and I think this has been part of the problem for those who are not happy with steel shot results. I recently went sage hen hunting with #6 steel shot (.110-inch diameter) and #5 steel shot (.120-inch diameter) using both with results not unlike those experienced with lead #5 shot. Consider BBB steel (.190-inch diameter) for

wild turkeys with body hits, and BB steel (.180-inch) for big birds where a higher pellet count is desired. Even large ducks can be dropped with #1 steel (.160-inch diameter). After you have the right size shot, make up your mind to become a better shotgunner with that shorter, narrower steel shot string.

Ballistics

A 12-gauge shotgun with $1\frac{1}{2}$ ounces of shot (when that much shot is allowed by the gunmaker) and 100 grains *volume* of FFg blackpowder or Pyrodex RS produces a muzzle velocity close to 1200 fps, challenging the modern shotshell. While not quite as powerful as a "baby magnum" 12-gauge round, this load is certainly not far behind it. A $1\frac{1}{4}$-ounce shot charge in the same shotgun with 100 volume Pyrodex RS or FFg blackpowder produces about 1300 fps MV. This gives a fairly good idea of blackpowder shotgun potency.

The Caplock Shotgun Nipple

The blackpowder shotgunner should obtain a nipple wrench and remove the nipples of his new shotgun before firing the gun. Some shotguns are fitted with nipples that have a very large vent, which causes a problem with blowback, as mentioned in an earlier chapter. Gas bleeds back through the orifice. Generally, this situation is not a problem because the large bore handles pressure nicely, and the comparatively small vent in the nipple does not allow an appreciable quantity of gas to escape through it. However, the base of the nipple can be a problem. The base should be flat with a pinhole in it. Some shotgun nipples are totally open at the base, and do not offer a seal at the nipple seat in

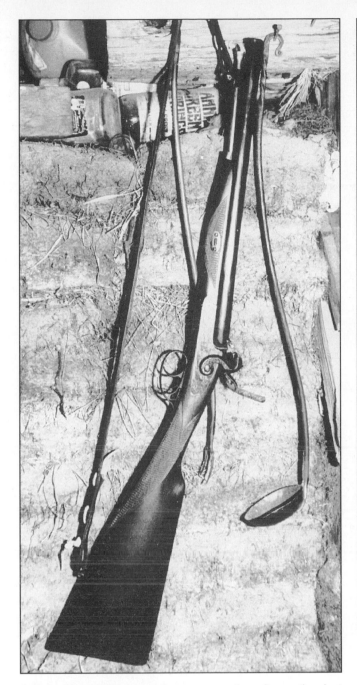

Called a "sawed off" shotgun because sometimes it was, the short blackpowder caplock was handy and quick to use.

Shot Sizes

Shot sizes have never been fully standardized. The following are for reference only, and are not to be considered the only diameters assigned to certain pellet sizes.

Shot Sizes

	Lead Shot			Steel Shot	
Size	Diameter (ins.)	No./Oz.	Size	Diameter (ins.)	No./Oz.
#12	.05	2385	#6	.11	315
#9	.08	585	#4	.13	192
#8	.09	410	#3	.14	158
#7½	.095	350	#2	.15	125
#6	.11	225	#1	.16	103
#5	.12	170	BB	.18	72
#4	.13	135	T	.20	52
#2	.15	90	F	.22	40
BB	.18	50			

Recommended for killing patterns: 140 pellets per charge for small ducks, 110 for medium ducks, 85 pellets per charge for large ducks, 55 pellets for small geese, and 35 pellets for larger geese.

Shot Energy
Lead vs. Steel Shot at 40 Yards

Lead #6 = 2.3 foot pounds
Steel #4 = 2.5 foot pounds
Lead #4 = 4.4 foot pounds
Steel #2 = 4.4 foot pounds
Lead #2 = 7.5 foot pounds
Steel BB = 9.0 foot pounds
Lead BB = 15.0 foot pounds
Steel T = 20 foot pounds
Steel F = 24 foot pounds

The above is not meant to be misleading. Of course, steel shot sizes show well energy-wise. They are of larger diameter in the comparisons above. For example, the lead BB is .18-inch diameter, while the steel T is .20-inch and the steel F is .22-inch. However, steel shot in larger sizes definitely carries reasonable energy for clean game harvesting. Loaded with the volumetric measure, of course, the weight of the steel charge will be less than an equal volume of lead shot. But with careful management, steel shot can be very effective.

the breech of the gun. Gas forces its way up the vent and out the cone of the nipple, causing a loss in pressure behind the shot charge and a loss in muzzle velocity. I have chronographed as high as 200 fps loss from this problem. If the nipples are of the baseless variety with a large vent, get rid of them. Replace with the proper nipple having a flat base with a pinhole in it and a more modestly sized vent.

The Round Ball in the Muzzle-Loading Shotgun

The .690-inch 12-gauge round ball weighs an average of 494 grains. While it is generally difficult to achieve much more than 1000 fps MV with the 12-gauge ball, even that modest speed brings about excellent close-range authority. The buck 'n' ball load, with a charge of shot in one barrel and a patched round ball in the other, suits hunters who may encounter birds or deer on the same terrain.

Having run extensive tests with buckshot in the blackpowder shotgun, I'm prepared to say that while it certainly has applications, I'm not in favor of its use on big game. That opinion is based on two things: First, I found that patterns were not good. Only a couple pellets hit the kill zone at even 30 yards. Second, each pellet carries low power, less than a 36-caliber squirrel rifle, which we would not use on big game. One #00 Buck weighs only 54 grains, lighter than the .350-inch ball for my 36-caliber squirrel rifle! This little pill is lucky to leave the bore at the speed of sound from the muzzle-loading shotgun, and not much more than 1400 fps or so from the modern shotshell.

The Readyload for the Shotgun

There are many ways to carry shot, powder and wads for the scattergun. The old flask worked fine, and powder and shot horns are good. And so are readyloads. The successful shotgunner should try each method until he finds the style he likes best. Readyloads can be prepared in many ways; a simple means is with plastic 35mm film containers. Pre-measured *powder* charges are put in one set of containers,

Today's blackpowder shotgun may have a recoil pad to tame the kick, as does this CVA 12-gauge. A slip-on pad also works well.

A Few Shotgun Safety Rule Reminders

1. After firing one barrel, be certain that the load in the other barrel has remained in place, especially when one barrel has been fired a few times without the other being shot. A load may creep upbore, leaving a space between powder charge and shot charge.

2. Don't use a powder horn to pour powder directly into the shotgun bore. A safer practice is to use readyloads, as with plastic 35mm film containers.

3. Uncap a loaded barrel before attempting to load a fired barrel. The logic behind this is clear.

4. Be fully certain that your shotgun bore is clear of plastic fouling if you use modern one-piece wads.

5. Use proper wad column materials only. Don't stuff newspaper, rag-cloth or other foreign bodies downbore in lieu of correct wads.

6. Be certain to cover and/or remove any powder container from the area before shooting. This goes for all muzzleloaders. I watched a trap shooter fire his frontloader right by a full can of blackpowder that had its lid off. Not a good plan.

7. Watch out for the old ones. There are original muzzle-loading shotguns in good shape, but you can't always judge condition by looking. At the least have an old gun checked by a blackpowder gunsmith, and checked by a metallurgist as with the Magna-Glow process to check for imperfections in the metal. Even after this precaution, consider an old muzzleloader a possible problem. There could be internal wear that cannot be located. Shoot originals with extreme caution. Fire your old-time shotgun by remote control before you try it from the shoulder.

8. Use only blackpowder or Pyrodex in the blackpowder shotgun—nothing else. But just because you use blackpowder does not mean that an old gun in poor condition is safe to shoot.

9. *Load by volume only.* Do not attempt to load by weights. Steel shot will weigh less per volume charge, but this is as it should be. Do not load steel, or any other shot by weight, nor the powder charge. Remember that a steel shot charge that weighs only $3/4$-ounce is recommended for ducks under thirty-five yards, while a 1-ounce charge of steel shot is recommended for ducks at thirty-five to forty-five yards distance. Even for geese and wild turkeys, a steel shot charge of only $1 1/4$ ounces is considered effective.

---

while pre-measured *shot* charges are contained in another. Wads can be carried in the hunting coat pocket. The method of working with these readyloads is all but self-explanatory. The gunner pops a top, drops a charge, runs a wad home on top of the powder, drops a shot charge from another plastic can, tops off with an over-shot wad and he's ready to shoot after capping the gun.

Recoil Problems

The successful blackpowder shotgunner manages his firearm with skill and grace. One way he can do this is by knowing that recoil of the gun is not going to cause trouble. It's no more than a push. Because blackpowder is not that efficient, and although it burns beautifully in the scattergun and gives nice patterns, that relative inefficiency means that *more* blackpowder is burned per load than the smokeless powder shotgun load requires. Since the *weight* of the powder charge itself is a factor in recoil, the blackpowder shotgun does offer some stout recoil, but it's manageable. If the shooter is bothered by the "kick," he may want to look at a slip-on recoil pad as offered by Michaels of Oregon and others.

Before leaving the blackpowder shotgun, consider the newest member of the clan, the modern muzzleloader shotgun. White Shooting Systems has two, the Tominator and the White Thunder. The Tominator comes with various choke tubes, including Super Full. Not long ago, I tested the Knight version of a modern muzzleloader shotgun, and it proved to be a powerhouse with an Extra-Full choke tube, 100 volume Pyrodex RS, and a full 2 ounces of shot.

For more versatility yet, the blackpowder shotgun may wear a sling. This 12-gauge CVA shotgun has built-in swivels.

The Versatile Scattergun

Successful blackpowder shotgunning is an all-encompassing enterprise, from choosing the right gun to loading it correctly, patterning, and practicing, especially to master steel shot. Each shot is a handload, and that's a plus because the shooter can alter his load for the field conditions at hand. I've taken a limit of grouse, for example, after which my smokepole scattergun was immediately loaded with a milder recipe for rabbits. And don't forget that the same gun can be turned into a fairly strong close-range deer-harvesting machine with a big round ball. The old-time blackpowder shotgun is a lot of fun to shoot. But it also offers the modern shooter a lot of options with good power and patterns.

●

chapter 36

The Blackpowder Sidearm

NO BLACKPOWDER BATTERY is complete without a sidearm—big-bore pistol, small-bore pistol, small-bore revolver, large-bore revolver, modern muzzleloader pistol, replica, custom, blackpowder cartridge revolver—the choice is yours, and there are plenty of guns to choose from. Basically, however, there are only two major sidearm types in the lineup. These are pistol and revolver. The pistol is usually a single shot, but can also be a two-shot double-barrel affair, or for that matter a multi-barrel piece capable of firing several times. A transition firearm between pistol and revolver was the pepperbox. It looked like an elongated cylinder taken from a revolver, but it was actually a collection of barrels. The barrels revolved into position one at a time to line up with the firing mechanism. I have no doubt that the pepperbox influenced inventors of the revolver.

Choosing Pistol or Revolver

This is not a difficult choice, really; the two are so different that a shooter easily knows which one he wants or needs. For example, if big game hunting with a blackpowder sidearm is your ticket, go for a pistol. Nothing in revolver form can touch a muzzle-loading pistol for sheer power. There are fine target pistols, too, some of which are extremely accurate. On the other hand, if you want to win the revolver match at the next blackpowder shoot, it's obvious that a wheelgun is your only choice, one with sights and accurate, too, something like the Ruger Old Army, perhaps. And if the Civil War period is your cup of tea, think revolver. Many revolvers of the era are back with us today in replica form.

Choice is one thing, but which type is truly better? Now things get tougher. Better for what? Below, a number of viable uses for the blackpowder sidearm are cited. It becomes obvious that for certain applica-

The Ruger revolver has what it takes to make a blackpowder shooter happy: accuracy and good sights. It's fun to shoot because you can hit with it.

tions, the pistol is better than the revolver, the reverse being true for other jobs. Better for firepower? That's the revolver. Power? The pistol wins. I don't like to play the weasel, but the fact is, neither is better. The pistol and revolver simply play different roles the majority of the time, whether for reenactment of a historical period, rendezvous service, collection, hunting, competition, or other use.

The Small-Bore Pistol

The small-bore pistol is almost recoilless with the proper load, and simply a lot of fun to shoot. With proper sights it's also a small game taker, and it can be used for target work. Dixie Gun Works' Mang Target Pistol in 38-caliber is an example of the latter. It fires a .375-inch round ball with 20

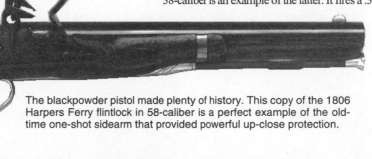

The blackpowder pistol made plenty of history. This copy of the 1806 Harpers Ferry flintlock in 58-caliber is a perfect example of the old-time one-shot sidearm that provided powerful up-close protection.

261

Pistol or revolver? Really, the choice is easy. A single shot pistol like this is certainly unlike a revolver. The shooter's choice is entirely his own, and for his personal reasons.

grains of FFFg for a MV of 902 fps—simply pleasant. A load of 30 grains of the same powder, same ball, delivers 1071 fps MV, while 30 volume Pyrodex P pushes the 80-grain .375-inch pill at 1112 fps MV. Energy runs around 200 foot pounds for the 30-grain charges, plenty for close-range small game hunting. There are other small-bore pistols fired mostly for plinking fun, or simply collected, examples being CVA's 31-caliber Vest Pocket Derringer, as well as the Snake Eyes double-barrel 36-caliber pistol from Mountain State Muzzleloading, and an Ethan Allen Pepperbox and a three-barrel 36 caliber Duckfoot from the same purveyor.

The duckfoot was another different kind of blackpowder pistol. All three barrels on this one fire simultaneously.

The Small-Bore Revolver

Surprisingly, the small-bore revolver was seen as a self-defense firearm, even a military weapon. There were a zillion of 'em around in the 19th century. Well, not that many, but more than a few, and far too many to list here. I believe that Wild Bill Hickok carried a 36-caliber revolver, and I know that the Civil War saw use of that size, although caliber 44 was certainly more popular in a wheelgun. A typical small-bore revolver is the 36-caliber Colt Pocket Police Revolver. It shoots an 80-grain bullet, .375-inch diameter, at 954 fps MV for a muzzle energy of 162 foot pounds. In my test, I stuffed in 24 volume Pyrodex P to get this energy, which is not much, but apparently sufficient for certain law and order, as well as unlawful, acts in its day.

And its day is not yet over. The small-bore revolver is a truly enjoyable plinker. Once again, lack of recoil must be cited as a major plus

with this one. It will, of course, take small game, but normally wears only the "point at 'em" type of sight, useful at close quarters when a large target threatens. Most enjoyable about this little revolver, aside from its plinking application and pure fun of ownership, is the replication factor. For example, Colt Black Powder Arms Co. offers several smallbores, including an interesting 36-caliber "Trapper Model" 1862 Pocket Police with $3^{1}/_{2}$-inch barrel and no loading rod. A separate $4^{5}/_{8}$-inch brass ramrod is included with this replica.

Colt also has a remake of the 1861 Navy, which I view as the most handsome cap 'n' ball gun of them all, along with the similarly-shaped Colt Cavalry Model 1860 Army with fluted cylinder in 44-caliber. The smaller 36 is graceful as a swan swimming on a blue lake. And if you thought 36-caliber was the smallest of the small-bore revolvers, it is not. Colt also remakes its 1849 Pocket Dragoon, which is only 31-caliber. And don't forget the Colt Paterson, the first commercial revolver, originally manufactured in Paterson, New Jersey. Navy Arms sells a beautiful replica of this 36-caliber revolver. It's easy to see that the small-bore revolver has a useful niche in modern blackpowder shooting.

The Big-Bore Pistol

This is a horse of a different color—in fact, it was sometimes called a horse pistol because it was packed on the old cayuse instead of in a belt holster. Big is the byword for these guns. In replica form, there's Lyman's Plains Pistol, calibers 50 or 54. In the latter, expect a 225-grain round ball to leave the muzzle at over 900 fps, which it will in front of 40 grains volume FFFg (954 fps MV). Muzzle energy is only 453 foot pounds, but a good shot at close range could certainly bag smaller deer

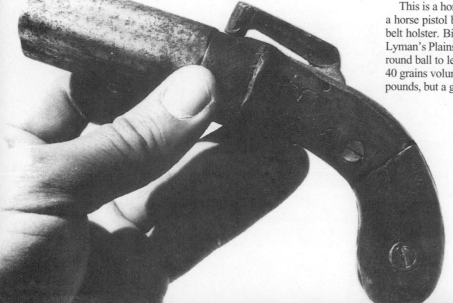

This is an original pepperbox pistol with revolving barrels. It's easy to see how it may have influenced the invention of the revolver.

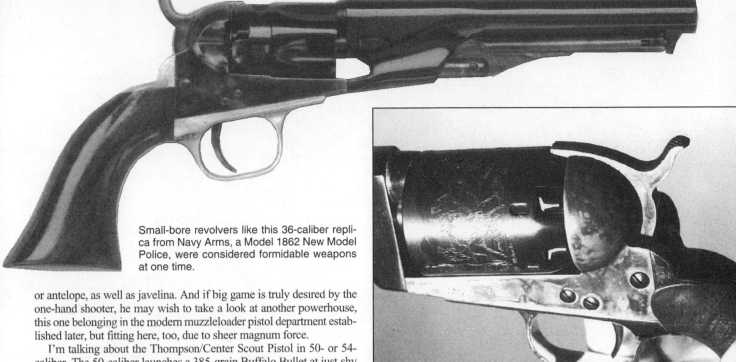

Small-bore revolvers like this 36-caliber replica from Navy Arms, a Model 1862 New Model Police, were considered formidable weapons at one time.

or antelope, as well as javelina. And if big game is truly desired by the one-hand shooter, he may wish to take a look at another powerhouse, this one belonging in the modern muzzleloader pistol department established later, but fitting here, too, due to sheer magnum force.

I'm talking about the Thompson/Center Scout Pistol in 50- or 54-caliber. The 50-caliber launches a 385-grain Buffalo Bullet at just shy of 1200 fps MV for a muzzle energy of around 1200 foot pounds. It does this with 100 volume FFg blackpowder. In my tests, the 54-caliber version propelled a 540-grain Maxi Hunter bullet at just shy of 1100 fps MV with 100 volume Pyrodex RS powder (71.5 grains by weight) for a muzzle energy over 1400 foot pounds. While this is only 200 foot pounds ahead of the 50, the heavy bullet can be counted on for great penetration. I'd not hesitate to hunt big game with the 54-caliber Scout within reasonable range limits. I don't think much more need be said. Clearly, the big-bore pistol can be a powerhouse.

The Big-Bore Revolver

Caliber 44 and up marks the big bore revolver, and there are plenty of them around. It's only fitting to mention one of the larger big-bore revolvers first, the famous Colt Walker. Many companies offer them, including Navy Arms, Traditions, Mountain State, and Colt. There is a factor concerning big-bore revolvers, however, that does need clarification. Even the big Walker—it weighs 73 ounces—does not produce the power generated by the bigger bore pistols. My best efforts in testing the Walker brought forth 1215 fps MV with a 141-grain .454-inch round ball for a muzzle energy of 462 foot pounds. "That ain't hay," as they say, but it's a mile down the road from the Scout with its 1400+ foot pounds of energy.

This Colt Whitneyville Dragoon is a caplock of modern manufacture from Colt. It's an exacting copy of the original, and it's extremely shootable.

The 44-caliber Navy Arms LeMat Army revolver fires its .451-inch diameter, 138-grain round ball at 687 fps MV for a muzzle energy of 145 foot pounds. But there's a little surprise that comes with the LeMat, an underbarrel in caliber 65, firing a .595-inch cast round ball at 571 fps MV for 230 foot pounds of muzzle energy. This is not a lot of snort, number-wise, but a big lead round ball is still a formidable missile. All in all, the big-bore revolver is represented by a whole army of six-guns, mainly in 44-caliber, and mostly producing 38 Special ballistics, roughly speaking. I do not consider any of them big game pieces, but as you can guess, big game has been taken with them. In the hands of a sharp pistoleer, I suppose that's all right, but most of us demand a little more than 38 Special or even 357 Magnum punch for deer and larger animals.

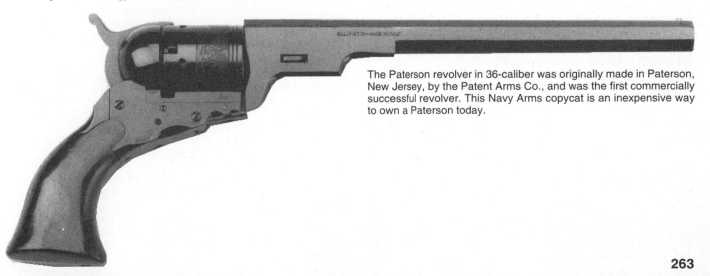

The Paterson revolver in 36-caliber was originally made in Paterson, New Jersey, by the Patent Arms Co., and was the first commercially successful revolver. This Navy Arms copycat is an inexpensive way to own a Paterson today.

Because of replication, a blackpowder shooter can experience handguns of old, like this LeMat 44-caliber Army model from Navy Arms, with its 65-caliber underbarrel.

Shooting Games

One of the many applications of the blackpowder sidearm is plinking, but this term has the wrong "ring" to it. Plinking sounds like just plain fooling around. I don't agree with that assessment any more than Abraham Lincoln would have when he went plinking with his favorite firearm, as he was known to do. Plinking is fun, sure, but it's also one fine way to get sharp. When I fired thousands of rounds annually from a 22 rifle I reached the point of what I'd call junior-grade exhibition shooter. Annie Oakley would have made me look silly, but I was able to light kitchen matches with my 22 rimfire rifle, and I could hit ten tin cans thrown in the air in a row. Not today. I got busy, quit plinking, and I'm much less the marksman for it. So I consider the plinking profile of the blackpowder sidearm far more than play. To me, it's serious practice while having a good time in front of a safe backstop.

Plinking is usually whacking tin cans and other inanimate objects set up against a dirt backstop. It may even be a roving game, much as the archer plays in his "stump shooting," where safe, well-distinguished targets are fired at when located in front of a proper backstop. Informal target shooting is different because it usually incorporates a more regular target, paper most likely, with a bullseye. Any blackpowder sidearm is up to par for this sort of work, and it's fine practice as well as enjoyable.

Shooting games are anything from ringing a gong to breaking charcoal briquets hanging on strings. Games are as varied as the imaginations of their inventors. There isn't a blackpowder handgun that is ill-suited to some sort of game, although for certain ones, only specific guns will do. You'll find that out if you try to ring a 12-inch gong at 200 meters with a point-at-'em sight arrangement, or a little gun meant for close-range work only.

Formal target shooting is self-explanatory. There are several matches held at rendezvous and other events in which formal blackpowder handgun shooting is held. The targets vary with the event. Rules are set down for each shoot, and within these rules lie the parameters for the specific handguns allowed. For example, there's often a 25- or 50-yard offhand match, generally one for pistol, another for revolver, but sometimes an open match where either sidearm is allowed.

Silhouette shooting is not yet established like the blackpowder cartridge rifle event, so with blackpowder handguns it is done informally. Hopefully it will develop into a national pastime of its own, as the cartridge rifle event has.

On the Trail

My favorite use of the blackpowder sidearm is on the trail. I enjoy backpacking, and in most areas small game is in season when I go. In the high country, mountain grouse and partridge are also allowed, and blackpowder handguns are legal for taking them. The type of handgun

Sam Colt's company introduced the world's most famous six-shooter, the single-action cartridge gun. But this revolver did not start that way. The first ones were blackpowder caplocks, every one, including many carried by now-famous shooters, such as Wild Bill Hickok.

is wide open, subject to individual choice. However, it better have good sights and reasonable built-in accuracy, or you won't get much game. I've taken, for example, blue grouse with my Thompson/Center Patriot Pistol. While 45-caliber seems a bit big for these fine birds, it's not, because only head shots are taken, and from very close range, usually ten yards or under.

A camp gun is any firearm that's handy, easy to manage, and reliable. I find the blackpowder six-gun all of those things and often carry a Ruger Old Army into camp. I've never needed it and probably never will, but you never know. A friend had a skunk stroll into camp. He didn't want to shoot it for the obvious reason of a gas attack, plus he had no desire to dispatch Mr. Stripes. He didn't know the skunk was rabid, and it ended up biting his nine-year old son. Quite an unlikely event, but it happened. A camp gun can come in handy.

Why shoot a blackpowder sidearm? Targets are one reason. This shooter plied his single shot pistol against a paper target at a rendezvous contest.

Why shoot a blackpowder sidearm? Another reason is for the fun of it.

Self-Defense

On the battlefield, aboard sailing vessels, as well as in saloons and streets all over 19th-century America, the blackpowder sidearm has accounted grimly for itself upon mankind. However, in spite of all this, I cannot recommend it for self- or family defense. It's outdated for the purpose, in my opinion.

Other Adventures

Completing The Period Outfit

Civil War reenactments demand proper outfits, including firearms, which means handguns as well as long arms. Other period dress, such as 19th-century fur-trapper regalia, also requires a proper handgun to complete the outfit. Study of the period reveals just what these guns are.

Rendezvous

Another worthwhile and important use of the sidearm is at rendezvous. These are held all over America. Some buckskinners, as rendezvous fans are called, don't get too excited about replicating period dress, but others do. Either way, blackpowder sidearms are important at rendezvous, if not for filling out the proper dress, then for the special shooting events.

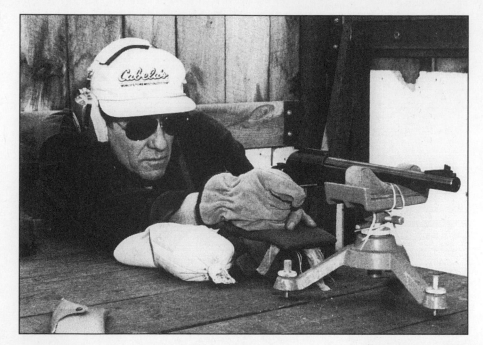

Why shoot a blackpowder sidearm? Hunting is another reason. This big T/C Scout pistol, when sighted-in properly, is entirely viable as a big game gun with the right load.

Collecting

One of the best things about collecting blackpowder pistols and revolvers is their size. You don't need a showroom to display your holdings. Furthermore, when not attended, the gun collection can be locked away in a modest-sized safe, whereas a rifle collection requires a lot more space. Collecting blackpowder sidearms can be as non-specific or as specific as you please. "Here's a collection of antique blackpowder pistols and revolvers," you might say, and the mix will interest anyone who likes firearms. Or, you can select a time period, such as the Civil War, and collect only handguns from that time block. Replicas are a good inexpensive way to collect outdated firearms, but, of course, the real thing is worth a lot more than the copycat.

For example, the Navy Arms replica of the Paterson Colt not only demands far less cash than the real thing, but try to find an original in the first place. Being the first production revolver made by Colt in 1839, you'd have to part with about $8000 to own one in Good condition, and around $18,000 for one in Excellent shape. None of the original blackpowder sidearms of important historical note is cheap. If you think the Paterson is expensive, try the Walker model. How about $75,000 for one in top shape? If you find one in Poor condition, expect to part with $20,000 for it. A Colt Model 1862 Police revolver could set you back $3000 or $4000, or close to it, for one in Very Good condition.

Custom Pistols

Custom pistols are within the ability of quite a few modern-day blackpowder gunmakers. Kennedy Firearms in Muncy, Pennsylvania, has specialized in hand-making good-looking pistols for many years now. Their American Revolutionary War Officer's Pistol, for example, is a flintlock with 8-inch octagon-to-round barrel, swamped, with wedding bands. It measures .890-inch to .750-inch to .800-inch diameter, and is in calibers 45, 50, or 54, smoothbore or rifled. It's an American adaptation of the English Georgian-style pistol built circa 1750-1780, and it's a beauty. The same company offers an English Dueler. I tested it for the sake of knowledge, using a 30 volume FFFg charge of powder behind a .490-inch 177-grain round ball for 776 fps MV. It was a lot of fun to shoot.

The Blackpowder Cartridge Revolver

How strange to think of the blackpowder cartridge revolver as a new wave, yet it seems to be shaping up as just that. Many workable replicas exist, with more planned. For example, Navy Arms is working on an "1875 Remington Style Revolver," and an "1890 Remington

A blackpowder pistol can be very powerful. This Thompson/Center Scout is capable of harvesting big game within its range limitation. It's made to shoot a heavy conical.

Style Revolver." Unless plans change, both will be available in 44-40 Winchester or 45 Colt, erroneously dubbed the "45 Long Colt," a cartridge that never was. The same company also has 1873 Colt-style revolvers, both blued and nickel finish, as well as an 1873 U.S. Cavalry Colt, and an 1895 U.S. Artillery Colt, both of these also offered in 44-40 or 45 Colt.

Other companies are planning comebacks of copycat blackpowder cartridge revolvers, and Colt has already been at the game for some time, offering their famous six-guns to the fan who wants to own a cartridge sidearm of the 19th century without having to pay collector's prices to own one. Navy Arms also has its Schofield revolver in two styles, the 1875 Cavalry model and an 1875 Wells Fargo model. The Cavalry model has a 7-inch barrel, while the Wells Fargo sports a 5-inch barrel. Both come in either 44-40 or 45 Colt chamberings.

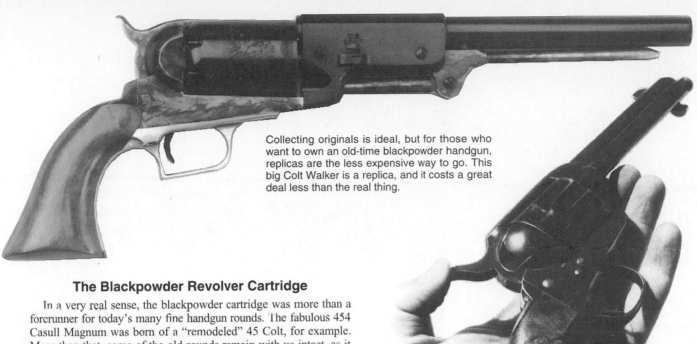

Collecting originals is ideal, but for those who want to own an old-time blackpowder handgun, replicas are the less expensive way to go. This big Colt Walker is a replica, and it costs a great deal less than the real thing.

The Blackpowder Revolver Cartridge

In a very real sense, the blackpowder cartridge was more than a forerunner for today's many fine handgun rounds. The fabulous 454 Casull Magnum was born of a "remodeled" 45 Colt, for example. More than that, some of the old rounds remain with us intact, as it were, with no change whatever, except modern loading in a current factory case with smokeless powder and a jacketed bullet. There are several such cartridges around. The four listed below are a few.

32-20 Winchester

This round is still alive. It doubled as a revolver as well as rifle cartridge. It never had a whole lot of power, and that fact in a very real way is its claim to fame. The little 32-20 is downright pleasant to shoot, and yet it will certainly put small game in the pot without a hitch, including the wily wild turkey. The 32-20 Winchester was introduced in 1882 for chambering in the company's Model 1873 lever-action rifle, and in the revolver before long. Colt's six-guns chambered for the 32-20 were quite popular. Bullets in the 100- to 115-grain category at below the speed of sound did quite a bit of work without undue destruction of tender edibles. Some shooters also relied on the little 32-20 for defense.

This original Colt cartridge six-gun is as collectible as guns get. Single-action blackpowder cartridge revolvers like this one, albeit mainly in replica form, are now hugely popular in the cowboy action shooting contests held all over North America.

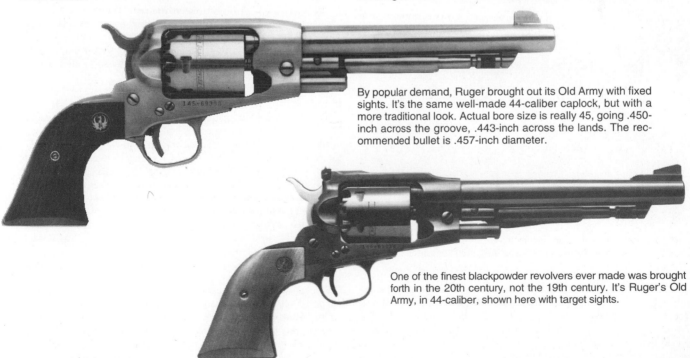

By popular demand, Ruger brought out its Old Army with fixed sights. It's the same well-made 44-caliber caplock, but with a more traditional look. Actual bore size is really 45, going .450-inch across the groove, .443-inch across the lands. The recommended bullet is .457-inch diameter.

One of the finest blackpowder revolvers ever made was brought forth in the 20th century, not the 19th century. It's Ruger's Old Army, in 44-caliber, shown here with target sights.

The Traditions Buckhunter Pro in-line pistol is a modern muzzleloader all the way, with adjustable trigger and thumb safety, plus fold-down rear sight. The 10-inch barrel is drilled and tapped for a scope mount. In 50- or 54-caliber, the Buckhunter comes in standard blue and an All-Weather model with C-Nickel frame.

The blackpowder handgun seems to know no bounds. This belt-buckle revolver from Freedom Arms is unique in style, and it shoots!

38-40 Winchester

The 38-40 Winchester is interesting because it is so close to the 44-40, right down to actual caliber size. It came along in 1874 and was based on the 44-40 case necked down. True caliber was .401, so the 38-40 could have been called the 40-40 Winchester. It fired lead bullets in the 180- to 200-grain class at modest velocities from the handgun, in the area of 900 fps. Nonetheless, it was well-liked, and some hunters stayed with the 38-40, even with blackpowder, well into the modern era. Of course, it fired enough bullet to serve as an effective self-defense round.

44-40 Winchester

The 44-40 Winchester also remains loaded to this day as another blackpowder cartridge that won't go away. The squat little cartridge has long been a favorite in the handgun and rifle, and it fires a 200-grain bullet at about the same velocity earned by a 180-grain bullet in the 38-40, or thereabouts. The 44-40 was the original chambering for the famous Model 1873 Winchester rifle, and it didn't take long to find its way into a sidearm. Actually, it will handle bullets of 250 grains, but the 200-grain is far more popular. As a sidearm to go along with a blackpowder cartridge rifle, either single-shot or lever-action repeater, the 44-40 is a good one. It's historically antiquated, yet not truly outdated.

45 Colt

The 45 ACP (Automatic Colt Pistol) was developed by John Browning in 1905 and it became the pistol round of the United States Army. The 45 Colt was developed way back in 1873 by Colt and did not go by the name of 45 "Long" Colt because there wasn't a 45 "short" Colt (45 ACP). Notes show original loads as a 255-grain bullet in front of 40 grains FFg blackpowder. I decided to test the 45 Colt in an original Colt Single Action Army revolver. I could not get 40 grains volume FFg into the case, but I did manage to stuff in 37 grains volume, which in this particular instance turned out to weigh 38.3 grains per charge. The result, with a 255-grain lead bullet was 773 fps MV for a muzzle energy of only 338 foot pounds. As always, the same reminder is given: a 255-grain 45-caliber bullet, even at low speed, is formidable.

Blackpowder handgun cartridges are loaded with modern smokeless powder in some firearms. Beware that *old blackpowder handguns can be blown up with smokeless powder.* Heed all information given. For example, a note I read recently provided with Accurate Arms' loading data book, current issue as this is written, page 317, stated that, "The SAAMI maximum average pressure for the .44-40 is 13,000

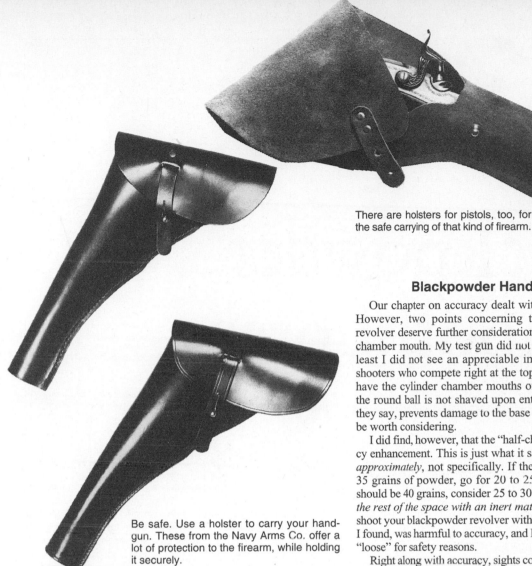

There are holsters for pistols, too, for the safe carrying of that kind of firearm.

Be safe. Use a holster to carry your handgun. These from the Navy Arms Co. offer a lot of protection to the firearm, while holding it securely.

C.U.P." Accurate Arms went on to say that their smokeless powder loads were maximum, running from about 11,000 to 13,00 CUP. Within this safety parameter, a 200-grain jacketed Nosler bullet earned 1008 fps MV with a pressure rating of 13,000 CUP. However, enough cannot be said about playing it safe with the old handguns. Stick with every aspect of the prescribed load, not only powder and powder charge, but exact bullet, primer, and case.

FFFg blackpowder and Pyrodex P are well-suited to the small-bore pistol and the blackpowder revolver. FFg works fine in many big-bore pistols, and can prove quite accurate in these. However, even in some of the big bores, FFFg is ideal, and suggested by the manufacturer. For example, the 50-caliber Knight Hawkeye pistol was very accurate with FFFg blackpowder.

The Modern Muzzleloading Pistol

Consider this trend new, too, for more are to come. Currently, I have tested only two modern muzzle-loading pistols. I have to call the Thompson/Center Scout Pistol modern because it's in-line ignition and up-to-date lines demand that. Data for this pistol has been given. The other modern muzzle-loading pistol I tested is the Knight Hawkeye, caliber 50. This is an in-line igniter, too, with plunger. It has adjustable sights and a good trigger. Barrel is 12 inches long, has a 1:20 rate of twist, and is engineered to shoot the pistol bullet/sabot combination. I got good accuracy with both the 240-grain 44-caliber jacketed revolver bullet, and a 260-grain lead pistol bullet with 50 volume FFFg blackpowder. Muzzle velocity for the 240-grain bullet was 1066 fps, 1051 fps for the 260-grain bullet. Both deliver well over 600 foot pounds of muzzle energy.

Blackpowder Handgun Accuracy

Our chapter on accuracy dealt with muzzloader bullet grouping. However, two points concerning the blackpowder cap 'n' ball revolver deserve further consideration. The first is chamfering of the chamber mouth. My test gun did not respond to this treatment, or at least I did not see an appreciable improvement. However, certain shooters who compete right at the top of cap 'n' ball competition do have the cylinder chamber mouths of their firearms beveled so that the round ball is not shaved upon entry into the chamber. This also, they say, prevents damage to the base of a conical, so the process may be worth considering.

I did find, however, that the "half-charge" was conducive to accuracy enhancement. This is just what it says, a half charge of maximum, *approximately*, not specifically. If the chamber of the revolver holds 35 grains of powder, go for 20 to 25 grains of powder; if capacity should be 40 grains, consider 25 to 30 grains of powder. But *do fill up the rest of the space with an inert material, such as cornmeal*. Do not shoot your blackpowder revolver with air space in the chambers. This, I found, was harmful to accuracy, and I do not like blackpowder loaded "loose" for safety reasons.

Right along with accuracy, sights count for a lot. A gun will shoot no better than its sights, so if the sights are crude, expect low level grouping. Many blackpowder sidearms come fitted with superb sights. My T/C Patriot, for example, holds a very pretty sight picture, as does my Ruger Old Army. As for sighting-in that was discussed in Chapter 30.

Triggers, too, have to be good for accurate shooting. That's a given, well accepted and understood. The factor plays no less a role in the blackpowder sidearm than any other gun. If trigger pull is too heavy for consistent let-off, or if it's cursed with creep, a competent blackpowder gunsmith can often help the situation. This is a job for a professional. Messing with a trigger can bring the disaster of a gun going off at the wrong time.

Cleaning the blackpowder pistol is not a problem. I'd be less than honest if I said the cap 'n' ball revolver was equally simple to care for. Fortunately, however, neither is that difficult to maintain. The revolver is so much fun to shoot that it's worth the little bit of extra effort to keep it "running smoothly." See Chapter 38 for cleaning details.

Sure, the sash gun was carried about the waist in a sash or belt, and mountain men were known to stuff a pistol into a similar place. However, super care must be taken to transport a blackpowder pistol or revolver safely. Holsters were invented for a reason, so use them for your revolver, and if you must tote a pistol in a belt, do so carefully.

For some shooters, the blackpowder handgun will be the most important firearm in the sootburning battery. These people like handguns! It's that simple. For the rest of us, though the pistol or revolver may not be number one in the lineup, it's obviously a very important part of blackpowder shooting. Nowadays, the choice of gun is broad, from old-time pistol to modern muzzleloader. Finding the right sidearm is only a matter of study and selection based on need and desire.

●

chapter 37

Blackpowder Smoothbores

THE WORLD OF blackpowder shooting was dominated by smooth-bores for a very long time. Even after rifled arms were widely available, some shooters preferred smoothbores. Nineteenth-century ivory hunter William Cotton Oswell was one who chose them for certain applications. He preferred his smoothbore for elephant hunting, for example, because it was faster and easier to reload with no rifling to capture fouling, plus it was easier to clean, especially in a camp setting. Oswell, who died in 1893, hunted tuskers as a young man. S.W. Baker said Oswell was absolutely the first white man to show up in certain parts of South Africa. And when Oswell did show up, he was carrying his favorite "rifle," only it was a smoothbore made by Purdey.

In Africa

It was a 10-gauge weighing 10 pounds and charged with "six drachms of fine powder" according to Baker, who borrowed the piece from Oswell for an African hunt. Incidentally, the term *drachm* was synonymous with dram. A drachm or dram equals 27.34 grains weight, so the load was about 164 grains of powder. Baker wrote that he enjoyed great success with Oswell's smoothbore. He said, "There could not have been a better form of muzzle-loader than this No. 10 double-barrel smoothbore. It was very accurate at 50 yards. . ." (*Big Game Shooting*, 1902). Of course, "very accurate" must be qualified. Baker was hunting very large animals at close range.

Oswell preferred pre-patched round balls for his 10-bore, wrapping them in either "waxed kid" (leather) or linen. The "object of the smooth-bore was easy loading," said Baker. The pre-patched ball was rolled tightly in cloth or leather with the excess trimmed close with "metal scissors" so that the wrapping became a part of the projectile. The powder charge was also pre-measured and carried in a paper cylinder, "the end of which could be bitten off," Baker noted. The whole package of powder—paper and all—was thrust downbore after the end was nipped off, followed by the pre-patched ball rammed home with a

Blackpowder is inefficient compared with smokeless, but with a big bore like this one, large bullets can be fired at reasonable velocity for good power. This smoothbore can handle round balls of about 500 grains weight, and that's a lot of bullet.

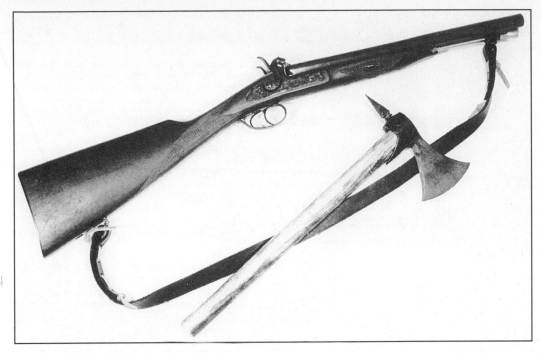

No one is trying to convince anyone that the smoothbore is better than a rifled arm, but smoothbores do have their place. This short-barreled percussion shotgun loaded with buck 'n' ball can do a great deal of work, up to and including the harvesting of a deer as well as a bird for the dinner table.

"powerful loading rod." In the name of honest reporting, I must add that Baker cursed smoothbores in print later on. He had admired his mentor's smoothbore to begin with, but found smoothbores lacking for long-range shooting compared with a precision-made rifled longarm, which is all Baker carried for his own big game hunting in Africa and Ceylon. In his book, *The Rifle and the Hound in Ceylon*, S.W. Baker verbally cuts the smoothbore to ribbons with an abrupt tongue-lashing. "Smooth bores I count for nothing, although I have frequently used them," said he.

Another well-known hunter of his era, J.H. Walsh, known as "Stonehenge" to his friends, also downplayed smoothbores for hunting. In his book, *Modern Sportsman's Gun & Rifle* (reprinted by Wolfe Publishing Co.), Walsh warned: "If, however, the six inch circle at 50 yards could be depended on, I should be ready to admit that for large game it [a smoothbore] is a most useful weapon; and with this view I have repeatedly tested smooth-bores by various makers, but the trial has invariably ended in disappointment. Sometimes the first or second, but oftener further on in a short trial, a wild shot occurred, and of course this wild shot may be the one to cost a sportsman his life, when charged by any kind of large game." (From p.7 of reprint.)

On the other hand, many military men of the past applauded the smoothbore. Our own General Washington often replaced rifled arms with muskets, believing the smoothbore a better tool for battle. The smoothbore musket was easier to keep in repair, simpler and faster to reload for rapid fire, and it carried a fixed bayonet better than a rifle (Washington believed). The Brown Bess smoothbore musket remained Britain's first choice of arms, too, for a very long time, firing a .753-inch ball (11-bore) with 70 grains of powder. General George Hanger, said to be the best shot in the British Army (he served with Hessian Jaegers during the Revolution), reported that "a soldier's musket, if not exceedingly ill-bored (as many of them are), will strike the figure of a man at eighty yards; it may even at 100. . .," but he concluded that "firing at a man at 200 yards with a common musket, you may just as well fire at the moon." (*American Rifleman*, August, 1947) Notice that Hanger thought smoothbores were more manageable on the battlefield than rifles, but he did not consider them accurate.

Certain American hunters preferred the smoothbore well into the era of rifled long arms. Many men traveling with Lewis and Clark into the Far West carried smoothbores. After all, that's what their "fusils" truly were: simply smoothbore arms, the word borrowed from the French, meaning either steel or tinderbox. Fusils, as shoulder-held firearms, were mentioned in print as far back as 1515 in French hunt-

This old smoothbore, with sights, had its good points. It was easy to clean (no grooves, because there's no rifling) and up close it could get the job done on anything from quail to moose.

ing ordinances, but still in use during the 19th century in one form or another. Sometimes, fusils were also noted as a "trade rifle quality" arm, meaning a fairly cheap gun used for bartering. Speaking without first-hand authority, I must conclude that the fusil was just another smoothbore, and certainly nothing special.

First Research

But I did become intrigued with these guns for two reasons. First, the historical pull was magnetic. If these muskets were so ill-firing, so worthless and inaccurate, why did they hang on for so long, even after rifled arms were widely available? Second, in some locales today, smoothbores must be used for certain blackpowder-only primitive hunts. Do hunters packing smoothbores have a prayer of cleanly dropping a deer, even at woods ranges of fifty or sixty yards with a hunting tool so maligned? After all, even the longbow outshot the common musket back in 1792 in a match on Pacton Green, Cumberland. The range was "over 100 yards" and the bowman placed sixteen arrows out

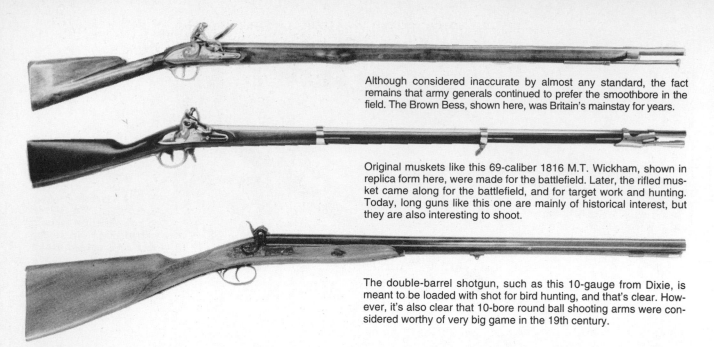

Although considered inaccurate by almost any standard, the fact remains that army generals continued to prefer the smoothbore in the field. The Brown Bess, shown here, was Britain's mainstay for years.

Original muskets like this 69-caliber 1816 M.T. Wickham, shown in replica form here, were made for the battlefield. Later, the rifled musket came along for the battlefield, and for target work and hunting. Today, long guns like this one are mainly of historical interest, but they are also interesting to shoot.

The double-barrel shotgun, such as this 10-gauge from Dixie, is meant to be loaded with shot for bird hunting, and that's clear. However, it's also clear that 10-bore round ball shooting arms were considered worthy of very big game in the 19th century.

of twenty on target (size not given). Meanwhile, the best musketeer only hit the target twelve for twenty tries. As Karl Foster (of rifled slug fame) said in the *American Rifleman,* "Round balls in smooth barrels have lacked accuracy since guns were first made." (October 1936 issue, p.23.)

In Scotland, in 1803, soldiers practiced to meet Napoleon by firing their muskets. However, they were content when ". . . every fifth or sixth shot is made to take place in a target of three feet diameter at the distance of 100 yards." (*American Rifleman,* August 1947, p. 8.) When I presented this quote to one of my "buckskinner" friends who had laid out a fat wad of greenbacks for a custom rifle with a smooth bore, he replied, "Must have been damn poor shots." And he went on to tell me that if ever I was in need of venison steaks for supper, just let him get within seventy-five yards of a buck and "We'll be in meat with my smoothbore." Then he added, "Of course, I stalk for close shots. You do remember stalking don't you? That's where you get close before you shoot," he said sarcastically.

I decided to do some research, to discover what other modern marksmen thought about smoothbores, and I also made up my mind that I would conduct a test of my own. The first part of my smoothbore study uncovered a piece by Harry Root Merklee, a well-known authority on blackpowder arms. His article resided in *Muzzle Blasts* magazine for April of 1961. Here is what Harry

learned about accuracy with military smoothbore flintlock muskets:

> Five men of military age assembled at a local range, each armed with a cal. 69 smooth bore flintlock musket. These were rifles of the Napoleanic wars and were in first class condition. Except for minor details of construction these muskets were the same as those used during most of the flintlock period which includes the American Revolution.
>
> The loads for these muskets varied according to their owner's preference but all used the same caliber round ball which would slide down the bore of its own weight; 'fall down' would be a better description, a rattling fit at any rate. No patch was used of any kind. Powder charges ranged from $3^{1}/_{2}$ to 5 drams of FFG powder. Regular shotgun wads of felt $^{3}/_{8}$-inch thick were used over both powder and ball.

That was the shootout. Note that no patch was used. I doubt that a patch would have made much difference with these muskets in this particular instance. A wad held the ball in place for safety, and while a patch can transfer the impetus of the rifling to the projectile, remember there was no rifling in these smoothbores. Here is what happened: Shooting from a sitting position, the marksmen kept most bullets within a 16-inch circle at 50 yards. Sights on these muskets were too crude to ask for much more in the first place, and trigger pull was referred to

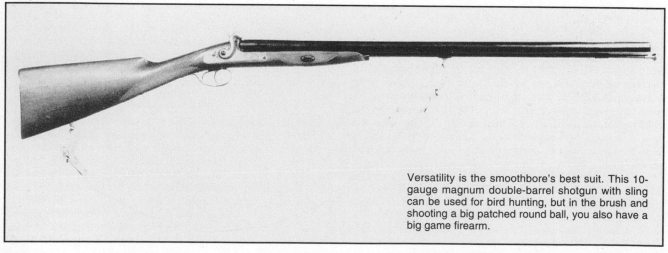

Versatility is the smoothbore's best suit. This 10-gauge magnum double-barrel shotgun with sling can be used for bird hunting, but in the brush and shooting a big patched round ball, you also have a big game firearm.

Hunt with a smoothbore? You can if you want to, or if the rules of a specific hunt demand smooth-bores only. The secret to success is getting close, which is entirely possible with the honing of hunting skills and careful stalking.

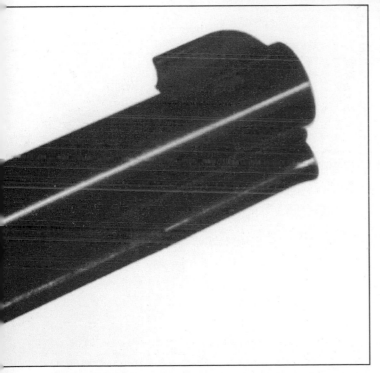

There is no way for a smoothbore to compete with a rifled arm, but put decent sights on a smoothbore and with good loads, harvesting big game at close range is no problem. Thompson/Center found this out with its own 56-caliber smoothbore.

as "horrible." Powder charges ranged from 3.5 drams to 5 drams, and recoil proved bothersome with the latter from the sitting position, mainly due to the poorly-designed stock. The 69-caliber round balls would have weighed about 494 grains. In grains weight, powder charges ran 96 to 137 grains. A 16-inch group at 50 yards is a poor showing, but all shooting was done sitting, not from a bench. What could be accomplished under slightly more favorable circumstances? I wanted to find out.

Early Trials

My first attempt came with a double-barreled shotgun shooting patched round balls. I was caressed by success and slapped by failure simultaneously—success in power, failure in accuracy. The shotgun was a 28-inch-barreled 12-gauge side-by-side caplock. Eighty grains of GOEX FFg provided a MV of 927 fps. That load was mild, even behind a .695-inch patched ball, which averaged 502 grains weight, so I progressed to a flat 100 grains of the same powder for a MV of 1190 fps. I did this because the heavier charge developed 1579 foot pounds, which would make a pretty strong showing on deer-sized game in the thicket. I was interested in close-range shooting, which would be the normal application for such a gun and load. A target 1-foot across was set up 40 paces away. The 12-gauge round ball had been tested on my bullet box and it penetrated a couple feet of media at 50 yards, bettering the "wound channel" I'd gotten from a couple 30-06 loads. Performance was fine, but accuracy was not. I failed to keep all bullets inside the 12-inch bull, even though I had a solid rest. Of course, sights were not conducive to good bullet placement, but I still expected closer clusters from a distance of only 40 yards.

While the gun only wore double shotgun beads for sights, these could be aligned for some semblance of accurate aiming. Nonetheless, no matter how carefully I aimed this double-barrel smoothbore, grouping was poor. The left barrel had a penchant for dropping its projectile into the black at 40 paces with some regularity, while the right often sailed its bullet right off target, missing everything, including the target frame. I tried the old trick of filing the muzzles to regulate the barrels (make them shoot to the same point of impact). Cutting the inside edge of the right-hand muzzle brought the ball over a little in point of impact, but nothing close to true bore regulation resulted. I gave up after the muzzle looked like a kid with a hacksaw had attacked it.

Perhaps, I said to myself, we must expect inaccuracy from an unrifled firearm. But in a way, shooting round balls, as I was, should allow at least something resembling deer-hunting accuracy at only forty or fifty yards. As noted before, a good round ball should fly relatively well, even when not spinning on its axis. The round balls I used were not lopsided. My test for this was crude, but not entirely without merit. I rolled and spun them on a flat surface, and the high-quality lead pills did not wobble off track. So I still did not know why the lead sphere was so inaccurate from a smoothbore, because I'd managed many 1-inch center-to-center, five-shot groups with patched round

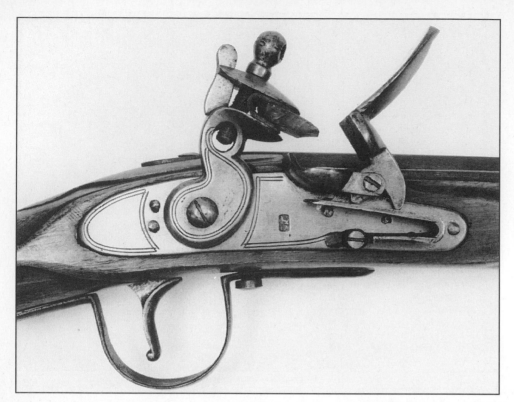

This is an original flintlock smoothbore trade gun of English manufacture, circa the mid-1700s to the early 1800s. This one is a 60-caliber with a 42-inch barrel, 57$\frac{1}{2}$-inch overall length.

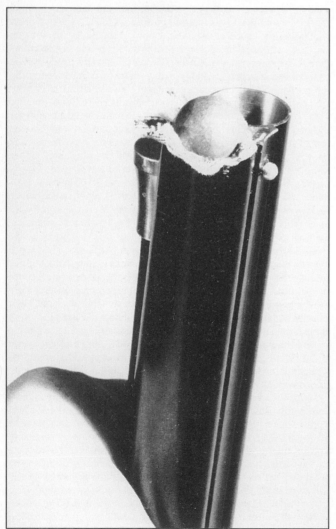

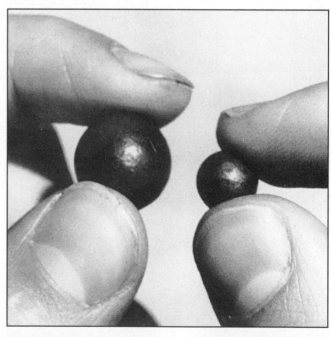

The strength of the smoothbore lies in its caliber. As we know, small round balls don't carry a great deal of energy, but a big round ball at even modest velocity is formidable.

Loading the smoothbore is rather simple. A patched round ball is the way to go.

Some wild animals are easier to get close to than others. A smoothbore is all that is needed for javelina, for example. They can be stalked to very close range if the hunter keeps quiet and the wind in his favor.

balls at 100 yards from scoped rifled long guns. I knew that the ball's center of mass rotated on its axis the same as a conical, and that imperfections in the ball itself would cause the projectile to leave the bore at a different angle of departure each shot. Because matter in motion moves in a straight line, the "heavy" part of the projectile would determine the initial line of flight of the round ball.

An imperfect ball would tend to travel on a tangent from the line of the axis. In other words, static imbalance would ruin accuracy. But the round balls I was testing were not that bad, and I began to wonder just what sort of static balance (the actual precision of the projectile in terms of mass distribution) was present in those old missiles of the past, where the boys felt lucky to infrequently hit a 3-foot target at 100 paces. I knew better accuracy potential resided in the smoothbore firing the good lead spheres I had to work with. More testing was in order.

I studied more background data first to see if a lead ball should do better. I learned that a sphere would be less sensitive to rotational stabilization than a conical. W.W. Greener said, "Rifling, therefore, is of greater importance when a conical or elongated projectile is used than when the bullet is spherical," (from *The Gun*). The principle of rotating an elongated missile for stabilization was a phenomenon tested hundreds of years ago. There are even relics of crossbow bolts which had been grooved to create a spiral motion. The big ball had mass going for it, too. The greater the mass, the greater the inertia. The heavier the projectile, the less rotation on its axis necessary to stabilize it, and for big game hunting with the smoothbore muzzleloader, missiles of *at least* $1/2$-inch diameter prevail. The .690-inch ball for the 12-gauge, for example, weighs 454 grains. Thompson/Center's 56-caliber ball is 252 grains heavy (on my scale).

Smaller Balls

So a smaller ball would gain more advantage from rifling than a larger ball, while the large round ball would be more inherently stable, if made right, as my test spheres were. More research brought me to Ezekiel Baker, the well-known court ballistician who wrote a gunnery treatise for His Majesty George IV. He said in his 11th edition of the work, "The Honorable Board of Ordnance being anxious to ascertain if rifling a large piece would have the same advantage over smooth barrels which rifles possess over muskets, and would be equally effective in carrying the ball, the experiment was tried at Woolrich [on May 15,

1806] with two wall-piece barrels of equal dimensions, one rifled, the other not rifled."

The barrels were 4 feet, 6 inches long, each weighing 20 pounds. The projectiles were 5-gauge round balls. The advantage of the rifled piece was not nearly as pronounced as it had been with smaller round balls of 20-gauge size. What was not tried, however, was very careful sorting of round balls in the 20-gauge and smaller smoothbore firearms. It was long known that balls "created by pressure," swaged, in other words, were the more uniform; however, I found no test in which the smoothbore was fired with very carefully weighed (sorted) round balls. In other words, would *balanced* spherical missiles make a difference in the smoothbore? Static stability would be improved, for certain, which should improve dynamic stability.

Years later, Dr. F.W. Mann concluded, after thousands of experiments, that precision of projectile was the most important single aspect of accuracy. The sphere, if perfect, would theoretically fly true, would it not? Even from a smoothbore. Round ball perfection wasn't possible. But precision was. Rifling vastly improved round ball accuracy because it "averaged" the imperfections in the ball on a common axis, such as a ball heavier on one side than the other. Rifling twist equalizes lopsidedness on the axis through rotation. So in my later tests, I presorted round ball projectiles, which improved accuracy in the shotgun mentioned above. But I still wasn't happy.

When I turned to the T/C factory 56-caliber ball, sorting was not necessary. The greatest variation in random sampling of ten balls was only .9-grain. The heaviest in the string was 252.1 grains, the lightest 251.2. The micrometer gave an average diameter of .552-inch. If the T/C ball was pure lead, it would weigh 253 grains. Weighing proved that the T/C ball was precise, and it was not an alloy. While homogenity was not proved by weighing, uniformity was. In short, these T/C round balls were darn good. Now I wanted to see what they would do from a well-made smoothbore like the Thompson/Center Renegade in 56-caliber with adjustable rear sight, which was created to give the blackpowder hunter a reliable firearm where law or desire called for a non-rifled bore. Sights at last! I had been spinning my wheels, even with a good benchrest, trying to remove extraneous variables while aiming with shotgun beads for sights.

First, the T/C rifle proved totally reliable, with 100 percent ignition using CCI No. 11 caps. The test run included the firing of eighty .550-

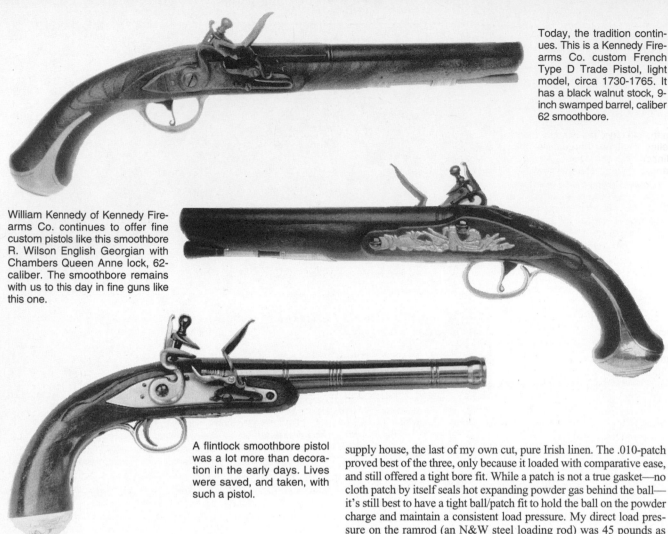

Today, the tradition continues. This is a Kennedy Firearms Co. custom French Type D Trade Pistol, light model, circa 1730-1765. It has a black walnut stock, 9-inch swamped barrel, caliber 62 smoothbore.

William Kennedy of Kennedy Firearms Co. continues to offer fine custom pistols like this smoothbore R. Wilson English Georgian with Chambers Queen Anne lock, 62-caliber. The smoothbore remains with us to this day in fine guns like this one.

A flintlock smoothbore pistol was a lot more than decoration in the early days. Lives were saved, and taken, with such a pistol.

inch T/C cast "265-grain" projectiles. Actually, they went closer to .551-.552 average diameter, weighing about 252 grains.

Encouraging Tests

Loads were selected from the T/C manual, *Shooting Thompson/Center Black Powder Guns*. Three were given, all using T/C patch material, a No. 11 cap, and Maxi-Lube. My tests included three patch types and three lubes. The shooting patches were .005-inch, .010-inch and .013-inch, the first two from Gunther Stifter's German supply house, the last of my own cut, pure Irish linen. The .010-patch proved best of the three, only because it loaded with comparative ease, and still offered a tight bore fit. While a patch is not a true gasket—no cloth patch by itself seals hot expanding powder gas behind the ball—it's still best to have a tight ball/patch fit to hold the ball on the powder charge and maintain a consistent load pressure. My direct load pressure on the ramrod (an N&W steel loading rod) was 45 pounds as maintained by a special tool a reader of mine built for me, which releases when 45 pounds pressure is reached.

The lubes were grease, cream and liquid, RIG, Young Country Lube 103 and Falkenberry Juice. All three worked equally in terms of accuracy. Initially, shooting from the 50-yard bench, the balls struck the black with a 6-inch center-to-center group. My early 100-yard tests were folly. Test-shooting with open iron sights, coupled with initially large groups, did not warrant 100-yard shooting, but that would change. Boring the reader with details of continued failures is pointless, so I'll get to the method which gave best accuracy.

Summary of Accuracy Steps

1. Sort round balls by weight and discard those that do not fall within the norm for the batch. Round balls are capable of excellent uniformity in weight, whether commercial or home-cast.

2. Ensure a consistent powder charge by following a set procedure when using the powder measure. The smoothbore needs every break it can get in order to achieve hunting accuracy, and charge-for-charge consistency always promotes accuracy.

3. Use a buffer such as hornet nesting material between the powder charge and the patched ball. This little step provides assurance of good patch condition, which in turn may promote accuracy.

4. Use good patches that take up the windage in the bore. Although the patch cannot translate the rotational value imparted by rifling because there is no rifling in the smoothbore, the patch is part of the load chain, and it can hold the round ball firmly on the powder charge with good pressure.

5. Be sure to wipe excess lube from the bore following loading. This step ensures the same bore condition from shot to shot, and such consistency never harms accuracy.

6. Review accuracy aspects as outlined in Chapter 23 just to be certain that you've done all you can do to upgrade smoothbore accuracy.

7. Choose large calibers. Large round balls have more accuracy potential than small ones because they have more mass and tend to stay "on line" better.

8. If possible, have sights fitted to a smoothbore intended for big game hunting. No firearm can be expected to shoot well without proper sights.

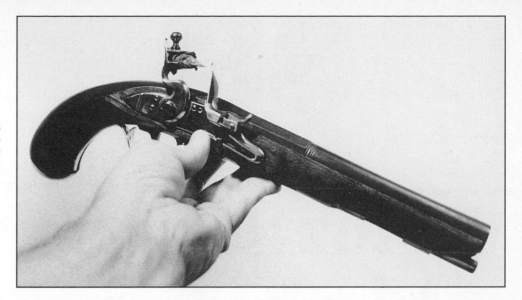

In the old days, many adventurers relied on smoothbores like this flintlock for protection. At close range, the big pistol was deadly.

No accuracy difference was found among the three allowed powder charges of 80, 90 and 100 grains of FFg, so the latter was used. It only developed about 6000 LUP with comparatively mild recoil, and muzzle velocity averaged 1366 fps. Thompson/Center tests averaged 1300; however, powder lots can vary, and I was shooting at an altitude of 6000 feet, with a temperature of 85 degrees F. The muzzle energy of this load was a bit over a half-ton at 1044 foot pounds, with a 252-grain projectile. At 50 yards, my chronograph showed a retained velocity of 1101 fps for the 56-caliber ball, with a 50-yard energy of 678 foot pounds. This doesn't sound like much, but coupled with an *entrance* hole over half an inch across it is certainly ample for deer.

Loading Consistency

The ball-shooting smoothbore proved amply accurate for deer hunting in woods and timber. Previous shotgun clusters shrank to consistent 100-yard 8-inch groups. That's with iron sights. A test-mounted rifle scope might reveal further accuracy potential. At 50 yards, 3-inch center-to-center groups were common. But good groups at any range were possible only after using careful regimen in the loading process. First, as above, I made sure that the missile was precise. T/C's commercial cast ball proved to be well made, so I stayed with it. Second, consistency of powder charge was maintained by over-filling the measure, tapping the barrel of the measure ten times, then swiping off excess kernels of powder by swinging the funnel section of the measure into line with the barrel.

Next, I used a hornet nesting material buffer between the patched ball and charge. As you recall, hornet nesting material does not catch fire inside the bore, thus saving the patch from burnout. Moreover, a buffer between patch and charge serves to absorb excess lube that might attack the powder charge. The fourth step in the accuracy process was wiping the bore free of excess lube after the load was seated in the breech of the rifle. I fired several groups with the bore untouched (damp) after seating the ball and several with the bore wiped with a cleaning patch after seating the ball. The latter were always better in my particular test rifle, plus point of impact remained constant with the dry bore method of loading. Final sight-in was accomplished with a lube-free bore.

The Thompson/Center New Englander, a single-barrel 12-gauge shotgun with auxiliary rifled 50-caliber barrel, tested for the Second Edition, was retested for this book. The smoothbore barrel was loaded with one .690-inch round patched ball using the sequence mentioned above with hornet nest buffer. The retest centered on Pyrodex RS powder with a charge of 100 volume (70.5 grains weight for the particular lot of powder). Accuracy was more than acceptable, especially considering the shotgun bead as an aiming device instead of true sights. Groups were better than before. Instead of 8-inch clusters with the big 69-caliber ball, 6-inch center-to-center groups were made at 50 yards. Carefully loaded, the New Englander could be counted on to strike the chest area of a deer at close range. And, after all, that's what we're looking for here, not real target precision.

Summary

The smoothbore long arm is capable of hunting accuracy, although these guns are not nearly as accurate as rifled arms. Where the law requires, or a blackpowder hunter, for his own reasons, chooses a smoothbore, effective close-range ball placement is totally possible on deer-sized game and larger.

●

chapter 38

Muzzleloader Maintenance

TRUE NON-CORROSIVE SHOOTING means firing the gun and putting it away without concern for its welfare. This is how modern-day non-corrosive ammunition works, as found in our 22 rimfire fodder, as well as in big bore ammo. Thus far, we have no muzzleloader propellant that promises the same non-corrosiveness enjoyed with modern ammo. Each propellant states that after-shooting cleanup is necessary. And even with non-corrosive ammunition, the fellow who never cleans his guns will pay the price, if not in eventual downright ruination, then in accuracy loss. I have doctored many sick centerfire rifles that "quit shooting accurately," according to their owners. "What did you do?" they ask when I return the rifle to its former accuracy. I tell them I cleaned it, really cleaned it, including the removal of copper fouling from the rifling. Is it any wonder that our blackpowder guns

shooting corrosive powder must be maintained through careful cleaning, when you have to clean guns that shoot non-corrosive primers and powders? Of course not.

The good news is that cleaning is easier and faster than ever, due to the modern chemicals and products outlined in Chapter 19. Keeping frontloaders in top shape is just not that much trouble anymore. So take heart. You can enjoy blackpowder shooting without too much after-shooting fuss.

If you think firearm maintenance will be a big stumbling block for you, a chore you really aren't going to like, even though you are highly interested in muzzleloaders, then by all means select blackpowder firearms that tend to clean up easily. There are a number of these guns. Often, it's a mere matter of barrel length that makes the difference. For

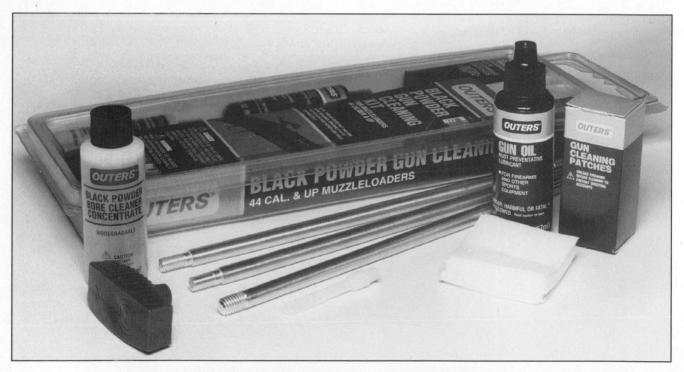

A blackpowder cleaning kit like this one is a good way to get started in muzzleloader maintenance.

example, a 22-inch carbine barrel is easier to clean than a 32-inch rifle barrel. There's less bore to scrub free for starters, plus it's easier to manage the shorter barrel in a sink or bucket, or for that matter, out on your back porch using a no-water, solvent-only cleaning method. The modern muzzleloader with removable breech is another option. Cleaning from the back end is easier and faster than cleaning from the muzzle, plus you can see when the bore is clean a lot better when you can look through it all the way, from breech end to muzzle. But don't choose your muzzleloader based only on cleanup, unless maintenance is vital to you, because with today's chemicals, even long toms are not that difficult to spiff up after shooting.

Today you can actually start cleaning before you start shooting, by using modern lubes made to break down blackpowder fouling. I conducted a little demonstration of my own using patches lubed with synthetic whale oil, and then patches lubed with a modern shoot-all-day product. The modern lube made after-shooting cleanup easier.

The simple water-cleaning method used eons ago got guns clean enough to preserve them, and there is nothing wrong with that method today. However, at least consider after-cleaning metal preservers following the water-only cleaning method. While the old way of simply pouring water downbore and cleaning with a patch on the end of a ramrod or wiping stick is workable, at the very least add a metal preserving agent to the process after the main cleaning is done. It'll pay off. Also, I want to make an admission here. If you take one of my frontloaders from the rack and run a solvent patch through the bore, the patch may not come out lily white.

I believe in getting 'em clean, and I'm not lazy about it, but I do not feel that a shooter must spend hours at the job. Perhaps a muzzleloader will last a couple hundred years with perfect cleaning, but I figure it will last at least a couple lifetimes with good honest maintenance as described below. So get your guns clean, by all means, but don't think you have to spend a whole evening doing the job. The oldtimers didn't have the products we have to work with, and a trip to a gun museum will turn up a good many blackpowder firearms of yesteryear that are still shootable. Some would be better off, to be sure, had they been cleaned as religiously as we intend to do the job, but nonetheless, they're intact.

A ring in the bore can build up when the bore is not cleaned properly. This ring represents a fouled area that did not get scrubbed clean. It is an etched circle that never goes away. Also, poor cleaning could

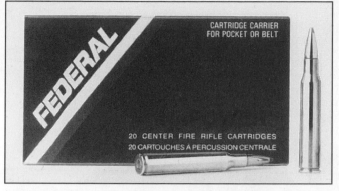

Nowhere on this box of modern non-corrosive Federal fodder does it say you must clean your rifle as soon as possible after shooting or the metal will be damaged. We have no blackpowder or blackpowder substitute at this time that can boast the same thing, and that is why cleaning a muzzleloader after shooting is imperative.

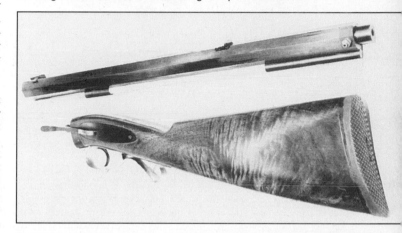

Some muzzleloaders are easier to clean than others. Barrel length and depth of the rifling grooves have a lot to do with this fact. The Storey Takedown Buggy Rifle shown here is one such rifle. It has a short 22-inch barrel, and it's a conical-shooter, so its groove depth is shallow.

Thompson/Center's All-Natural Muzzleloader Cleaning & Seasoning Accessory Kit includes Natural Lube 1000. Using "all-day" lubes is really the beginning of the cleaning process because such lubes keep fouling soft and easier to remove after shooting.

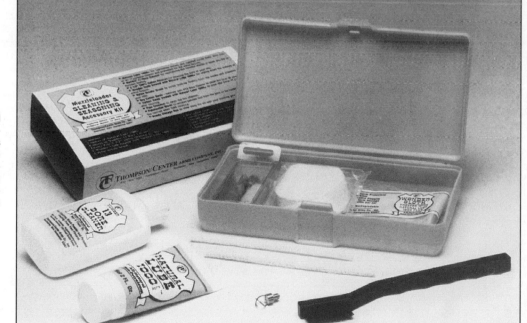

Every gun has its own takedown features. This Thompson/Center Scout pistol is easy to reduce into its main parts for cleaning.

cause the bore to pit. Pits are rust pockets. Furthermore, the interior of the lock may suffer, and fail to work properly. If the notches on the tumbler become damaged, the firearm could be dangerous and go off prematurely because the nose of the sear slips out of the full-cock notch or half-cock notch. A wooden stock can soften and decay from lack of after-shooting cleanup, but a pitted bore can mean reduced accuracy as well as loading difficulty. The nipple seat area may also deteriorate to the point where it fails to transfer the spark from the percussion cap to the breech (misfires and hangfires). The flintlock's touchhole will one day burn out anyway after years of use, but when left uncleaned, it can get damaged a lot sooner.

Corrosiveness

Chapter 14 dealt with the nature of blackpowder, and the fact that many different salts are left behind following combustion. These salts can cause damage. They are reduced (broken down) with polar solvents, such as water, as well as with many commercial products. Never forget that blackpowder is hygroscopic, meaning it attracts moisture. Think of a soda cracker left out in a high-humidity environment. Pretty soon it's soggy. Muzzleloader powders have a tendency to soak up moisture, too, and we all know what water can do to metal.

Sometimes the corrosive nature of blackpowder and Pyrodex is overblown. These powders do not eat up metal on contact. I have tested original muzzleloaders that were left loaded, perhaps for years. Where the load rested there was evidence of damage, but I've yet to find a gun eaten up by blackpowder. I have also left my muzzleloaders loaded on a hunting trip, with the percussion cap removed and the firearm put away for the night for safety's sake. I have never seen damage to the breech where blackpowder or Pyrodex rested downbore. On the other hand, a charge of muzzleloader propellant left in the firearm in a high-humidity situation may cause trouble, not only from the corrosive nature of the powder, but also from its hygroscopic nature. The moisture in the powder is a bigger bandit than the powder itself.

A sidebar at the end of Chapter 15 summarizes a long corrosion study I conducted for the purpose of testing bore wear with three different powders: GOEX blackpowder, Black Canyon Powder and Pyrodex. As part of this study, two cleaning methods were employed:

water only and solvent only. The reader may enjoy studying the results of the test as described in Chapter 15. Meanwhile, take a look at the two general styles of cleaning, plus a third method, which is a mix of the first two. I think you'll agree that maintaining the blackpowder firearm nowadays isn't that difficult.

Disassembly

In all three methods of cleaning, the firearm should be broken down before working on it. For our purposes, this means no more than removing the barrel from the stock, and then, if necessary, taking the lock from its mortice. Many modern muzzleloaders have breech assemblies that can be removed so that cleaning takes place from the breech end, which makes the job all the easier. Stocks that are pinned require special care. The pins must be driven out without damaging the stock itself. In many cases, the wise move is to have a blackpowder gunsmith do the job, because pins may be tapered, meaning they must be driven out from one side only, or the stock will split. You can tell a pin is tapered when one end is larger than the other end. Ordinarily, guns with pinned stocks should be cleaned without removing the barrel from the stock. Wrap a cloth around the muzzle to prevent water from seeping between the barrel and its channel in the stock. It's a good idea to occasionally rub down the stock with a little linseed oil on a soft rag. A little goes a long way.

Method One—The Water-Only Cleanup

Ordinary tap water is a perfectly acceptable cleaning agent because of its nature as a universal solvent. Water breaks down and dissolves most blackpowder residue. It works not only on the salts, but on carbon and sulfur, too. Water is cheap, so a shooter can afford to flush the bore with it, and that's a good way to get rid of fouling. Here's how:

1. Make certain that the firearm is unloaded. "The gun went off when he was cleaning it" may be a very old story, but it has happened. You know how to check for a loaded gun. See Chapter 8 for safety measures. As a review, remember to insert a ramrod to see if it bottoms out, or if it falls short of the bottom of the breech, indicating a load is present.

2. In order to get into the lock region or into the various breech parts

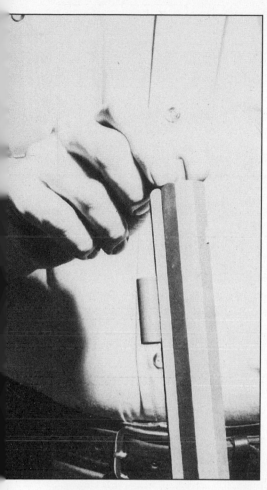

Be sure the firearm is unloaded before attempting to clean it. Here, Dale Storey checks for a load using a ramrod. He drops the ramrod downbore, notes how much remains protruding from the muzzle, removes the rod, and holds it against the side of the barrel to gauge it against the breech plug. If the rod bottoms out in the breech, the bore is unloaded.

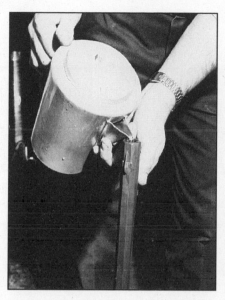

The bore can be flushed with cool water initially to reduce some of the major fouling. After that step, hot water is a good idea because it heats the bore, aiding the drying process later on. Water can be heated in a container like this one, or hot tap water can be used.

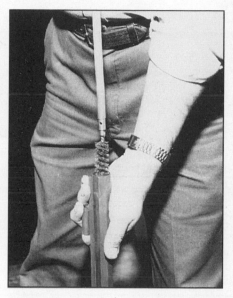

A bristle brush is important because the bristles get down into the bottoms of the grooves of the rifling to remove fouling.

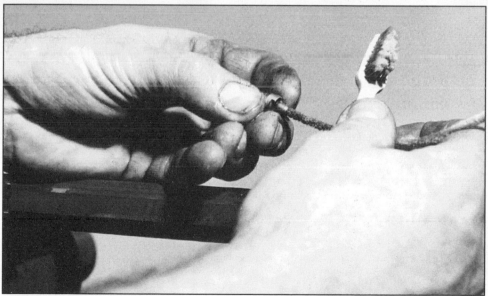

A pipe cleaner goes to work in the vent of the nipple to clear fouling and to dry up water or solvent.

of the modern muzzleloader, you must disassemble the firearm. If you don't know how to do this job right, find out before starting. Guns can be ruined when taken apart incorrectly.

3. Flush the bore with cool water. The theory, and it is no more than a theory, is that cool water will not "set" fouling in the bore. This step gets the process underway, breaking down and flushing out major fouling.

4. Now flush the bore with hot water. This is very important. Hot water heats the barrel, which in turn aids the drying process. It is impossible with some firearms to reach every little corner with a clean-

ing patch or pipe cleaner, but if the bore is good and hot, traces of trapped water evaporate away.

5. Use a bristle brush to scrub the residue from the rifling grooves.

6. Remove the nipple. If the firearm has a cleanout screw, remove it now. Some guns have this little screw; most do not. The screw allows a flow of water through the breech section. Whereas the nipple seat also affords this passageway, the cleanout screw opens a direct port from the breech. Flushing with the cleanout screw removed helps clear powder residue directly from the breech. Water squirts out vigorously through the cleanout screw hole when poured directly down the muzzle.

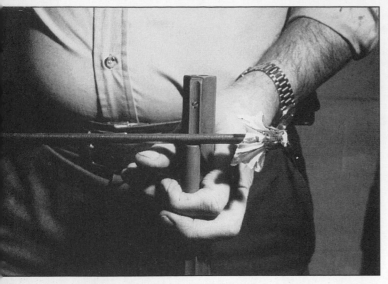

Cleaning patches on jags remain an excellent means of sopping up the "goop" created by brushing the bore.

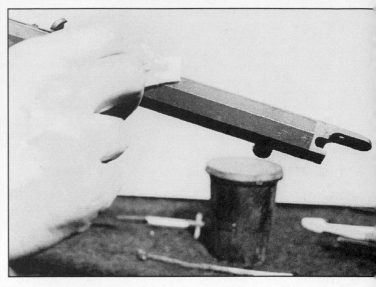

A little metal preserver after cleaning helps to thwart rust and corrosion.

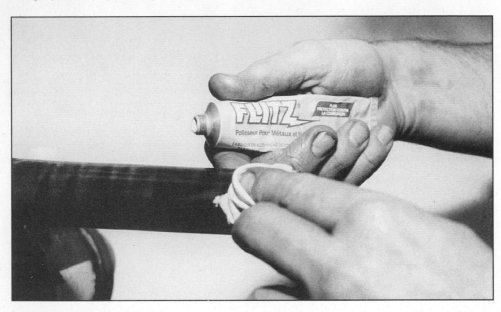

Metalwork, such as a brass nose cap, can be brought back to shiny with Flitz or other brass cleaner.

7. Leaving the cleanout screw hole open, work the bristle brush through the bore several times with hot water. Be careful to guide the stream of water from the cleanout screw hole away from the lock and stock of the firearm. If that dirty water gets into the lock, lock mortice, or barrel channel, it can promote rust and/or decay of the wood, especially if the mortices and channels are not well-finished.

8. Douse a toothbrush with hot water and scrub the hammer nose, snail area, metal parts around the lock, nipple seat, nipple threads and other exposed metal parts. A toothbrush does for the outer parts of the gun what a bristle brush does for the bore.

9. Dip a pipe cleaner in hot water and swab the vent of the nipple, as well as the nipple seat of the firearm. Use the same treatment to clean the touchhole of the flintlock.

10. Now it is time to sop up the moisture that is left in the bore. Cleaning patches pick up debris that has been knocked loose by the bristle brush, but has not been carried off by the water flush.

11. Repeat Step 10 with more dry patches. The first patches out of the bore may be quite wet. Since you used hot water, it won't take long to dry the bore. You may wish to continue wiping the bore with dry patches until one emerges dead white; however, that may take some time. A very light gray (almost a white patch) means that most of the fouling is gone.

12. Wipe all channels, cracks and crevices dry with pipe cleaners.

13. Dry the outer metal surfaces. In the process of cleaning the muzzleloader, some moisture and even blackpowder residue may have been transferred to the barrel and lock of the firearm, generally caused by pouring water downbore. Dry it with a clean cloth.

14. Wipe the wood to clear off water and residue then rub the stock with a clean cloth plus a few drops of pure boiled linseed oil.

15. Go over the entire firearm one last time with a soft clean cloth, free of solvent, oil, or any other product. This step picks up traces of residue, extra solvent or linseed oil from the surface of the gun.

16. Run a cleaning patch through the bore dampened with a little metal preserver. (Be certain to run a dry patch downbore before shooting the firearm again so that this preserver is removed.)

17. Now apply a light coating of metal preserving chemical to the outside of the barrel and lock.

18. Occasionally, the lock requires special cleaning. Remove it from its mortice in the stock. Blow dirt out of the workings with canned air or an air hose. Use a toothbrush to flick away stubborn dirt. Wipe all exposed parts clean. Oil it lightly.

19. Brightwork, such as brass fittings, may be cleaned with a little Flitz on a cloth. Follow with a clean, soft, dry cloth.

Tip: An oily bore may shoot wide of the mark, or it may produce overly-large groups. Sight-in with an oil-free bore and maintain an oil-free bore during shooting. It is perfectly all right to run a cleaning patch downbore after the projectile is seated on the powder charge. This may sound dangerous, but it includes no more risk than already experienced when running the bullet down onto the powder charge. The clean patch sops up any excess oil or lube left in the bore from the ball patch, or from grease on the conical. After storing your muzzleloader with a little preserving oil in the bore, be sure to remove that oil before firing the muzzleloader again.

Method Two—The Solvent-Only Cleanup

No water whatsoever is used in this method of muzzleloader cleaning. William Large, barrelmaker for over fifty years, "never touched water to barrels," to quote his admonition to me a number of years ago. He didn't believe in water. He used solvent alone to keep his barrels in perfect shooting condition. Since his barrels won literally hundreds of muzzleloader matches, I'd say Mr. Large knew what he was talking about. Nonetheless, the worn cliche "to each his own" continues to carry a lot of weight, and I feel a shooter should use the method he prefers, whether it is all-water cleaning, all-solvent cleaning, or a combination of the two. Furthermore, some shooters report that in regions of high humidity, the hot-water flush does the best job of getting the gunk out, while allowing the bore to dry thoroughly. All water? No water? A mixture of the two methods? The choice is yours. But for now, here's a look at the solvent-only method of maintaining a frontloader after the shoot:

1. Once again, ensure that you're working with an *unloaded* firearm before attempting any cleanup or disassembly.

2. Attach a cleaning patch to a cleaning rod via a jag. The jag should hang onto the cleaning patch during swabbing. If it doesn't the patch is the wrong size or the jag is wrong for that specific firearm. Soak the patch with solvent using one of the many good chemicals available today. Now run the solvent-soaked patch through the bore several times to loosen fouling.

3. After swabbing the bore with solvent, remove the nipple and cleanout screw (if the gun has one) and repeat Step 2. The reason for waiting to remove the cleanout screw until this step is to help loosen it. Solvent runs into the threads of the nipple and cleanout screw, making both easier to remove.

4. Scrub the bore with a bristle brush soaked in solvent. Depending upon how much powder was burned in the bore, anywhere from only a few passes with the brush to dozens are required to loosen fouling. The bristle brush is very important to the no-water cleaning method. It is the brush that finds its way into the grooves of the rifling to scrub out fouling.

5. Now sop up the goop in the bore by running dry patches through. If a lot of powder was shot, and in spite of cleaning between strings of shots, it may take a few patches to soak up all of the dark liquid in the bore after the bristle brush and solvent have gone to work.

6. Repeat Step 4—more solvent, more bristle brush work.

7. Repeat Step 5—more soaking-up cleaning patches. At this point, patches should emerge from the bore at least gray instead of black or dark brown.

8. Now refer to Steps 12 through 19 in the water method, but forget the water. Use only solvent on your cloth to clean exterior metal. Use solvent on a pipe cleaner to reach cracks and crevices. Be certain to dry every part well. Don't fail to use the pipe cleaner or toothbrush. These tools work well with any cleaning method.

Method Three—Water/Solvent Combination

This may be the best method of all, because it incorporates the hot water flush, plus scrubbing the bore with a solvent-soaked bristle brush. It goes like this:

1. Make sure that the muzzleloader is *unloaded* before attempting to clean it. Knowing it is unloaded, you may take it apart.

2. Flush the bore with cool water.

3. Flush it with hot water.

4. Run a solvent-soaked cleaning patch through the bore several strokes.

5. Make several passes with a solvent-soaked bristle brush.

6. Remove the nipple and cleanout screw if there is one.

7. Run a solvent-soaked bristle brush through the bore several strokes with the nipple and cleanout screw removed.

8. Sop up all the liquid from the bore with cleaning patches.

9. Clean the nipple seat and cleanout screw hole with a pipe cleaner dipped in solvent.

10. Wipe all channels with the toothbrush and pipe cleaners using only a little solvent. Over-using solvent on the toothbrush or pipe cleaner causes the solvent to run into mortices, etc.

11. Wipe the outer portion of the firearm, metal and wood, with

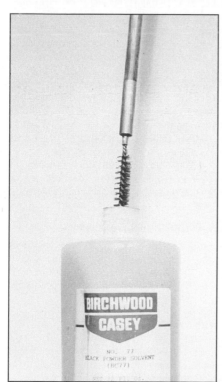

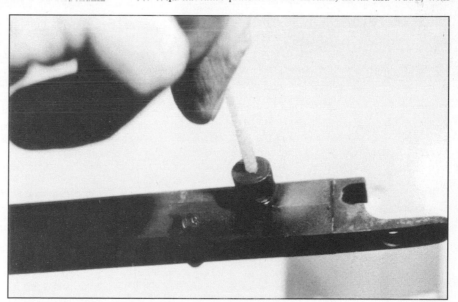

Here, the cleanout screw has been removed on this drum-and-nipple type percussion rifle. Water or solvent can now flow through the area. Also, with the cleanout screw removed, a pipe cleaner can be worked into the breech as shown here.

The no-water method uses solvent like this to break down blackpowder fouling.

When a dirty rifle is truly stubborn, it may be necessary to remove the barrel and place it in a vise, as Dale Storey does here, to really apply the "elbow grease" to it.

(Below) Thompson/Center's Expediter cleaning kit is a good barrel-flushing unit that works on percussion firearms, including modern muzzle-loaders like this one.

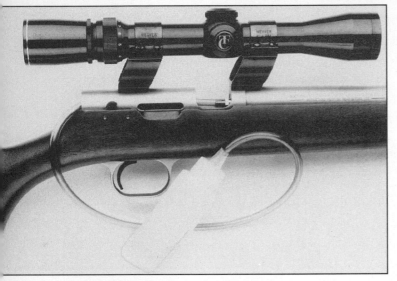

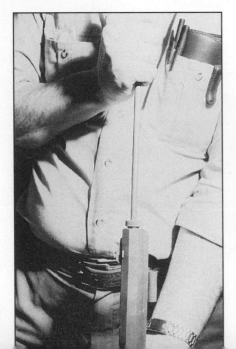

A muzzle protector keeps the cleaning rod centered in the bore so that the rod does not scrape the lands of the rifling, especially at the crown of the muzzle.

clean rags, using a little boiled linseed oil on the stock and a little oil on the metal.

12. Protect the bore with a light coating of rust inhibitor or other metal preserving agent.

13. Attend to the lock if necessary. Remove the lock and clean it inside with a cloth and solvent. Use pipe cleaners to get into the crevices. Lightly oil all parts of lock.

14. Attend to the brightwork with a little Flitz on a cloth.

Cleaning Hints

The nipple can be cleaned by dropping it into a small container with solvent. Soak the nipple in solvent while working on the rest of the gun. Then run a pipe cleaner through the nipple vent. Clean the threads of the nipple with a toothbrush. Dry it with a new pipe cleaner and a rag, then replace it into the nipple seat of the firearm.

Continued shooting of one-piece plastic wads in the shotgun, or plastic sabots in rifles or pistols, can leave a slight deposit of plastic in the bore. This may be removed with a modern solvent, such as Shooter's Choice, using a bristle brush. Follow the instructions on the container.

After two or three seasons of heavy use, it's a good idea to treat the muzzleloader to a real checkup. There are two ways to do this. You may take the firearm to a blackpowder gunsmith and have him go through it. Tell him you want the firearm debreeched. When he removes the breech plug, he'll be able to get a good look at the breech end of the bore to determine if there is a ring, pitting, or other problem. Also, with the breech plug out, the gunsmith can do a heavy-duty job of bore cleaning. Another alternative is doing this advanced maintenance yourself. See *Black Powder Hobby Gunsmithing*, a DBI book, for instructions.

Shotguns, Pistols and Revolvers

Shotguns and pistols are cleaned very much like the muzzle-loading rifle. All in all, the same goal pertains: get rid of fouling. The revolver is different. A few instructions are provided here. A major difference in revolver cleaning is disassembly. Blackpowder fouling can get into the workings of the revolver, binding up the action.

Cleaning a Revolver With Water

1. Ensure that the gun is unloaded before cleaning it.

2. Disassemble it appropriately, including removing all nipples from the cylinder. If this is a routine cleaning after firing only a couple cylinderfuls, the revolver may not require full disassembly.

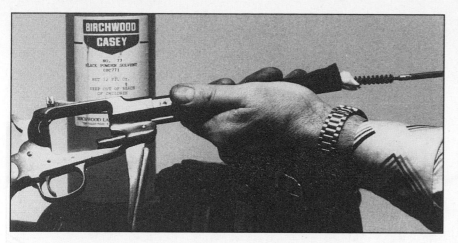

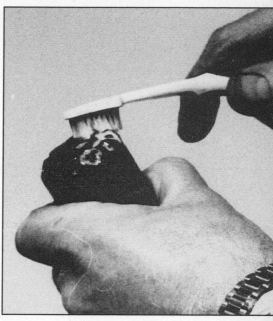

When a cleaning jag won't work, a patch can be retained with a bristle brush, as shown here. The bristle brush will not leave the patch downbore or allow it to get "lost" mid-bore.

A toothbrush loaded with solvent is superb for cleaning away fouling, as shown here with the cylinder from a cap 'n' ball revolver.

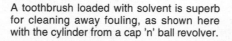

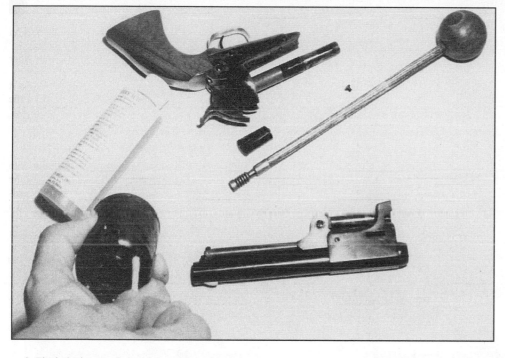

Drying all parts to soak up leftover water or solvent is an important step in maintenance. A pipe cleaner is used here to reach hard-to-get-at spots.

3. Flush the bore and chambers with cool water.

4. Flush the bore and chambers with hot water.

5. Scrub the bore with water on a bristle brush.

6. Scrub the cylinder and chambers with a toothbrush.

7. Run cleaning patches and water through the bore and chambers.

8. Dry the bore and chambers with patches.

9. Use pipe cleaners and water to clean the nipples.

10. Dry all the parts. If the revolver was not fully disassembled, be certain that no water has invaded the workings of the gun. If you suspect that water has gotten into the working parts, the revolver will have to be fully stripped and all moisture removed. Parts may be dried in an oven on low heat.

11. Wipe the revolver down. Metal parts are wiped with an oily rag. Wooden grips may be wiped down with a rag and a touch of linseed oil.

12. For storage purposes, leave a light trace of oil on all metal parts, on the working parts as well as on the frame, barrel, and bore. Revolvers with brass frames may be brought back to bright with brass cleaner.

Important: The revolver will require full disassembly if several cylinderfuls of blackpowder or Pyrodex have been fired. If you are not fully versed in the takedown procedure, get help, but be sure to observe the steps so you can do this job yourself next time around.

Cleaning a Revolver With Solvent

Basically, the steps are the same as for the water-only method of cleaning the revolver, with the exception that solvent is used in place of water.

Remember that the revolver will demand a full takedown if several cylinderfuls have been fired. Furthermore, full takedown is necessary from time to time even when the blackpowder revolver is fired only a few times.

As noted at the beginning of this chapter, muzzleloader maintenance is easier than ever. It does take a little time and effort, plus a dab of patience to do a good job, but the fun of blackpowder shooting is well worth the little bit of effort required to properly maintain your firearms.   ●

Blackpowder Corrosion Tests

MOST "HOMEGROWN STUDIES" outside of sophisticated labs are more demonstration than scientific discovery. However, in an attempt to see what really happens to bores when subjected to muzzleloader propellants, I ran a carefully planned sequence over several months that I believe resulted in reliable information.

The Plan

Five Thompson/Center 50-caliber Hawken barrels with sequential serial numbers were used with one T/C stock. The bore dimensions before and after testing were measured to indicate wear, and are given below. Shooting commenced in February and ended in October. There were a total of twenty-six sessions with a rest period in between to allow metal deterioration to take place if it was going to. The barrels were stamped A through E, as were the breech plugs, so there was never a mixup.

Powder Charge

Three powders were used: GOEX FFg blackpowder, Black Canyon, and Pyrodex RS.

- GOEX FFg - 100 volume for 100 grains weight
- Pyrodex RS - 100 volume for 71.5 grains weight
- Black Canyon - Volume for 100 grains weight

All powders received the same seating pressure of 60 pounds, using a spring-loaded regulator.

Cleaning Methods

Two cleaning methods were used, water only and solvent only. The solvent for cleaning was Hodgdon's Spit Bath. The barrels were treated thusly:

Barrel A: Pyrodex RS/cleaned with water only.
Barrel B: GOEX FFG/cleaned with water only.
Barrel C: Black Canyon/cleaned with water only.
Barrel D: Pyrodex RS/cleaned with solvent only.
Barrel E: GOEX FFg/cleaned with solvent only.

Velocity Check

For the sake of added information, a chronograph was used to record velocities. Here is a typical velocity check with the 28-inch

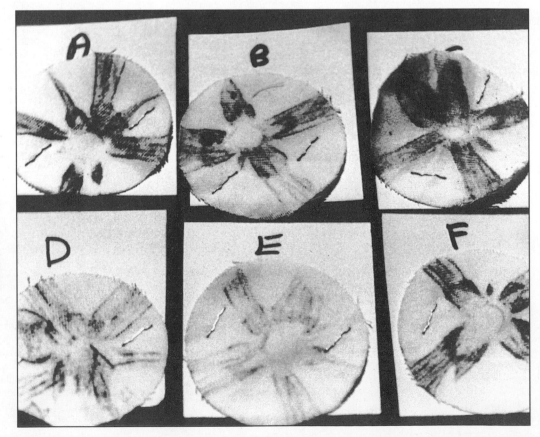

Typical appearance of cleaning patches after firing three shots during corrosion testing. These are for Barrels A, B, C, D, and E, with an extra trial run for a special test barrel, Barrel F, which was fired with GOEX FFg and a patched round ball for comparison purposes only.

barrel T/C Hawken and the 385-grain Buffalo Bullet taken from two different sessions, three-shot strings for all tests:

Test Session No. 5		Test Session No. 17	
Barrel	Muzzle Velocity (fps)	Barrel	Muzzle Velocity (fps)
A	1456	A	1556
B	1425	B	1402
C	1109	C	1107
D	1438	D	1484
E	1430	E	1434

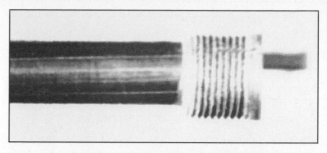

A closer look at the breech end of the barrel with breech plug removed reveals no metal deterioration after the test.

Conclusions

1. No discernible corrosion damage was experienced with Pyrodex RS, GOEX FFg, or Black Canyon powder in the course of the study.

2. All three powders left residue after shooting. Residue was successfully removed with normal cleaning practices.

3. All barrels cleaned with water only or with Hodgdon Spit Bath solvent only, showed no traces of etching, pitting, discoloration, or other evidence of bore damage.

4. Temperature and humidity variations did not cause recognizable differences in velocity during the course of the study.

5. GOEX blackpowder, Black Canyon powder, and Pyrodex RS are all viable propellants in muzzlcloaders in terms of bore corrosion. With a reasonable maintenance schedule, bores showed no damage with these powders. Nothing was indicated in after-test inspection to reveal bore wear.

6. All bores were "miked" about one inch into the muzzle before and after testing. No measurable wear was revealed.

7. The five breech plugs were visually studied after the test using 8x magnification. Conclusion: no etching, pitting, discoloration, or

(Above) All breech plugs were removed before barrels were cut lengthwise after the corrosion test. Breech plugs were not damaged. This includes the nipple seat.

After the corrosion study, all five barrels were cut in half lengthwise for inspection.

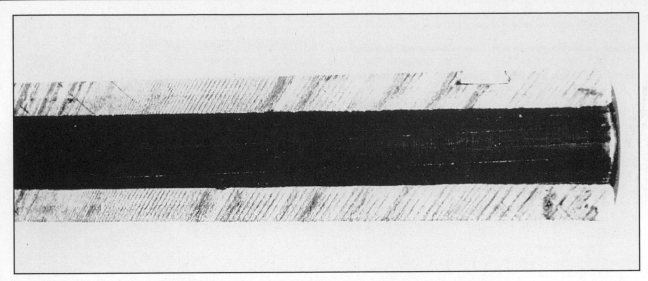

This is a close look at one of the barrels after the study when the barrel was cut in half lengthwise. There is no indication whatsoever of etching, discoloration, pitting or other metal deterioration.

other traces of damage due to corrosion were noticed. However, breech plugs D and E (solvent only cleaning) had shiny cones, while breech plugs A, B and C (water cleaning only) showed slight remaining residue. No attempt was made to determine the nature of this residue, as it was considered more important to leave the cones of the breech plugs, as well as the bores of the barrels, in an "as is" condition after shooting.

8. Hammer noses revealed some corrosion. This fact leads to the conclusion that this area requires more cleaning attention. Throughout testing, the nose of the hammer showed that normal cleaning with toothbrush and water or solvent was not sufficient to attend to this area. It is recommended that shooters clean the nose of the hammer twice after shooting, rather than once, and that they check the hammer nose periodically to determine that it is corrosion-free.

Notes: Cleaning with water-only and solvent-only methods were conducted normally with no attempt to be extra careful or to use overly-long cleaning periods. On occasion, bores were pur- posely abused by leaving them dirty for a week rather than cleaning them right after shooting. This abuse factor was included to make maintenance more "realistic," assuming that shooters do not always clean their muzzleloaders the same day after firing them, even though that has always been the most desirable method for maintaining this type of firearm.

After every shooting session, the bores were observed with two bore lights, a CVA unit that drops down into the breech, as well as a handmade miniature light passed from breech to muzzle.

After the final test, all five barrels were cut in half lengthwise for study. An 8x loupe was used to investigate the bores. No damage was revealed. The bores were studied by the machinest who cut the barrels, as well as by a gunsmith and by the author.

It is justifiable to say that water-only and solvent-only cleaning methods are both viable. A combination of the two could be ideal, with a water flush to remove most of the fouling, followed by a solvent cleaning.   ●

chapter 39

Troubleshooting The Frontloader

NO MATTER HOW fine the blackpowder gun, or how well made, things can, and occasionally will, go wrong. That's what this chapter is about: finding out what's wrong and hopefully fixing it if we can, and if not, getting our pet to someone who can—our friendly blackpowder gunsmith.

Scoped Rifle Won't Stay Sighted-In

Although there may be other problems at hand in this situation, odds are one of two things is wrong: the scope reticle is wandering or the mount is loose. To check for the first, secure your scope sturdily, as in the cutout of a cardboard box, and move the dials. Sometimes you'll actually see a crosswire jump, rather than cleanly move over its allotted distance. To check for the second problem, simply remove the scope and test base and ring screw tightness with a screwdriver. If a rifle "shoots loose," that is, stays sighted for a short while, then has trouble again, a mount may have to be secured with Loctite. There is a milder formulation of Loctite that is just right for this job. It will help secure the scope mount, both rings and base, while permitting easy removal later.

Accurate Muzzleloader Goes Sour

This problem exists more with modern cartridge guns than frontloaders, but an accurate blackpowder gun can also start shooting fatter groups all of a sudden. With modern arms, a good cleaning to get rid of copper fouling often helps. With muzzleloaders, a good cleaning to rid the rifling of stubborn fouling can also do the trick. In severe cases, a gunsmith will have to remove the breech plug from the firearm so that the bore can really be gotten to with a stout cleaning rod and a series of bristle brush scrubbings. When putting the gun back together, ensure that everything is tight. If loose bolts or screws were part of the problem, this little step will cure it.

If a good gun goes bad and cleaning does not help it, look for crown damage next. If the rifling is harmed at the crown of the barrel, accuracy can go astray. After all, the crown is the last area of control that the rifling has on the projectile. If the crown is damaged, a gunsmith can recrown the barrel. If a damaged crown is not the problem, check for patch damage with a round ball shooter. Naturally, we're assuming that when an accurate frontloader begins to shoot inaccurately, that it

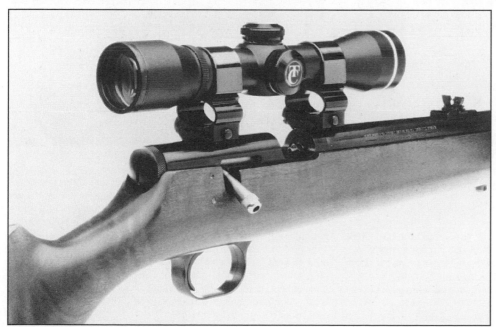

Did the scope move? Good solid mounts like these from Thompson/Center fairly well assure that a scope will remain in place. But even the finest mounts require tightening from time to time.

When a blackpowder firearm loses its accuracy edge, a look at the crown of the muzzle is in order. Damage here can cause a deterioration in bullet grouping.

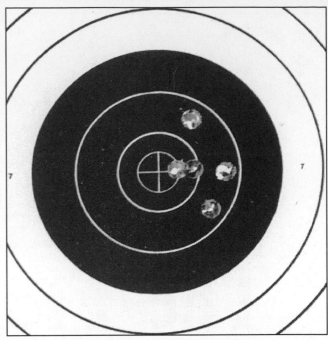

In troubleshooting a gun for accuracy problems, the shooter must first know what the firearm was originally capable of, otherwise it's impossible to know if there has been deterioration. Keeping typical targets, well-marked, is one way to know when a muzzleloader "goes sour."

(Above) Sometimes a barrel must be debreeched in order to find a problem. Here, the breech end of a barrel is shown with its breech plug (far right), drum and nipple removed.

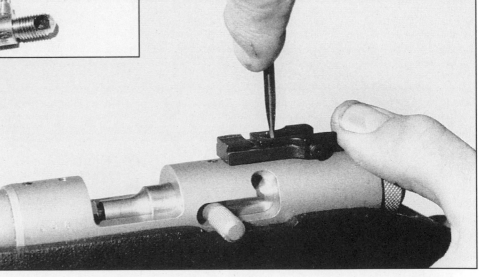

Something as simple as tightening a screw or bolt can make a rifle shoot well again.

does so with a load chain that used to shoot well. Sometimes the gun gets the blame when in fact the shooter has changed the recipe that used to work, especially the projectile.

Round Ball Inaccuracy

Suppose that your muzzleloader has gone sour and you suspect the round balls you're shooting. Checking the round ball out is easy. Scale-weighing should have been done when the balls were cast, but if not, do it now. You may find air pockets causing some light missiles. Also, check for damage. While a slightly out-of-round ball will fly pretty straight, nose damage especially seems to throw the sphere off course, and any damage at all is bad for accuracy.

Conical Inaccuracy

Various forms of conical bullet damage can do various degrees of harm to accuracy. Minor aberrations are not serious; however, it's just as well to protect the muzzleloader missile from dents and flat spots. The original box is good for regular storage, but for travel from home to range or hunting field, a metal box is better. Any tin box will do. Take up excess space in the tin container with a soft cloth or paper towel to prevent the soft lead missiles from crashing into each other en route to the shooting site. Seal the box, as with a layer of plastic wrap, if the bullets are going to be stored for a while. Conicals seem to suffer badly from base damage. If the base or skirt of a conical is damaged, chances of best accuracy are slim.

The Muzzle Mitt is a water-proofer that covers the muzzle of a blackpowder gun to prevent invasion by rain or snow.

(Below) The Muzzle Mitt in place over the muzzle of a rifle ensures that no water or snow will flow downbore. A small rubber band is used for extra insurance against the Muzzle Mitt's falling off.

Troubleshooting an accuracy problem should include matching the right bullet to the gun, as well as using only good, undamaged round balls or conicals. These Thompson/Center conicals belong in a firearm with a conical twist, and they are obviously of high-quality manufacture.

Normal Group Hits Left, Right, High, or Low

An oily bore can cause this problem. Troubleshooting it is very simple. Wipe the bore free of excessive lube with a single cleaning patch after the gun is loaded.

Muzzleloader Failed To Go Off in Damp or Rainy Weather

Take it from a fellow who learned the hard way, forgot his lesson, and had to learn the hard way again. My Colorado guide, Steve Pike, called a large six by six bull elk up to 12 paces. I know, because after the incident, I paced the distance off so I could kick myself. The big bull stood and stared. I aimed. Ftttt! It had been raining, and in my haste to leave town, I left all my waterproofing stuff behind. Worst of all, I know I would have gotten that bull, because when the rifle misfired, the sights were lined up perfectly in the center of the neck and there the sight picture remained. Don't let this happen to you. There are three general areas that can cause trouble. The least likely of the three is the lock, but a lock can leak water and the water can find its way into the pan on a flintlock or perhaps into the nipple seat via a circuitous route. Although lock leakage probably won't cause a problem, here is a way to make the lock at least moisture-proof if not waterproof.

Waterproofing and Foul-Proofing the Lock

Saturate a thin piece of paper, such as onionskin writing paper, with gun grease. There are many good products on the market. Just make sure it's a true grease that will stick, and not a runny oil. An alternative is the pre-greased protection paper that comes with new firearms. Sometimes your local gunshop can supply a little of this paper. Remove the lock from its mortice in the stock. Using the mortice as a pattern, cut the greased paper to fit, but with overlap. The paper is going to form a gasket between the lock and the interior of the stock, helping to keep moisture out of the important tumbler area when the lock is refitted to its mortice. The greased paper gasket will also discourage blackpowder fouling from attacking the inner lock workings. Don't overdo it. One thin piece of greased paper will do. Excess paper might get caught up in the workings of the lock.

Serious Rainproofing

The other two areas that demand waterproofing are much more serious than the lock. The muzzle can, of course, allow water to enter. These days, there are specific rubber protectors that slide right over the muzzle. In the past, we used small party balloons, which will still work. A commercial protector, or a balloon, will prevent water from entering the bore. The touchhole of the flintlock and vent of the nipple, as well as the nipple seat, are real problems. I'd say that the vast majority of failures on rainy days come from these areas, and not from the lock itself or water getting in the muzzle. The flintlock can wear a boot,

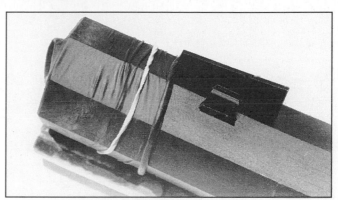

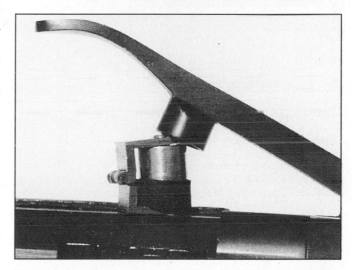

(Above) Waterproofing a muzzleloader can be one of the most important things a hunter does. The Cap Locker is a device that covers the nipple entirely, not only for waterproofing, but also for safety. Spring-loaded, the top pops out of the way when the gun is cocked. The nipple is surrounded by a rubber gasket.

Here, the Cap Locker is shown with the hammer of the rifle cocked. The cover of the Cap Locker has jumped out of the way so the rifle can now be fired. Note the rubber gasket around the nipple.

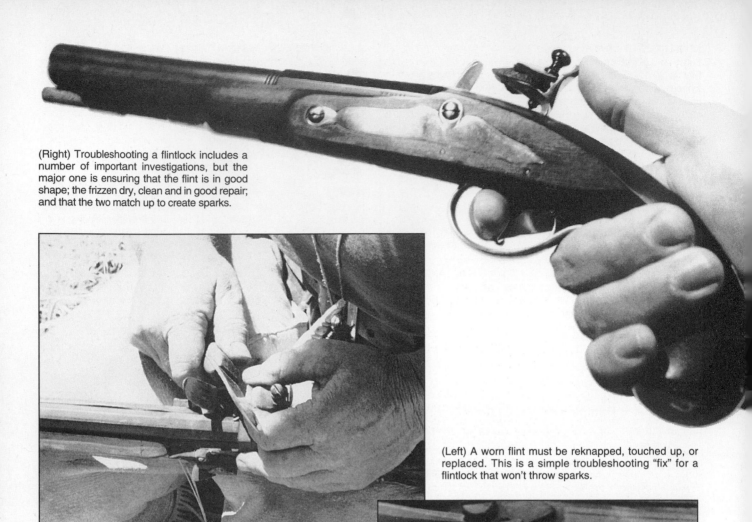

(Right) Troubleshooting a flintlock includes a number of important investigations, but the major one is ensuring that the flint is in good shape; the frizzen dry, clean and in good repair; and that the two match up to create sparks.

(Left) A worn flint must be reknapped, touched up, or replaced. This is a simple troubleshooting "fix" for a flintlock that won't throw sparks.

which can be found at your local blackpowder shop. The boot is not entirely handy, but it beats letting your lock get all wet. The nipple and nipple seat can be waterproofed very nicely with devices that do the job. Again, check your local blackpowder shop. If I had my Kap Kover on Old No. 47, a big bull elk would have been mine. Another good nipple waterproofer just came to my attention as this edition was underway. It's from Big Bore Express, and it's spring-loaded. While hunting, hammer nose pressure keeps the cover down firmly on the device, but when the hammer is cocked, the cover snaps out of the way so the rifle can be fired.

Flintlock Problems

A Poor Spark

Be sure that the bevel of the flint is downward. Make certain that the edge of the flint mates squarely with the face of the frizzen with uniform contact. If the flint is ragged and it will not make full contact with the face of the frizzen, get a new flint, reknap the old one, or have it reknapped. If your flint is brand new, let it drop against the frizzen a few times. This will help to mate the edge of the flint with the face of the frizzen. Be sure that the flint and the face of the frizzen are both clean and not oily. If the flintlock firearm still throws poor sparks, check the hammer throw. A spring may have gone weak, failing to provide a good whack of the flint against the face of the frizzen.

Also, check the frizzen itself. These can eventually wear out, demanding replacement. When the flint wears down, it may still have some life in it. You may wish to reangle it in the jaws of the hammer just a little bit. Then let the hammer fall a few times to see if the flint will reknap itself on the face of the frizzen to some degree. Also, move the flint sideways

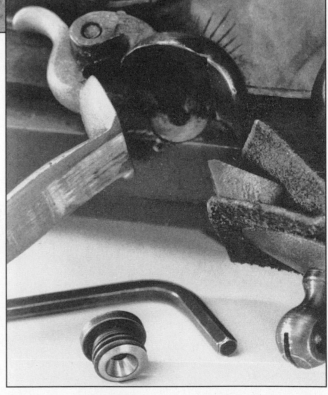

Sometimes a flintlock will act up because the touchhole is worn. In such a case, troubleshooting may mean installing a touchhole liner, like this one from Uncle Mike's.

just a little bit to create a new contact point between flint and frizzen. Sometimes this puts a bit of new life into a used flint. Finally, move the flint forward in the jaw so that it makes fuller contact with the face of the frizzen. Also try turning the flint over so that its bevel points up instead of down. Be sure to remount it in the jaws of the hammer so it sticks out pretty far. Adjust it until only the edge of the flint makes contact with the frizzen when the hammer is allowed to fall forward under control of your thumb. Let the hammer fall freely several times. This will help to remove the "bad" part of the flint, bringing into contact some new material. After you have done this process, remount the flint in the bevel-down attitude it was in before. The new edge you have created with the turn-over method may allow a flint to give considerable extra service.

Flash in the Pan

There are many reasons for a flash in the pan, but consider that the flint did deliver sufficient sparks from the frizzen to ignite the pan powder, so looking to the flint in this case would be wrong. A damp or oily pan can surely cause trouble. So can a touchhole packed full of powder. While blackpowder ignites very easily, when it is packed into such a small channel, which the touchhole certainly is, a fuse may be created. A fuse can slow ignition down, or even cause a flash in the pan. See Chapter 20 for more details.

Caplock Fails To Go Off (Gun is Dry)

Modern caps are well made and very reliable. I recently tested a box of RWS caps purchased from the Navy Arms Co. by randomly selecting twenty caps from the box and placing them in a high-humidity environment on an open piece of paper. After 48 hours of running a humidifier in a closet with these caps, I fired the caps with the Dixie Screwbarrel Pistol, barrel removed. All twenty went off. Nonetheless, caps can be inundated by moisture. By firing a cap or two on the nipple prior to loading, damaged caps can usually be detected. If a cap does not give a good bang on the nipple during the clearing process, beware. Cap debris can also cause a misfire. This fact is also broached in Chapter 20. A damaged nipple can be a problem. The wrong nipple can cause trouble. Naturally, the cap and nipple may not be the root of the misfire at all, if the powder charge got damp.

Brass Furniture Gets Dull

Actually, some people prefer the tarnished look over the bright appearance, and oxidation won't hurt brass furniture in the first place. But for those who prefer the bright look, there is a way to slow the oxidation process. Apply a coat of beeswax to brass furniture, such as nose caps, patch boxes, buttplates, toe plates and other metalwork. A good product for this is Birchwood Casey Gun Stock Wax. The wax coating forms a modest, but helpful, barrier between metal and the atmosphere, which keeps tarnishing to a minimum.

Lead Bullets Look Powdery or Moldy

I'm not sure that either term is accurate. But we have all seen lead projectiles that had some sort of crud on them, sometimes a powdery frosting, sometimes a moldy look. These bullets can be restored, if only for the sake of appearance. I'm not certain that minor coatings really harm accuracy, especially with the ball, which is wrapped in a cloth patch to begin with. But I suppose that extremely minor changes in flight characteristics could occur with moldy bullets. Coating can be thwarted by spraying lead projectiles with WD-40 or a similar product. Also, moldy missiles can be cleaned with solvent, then treated to a little WD-40. Dry all bullets thoroughly before shooting.

Ramrod Problems

Stuck Downbore

My N&W loading rod seldom gets stuck, but if it does, the brass knocker pulls it free. However, should a ramrod become stuck in the

(Above) In the spirit of prevention instead of having to troubleshoot later, the Thompson/Center synthetic ramrod was designed to take a terrific beating. It's shown here beneath the tire of a vehicle. If this ramrod gets stuck, it won't break during removal.

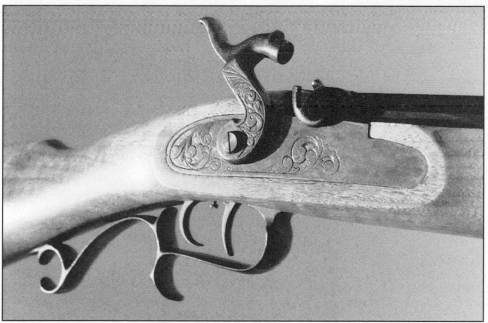

Troubleshooting cap debris is simple enough: Fire one and see what the cap looks like. Is the body of the spent cap intact or fragmented? Can you pick up cap debris from the vent of the nipple using a pipe cleaner?

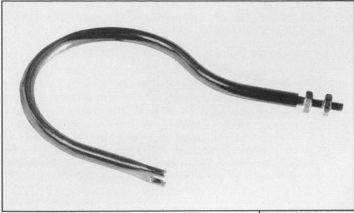

The Blue Ridge Tug 'N' Pry is shown here unattached to a ramrod.

(Below) One way to remove a stuck patched ball that was put down without a powder charge is with Thompson/Center's Silent Ball Discharger.

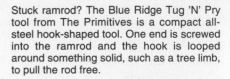

Stuck ramrod? The Blue Ridge Tug 'N' Pry tool from The Primitives is a compact all-steel hook-shaped tool. One end is screwed into the ramrod and the hook is looped around something solid, such as a tree limb, to pull the rod free.

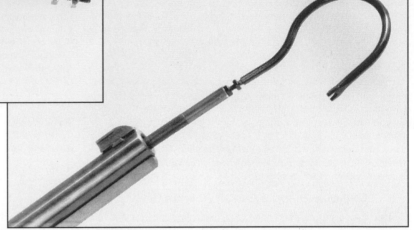

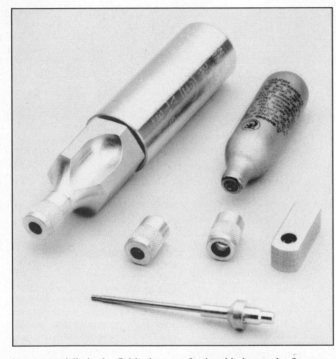

(Above) Another prevention measure is the use of high-grade chemicals like these from Jonad Corp. They help prevent rust and other problems that lead to the necessity of troubleshooting.

bore, especially in the field where professional help may be far away, try a leather thong to remove it. Tie the thong in a clove hitch around the extended ramrod shaft. Secure the end of the thong to a stationary object, such as a tree. Pull carefully and slowly, but steadily and strongly, on the rifle. If a stuck cleaning patch will not come free, run some liquid patch lube down the bore and try again. Be sure to clean the bore thoroughly before the next load is run home. Oil left in the bore can cause bullets to go astray.

Keeps Breaking (Especially in A Dry Area)

Soak your ramrod in coal oil, kerosene or neat's-foot oil and it will become more supple, yet not so pliable that a patched ball can't be rammed downbore. A piece of tubing can be corked on one end, and the ramrod placed inside the tubing and within the liquid. Ramrods can also be treated with boiled linseed oil to help in the prevention of drying out and cracking.

A Ball Stuck Downbore

The Thompson/Center CO_2 Magnum Silent Ball Discharger will literally blow a stuck ball free. However, lacking this device, a stuck ball can usually be removed the old fashioned way, with a screw (not a worm). The screw should be used with a muzzle protector because the protector will center the screw in the bore, preventing its contact with the bore walls and also delivering it to the center of the stuck ball where it belongs. A cleaning rod with a knocker is excellent for removing a stuck ball. If your rod does not have a knocker, the thong method, noted above for freeing a stuck ramrod, may have to be employed to encourage a stuck ball's removal.

Seated Ball, No Powder

It happens to the best of us. We're thinking about that next shot and the patched ball is rammed home without a powder charge in the

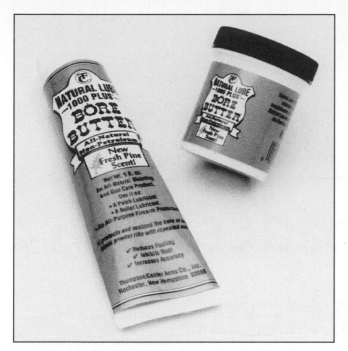

By using modern "all-day" shooting lubes, such as Thompson/Center's Bore Butter, chances of a stuck ball and other problems are reduced.

A rifle suddenly hits wide of the mark. How do you troubleshoot it? Check the sights. This Cabela's modern muzzleloader has a hooded front sight for protection. But could the rear sight have been moved somehow?

breech. Now what? The stuck ball method is the best way to get the patched ball back up the bore, using either the CO₂ device or the screw attached to the end of a metal loading rod fitted with a muzzle protector. The ball can be safely shot free with a small powder charge, but if it is not driven all the way out of the bore, it should be reseated before once again trying to blow it free.

In order to insert a little blackpowder behind the patched ball, remove the cleanout screw and trickle powder directly into the breech. Then replace the cleanout screw, cap the firearm and shoot the ball out in a safe direction. Sometimes powder can be trickled into the touchhole of the flintlock rifle for the same purpose. If there is no cleanout screw on the caplock firearm, the nipple can be removed and powder can be introduced down into the nipple seat area. Then the nipple is returned to its seat and the ball is fired clear. You may have to clean the cleanout screw channel, nipple seat or touchhole before powder can be trickled into the breech. Use a pipe cleaner for this, or a nipple pick.

Stored Muzzleloader Attracts Rust

This is simply solved with one of the many fine rust inhibitors of the day. Modern metal preservatives, such as Jonad's Accragard Metal Protector, thwart rust.

Patch Lube Good in Summer, But Not in Winter

Check it out for yourself. Put your patch lube in the freezer on a patch. Does the patch harden into a brick? There are several modern patch lubes that do not harden up, or at least not enough to cause trouble. These lubes remain functional in cold weather. Purists may wish to look to whale oil for its good cold weather properties. Others should read the statement on the lube container, as well as freezing treated patches as a test. Once loaded downbore, a lube-hardened patch is not necessarily a big problem, but loading stiff patches can be tricky in the field.

Troubleshooting the Correctly Seated Load

Put a witness mark on your ramrod after you've decided on a pet load for a certain muzzleloader. It's easy to do. Drop your powder charge. Ram the patched ball or conical downbore and seat it correctly. Do not remove the ramrod. Instead, make a mark to coincide with the ramrod's position in the bore. Indelible inks will permanently mark most rods. The mark, which is placed where the ramrod meets the crown of the muzzle, will forever reveal the correct seating depth of your load. If, for example, the mark rests well above the muzzle, you know that the load is not properly seated. Perhaps there is heavy fouling in the breech, or a patched ball is stuck off of the powder charge. The marked ramrod will let you know that things aren't right. In the name of consistency, mark the ramrod so that each pet load is delivered to the breech in very much the same position as the load before it.

Misses Caused by a Moved Sight

Put a witness mark on your sights. This way, if the gun gets bumped, you will immediately know if the sights were knocked out of line. Once the front sight and rear sight are properly aligned in their respective dovetail slots following the sighting of the frontloader, make a mark (or have your gunsmith do it for you) that will prove that the sight has not moved. This tiny line extends only a fraction of an inch from the sight to the barrel. Should a sight move, the movement will be readily spotted, because the line that extends from sight to barrel will be broken.

Stock Gets Burned from Ignition

The best prevention is a flash cup for the percussion firearm. A flash cup diverts sparks away from the wood. Also, a bit of masking tape around the nipple area or the lock area of the flintlock will save the wood from getting burned.

Round Ball is Damaged when Seated

The nub end of some short starters can put a deep dimple in the round ball during the initial start into the muzzle of the firearm. The nub of the short starter can be made less offensive to the lead ball by attaching a small piece of electrician's tape to it.

This has been a little look at a few possible problems that can be dealt with by applying proper troubleshooting techniques. ●

chapter **40**

Muzzleloader Modifications

MUZZLELOADERS CAN BE modified in a number of ways, some of them requiring a great deal of skill at a professional level, others demanding no more than the simple installation of a part. The hobbyist interested in working on his own guns may wish to consult *Black Powder Hobby Gunsmithing*, a DBI text devoted to building muzzleloader kits, creating numerous interesting and useful muzzleloader modifications, as well as building a rifle from scratch. Why bother talking about modifications here that demand expert attention? For a good reason: If the reader knows what can be done to alter his charcoal burner, he can elect to have the work accomplished for him. If he is oblivious to such possibilities, he will not even know what can be accomplished.

For example, a shot-out or corroded bore can be replaced with a liner. Knowing that, the gun owner can take his favorite frontloader to an expert who can bring Old Betsy back to shooting life, as it were. Not knowing that this can be done, the gun owner has a wallhanger in his possession, and nothing more. Of course, lining a bore is a job for a pro, and not just any gunsmith either, but one with the knowledge and tools to accomplish this specific task. On the other hand, anyone can purchase a drop-in replacement barrel for his muzzleloader (provided it is a gun for which drop-in barrels are made) and by that simple act he can replace a bad barrel, or add to the versatility of that specific firearm. For example, certain frontloaders can be both shotgun and rifle by merely switching barrels in the same stock. Here are a few modifications possible with blackpowder guns:

Tuning a Lock

Locks can indeed be tuned to function more smoothly and more reliably. Engagement or contact points within the lock can be polished. A frizzen can be hardened. Only a few tools and some fine grit paper are required for the first chore, while the second demands a good source of heat, a safe place to work, and some sort of cooling substance, such as a container of quenching oil.

Full details are given in *Black Powder Hobby Gunsmithing*, but anyone interested in hardening a firearm part, such as a working piece of a lock or a frizzen, should be aware of the possible dangers of working with torches. For the advanced hobbyist who has the equipment and the place to work, however, parts hardening is a worthwhile modification to a muzzleloader, for harder parts may improve the function of the gun.

Flint to percussion conversions were done long ago and can be done today. However, this is definitely a job for the expert.

The fly allows the nose of the sear to override the half-cock notch for

Is it a job for you? Perhaps not, but have you read *Black Powder Hobby Gunsmithing*? This DBI book (same company that publishes this text) by Fadala and Storey, is loaded with how-to muzzle-loading gunsmithing projects.

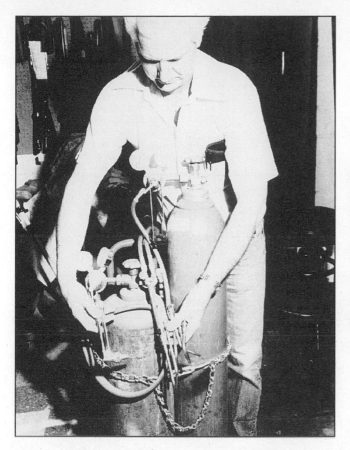

(Left) Heat-treating parts can harden them, which makes them last longer and sometimes work more efficiently. However, in many cases, sophisticated equipment is necessary. This equipment can be dangerous to operate.

(Above) Showing the hardening of a lock part by heating and then quenching the hot part in oil. *Black Powder Hobby Gunsmithing* goes into this process in more detail.

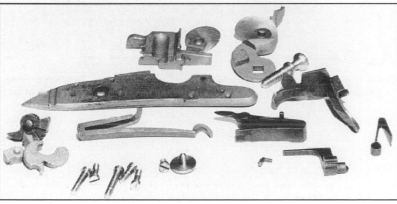

Not all modifications are suitable to the home workshop. Tuning a lock may be one of those, depending upon the gun owner's level of expertise. But a lock can be tuned to function like a Swiss watch. The tools needed are knowledge (first and foremost) and a set of fine hones.

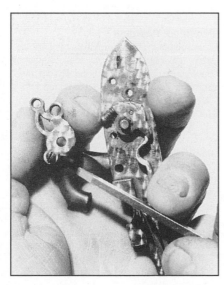

Fitting a fly to the tumbler is, in my opinion, a job for an expert blackpowder gunsmith or at least a very advanced layman. I'm not so sure that even an expert modern gunsmith would know the routine. But a fly can be installed so that the nose of the sear overrides the half-cock notch for positive action.

smooth operation. A fly or detent can be installed by a professional gunsmith. The fly may be a screw fitted by drilling and tapping a hole. The screw can be held in place with Loctite or simply bottomed out.

Trigger Adjustment

Muzzleloaders have many different types of triggers. Most of these triggers demand the attention of a gunsmith for tuning, not because they are so terribly complicated, but because of safety. After all, it's the trigger that touches off the shot. If the trigger is not properly adjusted, the shooter may lack full control of his gun's function, which can be dangerous. Double-set triggers with a set screw can be adjusted by the shooter, but only with tremendous care. No trigger should be adjusted so light that shooter control is compromised.

The Modern Muzzleloader Trigger

These triggers are very often the same models used in today's cartridge rifles. Yes, they do offer adjustment; however, they can be tricky. I've seen professional gunsmiths work on a modern trigger for a solid hour before announcing that it was both improved and safe. A light trigger promotes hitting the target, but as noted above, too light

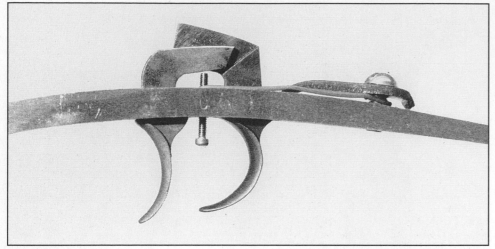

Trigger adjustment can be difficult or easy, depending upon the firearm in question. The knowledgeable gun owner can certainly adjust the set screw on a multiple-lever double trigger system, while he is well-advised to have an expert work on a more complicated trigger system. The set screw between the two triggers shown here can be adjusted by the gun owner, but super care is demanded for safety, even with this rather simple arrangement.

(Right) Many modern muzzleloaders use triggers that come straight from the world of cartridge rifles. These triggers are best adjusted by a professional gunsmith for reasons of safety.

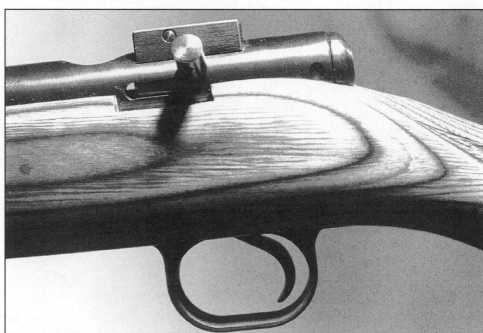

(Below) This is a trigger from a modern muzzleloader. It is adjustable, but the gun owner is well-advised to have an expert do the job.

can be dangerous. Leave most trigger adjustment to the experts, but if you want your trigger reduced in creep with a crisp letoff, take your blackpowder firearm to a gunsmith and tell him what you're after. The improvement can be dramatic.

Bending a Hammer

Sometimes a hammer will not fall exactly where it should. The nose of the hammer, in this case, does not make flush, full contact with the topmost part of the nipple cone. By bending the hammer, however, the angle can be changed so that the hammer nose does make full contact. This process requires heating and is best accomplished by a gunsmith. But it can be done. That's the important thing to know.

Barrel Improvements

The Barrel Liner

The barrel liner, or bore liner, is a barrel that fits within the existing barrel. The liner replaces a bore that is shot out or corroded. The original barrel is carefully drilled out to accept the liner. This job is only for the competent gunsmith. In fact, it is only for those smiths who have specific barrel-liner knowledge. I have fired only two rifles that were treated to barrel liners. Both were entirely shot out, and neither could be counted on to hit a target, and were not much better than smoothbores before the job. The same gunsmith lined both of these rifles. With their new liners, both rifles shot well again.

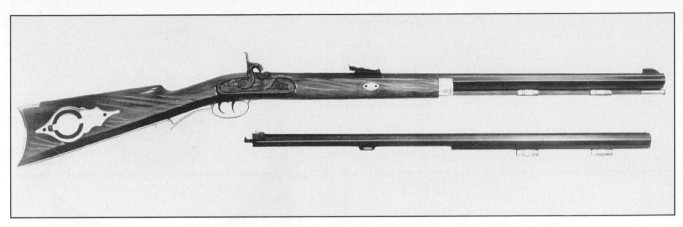

Adding a barrel can be a good modification for muzzleloaders, and it takes no more savvy from the gun owner than simply buying the correct new barrel and installing it. There are a number of muzzleloaders that have extra barrels offered for them. These barrels simply drop into place in the stock, and the shooter has added another dimension to his firearm. This Armsport Model 5114 takes a 45-caliber barrel, a 50-caliber and a 20-gauge, all on one stock/lock.

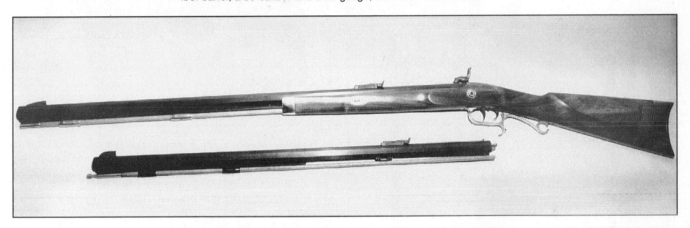

Many different drop-in barrels are offered. In this illustration, the Thompson/Center Hawken rifle is shown with two drop-in barrels made by Green Mountain Rifle Barrel Co., Inc., of New Hampshire. One is 50-caliber with a 1:70 rate of twist, and the other is 36-caliber with a 1:48 twist.

Adding a Drop-In Barrel

This is an extremely simple matter, for it entails no more than buying another barrel for an existing muzzleloader, provided of course that there is a drop-in barrel offered for that particular firearm. Some muzzleloaders come ready to do double duty. For example, the Thompson/Center New Englander is a shotgun or a rifle depending upon the barrel in place. Green Mountain company also offers several drop-in barrels that can change a muzzleloader's character in a few seconds. For example, the Thompson/Center Hawken has 36- and 50-caliber replacement barrels available, both ball shooters. The 36 has a 1:48 rate of twist, while the 50 has a 1:70 rate of twist. How simple it is to modify a muzzleloader with a mere switch of barrels for those guns that offer multiple-barrel options.

Rebarreling

Without a doubt, this is a job for the expert, but it's surprising how few shooters take advantage of the rebarreling process. When a bore is shot out or corroded, the firearm is rendered unshootable, at least where accuracy is demanded. It's cheaper to rebarrel than to throw the gun away and start over, except with cheap guns, in which case it costs more to rebarrel than to buy a whole new firearm. The obvious advantage of rebarreling is returning a muzzleloader to shooting condition. However, that's hardly the only reason. Consider the competition

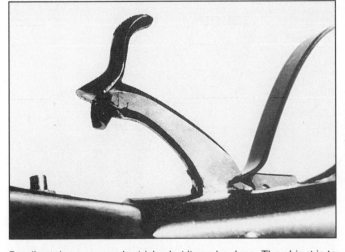

Bending a hammer can be tricky, but it can be done. The object is to better align the hammer nose with the face of the nipple cone, thereby promoting a direct and clean blow upon the percussion cap. A professional will heat and bend the hammer to accomplish this goal. This underhammer rifle has had the hammer bent so that the hammer nose strikes flush on the face of the nipple cone.

A barrel liner is another excellent means of fixing a bad barrel; however, this task is definitely one for the trained expert with the proper equipment and know-how. When done properly, the result is a new bore in an original barrel, which can be very important when the firearm itself is an original.

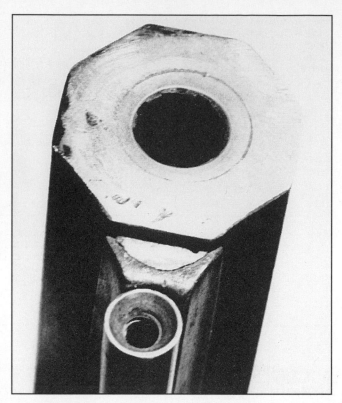

Rebarreling a firearm is definitely a job for a professional blackpowder gunsmith, but it is also a wonderful modification for a firearm that is otherwise good, but has a shot-out or corroded barrel. Dale Storey poses with an original musket that he rebarreled.

shooter who felt he could win more metallic silhouette matches with his single shot breech-loading blackpowder cartridge rifle if only the rifle were capable of better accuracy.

He liked the firearm, and it wasn't a cheap model. So he had the rifle fitted with a top-grade barrel. The work was done by a professional who performed an extremely careful job with close tolerances. The result was definite accuracy improvement. I don't know if the man won more matches, but I do know that his rifle shot much tighter clusters, because I was the one who pre-tested the breechloader before rebarreling, and then shot it again after the new barrel was fitted. It shot a lot better with the new barrel. So rebarreling can replace a worn-out bore, and it can enhance accuracy if a top-grade barrel is used for replacement. Incidentally, I know of one original muzzleloader that was rebarreled simply because the owner wanted to shoot the piece. He kept the old barrel, intending to replace it if he decided to reinstate the rifle as an "official" collector's item. So this is another possible reason for rebarreling: to shoot an original muzzleloader.

A Burned-Out Touchhole

In time, the fire flying through the touchhole of the flintlock firearm causes burnout. Then, misfires and hangfires may occur. However, the situation can be remedied with a touchhole liner. Once again, the pro is called upon, because this job entails precision hole-drilling, followed by correct tapping of the hole so that the threaded portion of the liner screws firmly into place. The Uncle Mike's touchhole liner is readily available, easy on the pocketbook, and its installation can give a flintlock a new lease on life as well as improved ignition.

Chamber Polishing

The chambers of the cap 'n' ball revolver can be polished using fine-grit paper on a dowel or other well-fitted cylindrical tool of the correct size. A smooth, polished chamber is easier to clean and keep clean. Polishing should be done by hand. The use of a drill can speed the process, but it also increases the possibility of taking off too much metal. Also, angling the bit while polishing may damage the uniformity of the chamber. Turning the polishing medium by hand is much slower, but much safer, with less chance of taking away too much metal or polishing off-center.

Here is a severe modification—changing a flintlock to percussion. The touchhole has been drilled and tapped to accommodate a drum and nipple, definitely a job for a person with the right tools and knowledge.

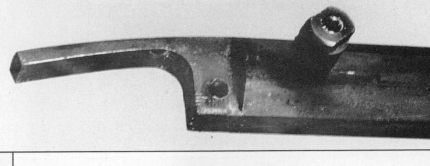

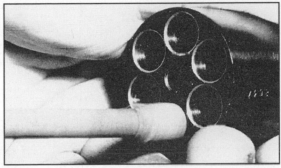

Polishing the cap 'n' ball revolver's chambers can be accomplished by the gun owner if he uses care. He centers a rod in the chamber, the rod holding fine-grit emery cloth for a nice polishing job. A polished revolver chamber is a bit easier to clean and keep clean.

Chamfering the mouths of the blackpowder revolver's chambers creates a bevel which allows seating the round ball without shaving lead from it. Some shooters feel this promotes accuracy by keeping the ball intact. Here, a chamfering tool is being used.

Chamfering the Revolver Chamber Mouth

The mouths of a blackpowder revolver's chambers can be chamfered, or angled, so they form a sort of mild cone shape. A chamfering tool is used. Once again, turning by hand is appropriate. There is no need for speed here, but there is a need for uniformity. By turning the tool by hand, the mouth of each chamber can be chamfered neatly and uniformly. The end result is no more shaving of the lead ball as it is forced into the mouth of the cylinder chamber. Some shooters believe this promotes accuracy by preserving the shape of the round ball, although not all pistoleers agree with this assessment.

Fitting a Scope Mount

This is normally a very simple procedure these days. That's because so many muzzleloaders come from the factory already drilled and tapped for scope mounts. The gun owner needs to know little more than how to use a screwdriver or Allen wrench. Of course, he must have the correct scope base and rings for the firearm in question; and in using that simple screwdriver or Allen wrench, he will want one that fits perfectly so that the screws of the scope mount are not "chingered," as the term goes. For those muzzleloaders that are not drilled and tapped for scopes, the gunsmith comes in. He must find a mount that

A great deal of woodwork can be accomplished by the gun owner, provided he has a place to work, a good body of woodworking knowledge and, of course, the tools, such as these chisels and gouges.

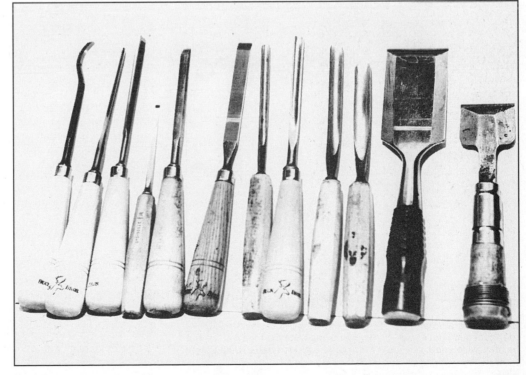

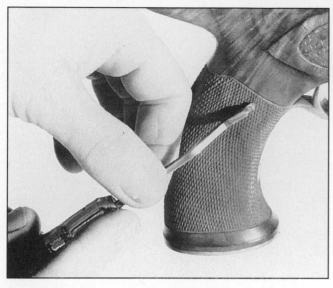

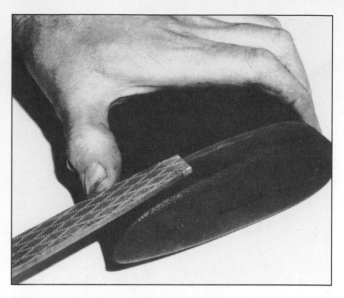

Fine checkering is normally accomplished by a professional; however, some gun owners can do it themselves, given the correct tools.

Adding a recoil pad is a modification that is tricky, yet can seem overly simple to accomplish. On the other hand, a pad can be attached very easily if the gun owner is not concerned about a 100-percent perfect fit.

A sling can also be added when no hardware is available. The Uncle Mike's Slinger, shown here, takes care of the problem.

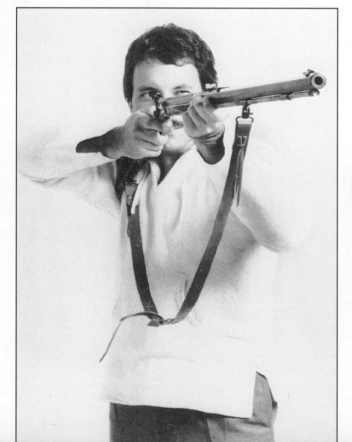

While hot bluing at home is out of the question for the average black-powder shooter, browning is not, and it is an authentic old-time metal finish that is handsome and practical.

Adding a sling can be a very simple process when the firearm comes with sling swivel eyes already attached. This one is Uncle Mike's Full Grain Carrying Strap.

Refinishing a gunstock (or finishing one) is a job that the home hobbyist can do. See *Black Powder Hobby Gunsmithing* for full instructions. Final coats of oil go on here.

will fit the contour of the gun, after which he drills, taps and fits the mount. Sometimes a mount must be modified to fit a muzzleloader for which there is no existing mount. All of this requires professional work, but it can be accomplished.

For example, I had a Bausch & Lomb 6-24x variable scope mounted on a custom Dale Storey muzzleloader for experimentation. Mounting a scope on this handsome traditional-style sidehammer custom rifle was a terrible thing to do to a rifle intended to fit the old scheme of things, but what was learned was worth the risk. Proved was the fact that a fine blackpowder rifle with top-grade barrel can group into a minute of angle, which this rifle did, but only after a scope was mounted so that the rifle's true potential could be realized. Those shooters who require a scope sight due to aging eyes, or to hit small targets, or to promote tighter grouping may want to mount a scope sight. As stated before, scoping is a personal matter. Occasionally it may be a matter of law. Scopes are not legal on muzzleloaders for certain hunts in certain areas. So before mounting a scope on your frontloader, check the rules.

Woodworking

Woodworking of all kinds can be accomplished by the amateur smith, including fixing a cracked stock, as described in *Black Powder Hobby Gunsmithing*. Chisels and gouges are needed for inletting. Stocks can be carved, too. Many embellishments can be created. A dent can be lifted, and of course, a stock can be finished or refinished.

In some cases, stocks can be shortened by lopping off a bit of the buttstock and refitting the buttplate. The latter can be tricky, however, and stock shortening may best be left to the pro. A stock can be lengthened, too, by adding wood to it or by installing a recoil pad. Checkering is possible as well, although top-grade checkering is accomplished only by the experts. A cheekpiece can be modified, too, even heightened as well as reduced.

A sling can be added in one of several ways. The task may be as simple as attaching the sling with pre-installed swivels; swivels can be installed, or a leather sling requiring no swivels may be selected. The options are many.

Muzzleloader modifications are limited only by the imagination. Decorative tacks, for example, can be installed on a frontloader, as they were in days of old. While tacks are generally a crude way of embellishing a firearm, they can be fitting on some pieces. Finish can be applied to the lock mortice to make the wood more impervious to blackpowder fouling, as well as liquids that leak past the lock during the cleaning process. Finish can likewise be applied to the barrel channel in the stock. Bluing and rebluing should be left to the expert with the tanks to do the job, but metal browning is quite possible in the home workshop, as *Black Powder Hobby Gunsmithing* explains. Modifications can improve the looks of a muzzleloader, or the way it shoots, or both. What would you like to change on your favorite frontloader?

•

Modern Old-Time Gunsmiths

THE OLD-TIME GUNSMITH made an indelible impression upon the fabric of North America. His was a responsible business, conducting a trade that included teaching his skills to apprentices. His guns helped bring food to the table, and they went west with the fur trappers and pioneers. He was also responsible for art that still exists to this day in collections. Surely, the old-time gunsmith is gone, with his simple one-room operation. And yet, others have taken his place. Today we still have old-time gunsmiths. They continue to build the muzzleloaders of yesteryear, and many of them compete favorably with the masters of the past.

In order to understand the modern blackpowder gunmaker, it's important to look to his predecessor. While it is true that there were barrelmakers and lockmakers who built parts for early gunsmiths to work with, the fact remains that smiths of early America were multi-gifted craftsmen. They had to be many things at once: blacksmiths, woodworkers, toolmakers, sculptors, ironworkers, inlay artists, carvers, and more. Proficiency was demanded at each skill, and so the old-time gunsmith was a master at his trade. The community at large depended upon him because a firearm was a staff of life. As stated above, and worth repeating, the rifle helped open a frontier. It fed the family, thwarted enemies, protected life and limb. You didn't last long in untamed America without a shooting iron. And guns, being mechanical instruments, required not only building, but repairing. A territory without a gunsmith was like a house without a foundation.

Mentioned in preceding chapters, gun history is a winding mountain path with countless off-trails. Dates and places are arguable. However, it's fair to say that early gunmaking in this country began in earnest with German (and Swiss) immigrants who brought with them the Jäger-type rifle from the Old World in the early 1700s. These were stout rifles, often of large caliber, but they burned comparatively small powder charges. The Jäger was fine for close-range shooting associated with the European countryside. But the gunmakers of the New Land saw early on that the Jäger was not perfect for America. A smaller ball would get the job done. The barrel could be longer. Greater accuracy would be appreciated.

A truly American firearm was born of these requirements. It came to be known as the Kentucky rifle by most people. Not that the rifle was made in Kentucky. The name may have come from a popular ballad relating to the Battle of New Orleans. The song was called "The Hunters of Kentucky," and the rifle mentioned in it was decidedly an American long arm—hence, "Kentucky rifle." Of course the progression from Jäger to Kentucky had a third branch—

the plains rifle. However, the artists in wood and metal mainly built the middle model—the Kentucky, or Pennsylvania rifle—and not the stout rifle of the mountain men, as discussed in Chapter 48 on the Hawken rifle.

The Pennsylvania/Kentucky rifle had specific characteristics that set it apart from anything that came before, as well as any firearm that followed. Some argue that no more beautiful guns were ever made at any

To get a handle on today's blackpowder gunmakers, it's a good idea to look at some from the past. George Schalk was one of many great gunmaking artists. He lived in the 19th century and produced hand-made rifles of high-grade accuracy.

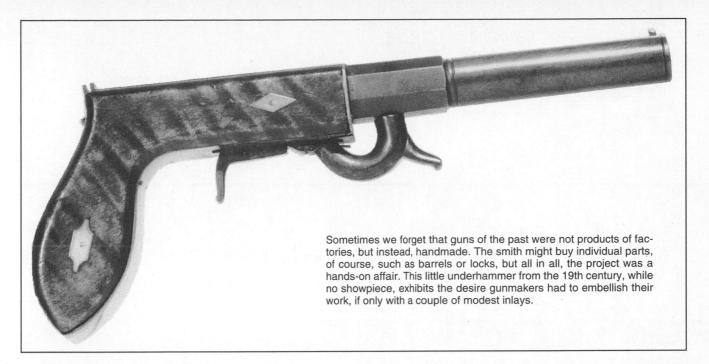

Sometimes we forget that guns of the past were not products of factories, but instead, handmade. The smith might buy individual parts, of course, such as barrels or locks, but all in all, the project was a hands-on affair. This little underhammer from the 19th century, while no showpiece, exhibits the desire gunmakers had to embellish their work, if only with a couple of modest inlays.

time, by anyone. Graceful, with slim wrists, the rifles were often embellished with wire inlay, carvings, patchboxes and other artistic touches. Their long barrels did not promote accuracy directly. But the consequently long sight radius provided a sharp sight picture. Up-front weight also helped the shooter hold his Pennsylvania rifle steadily from the off-hand position.

These rifles were often made in one-room gunshops. While the Pennsylvania rifle had a recognizable style, it was a "custom" all the way. No two were identical. Furthermore, there were *schools* of design. A rifle of the Lancaster school was slightly different from a rifle of the Bedford school (both counties in Pennsylvania). Russell E. Harriger's fine volume *Longrifles of Pennsylvania* discusses these various schools. Volume I includes Jefferson, Clarion and Elk counties of Pennsylvania.

Here is a gunshop of the 19th century. The only feature that reveals the work accomplished here is the large rendition of a firearm hanging over the door. This shop belonged to the fine gunmaker, George Schalk.

Usually forgotten about the old-time gunsmith is his inventive nature. Not all of George Schalk's guns were handsome heirlooms. This percussion pistol, with what amounts to a type of falling block, was one of his inventions.

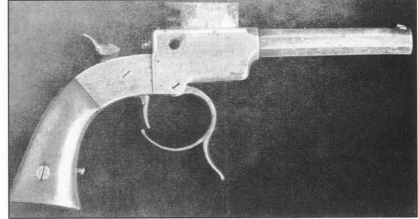

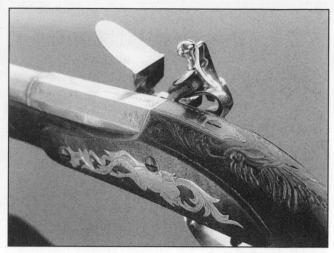

William Kennedy's handwork is exhibited in this fine flintlock pistol.

(Left) Reminiscent of gunshops past, this is William Kennedy's shop in Muncy, Pennsylvania. The handwork that comes out of this door rivals anything accomplished by past masters.

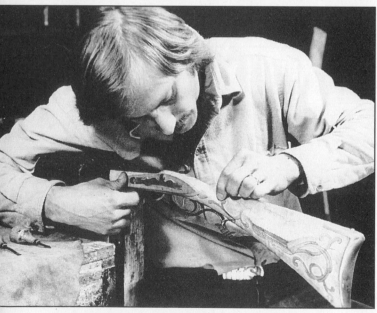

Kennedy is an old-time modern-day gunsmith specializing in Kentucky/Pennsylvania rifles—in every detail, including the finest points of embellishment.

Kennedy plies his trade in the old tradition. His wonderful handmade blackpowder arms are mainly a product of handwork.

While hand skill is an obvious criterion of professional-level custom muzzleloader making, planning every move is also vital. No fine gun is produced without careful thought. Here, Kennedy follows his plan for embellishing another of his handsome copies of a long rifle of the 18th century.

Today's full-time blackpowder gunsmith often works with modern arms as well, as does Dale Storey, shown here. While Dale puts a tremendous amount of hand labor into his work, he also enjoys the use of modern tools, such as his lathe.

(Above) We sometimes fail to recognize that the modern-day old-time gunsmith, like Dale Storey, shown here, works in metal as well as wood, and with many different types of tools.

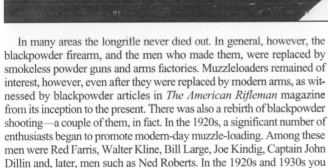

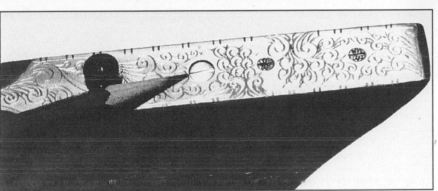

Everything has to be correct. Is this screw perfectly aligned? If not, it will be when the master craftsman is finished with his work.

In many areas the longrifle never died out. In general, however, the blackpowder firearm, and the men who made them, were replaced by smokeless powder guns and arms factories. Muzzleloaders remained of interest, however, even after they were replaced by modern arms, as witnessed by blackpowder articles in *The American Rifleman* magazine from its inception to the present. There was also a rebirth of blackpowder shooting—a couple of them, in fact. In the 1920s, a significant number of enthusiasts began to promote modern-day muzzle-loading. Among these men were Red Farris, Walter Kline, Bill Large, Joe Kindig, Captain John Dillin and, later, men such as Ned Roberts. In the 1920s and 1930s you didn't run to your local gunshop to buy a factory-made blackpowder rifle. You either found an original or someone had to make a firearm for you. This fact promoted the modern old-time smith.

It's a marvel that the art remained alive, especially with rifles of such superb quality, many as fine as the finest arms handcrafted when America was in its infancy. The names of the world's best muzzle-loader makers are well known to blackpowder enthusiasts. Hacker Martin, who died in 1970, built many beautiful and functional longarms and did much to promote modern old-time gunmaking. William Large made some of the finest blackpowder barrels ever seen. He did so for over half a century. In more recent times, Homer Dangler, Ted Fellowes, Dennis Mulford, Andy Fautheree, William Kennedy and many others have kept alive the tradition of fine art in custom-made muzzleloaders of the old schools. Some builders lean more to art than function. One of my friends, upon buying a beautiful hand-made muzzleloader, declared, "I'm not going to shoot it. I'm just going to appreciate it." On the other side of the coin, another shooting associate had a

handsome and highly functional frontloader built for hunting. The custom gunmaker, Dale Storey, did a fine job, but the rifle was as rugged as it was pretty, and it was highly accurate and powerful to boot.

While our country's original guncrafters used hand-powered tools because there were no power tools (many modern armsmakers still build muzzleloaders truly by hand) today's artisans use both hand-powered and electric-powered tools. Bow drills and other old-time tools have been superseded by modern machine shop implements, including lathes. But handwork remains the cornerstone of modern muzzleloader custom building. Today's custom long rifle has more handwork in it than power tool work. I know of one gunsmith who actually draw-files his round barrel stock into octagonal form. He feels this touch gives his rifles a true handmade quality.

Many blackpowder gunmakers ply their trade as a part-time effort. However, others are so skilled in what they do that the call for their work has lured them away from "a job" into a full-time shop. Dale Storey decided to go full-time when, as he said, "I was having to work every weekend and every night to keep up with the demand. Finally, I had to make a decision. I could either turn down further requests for custom rifles, or I could open up a full-time shop." He opened a shop because part-time effort could not come close to fulfilling his customer's needs. Ted Fellowes went full-time in 1964. Requests for his fine rifles could never be met by part-time effort. His Beaver Lodge gunshop in Seattle was busy all the time, with customers waiting a year or longer for one of his rifles, while declaring the wait worth it. Fellowes makes "Everything from Hawkens to Southern Mountain rifles, English sporting guns and some schools of Pennsylvania long rifles,"

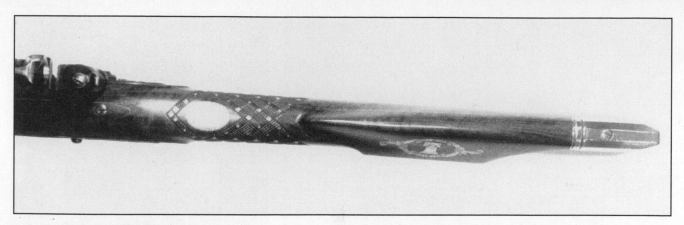

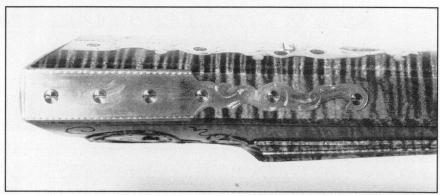

Andy Fautheree is known for his accurate renditions of guns past. He does not try to duplicate an old gun exactly in all of his work, however. He adds his own master touches. Here is a top view of one of his Western Pennsylvania rifles. The wrist exhibits a skipline checkering pattern with silver dots, the latter a Fautheree trademark.

This J.P. Beck rifle is one of Andy Fautheree's fine modern-made rifles on a past design. Here he has copied not only the toe plate pattern accurately, but also the engraving, using *Thoughts on the Kentucky Rifle in its Golden Age* by Joe Kindig as a photographic guide.

he said. William Kennedy is another who turned gunmaking into a full-time effort. His Kennedy Arms Co. now makes some of the finest custom muzzle-loading pistols in the world.

The quality of the better modern custom blackpowder rifle is without parallel, plus many of the rifles replicate originals amazingly. Homer Dangler, a name associated with superb handmade long arms for many years, produced a great number of high-grade rifles with every sort of embellishment of original style. Dennis Mulford built Jägers for people who knew they'd never find, or could not afford, an original version of this historical rifle. Dale Storey worked with sidehammers. Andy Fautheree always prided himself in the correctness of his Pennsylvania rifles, as well as his actual copying of guns from books. For example, Andy accurately copied a J.P. Beck rifle, including the engraving, from Joe Kindig's famous book, *Thoughts on the Kentucky Rifle in its Golden Age*.

The professional modern old-time gunmaker is a unique individual. He generally works alone. He has learned, as his forebears had to learn, how to do many things from inletting to finishing wood and metal. He has a great deal of knowledge with talent to match. The master muzzleloader builder is an artist in wood and metal. There's is nothing quite like a correctly built muzzleloader. And the modern custom longarm wouldn't exist but for the men who have devoted themselves to its correct manufacture—by hand. ●

chapter 42

Big Game with Blackpowder

THE MAJOR REASON modern hunters take up a smokepole when they hit the big game trail is opportunity, clear and simple. Special blackpowder-only "primitive" hunts abound, almost every state offering several annually. These hunts are held, sometimes, on grounds unsuitable to long-range modern guns. Or, since the blackpowder challenge normally results in a lower harvest, special blackpowder hunts are allowed before or after regular seasons, when conditions are more prime due to better weather (early hunts) or rut (later hunts). I suspect that the blackpowder bug bites most hunters who try front-loader-only seasons, and then these men and women go because they love it, not because they know it's a good deal. The best writer in the world would not be able to put on paper the enjoyment experienced going for big game with a charcoal burner, nor the thrill of "making meat," as they used to say, with the old time shootin' iron. It's like winding the hands of the clock back a hundred years, stepping out of your time machine, and setting foot in a whole new world of double challenge and triple reward.

In North America, the white-tailed deer is number one, but hardly the only big game hunted with blackpowder guns. Mule deer are prime, too, and pronghorn antelope, wild boar, bison, elk, javelina, black bear, grizzly, polar bear, black-tailed deer, Coues deer, wild sheep, mountain goat, and more. Abroad, there are hundreds of species open to smokepole hunting. Africa alone offers dozens of plains game animals, not to mention the beasts of the bush and jungle. This is not the forum to list blackpowder big game huntables of the world. Suffice that even Italy, not known for hunting, let alone muzzleloaders, has blackpowder big game hunting. A law set down in 1993 helped liberalize the ownership of muzzleloaders and their use in Italian hunting.

Which Guns?

Muzzleloaders get the special hunts, but blackpowder cartridge rifles, loaded with blackpowder or Pyrodex, are on the move, too, for big game hunting. As for gun types, the reader knows there are many good blackpowder big game choices. Put me in a tight thicket for white-tailed deer with a short-barreled double smoothbore (shotgun) loaded with patched round balls and I'll be happy. Or place me in a more open setting with any muzzleloading long gun, such as a rifled musket, a Kentucky, a plains rifle—any will do. A Sharps, Remington, or other blackpowder cartridge arm would be fine for hunts during regular seasons, too. These guns are normally disallowed for special primitive hunts. And it doesn't have to be rifles only. As clearly stated before, there are a few blackpowder pistols that develop big game

The white-tailed deer is the number one big game animal in North America. One way to hunt this deer is from a treestand, which can provide a close shot.

power, the Thompson/Center Scout coming first to mind, with big 50- or 54-caliber conicals at good muzzle velocities.

Every big game hunter wants the same thing: a fast and clean harvest. This is exactly what the big bore muzzleloader or blackpowder cartridge rifle is capable of. Plenty of information has been passed down in this book on calibers, loads, guns and more. See Chapter 22 on ballistics, for example. Clearly, blackpowder power comes from

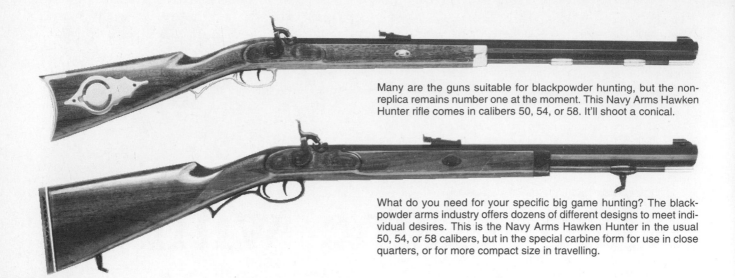

Many are the guns suitable for blackpowder hunting, but the non-replica remains number one at the moment. This Navy Arms Hawken Hunter rifle comes in calibers 50, 54, or 58. It'll shoot a conical.

What do you need for your specific big game hunting? The black-powder arms industry offers dozens of different designs to meet individual desires. This is the Navy Arms Hawken Hunter in the usual 50, 54, or 58 calibers, but in the special carbine form for use in close quarters, or for more compact size in travelling.

Nick Fadala harvested this white-tailed buck with a 50-caliber CVA muzzleloader using a Buffalo Bullet conical. There are many ways to hunt with a muzzleloader. Still-hunting got this buck.

good-size bullets, since velocity falls into the medium realm only. There are no 3000-fps frontloaders. Round balls that travel at 1800 to 2000 fps, conicals running 1500 fps or so, and pistol bullets in sabots falling in between the two, constitute muzzleloader bullet speeds. Add bullet mass (weight) to these velocities, and you have big game authority, provided you get close to your quarry. Although there's plenty of talk currently about 200-plus-yard blackpowder big game shooting, I don't buy into it. I say let's forget about developing the 200-plus-yard muzzleloader and work instead on promoting the *minus* 100-yard-hunter, the man or woman who can close in for that one perfect shot and make it count.

Any big-bore muzzleloader allowed a reasonable powder charge and fed the right projectiles will take big game cleanly from close range, which, to me, means 100 yards or under, and preferably not farther than 125 yards. But there are a number of special big game charcoal-burning rifles intended to give an extra power edge for big game hunting. The Zephyr is one. From the Pacific Rifle Company, this 62-caliber round ball shooter is allowed a hefty powder charge for decent muzzle velocity. The big round ball does the rest. The Storey Takedown Buggy Rifle is another. In my version of this heavy-barreled rifle, 50-caliber, I burn a big dose of FFFg blackpowder behind a big bullet, such as the 600-grain White Superslug. Gonic's 50-caliber Magnum modern muzzleloader shoots big bullets, too, like the company's own 486-grain Copper Nose. Large powder charges are allowed in the Gonic Magnum, such as 150 grains of Pyrodex RS. Combine 150 grains Pyrodex RS with the 486-grain Gonic bullet and the result is 1610 fps muzzle velocity for an energy rating close to 2800 foot pounds.

Sam's Bullet Box

Over the years, I've had great success with a simple wooden box, compartmentalized. In the compartments, different media are placed to represent not bone, bodily fluids or muscle structure so much, but rather obstacles that challenge bullets. My obstacles may, in a way, reflect upon animal tissue and the like, because I do see a high correlation between good results in the box and good results in the field. But I still lay no claim to making a working model of a real big game animal with my bullet box. Use of water-filled balloons, however, do challenge a bullet, as liquid is prone to do, while wooden boards—not really related to bone—supply hardness. And clay, while not akin to muscle, does give an energy-absorbing medium that bullets have to struggle through. The large department store catalogues I use are tortuous to bullets, too. All in all, if a bullet performs well in my bullet box, it'll do likewise in the field.

I look for many things from my bullet box tests: penetration, bullet upset (mushrooming), wound channel, retention of bullet mass, and so forth. Comparing modern bullets from modern arms has always been interesting. For example, the 7mm Remington Magnum, a wonderful big game number, with a 175-grain bullet at 3040 fps muzzle velocity, having an impact velocity of 2300 fps for this test (representing about 300-yard shooting) went through a large Sears catalog, a couple of inches of clay, a water balloon, and three-fourths of a second catalog before coming to rest. The recovered bullet weighed 69 grains. At an arrival velocity of 2512 fps, representing a 200- to 225-yard hit, results were almost identical, telling me that that particular bullet was about burned out after the book, clay, water, and three-fourths of book number two. Incidentally, the recovered weight of the faster-striking bullet was 66 grains, amazingly similar to the result with the slower-striking bullet.

The same bullet box with identical fresh media was struck by muzzleloader bullets for comparison. A 58-caliber 625-grain Minie hitting at 1200 fps impact velocity travelled through a book, the clay, the water, another full book, and yet another full catalog, coming out of the back of the bullet box. Do not misunderstand. This fact is not presented to show that blackpowder guns have more power than modern guns. Not so. It simply shows that a big lead bullet, even at slow speed, can do some work. The recovered weight of the 625-grain Minie was 508 grains, or over 80 percent of its original weight. The 7mm bullets retained about 40 percent of their original weight. Again, this is not a blow against the great 7mm Magnum. In fact, some would say it favors it, showing that the jacketed bullet probably left more energy within the target than the big lead bullet did. But the bullet box test was convincing concerning the fact that all-lead bullets tend to remain intact, rather than fragmenting.

The bullet box also proved that little round balls, wonderful as they are for small game and such, are no match for big lead bullets when it comes to penetration. The little 36-caliber squirrel rifle with .350-inch 65-grain round ball only made it halfway through the first book. And that round ball took a beating, too, losing half its weight in the catalog. A .490-inch round ball did far better. Striking at 1200 fps, it lost 40 percent of its original weight, but somehow made it through the first book, the clay, the water, and one-third of the second catalog. A 58-caliber round ball did better yet. Of 280 grains, it struck at 1100 fps, losing only 20 percent of its original weight and passing through the first catalog, the clay, the water and completely through a second catalog, stopping on the cover of a third catalog.

Chapters 16 and 17 contain numerous facts about round balls and conicals respectively, and that data will not be repeated here, because it's the same old story: large-caliber round balls make for power; heavy conicals likewise.

We've been talking about guns and bullets for blackpowder big game hunting. It boils down to personal preference pure and simple, but the hunter must be prepared to accept the challenge he puts upon himself. I've had wild success with the patched round ball, finding it absolutely deadly. But I don't try to shoot from my county into yours. I've also had super luck with conicals, but I don't expect a solid-nose lead bullet to drop a game animal in its tracks with the universally-accepted chest hit. Sometimes that's just what happens. But without bullet expansion, which may happen because of the lack of bone in this region, the deer, elk, bear, whatever, may stroll a little distance before toppling over.

The Wound Channel

The wound channel is the hole created by a projectile, in this case, in a big game animal. The process of creating that channel is far more complex than meets the eye. Bullet upset, or mushrooming, dispenses energy, but it also changes the shape of the wound channel. Everyone who has experience harvesting big game with modern cartridges has seen an exit hole far larger than the size of the bullet. This is extremely evident with modern high-speed mushrooming-type big game bullets. A 2-inch exit hole made by a 30-06 Springfield or similar cartridge is not at all unusual. Did a 30-caliber bullet really expand to 2-inch diameter? Not at all. If it doubled its original diameter, that would only be .60-inch, or a bit over half an inch.

One season I watched through a spotting scope as my son took a buck antelope. I had the 20x glass steadily on the animal when the bullet struck. The hide flew out and away from the animal as if a balloon were blown up. My son's 30-06 had created that 2-inch exit hole noted above, and yet, the bullet was trapped against the hide of the pronghorn. What pushed the hide out? What made the 2-inch exit hole? The bullet did all of this, of course, but not directly. There is a shock wave in front of the projectile, and it is a product of force (energy). It can do an immense amount of work. Wound channels result indirectly from bullet energy of course, but not directly applied bullet energy.

This whole business of wound channel can be immensely helpful to the blackpowder hunter interested in learning more about his craft. I speak of he who wants to *understand* what's going on. By studying

Sam's Bullet Box is simple but effective. It's merely a compartmentalized unit capable of holding various test media, such as water balloons, clay, and large department store catalogs.

actual wound channels, the hunter learns how his blackpowder bullets work, and they certainly do not work the same way with varying bullet styles. For example, I've seen buck deer take a strike directly behind the shoulder from a 45-caliber 500-grain lead conical, and yet the animal walked 100 yards before falling over. Why? Because very little energy was imparted from bullet to game. On the other hand, I've witnessed the lowly round ball drop a buck deer in its tracks, and yet, the round ball arrived with far less authority than the big conical.

Why did the lead ball do so well at close range? Because it imparted its energy through bullet upset. If nothing else is agreed on in this chapter, let one important point stand: With bullets that do not tend to expand, a hit in the "boiler room" may not show much effect. Please follow up on all shots taken at game to make certain that you missed, if indeed you did miss. You may find your game in the brush defunct, but a short distance away. If you're shooting a solid lead conical, you may wish to aim just a bit forward of the chest region, into the scapular area. This does not imply far forward into the point of the shoulder, but rather into the shoulder blade. It's a deadly aimpoint with solid lead conicals.

Bullet Placement

We devoted a whole chapter to sighting-in, Chapter 30, so that trail will not be trod again here. However, it's important to know your gun's trajectory, be it a muzzle-loading rifle, pistol, smoothbore or blackpowder cartridge rifle. Knowing trajectory as well as the basics of wind drift can make a big difference in bullet placement. Know the trajectory of your hunting instrument so you'll know how to aim for perfect bullet placement at various ranges.

See for yourself. Place a target behind a good brush screen and fire at the bull's-eye from 50 yards or so. Bullet deflection will be shown.

Take a rest. If at all possible, rest that rifle for the best possible shot. Here, Fadala rests his Hawken replica across a boulder, with his hand under the forend to protect the wood.

Uphill and Downhill Shooting

How many big game animals are missed every year because hunters will not believe shooting angle makes a difference in bullet trajectory? It's not practical to go into details on uphill and downhill shooting, because no one can judge range exactly, let alone degrees of elevation angle. The practical way to deal with this phenomenon is simply this: hold lower than you normally would hold when your target is at a steep angle *either uphill or downhill* from you. We are not talking gravity directly. We are talking angle. It is angle that counts, either angle up, or angle down. So if that deer is steeply downhill from you at what you consider 100 yards, and your rifle is sighted in for 100 yards, *hold on the deer, but a bit lower than you would if the deer were on the same level as you. Likewise if the deer is steeply uphill from you.*

Brush Shooting

What a surprise I had. I stalked a buck successfully and left myself a 20-yard shot. The buck was feeding in a brush patch on the side of a steep hill and the approach was a cinch. I sat for the shot. The deer, feeding away from me, didn't even know I was there. When he turned broadside. . . Boom! He looked at me as if to say, "You helpless fool," and then he ambled away as I sat speechless. I spent two hours trying to spot so much as one small sign of a hit. There was none. Finally, I walked over the ridge and glassed. The same four-by-four was feeding across a canyon. It was him, all right, and in perfect health. I went back to the scene of the . . . joke on me? There it was. The round ball had met up with a branch, shattering it. That branch had deflected the 54-caliber lead pill for a clean miss on a deer's broad chest at only 20 paces.

Since that time, I've spent hours trying to learn about shooting in the brush, and here's my conclusion: avoid it like the measles if you can.

Wait for any opening and guide the projectile through that. You'll be better off for it. I've had fellow shooters mock my approach to brush shooting—until they've tried my little test. It's simple enough. Place a standard paper target with a nice big bull's-eye behind a screen of brush about 50 yards away and shoot at the target. If the bush has only finger-thick twigs, you'll probably do OK. If twigs are much larger, be ready for disappointment. While I still think a big-caliber, blunt-nosed, not-too-fast bullet makes more sense in the brush than a small fast missile, the truth is, I've had about the same results with either.

In one of my little brush shoots with a 243 Winchester, a 458 Winchester, a 54-caliber round ball rifle, and a 58-caliber conical-shooter, not one of these guns made a truly decent group on target through brush at only 50 yards. Every bullet was deflected, even the 500-grain round-nose from the 458 Winchester. Now you know why I prefer trying for a path in the brush rather than counting on a bullet to "buck" the brush. In that little four-rifle test above, the darn 243 did about as well as any of the other guns. The round ball was worst. It simply flew off course. The farther back in the brush the target was set, of course, the worse things got, because the deflected angle of the bullet was more pronounced.

Taking the Shot on Big Game

An offhand shot is sometimes the only one you'll get, and you simply must be in practice to do this shot well, but there are many times when a rest is entirely possible. But do we always take it? No, we don't. And we should. I trained myself to rest my muzzleloader on my walking stick, or across a tree limb or on a boulder before firing. What a difference this makes. Also, how many times do we fire without ensuring a perfect sight picture? Finally, what about trigger squeeze?

This Loc-On All Steel Folding Ladder Treestand offers one way to hunt with blackpowder. Whitetail hunters all over North America use treestands these days.

Kelly Glause builds box blinds, which he sets up near waterholes. They work supremely well for blackpowder hunters as well as archers.

We know that if we yank off a shot at the bench, the bullet will either go wide of the mark or fly out of the group, and yet on game, we'll jerk the trigger. So get a rest if you can. Make sure your sight picture is perfect. And squeeze, don't jerk, that trigger.

A Few Methods of Blackpowder Hunting

The following are a few suggestions for bagging big game with a blackpowder gun. There are many other ways to get the job done.

The Whitetail Hunter's Treestand

Take a tip from the modern whitetail bowhunter. Go climb a tree. But don't just climb a tree. Use a treestand. Whitetails tend to pattern pretty well. There's no guarantee that a buck will come by your treestand, even though evidence shows one travelling the trail you're set up by. But there's a darn good chance that if you sit and watch long enough, your buck, or another, will show up. It happens every year to thousands of bowhunters. So think about a treestand for whitetails.

The Ground Blind

Ground blinds come in various forms. Kelly Glause, a guide, builds three kinds of wonderful blinds. The first is a pit dug right in the side of a dirt bank overlooking a trail or waterhole. The second is a little rock corral, usually constructed by a game trail. And the third is a wooden box near a waterhole. I've also had good luck with simple blinds made of cheesecloth (the material game bags are made of) and bits of brush. The cheesecloth is draped over a couple of bushes near a game trail or waterhole. Then, leaves and bits of brush are stuck on the cloth to form a blind. This type of ground blind makes a

A ground blind can be built with cheesecloth like this. After stretching the cheesecloth out and pinning it on the brush, leaves and twigs are stuck to the cloth to make it a good hideout.

Very skilled still-hunters can close in on game, such as this feeding buck.

Sometimes going guided is the very best way. Fadala took this buck with his Storey Buggy Rifle at close range while hunting with guide Kelly Glause.

Sign means tracks and droppings, but also evidence of feeding. Javelina fed on this plant. Knowing that the animals are nearby, due to this fresh sign, the blackpowder hunter can start looking for the animals themselves.

pretty good hideout. Just be sure that you can get a shot off without exposing yourself too much.

Still-Hunting

Still-hunting does not mean standing still. It means moving quietly, with the wind in your favor, through game habitat, and with a plan, not haphazardly. Still-hunting is probably the number one deer-hunting method used in North America. It does, however, demand skill. Aimless wandering is not still-hunting. And while the hunter with a scoped long-range rifle might do all right bumping game off across the canyon, the blackpowder shooter won't have much luck with this plan. So go slowly. Look a lot. Keep the wind in your favor. Try to be quiet. And know the habitat so you'll be in the correct field position for a shot, rather than stuck in the center of a thicket as deer burst out of the brush on all sides, heard, but not seen.

The Drive

A well-devised deer drive is deadly. Likewise for elk in some areas, depending upon the lay of the land. The idea is to post a hunter or hunters, while others move toward the stationary men or women on stand. The wind should be at the backs of the drivers, and in the faces of those on stand. Ideally, game will be pushed from drivers to shooters. This can be accomplished with as few as two hunters. When only one hunter does the driving, he should be double sure to have the breeze at his back so that game scents him coming, sneaking away from him to his partner.

Cutting Sign and Scouting

Big game blackpowder hunters do well to scout an area, searching for sign of game. This sign is generally in the form of tracks and droppings. In time, a hunter learns to determine the age of such sign. By scouting an area before season, when possible, hunters don't waste time where the game is not. That's the whole point of it. Of course, it's difficult to scout an area that's far from home. Sometimes, a hunter needs a guide.

In stalking, move only when the animal is feeding, as this buck is here, or otherwise has its attention diverted.

Stalking for the blackpowder shot is a matter of using what the terrain provides to hide behind, as well as keeping the noise level down and the wind in your face. See the javelina (arrow)?

Going Guided

The past few years, as work obligations increased, I found myself scouting less, even in my own district. Balancing the cost against the results, I got some help. Three super guides have provided great opportunities for me. Kelly Glause does a fine job with blackpowder hunters who want to get into desert mule deer and pronghorn antelope. A trapper by trade, Glause is a marvel at finding game by sign, then locating it in the flesh, and finally putting the blackpowder hunter close for a shot. A high-country Colorado super elk guide is Steve Pike. He guides on the special blackpowder-only elk rut hunt near Gunnison, Colorado. This guy can talk an elk into slingshot range, let alone smokepole distance. Third is a father-and-son team who hunt very interesting mountain lion country in Northern Wyoming, along with excellent elk, mule deer, and whitetail habitat. These men are Dean and Daylen Carrell. If these men like anything better than finding game for themselves, it's locating nice trophies for other people.

Spot and Stalk

This is a very basic way of hunting, and yet it can be a fine way to go with a muzzleloader. Spot and stalk is just that: you cover ground until you spot something and then you stalk it. In open country, as well as in creek and river bottoms, I always search with binoculars, often finding bedded or feeding animals. It's the old rule again: game found before it sees the hunter can be stalked for a close shot.

Tracking

Tracking is a wonderful art few of us do well. I watched a fantastic tracker in Africa follow a single animal through a maze of trails. I did not believe in him, but he proved that he was an unbelievable tracker. While I've never seen a tracker in my own country that great, we can do pretty well in snow. Even then, however, a big mistake is made when a hunter "dogs" right on the trail of a deer or other game. Such game watches its back-trail. It's better to leave the direct trail from time to time, circling ahead with the wind in your favor. I've caught deer

Steve Pike guides blackpowder hunters on Colorado's special bugling elk hunt in the high country.

A guide can be invaluable for a blackpowder hunter who is unfamiliar with the region. Kelly Glause is a Wyoming guide who hunts thousands of acres. His success ratio is very high on mule deer and pronghorns.

"Staying over" is the author's favorite way to hunt. It really doesn't take that much gear, but it does require good, well-organized equipment and supplies for safety and comfort on the backtrail.

(Above) Calling game is one way to get that close shot. Here, Steve Pike, Colorado blackpowder guide, calls from his horse to see if he can get a reply from the black timber in the background.

This whitetail is testing the wind. In this situation, scenting an approaching hunter is the animal's best line of defense. That's why keeping the wind in favor is so important, as well employing deodorant tablets and other means of cutting human aroma.

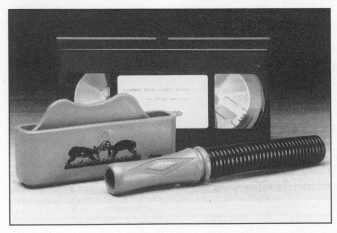

Lohman has an artificial rattling box, plus grunt call, with a video to teach a hunter how to call deer in close for a good shot.

Burnham Brothers calls have been around for a very long time. This is one of them.

Organizing hunting gear is very important, because the well-planned trip is the successful one. This Uncle Mike's large equipment bag is rugged, and it has pockets for sorting things.

Some camps have it all, right down to a hot shower, even in the middle of nowhere. The large tanks on top of the canvas enclosure have both cold and hot water, the hot water provided by the barrel stove.

Pitching a camp in a remote setting can make a huge difference in success, because wild animals aren't as wild when they reside off the beaten path.

looking at their back-trail, sensing something was there, while in fact that something was me looking at them.

Calling and Rattling

I watched Steve Pike call a bull elk within twelve paces of us. I have had good luck with a Haydel's deer call, too, bringing deer in close. There are many different big game calls around these days, and audio tapes to teach their proper use.

Antler rattling is a very old art. It, like calling, is rut oriented. But it can bring fine results for the blackpowder shooter who wants a close shot at a buck deer.

Staying Over

Staying over is simply my favorite way to hunt. My outfit consists currently of a Remington pack and frame for long hauls, along with a very fine Pack Idaho fanny pack with straps. I also carry a lightweight tent, sleeping bag, and other necessary gear. The object is to camp with the game instead of heading for camp in the afternoon. In the morning, you wake up where the wild animals live, usually in a remote spot away from the beaten path.

Hunting Tricks

Buy a little plastic squeeze bottle of archery glove powder from your local archery shop. This fine powder not only shows the general

direction of the wind, but also reveals the pattern of the wind. It works remarkably well for stalking.

Nullo Ghost Scent works from within to reduce human body scent. Since deer and other animals often detect human presence by nose, the plan here is obvious. Get a waterproof parka, good boots, and all good clothing not only for a safer hunt, but also because a comfortable hunter hunts better.

I have cost myself some unwanted misery because my gear was not truly packed and ready to go. Well, not any more. Two fine bags that are working for me today are an Uncle Mike's large gear bag and another fine product from Bagmaster. The latter is really a shooting bag with a padded interior, but I use it for special gear, not only hunting items, but cameras as well.

The Black Powder Hunter's Camp

Good shelter provides rest. A rested hunter is a safer and better outdoorsman. So make the best camp possible. Sometimes it's the smallest thing that counts. I recently tried the Coleman North Star Dual Fuel Lantern, for example, and it proved a real boon. I like a well-lighted camp, and this fine lantern turned the area into daylight. A little thing? Not to me.

There is no way to capsulize the broad world of blackpowder big game hunting into a book chapter, but I hope the few ideas laid down here will pay off.

●

chapter 43

Small Game and Varmints with Blackpowder

SMALLBORE SQUIRREL RIFLES in 32 and 36 calibers make the most sense for little edibles and non-edibles, with up to 40-caliber round balls for varmints in the coyote category. It's the same old tale told by a different campfire: smokepoles add an entirely different dimension to harvesting small game or controlling wildlife on the non-game list.

Small Game

The number one small game animal in North America, and the world, I guess, is the cottontail rabbit or relatives thereof. Powderpuff Tail lives just about everywhere from the sunny deserts of the Southwest to the frigid winter forests of the Northlands, and he's just wonderful to hunt. The tree squirrel is second on the small game list and, if anything, even more cherished for the challenge. Prepared correctly, it too is a fine dish on the table. Then there are dozens of small things that are also huntable and edible. The bullfrog comes to mind. Also the raccoon, opossum, porcupine, rattlesnake, turtle, woodchuck, rockchuck, prairie dog, muskrat and beaver.

Non-Game Animals

There is considerable crossover in categories. The raccoon, for example, is classified as a non-game animal, but can be sought after for its pelt or tail, and certainly the raccoon has been, and will continue to be, cooked for the supper table. Muskrats have fed thousands of trappers and pioneers, but they are generally classified as furbearers, and therefore non-game animals. Beavers likewise. Coyotes and bobcats are not normally food for man, but they cross over, too, as both varmints and furbearers, and therefore non-game or varmint species.

Varmints

In the politically correct modern context, game departments are going away from the quaint and accurate term "varmint" to "non-game species." But everybody has a varmint. Two across-the-highway neighbors of mine who don't hunt, and truth be known are against the sport, take care of injured birds. They've spent the past twenty years at this. Now in retirement from their jobs, they are full-time caretakers of injured eagles, owls, falcons, hawks, and for that matter, crows, jays and any avian that may have flown into an obstacle or was illegally shot at and wounded by a "hunter."

One day I brought an injured sparrow hawk (got caught in a fence) to the pair. As we took the bird to its cage, I noticed something. "What's that?" I asked. I had never seen a mousetrap that large.

"Oh," the lady told me, "that's our rat trap. Rats come in and eat our bird food, so we trap them."

Varmints, or non-game animals if you prefer, make interesting objects of the chase for blackpowder shooters. Of course, in no way will the muzzleloader control the numbers of unwanted damage-causing species. Admittedly, modern arms cannot get that job done either. I recall a ranch that was infested—and that word is intentionally chosen—with prairie dogs. The pastures were riddled with holes. The area was like the underground of Tombstone, Arizona, where miners tunneled everywhere for precious metals. The rancher called upon hunters to thin prairie dog numbers. Numbers were thinned, but the overall population of rodents remained high—until a poisoning program went into effect. That did it. While there has been some repopulation, prairie dogs on that ranch are definitely fewer in number than they were—but not from hunting. And yet, hunting can reduce overall numbers to some degree.

Controlling numbers of non-game animals is not a big factor, so why hunt varmints? For one reason, there is crossover. Young prairie dogs, for example, have been food for man as long as man has lived on the plains. They still are food. Porcupines, while not my idea of tasty fare, have also been eaten, and yet when these animals are in too great numbers, they can destroy many valuable trees, so they're a varmint. Their quills have also been used for decoration over the years, and still are. The blackpowder hunter goes for varmints because they represent an interesting animal, and they can provide food in some instances, or pelts. Meanwhile, there may be some overall benefit. Coyotes eat many antelope fawns where I live. Taking but one coyote can mean a few more pronghorns on the plains, which is important to many.

Some varmints are just as great a challenge as big game. For example, you have to be pretty darn good to consistently bag coyotes, espe-

The number one small game animal in the world is the cottontail rabbit. It's small, but it represents thousands of hours of wonderful outdoor activity and literally tons of superb food every year. Going for Mr. Ears with a muzzleloader turns the sport into an adventure.

The number two small game animal is the tree squirrel. There are hunters who would rather spend a day after tree squirrels than a day after deer. Hunting the Chatterbox can be addictive, especially when blackpowder smoke is added to the scene.

cially with your smokepole, where the 300-yard shot is out of the question. I've had a tough time on marmots, too, in this case Western marmots or rockchucks. And I have correspondents who write to me when they bag an Eastern woodchuck with a smallbore muzzleloader. The same hunters wouldn't bother telling me that they got a whitetail or a wild turkey. The blackpowder challenge, however, is a lot more fun that it is work.

The Advantages of Small Game and Varmint Hunting

Small game, and some varmints as well, are often just a little distance down the road. There is no extensive travel involved in hunting them, nor big outlay of cash for hunting tags and licenses. Trespass fees are generally unheard of, too, on private lands where deer, antelope or other game command folding green if you want to hunt private

Varmints with a muzzleloader: What a challenge and what an experience. This coyote came on the run to a Burnham Brothers call. The Burnham Brothers gave the varmint hunter some of the world's first and best nationally known calls. Since getting close is the name of the blackpowder shooting game, calling makes a lot of sense—bringing 'em in close to you, instead of you stalking to get close to them.

Everybody has a varmint. The kindest granny in the world will hurl a cup at a tiny mouse on her kitchen floor. Prairie dogs do incredible range damage, and ranchers are happy to see hunters remove a few.

Add a flintlock 36-caliber squirrel rifle to the hunt and going for bushytails becomes a prime outdoor experience.

property. In many areas, going for small game or varmints is no more expensive than getting in your car and driving a few miles. And for that, you could come home with several pounds of high-quality food or, on inedibles, an interesting day in the field, and maybe even a little healthful exercise.

There's also economy in the shooting, especially if a smallbore squirrel rifle is put to work. As we know, a 32- or 36-caliber round ball rifle can be fired for less cash outlay than a 22 rimfire, provided the shooter casts his own projectiles and perhaps even makes his own percussion caps with Forster's Tap-O-Cap tool. You'll get about 155 32-caliber round balls, .310-inch diameter, from 1 pound of lead. That's at 45 grains per ball, 7000 grains to the pound of lead. Find scrap lead, and you're really in business. And at 10 grains volume per shot, count on 700 pops for 1 pound of FFFg blackpowder. The 36-caliber, at 65 grains per ball, does not work out as cheaply, at about 107 balls per pound. But that's still a lot of shooting for 1 pound of metal, and the 36 gets good ballistics with mild powder charges, too.

Cottontail rabbits are so high in protein that a diet of that meat alone would not be wise. However, cooked right, these rabbits are as tasty as chicken, and with the proper accompanying vegetables and other foods, rabbit makes a healthy meal. Par-boiled/fried cottontail is tender and tasty. Squirrels are equally fine, although they require their own cooking style. Many other small edibles can grace the table, not only to the delight of the palate, but they're also worth cash, as anyone who has purchased rabbit and similar meat knows. For a whole lot of information on game care, plus dozens of good small game recipes, look into *The Complete Guide to Game Care and Cookery*, a DBI book.

Both small game and varmint hunting with a muzzleloader have a tremendous amount of transfer value to the big game field. The hunter practices his finding and stalking techniques, not to mention learning about new gear from boots to coats. He also gains a great deal of field experience with loading and shooting his charcoal burner. One afternoon, I spent four hours in an area terribly overpopulated with ground squirrels. Did I get any practice? I should say so. I loaded and fired my sootburner many times.

The Shotgun for Small Game and Varmints

The muzzle-loading shotgun, flint or percussion, single or double barrel, is a worthy small game instrument. I think of varmints as rifle

The marmot is a major huntable varmint in North America. In the East, he's a woodchuck. And in the West, he's a rockchuck. In either place, the 'chuck is a challenge to the hunter with a blackpowder rifle.

candidates, but a shotgun can be a small game delight and, in some circumstances, better than a rifle. For example, I've hunted cottontails that were wilder than first-grade kids on the last day of school. Jump and run was their game. Under those circumstances, a shotgun turned the trick. I've tried as little as 28-gauge and as large as 10-gauge on small game, and due to the wonderful versatility of the shotgun, every size worked. A mere half-ounce of #6 shot can be enough when cottontails are taken almost underfoot, which is sometimes the case, and on the other hand, more shot may be the best bet when rabbits are jumping from 30 yards or so, with the target and the shot charge coming together at maybe 40 yards by the time the hunter gets off his shot.

As with quail hunting, I've learned that, for close-range encounters with small game animals, a shotgun loaded with a sufficient charge of shot and a reduced powder charge can be ideal. One afternoon, I ran across a bunch of truly wild rabbits in a field of heavy brush. I had a 32-caliber round ball rifle with me; spot and stalk just didn't work. I couldn't find sedentary rabbits in the thick brush. The only opportunity to make meat was with the scattergun, and my short-barreled 12 was in the truck waiting to be put into action. This chokeless double is good for short-range work and, with one-piece plastic wads, puts out a fairly decent pattern out to about 30 yards. I had my favorite rabbit shot along, #5s. I dropped 70 volume FFg down each bore, followed by one over-powder wad and one cushion fiber wad, then an ounce of #5 shot with one over-shot wad. The limit at the time was five cottontails (now boosted up to ten in the same area). I had my five inside of an hour.

The Rifle for Small Game and Varmints

For small game, the 32- or 36-caliber squirrel rifle is ideal. If forced to choose one over the other, the 32 gets the nod, simply because it's large enough, with its 45-grain round ball, to get the job done and even more economical to shoot than the easy-on-the-pocketbook 36. I can't imagine needing anything more than a 32 for small game hunting. Of course, only head shots count anyway, so this prescription for a 32 or 36 is not chiseled in concrete. If a hunter has a fine 40 or 45 that he shoots well, let him seek small game with it, but I've found larger-than-36 too destructive when other than a perfect head shot is made. For example, I've used a 40 on cottontails, and when the ball hit more neck than head, the shoulder meat was damaged.

However, for varmints, larger-than-36 is OK, although I can't imagine when over-40 would be necessary, since the largest varmint normally encountered is the coyote, and one 40 ball through the chest region on the desert dog will humanely do the trick. For comparison purposes, here are a few figures to work with. The ballistics below belong to four different rifles. Remember that a 22 Long Rifle round with a 40-grain bullet at 1250 feet per second muzzle velocity gets 139 foot pounds of muzzle energy, and at 100 yards, velocity is around 1000 fps and energy is 89 foot pounds.

While the smallbore squirrel rifle is ideal for taking cottontails and other little edibles, in some situations a shotgun is preferred. The rabbits in this setting were simply wild, offering no standing shot. However, a blackpowder double-barrel shotgun proved perfect.

Sample of Small Game/Varmint Rifle Ballistics

Powder Charge (grs. volume)	Muzzle Vel. (fps)	Muzzle Energy (fp)	100-Yard Vel. (fps)	100-Yard Energy (fp)
Dixie Tennessee Squirrel Rifle, 41.5-inch barrel, 32-caliber, (.310-inch 45-grain round ball)				
10 FFFg	1263	170	720	55
20 FFFg	1776	336	852	77
30 FFFg	2081	462	936	93
Thompson/Center Cherokee, 25-inch barrel, 32-caliber (.310-inch 45-grain round ball)				
10 FFFg	1120	125	538	29
20 FFFg	1649	271	775	60
30 FFFg	1871	350	879	77
Hatfield Squirrel Rifle, 39.5-inch barrel, 36-caliber (.350-inch 65-grain round ball)				
20 FFFg	1471	312	794	91
30 FFF	1799	467	882	112
40 FFFg	2023	591	956	132
Ozark Mountain Arms, 36-inch barrel, 40-caliber (.395-inch 93-grain round ball)				
20	1294	346	828	142
30	1584	518	927	177
50	1993	820	1017	214

Small game hunting is much more than strolling around hoping to bump into something little, edible and in season. It's a learned skill. Here, binoculars are used to spot cottontails hiding in the grass of this old corral.

One way small game escapes is by doing nothing, as this rabbit is doing here. Just staying still until the hunter goes away.

One dash forward and this fellow is in its rock den. It's up to the hunter to spot the game before the game spots him.

We can agree that the little 32-caliber, with its 45-grain round ball at close to 1300 feet per second muzzle velocity, is just about ideal for rabbits, squirrels and similar small game. Even when barrel length drops from 41.5 down to 25 inches, a muzzle velocity of more than 1100 fps is achieved. Considering that rabbits and squirrels are often taken at 20 yards or so, these velocities, with energies ranging well over 100 foot pounds, will do the job—and with only a 10 volume powder charge. The 36 is a dandy, too, but admittedly, it does burn more powder. The 40-caliber long rifle is really more than necessary for small game, but on varmints, it's an excellent caliber. For coyotes and foxes, I'd choose the full 50-grain volume FFFg charge for close to 2000 feet per second muzzle velocity, giving a 100-yard energy figure more than ample for the largest "prairie wolf."

A Word About Pistols and Revolvers

Nothing said above is meant to downgrade pistols or revolvers for either small game or varmints. The criterion for using either is simple: Can the hunter hit his target? Pistols and revolvers have all the authority necessary to bag small game, as well as varmints. See Chapter 36 for more information.

In the Field

Once you've decided to hunt small game, then have gone out and picked the proper gun, you need to know what to do when in the field. There are various methods that work well when toting a smokepole.

Still-Hunting

Small game and varmints can be still-hunted successfully. Slow-motion walking is the ticket, whether in a field for rabbits or a forest for tree squirrels. Still-hunting is perhaps the most interesting way to chase small game or varmints because it keeps a person on the move, seeing things, as well as gaining a little exercise.

Jump and Shoot

Shotgunning for cottontails is really a jump and shoot endeavor. It's not very sophisticated, but it works. Walk. Scare something out. Shoot as it runs off.

Spot and Stalk

This is very much like still-hunting and, in fact, could be construed as pretty much the same hunting style. If there is a difference, it lies in covering a little more ground a little faster, relying heavily on binoculars for spotting game in the distance and then stalking close for a perfect shot.

Spot-Dash-and-Tree Squirrel Hunting

This manner of squirrel hunting really works, but before going into it, a warning: **Do not run with a loaded rifle that is either primed (flintlock) or capped (percussion).** There is plenty of time to drop a little pan powder in or fix a cap on the nipple after the dash part of this squirrel hunting strategy is over. Squirrels do not live in the trees all the time. They work the ground considerably. The hunter hikes a lot with this method, in prime squirrel habitat, looking for a bushytail on the ground. When he spots one, he makes a mad dash straight at the little rodent. Naturally, if the squirrel is close and a good shot is possible, the hunter can ready his rifle and shoot. But what usually happens is this: The hunter slowly approaches and the squirrel dashes off over the ground, quickly getting out of sight.

On the other hand, the mad dash provokes Chatterbox to climb the first tree it sees. Now the squirrel is isolated. True, he may head for his den, but usually that's not what happens. Instead, the squirrel climbs up

A lot of information is gathered by observing dens like this one. What might live here? A burrow like this could be the home of a fox or bobcat.

A couple tree squirrels on a Game Tote from the Lohman Co. suggest a fine upcoming meal. For small game recipes, check out *The Complete Guide to Game Care & Cookery*, by Sam and Nancy Fadala, a DBI book.

into the tree, sometimes lying flat on a branch. Squirrels in this setting are very hard to see, so I use binoculars to separate limb from old Umbrella Tail. Once the squirrel is spotted, the hunter maneuvers into shooting position, his rifle recently primed or capped. And now it's time for marksmanship to prevail on a tiny target amidst branches and sometimes leafy foliage.

Calling Squirrels

Contrary to popular notion, and in spite of the squirrel's nickname of Chatterbox, squirrels can be quiet and often are. On the other hand, they have a good voice and love to use it, especially for scolding. Because squirrels do like to talk, a call works well in hunting them. I have a Lohman squirrel call that has brought bushytails from out-of-sight hiding amongst leaves or branches, right out onto a limb in a fighting pose. The squirrel hunter interested in learning more about calling should invest in Lohman's fine video on the subject, called "Squirrel Challenge," and it instructs on aggressive calling, but contains much more than that about squirrels. Calling squirrels is a great way to locate them. It's also a good way to get a squirrel in a tree to show himself.

Calling Varmints

Calling varmints is quite different from calling squirrels. Here, your goal is bringing that fox, coyote or other critter right into smallbore blackpowder range. There are dozens of great varmint calls these days, not to mention the electronic devices available. Whole books have been written on the art of varmint calling. There can be many tricks to this trade. Interested? Do three things: Look for some good books on the subject, certainly, but also head to the video rental store for some good videos on using varmint calls. And, number three, when you buy your call, insist on an audio tape with it, and listen to that tape several times. Practice!

Not all calling is by mouth. These Johnny Stewart electronic calls are ideal in some situations.

323

Where to look? That's the first question. This woodpile proved to be home to cottontails.

Reading the signs on the ground is a great way to learn what small game or, in this case, varmint lives in the area.

Walk a little. Look a lot. That's the method used here to locate a bunny or two for supper.

Snowshoes

There's double meaning here. By "snowshoes" I mean both the hare and the things you wear on your feet to get around in the snow. I've lumped them together because I've had good luck along logging roads in winter for snowshoe rabbits while wearing showshoes on my feet. The snowshoe is not, to my taste buds, the treat the rabbit is. Snowshoes are hares, like jackrabbits. They are edible, but they can demand some cooking tricks, as noted in *The Complete Guide to Game Care and Cookery*. But they're also fascinating to hunt, and in many places, seasons are open all winter long. So put snowshoes on your feet and go out for snowshoe rabbits with your muzzleloader. Of course, watch the winter weather and always be prepared to stay safe, warm and fed should a storm come up.

As for the snowshoes themselves, you don't have to look only for snowshoe hares while wearing them. I learned the value of snowshoes during my residence in Fairbanks, Alaska, many years ago. When the snow piled up, I sure didn't want to be housebound, and so my friend Kenn Oberrecht and I donned our snowshoes and took off after cottontail rabbits. Our sandwiches froze into white bricks in our pockets, but we didn't let that deter us from having a good time. While the old-er-style wooden-frame snowshoe is a dandy, I've turned to Sherpa metal-frame shoes these days. The ones I have are the Lightfoot model, only 31 inches long, and they get me around on those logging roads just fine.

Jack the Rabbit

Jackrabbits are so special I save this varmint, non-game animal or—believe it or not—predator for last. That's right, my state, Wyoming, lists the jackrabbit as a predator. Classify him where you will, but admit that the jackrabbit has done a great deal to get newcomers started in hunting. My first four-footed conquest was at a very tender age in the then-unsettled desert regions outside of Tucson, Arizona. The prize was a jackrabbit taken with a homemade bow. Jacks can be difficult to approach at times, so the 40 round ball is good, with its greater reach-out than the 36 or 32. But the latter two can be plenty for most jackrabbit hunts.

Reading Sign

Small game and varmint hunting can be taken as seriously as the hunter wishes, right down to learning new areas by studying sign. In wintertime, it's easy to spot rabbit traffic. Trails in the snow, especially along fence lines, give their presence away. Also, crusty snow holds droppings in a starkly contrasting view. Varmint sign can also be studied. After all, just because it's small game or varmints does not mean it isn't full-blown hunting. So use all of your skills and enjoy the small game or varmint hunt as much as your quests for larger game. Small game is good food, and there's something you can do with just about every varmint, from fly-tying materials to pelts. Add a muzzleloader to these hunts, and you've just escalated the experience several notches. Try it and you'll agree. ●

chapter **44**

Wild Turkeys and Other Birds with Blackpowder

WILD TURKEYS ARE so important to hunters these days that there is a federation of enthusiasts who study them, write about them, transplant them, and hunt them. But Ben's Bird—so-called because Benjamin Franklin wanted the wild turkey as our national symbol instead of the fish-eating bald eagle—is not the only avian known to the blackpowder hunter, even though many consider it the most important. Every upland bird that has an open season is legal to the blackpowder shotgunner. And out west, a number of mountain birds are hunted not only with shotguns, but rifles as well. If hunting the wild turkey with muzzle-loading shotgun or squirrel rifle is a king's sport, surely the rest of the birds qualify with similar royal status.

Let's get this business out of the way early on, so we don't have to stumble over it later: The breech-loading blackpowder shotgun can be used on all birds, from doves and ducks to wild turkeys and geese. But it is not very popular these days, and will not be mentioned again in this chapter.

Some states have both fall and spring wild turkey seasons. That's pretty good, considering that by the early 1900s wild turkey populations were in the basement. Indiscriminate year-round hunting, avian diseases gifted to the wild birds by settlers' barnyard fowl, and lost habitat almost spelled doom for the wild turkey. By 1930 only twenty-one states had a grand total of around 20,000 birds. Now forty-nine states (all but Alaska) are home to our largest upland game bird with over two million of them running around. Far-thinkers decided that one way to preserve the gobbler was by transplantation. For example, a number of birds were introduced to the Rockies. Wyoming got some in 1935, when a few sage grouse were traded to New Mexico for fifteen Merriam's turkeys. About a decade later the flock grew to around a thousand birds. In the early 1950s, forty-eight turkeys, fifteen more from New Mexico plus thirty-three Wyoming residents, were turned loose in Wyoming's northeastern Black Hills region. They thrived. So did birds in other parts of the Rocky Mountain chain. While Arizona, New Mexico, Colorado, Montana, Idaho and other western states are not considered major wild turkey regions, they now offer thousands of opportunities for hunters.

The largest of the upland game birds, the male of the species struts in the spring and can be called into range.

325

The box-type turkey call can have excellent tone with a wide range of sounds. However, it takes know-how to get the most from it.

Wild Turkey Hunting

Wild turkeys live very different lives from area to area. A friend of mine owns a little ranch in South Dakota. He can count on seeing wild turkeys at a given waterhole during all but the winter months. These birds must wear wristwatches, because they arrive for a drink about the same time every day, depending on the season of the year. In the Rocky Mountains, where I hunt my birds, seasonal movement is very important to know about. Wild turkeys would be nearly impossible to locate in the vast western reaches that house them if hunters didn't know where to start looking and what to look for. Part of knowing where to look lies in the turkey's daily routine, its roosting sites, eating habits, food preferences, watering needs, communication calls, and strutting grounds. Sign-finding is very important, since the birds do move around.

Turkey Lifestyles

Since I hunt the western gobbler, it only makes sense that I talk about the bird I know, but information is transferable to a large degree. My birds live a rather simple routine. They get up early in the morning, feed, maybe go to the watering hole, belly up to the table again for a few hours, and then possibly lay up for a while at midday. If the mating season isn't underway they dust a little, scratch the ground here and there, feed a bit more, and finally head back to the roost in late afternoon. They perch for the night high above any four-legged predators prowling on the ground. This daily action constitutes a lifestyle: feeding, watering, scratching, dusting, roosting, and mating (in the spring), with plenty of meandering all the while. It's knowing these aspects of the wild turkey that make him huntable.

Wild turkeys move a lot and eat noisily. Most of the day, the birds are on the roam. That's good. If they brushed up, like whitetails, turkeys would be even more difficult to find than they already are. A band of birds may move two miles or more during a daily feeding romp. Feeding turkeys are noisy, and that's good for hunters. More than once I've cupped my hands around my ears to improve my all-too-human sound-sensors as I listened for them. I've also used sound-enhancers to listen for the birds exhibiting their bad table manners. Now the advantage swings a little bit to the hunter, who may be able to stalk the noise and get a shot.

The wild turkey is omnivorous, eating whatever providence sets in its path, from small beasties to a host of vegetable matter. Three years

running I took a tom in northern Arizona because I knew of a special field filled with grasshoppers. Every afternoon several flocks of Merriam's turkeys stepped out of the woods and onto this grassy dinner table. I had my pick of birds and deserve no credit for harvesting them, even with bow and arrow. They were only twenty paces away, busily gobbling up 'hoppers. I never saw another hunter in that particular

Videos can be very helpful in many aspects of shooting and hunting. This Lohman video instructs the wild turkey hunter.

The wonderful wild turkey has made a terrific comeback, thanks to sportsmen's dollars, game departments' hard work and, these days, the interest of the National Wild Turkey Federation, Inc.

region, which is too bad. Most of those fine-eating birds fell to winter snows and old age instead of a swift hunter's harvest.

Spiders, grubs, snails, crawdads, worms, ticks, millipedes, centipedes, beetles, salamanders, frogs, grasshoppers and just about anything else that skitters through the grass is fair game for the wild turkey. They also eat grass, berries and acorns. Leaves of many green plants, and numerous cultivated crops are also eaten. Wise hunters hold this image of the western wild turkey: a plucky fellow, unafraid to attack anything not large enough to attack it back. It is fond not only of live food, but if it's green, nutty, seed-like, or just about anything else that smacks of vegetables, the bird will go for it. The turkey gobbles food down whole, snatching it up, pecking at it, picking at it, stripping it away (such as grass seeds), clipping it off, or scratching it out of the earth. Paint this picture of your wild turkey, rather than a barnyard vision of a bird that walks around with its head down pecking for grain all day.

The turkey is a creature of action. The wild turkey can run like a race horse, but when feeding, its walking speed is around two miles an hour, except for occasional mad dashes at living things. The flock can easily move two miles a day through its outdoor grocery store. I have found most of my birds because of these mobile feeding habits. I climb up to any lookout spot, especially a bluff or bank above a waterway. (Note the combination: a high spot coupled with water.) Then I glass, looking for the feeding flock. I listen too, for clucks, gobbles, squawks and sometimes what sounds like a Chihuahua lap dog lost in the woods. Find. Stalk. Shoot. Of course it's not that simple, but the find-and-stalk approach works well for gobblers, especially around water. The birds don't necessarily drink every day, but you may have luck hanging around a watering station. A flock might just come in to drink.

Turkey Sign

Even though I never got a bird at its watering site directly, water is extremely important in turkey hunting. Check for sign around any waterhole, also along stream trails. When camping out on a turkey hunt, I search for water in the area even before looking for a roost. If there's a pond close by, I "dust it." This means taking a fallen branch and using it as a broom to clear away all tracks around the water's edge as best as possible. Next morning, the usual hunting plan is followed, but by afternoon I'm back to the waterhole to see who came to visit that day. Every track there, turkey and otherwise, will be fresh since all old sign was obliterated. If turkeys are using the water site,

I'll hunt hard in the immediate region to see if I can find the flock or its roost. What sign to look for? You don't have to be Sherlock Holmes to deduce what a fresh turkey feather lying along a trail means. And a turkey track on the trail is somewhat more impressive than a sparrow's.

Droppings are telltale. Amorphous blobs are left by the female. J-shaped rods belong to the male bird. Dusting sites may be found along trails and often on dirt roads. I located a particularly handsome dusting site on an old logging road one season and that single clue put me on that road for two evenings, leading me to a bachelor flock of tombirds. Scratchings also mark turkey activity. Hen scratchings are not always well-defined, but toms may clear a large area under a tree. Look for a V-shaped digout about a foot and a half long, centralized and well-defined. Sometimes the V-shape loses its form, with the entire base around the tree raked away. All the duff from beneath a pine tree, for example, may be removed. If a deer, elk or bear did this work, its tracks will say so. If the tombird did it, toe marks will prevail, deeply channeled into the ground by the bird's strong feet.

The Roost

The hunter who finds a "hot" roost has an excellent chance of bringing home a holiday bird. Remember that the wild turkey roosts every day, so finding his special tree is like locating his home address. Turkeys fly well. They can lift straight up from the ground to the roosting tree, but some roosting sites are below a dirt bank or cliff. I have seen the flock use these like a hang gliders's take-off strip. The birds hit the tree with the grace of a hippo at a tea party, hopefully clinging to a branch that will be its night's repose. These special trees are not always easy to find, so I look for them during scouting sessions.

Fall Calling

In the fall of the year, the call serves to locate a flock. I've never gotten a gobbler to come my way by calling him in during this time. Sometimes a few blasts on a crow call will bring a "gobble-gobble" that can be followed up, too. I have never used an owl call, but I'm told that it works the same as a crow call to bring a response. Once the birds give themselves away, they can be stalked. Although wild turkeys are not terribly smart, they are extremely wary. So stalking means slow-going with great care that you don't announce your progress with the crack of a twig.

A 36-caliber Mountaineer rifle from Deer Creek Products of Waldron, Indiana, accounted for this bird taken in snowy conditions.

(Below) One of the author's favorite wild turkey hunting tools is the Gobbler from Bracklyn Products. Sold directly or in sporting goods stores and archery shops, these high-quality hearing aids amplify the noisy feeding of wild turkeys as well as their gobbles and clucks.

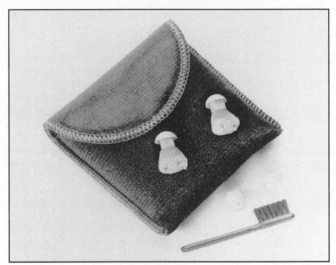

Spring Calling

Springtime calling is an entirely different matter. Amour is in the air and a tombird can be lured into your lap, provided you have a modicum of talent as a caller. I'm not very good, so I don't overdo it. When I get a response, I wait a minute, then answer briefly. I've found that where I hunt, toms won't come in after mid-morning, but they will answer a call all day, and that can be almost as good.

Strutting Grounds

In my state, only toms are legal in the spring, whereas both sexes can be hunted during the fall. So in the spring, I look for a strutting area. These can be traditional, used from season to season. I've been lucky on a couple of timbered ridges that lure springtime tombirds to strut for the ladies. One spring my partner and I slowly worked along one of these ridges. I had shot a bird, but he still needed his tom. I walked slowly, stopping to glass ahead every few steps. The sight that presented itself in my binoculars won first place in my memory for that season. I lifted the glass, and as if someone had turned a movie camera on inside of my binoculars, there was a big fan turning from side to side. I signalled my pard. He stalked and found several strutting toms ahead. One well-placed shot from a 32-caliber muzzleloader, and he had his prize.

Sound Enhancers

I use two sound amplifiers for turkey hunting. (I've used both on deer stands, too.) One is the Team Super Ear Personal Sound Amplifier by Silencio (sold in sporting goods stores). It uses earphones. The other is Bracklyn's Gobbler sold by Bracklyn Archery Products Co. These are tiny in-the-ear hearing aids, and they work remarkably well. In fact, the first time I used them, I could hear so much better it was almost annoying. However, I also heard a flock of birds I never would have located without the hearing boosters. See your archery shop.

Blackpowder Turkey Guns

For western turkeys, I like a 32- or 36-caliber squirrel rifle, loaded with a patched ball to about 1800 fps or so. The 32 normally achieves this, depending on barrel length, of course, with about 30 grains of FFFg or Pyrodex P, while the 36 usually gets this speed with about 40 grains of these powder. Often my chance comes from across a draw, not very far, but a bit long for a shotgun. With a squirrel rifle, I can put a bullet on target at close range, should a bird be called in, or at longer range if I can't get closer. One 32- or 36-caliber round ball cleanly

drops a gobbler with little loss of meat. A 32 or 36 loaded as suggested here is a bit more potent than a 22 WMR at close range.

Shotgun-wise, a well-loaded 20-gauge with Full choke can take turkeys, especially when the birds are called in close. But in all fairness to the birds and the hunters, I have to go with 12- and 10-gauge guns. This is personal, but I've found a mix of shot to be effective out of the blackpowder shotgun, so a charge may be about two-thirds #5 shot, with the rest BB size. The few birds I have taken with the shotgun (I hunt turkeys mostly with the squirrel rifle) were close, and the shotgun did a fine job. My tests of the Knight shotgun (modern muzzleloader) prove it to be a heck of a turkey gun. The Knight 12-gauge with Remington screw-in Extra Full choke tube was allowed a full 2 ounces of shot with 120 grains by volume of Pyrodex RS for a muzzle velocity of 1011 fps, plenty fast enough with all that shot.

The 25-20 and 32-20 rounds loaded with blackpowder are also good wild turkey takers. Loaded with FFFg or P, either one will put a bird down pronto, and without terrible meat loss.

The successful turkey hunter scouts prospective hunting areas before the season opens, when he can. He looks for sign around streams and ponds and he tries to find a roost. The wise turkey hunter also packs a pair of binoculars and uses them faithfully. He covers plenty of ground.

Other Birds with Blackpowder

The Mourning Dove and Whitewinged Dove

Pass-shooting, especially in or near fields or by waterholes, is the usual way to hunt these fine dark-breasted birds. The action is fast, the target small. Near waterholes, where birds are slowing down from their high-speed flight, an Improved Cylinder pattern is fine, but for longer-range shooting at faster birds, a $1^1/_4$-ounce charge of #$7^1/_2$ or #8 shot in front of a full complement of blackpowder or Pyrodex, plus at least a Modified, if not Full choke provide the type of pattern necessary to harvest a limit. The big trick is lead. One way to get enough is to pull well ahead and make certain that the trigger is pulled *while the gun is in motion*. This provides a sustained lead.

To stop the gun's motion and fire means to shoot well behind the target. Remember that shot is in a string. If you lead too much, you may miss with the first part of the string out in front of the bird, but the latter part of the string will make contact. But if the first part of the string is behind the birds, where else can the remainder of the pattern go but behind the bird as well. Leading too much on mourning dove or

whitewings is certainly possible, but not as prevalent as shooting behind the birds. Rifles are not allowed for dove hunting.

Quail

All kinds of quail can be hunted with the blackpowder shotgun. Rifles are not allowed for quail hunting. Bobwhites, scaled quail, Mearns, California, Gambel's—there are many kinds of quail in America. The open-choked muzzle-loading shotgun does a good job on these birds most of the time, using small size shot, as they tend to jump close to the hunter. This is mostly walk-and-jump hunting for those without dogs. Dogs, of course, add greatly to the hunt by locating birds.

Waterfowl

All waterfowl from the smallest duck to the largest goose can be hunted with blackpowder. No rifles allowed. For my money, a 12-gauge or 10-gauge double caplock with Full chokes is the way to go. Ducks can be puddle-jumped, of course, in which case shots may be close, but not always. I still like a 10 or 12 for puddle-jumping or stalk-

Waterfowl are definitely candidates for the blackpowder shotgunner.

(Below) Some birds don't jump until nearly stepped on, and for these the open bore muzzleloader is ideal. The author still prefers a fairly heavy shot charge, even for quail and partridge, but with a light powder charge for close-range shooting.

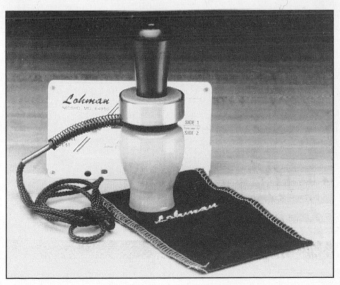

Calls are now scientifically produced to make realistic tones. This Lohman Sweet Talker Premium Duck Call is a perfect example of the new technology.

The blackpowder shotgun is capable of both good patterns and good power. This 12-gauge is allowed up to 1½ ounces of shot and a decent powder charge for a load that is extremely close to the modern 2¾-inch "baby magnum" 12-gauge shotshell.

ing along waterways where ducks are floating. Shooting over decoys can also mean very close work, but I find no problem with the tighter choke here. Pass shooting at ducks or geese demands Full choke and a good load.

Shorebirds

I include in this classification the fantastic sandhill crane, a big bird that can stand about four feet tall. Wings may span eighty inches or so, and bigger cranes may run about 12 pounds. Naturally, the 12- or 10-gauge gun is called for here, with a top load. Other shorebirds are much smaller, such as the different rails, snipes, and woodcocks. These littler birds call for smaller shot sizes and dense patterns.

Pheasants

Pheasants can be very well hunted with the blackpowder shotgun. The Full-choke gun is obviously the ticket for skittish birds that jump at longer distances. Modified choke works in other situations. I see no practical reason for Cylinder or even Improved Cylinder patterns on

This ringneck pheasant is sneaking along ahead of hunters. Without the aid of dogs to locate him, he'll probably disappear in the brush. That's all part of the upland bird hunting game.

This sage grouse is lifting off at a higher rate of speed than meets the eye. Even though it's a big bird, lead will be required.

Fadala's double-barrel blackpowder shotgun accounted for these sage grouse. With a choked gun, birds like this are no match for blackpowder shotgun patterns and power.

pheasants. I've had my best luck with smaller-size shot, but that's a personal preference. Number 6 shot has worked well for me, and in certain close-jumping situations, I found #7½ ideal. I try for head shots for the most part, or at least try to keep most of the pellets in the head/neck region.

Grouse, Ptarmigan & Partridges

In the mountains of the West, rifles are legal for some of these birds. *Be certain to check local regulations before hunting.* I much prefer the 32- or 36-caliber round ball rifle where a rifle is allowed, and when the birds are not disturbed. I like light loads, down to 10 or 15 grains if the rifle will shoot them accurately. My own mountain grouse and partridge hunting usually consists of very close shots, usually at 20 or 30 feet. Shotguns, of course, work just fine, too.

If a bird is legal and practical to hunt with a shotgun, a smokepole scattergun can do the job. Sometimes a smallbore rifle is even better. Either can add an extra dose of pleasure to bird hunting, giving the experience an extra dimension of enjoyment. ●

chapter 45

The Amazing 45–70 Government Cartridge

WHY DEVOTE AN an entire chapter to a single blackpowder cartridge? Because the 45-70 Government round is number one of its kind and king of the single shot breechloader silhouette game (see Chapter 50), outnumbering other rounds two to one in the competition. It's also *numero uno* in the hunting field. More rifles are currently chambered for the 45-70 than any other blackpowder centerfire cartridge, not only in single shot breech-loading rifles, but also in lever actions, a few bolt actions, at least one revolver, in pistols, and in the excellent short-barreled Thompson/Center Contender carbine. That is part of the reason I call it an amazing cartridge—for its staying power. The other reason is its ballistic impetus, with potential to outpower quite a few modern rounds.

The 45-70 Government cartridge, generally known simply as the 45-70, was brought on board as the official United States Army round chambered in the single shot Springfield Model of 1873, fondly known as the Trapdoor because its action was remindful of a door hinge. It had three-groove rifling, with a twist of 1:22. Its adoption followed an official act approved by Congress on June 6, 1872, authorizing the selection of a breechloader in both musket and carbine form to be carried by soldiers of the U.S. Army. Brigadier General A.H. Terry was senior officer of the deciding board and the Springfield got the nod, the 45-70 round being its cartridge. Of course the 45-70 was a blackpowder cartridge through and through, and it was christened with blackpowder nomenclature, such as 45-70-405—caliber 45, 70 grains of blackpowder, with a 405-grain cast lead bullet. The Trapdoor rifle was capable of withstanding something like 25,000 psi, so 70 grains of Fg blackpowder in the 45-70 was perfectly safe in that action with any normal 45-caliber bullet.

The 45-70 saw action as a military cartridge for many, many years. Not until 1892 did an ordnance board officially declare it finished when the Krag was adopted with its smokeless powder 30 Army round, known today as the 30-40 Krag cartridge. Even for the Spanish American war, many soldiers were armed with the Springfield Trapdoor, and a few Trapdoors remained in service until World War I. Add it all up, and the 45-70 saw some type of military use for about 45 years, even though many students of military history limit the official

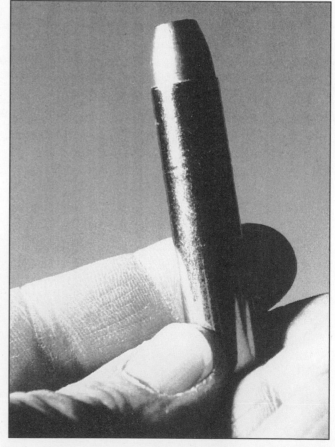

The 45-70 Government cartridge was designed over a hundred years ago, but remains a viable round, not only for blackpowder silhouette matches, but for big game hunting as well.

A 45-70 blackpowder round is shown here fully chambered in a single shot breech-loading Sharps rifle. The breechblock rises into battery behind the cartridge to prepare for firing.

The 45-70 Government was first used in single shot breechloaders like this Sharps rifle. It is still a big favorite in that type of blackpowder cartridge rifle, not only for silhouette shooting, but for hunting.

life of the cartridge to only two decades. Remember, too, that surplus 45-70s were offered to civilians for prices as low as $1.50 plus shipping. Now add the Numrich conversion kit for under $50, which turned Remington rolling blocks in 43 Egyptian and 7mm Mauser into 45-70s, with Navy Arms soon following with a ready-to-go 45-70 rolling block, plus the Martini and the Siamese Mauser in 45-70, and—you get the picture.

The Cartridge

The 45-70 Government was, by design or pure chance, a well-balanced cartridge. While the later 45-90 Winchester, for example, was indeed a bit more powerful, it wasn't really that much stronger than the older round. The 45-70 is a straight-walled rimmed case. Overall maximum case length runs 2.105 inches, including the rim. To give this fact visual meaning, consider that the familiar 30-06 Springfield case is 2.494 inches long, but that includes the neck. Take the neck away and the '06 case runs 2.111 inches to the base of the neck and 1.75 inches for the body length minus the head. Compared with some of the old rounds of the same era, the 45-70 was not really that large, certainly nothing like the 50 Sharps, for example. Across the rim, the round measured .608-inch. Leaving the rim out, the case head of the 45-70 is .505-inch across. The mouth of the case is .480-inch across. That's the 45-70, large enough to contain a good dose of powder, but not an immense case by any means.

Records show that at the outset, the 45-70 was loaded with 70 grains of Fg blackpowder for the rifle, but only 55 grains of the same powder for the carbine. In either case, the bullet was listed at 405 grains, which remained a popular weight for the 45-70 over many decades. Although the 405-grain projectile worked just fine, the government did load 500-grain bullets in the cartridge in the early 1900s. Many other bullets were also loaded for the 45-70, proving that it had more versatility than might be expected. Even a 45-caliber round ball is shown, listed at 140 grains. I found no data to support its construction or exact diameter. But assuming pure lead, the 140-grain round ball would have gone about .450-inch in diameter, since a pure lead ball of .450-inch weighs 137 grains and I imagine the 140-grain notation was rounded off.

Keeping It Alive

In 1931, amidst the writhings of that financial reptile called the Great Depression, Winchester was the last company to chamber the 45-70 in a lever-action rifle at that time (their Model 1886). That should have done it for the old girl, but it was not to be. You see, while chambering was discontinued, ammunition manufacture was not, and a shooter could still purchase 45-70 fodder and fire away. It's difficult to say why the round didn't perish. After all, it was totally outmoded by the smokeless-powder cartridge. In truth, velocities were maintained at about 1200 to 1300 fps for a 405-grain bullet, even with smokeless powder. Why so modest? Because there were plenty of Trapdoors out there and manufacturers did not want to risk hotter ammo in the actions of these rifles.

Also, I'm not really sure why so many of us just had to hunt with a 45-70, but we did. I had a Model 1886 Winchester lever action that I had swapped for. I got a couple of deer with it, and I can safely say that the 405-grain Winchester or Remington bullet from factory ammo whistled through deer like a draft through an open window. While my experience with the 45-70 on game is limited, add to it a number of notes I compiled from friends who also tried the old blackpowder cartridge in the hunting field, most generally for deer. They reported the same findings: bullets in the chest region zipped through. The deer, however, never got away. They ran 50 to 100 yards, perhaps, but always folded up. Hit *through* the shoulder instead of behind, the same-size buck collapsed on the spot like an imploded high rise.

Although there were no major companies chambering the 45-70 following the end of the Great Depression, the old-timer was destined for new life in many different rifles. In the early 1970s, for example, around ten manufacturers offered firearms for the cartridge, including Harrington & Richardson (H&R) with a remake of the Trapdoor. Navy Arms was busy offering Remington rolling block actions as well as finished rifles, and there were various Sharps rifles chambered for the 45-70 as well. Marlin got on board with its fine Model 1895 lever-action rifle chambered for the 45-70; Clerke company had one too, Ruger chambered both its No. 1 and its No. 3 single shot falling blocks

The shooting world would not let the 45-70 Government round die. It has been chambered in numerous rifles, and even in pistols and revolvers. This is Browning's version of the Model 1886 lever action chambered for the 45-70.

for the old cartridge; H&R offered it in a Shikari, which was essentially that company's break-top Topper action, and Browning came out with 45-70s in a handsome single-shot, as well as in a remade Model 1886 lever-action rifle. Numrich offered kits to convert original Remington rolling blocks that were not in caliber 45-70 to that round.

Ballistics

Records show that 45-70 factory ammo with 405-grain bullets was loaded to an advertised 1310 fps. Years ago, nobody had a chronograph, so if the company said 1310, that was fine with us. Having finally gotten around to chronographing some of the old ammo, I find that it ranged from around 1200 to 1300 fps that from old ammo picked up at gun shows in Winchester and Remington brands. The 45-70 is not

ballistically fabulous compared with modern smokeless rounds, as we see in our loading data. At the same time, there is nothing to pity about a 400- to 600-grain bullet leaving the muzzle at the speed of sound as pushed by a dose of blackpowder.

Although our interest in this book is strictly blackpowder and Pyrodex, we are obligated to look at smokeless-powder loads for the 45-70 just to be fair to the cartridge. After all, when we say that the 45-70 is not on a par with, let's say, the 30-06 Springfield, we're comparing onions with bananas—blackpowder against smokeless. So we need to look at smokeless loads in the 45-70 *intended only for rifles that can handle them,* of course. Heavy smokeless loads shove the old round right into the 21st century. For example, consider the special ammunition loaded by Garrett Cartridges, Inc. (Note: Not all of the loads listed and Browning single-shots." This is special ammo for only specific firearms. We include them to show the potential of this old-time round when loaded "hot" with smokeless powder. Meanwhile, Winchester has a very good factory load firing a 300-grain jacketed hollowpoint bullet. I chronographed it from a short 16-inch barrel at very close to 1800 fps for over 2100 foot pounds.

Blackpowder Power

The 45-70 is capable of freight-train performance when loaded with smokeless powder, but we are going to use Pyrodex or blackpowder, so what can we expect? We can expect to match or even slightly improve upon the ballistics associated with standard factory cartridges. Remember, of course, that Winchester and Remington were forced to load this ammo down just in case the cartridges found their way into an old Trapdoor rifle or some other action of low pressure capability. In fact, factory 45-70 ammo is rated at about 16,000 psi. Looking at data from *The Gun Digest Black Powder Loading Manual,* a DBI book, we find an Armsport Sharps with 28-inch barrel getting 1301 fps with a 420-grain cast bullet and 70 grains of Goex Cartridge-grade blackpowder. The actual weight of this charge was 70.8 grains. That gives a muzzle energy of 1579 foot pounds.

Browning's replica Model 1886 lever action in 45-70 was tested with a 322-grain cast bullet and 60 grains of Pyrodex CTG (cartridge granulation) for a muzzle velocity of 1437 fps. The same rifle produced 1205 fps with 65 grains by volume of CTG and a 420-grain cast bullet, and 1103 fps with a 490-grain cast bullet and 60 grains by volume of CTG. Muzzle energies ran 1477, 1355, and 1324 foot pounds, respectively. A C. Sharps Model 1855 High Wall in 45-70 with a 28-inch barrel earned 1313 fps with a 405-grain cast bullet and 70 grains of Goex FFg powder (69.5 grains by weight) for 1551 foot pounds. So when we say that blackpowder ballistics are very much on par with the light smokeless-powder loads offered for years by the ammo companies, we're pretty much on the money. The 45-70 with a 500-grain bullet and a full dose of blackpowder or Pyrodex remains big game worthy to this day for those willing to stalk for a decent close shot—even on elk and moose.

Kinetic Energy

Do not be disappointed by the fact that the 45-70 produces low figures on the energy chart. For example, a 243 Winchester with a 100-grain bullet at 3100 fps delivers 2134 foot pounds of muzzle energy, while a 45-70 with a 500-grain bullet at 1200 fps shows only 1599 foot pounds. While the old round does come up short on paper, it's better than it looks in actual practice. This is not a jab at the Newton formula, for it works pretty darn well. But the big lead bullet from the 45-70 penetrates well on big game and has proved itself effective for years.

In my own tests with a bullet box, the 45-70 with heavy lead bullets has always penetrated well. Years ago, when I used pine boards to test penetration, the round also shined. For example, from a 26-inch barreled rifle a 45-70 500-grain bullet at 1170 fps penetrated eighteen soft pine boards, each one $7/8$-inch thick, the bullet fired from 15 feet. The same bullet can be counted on to shoot through the rib cage of an elk or moose.

Getting the Range

The 45-70 with blackpowder should be a short-range round because of its comparatively low velocity and high trajectory. Yet, I watched a group of shooters consistently hit a buffalo silhouette at 900 yards with single shot *iron-sighted* breech-loading cartridge rifles chambered for the 45-70. But, you're going to say, surely the rounds were loaded with smokeless powder for best velocity. No they were not. Smokeless powder was not allowed at this event. Every shot was taken with blackpowder or Pyrodex loads. So in spite of low velocity, the old round is capable of hitting at long range, but only when the shooter has a good idea of the distance, and he knows his sight picture and how to adjust his sights.

In one demonstration, a 500-grain round-nose bullet from a 45-70

For some reason, interest in the 45-70 cartridge did not die. Even when there were no new rifles chambered for it, shooters found old ones. The author carried an original Model 1886 rifle in 45-70 in his younger days.

here are still available from Garrett, as the company is always upgrading.) These loads prove the point that the 45-70 is capable of big power. How about a 415-grain bullet at 1730 fps for well over 2700 foot pounds of muzzle energy? Of course, this ammo is *not* for every 45-70 rifle. In fact there is a clear warning written on the box: "Garrett's 45-70 Government cartridges safe only in modern rifles & T/C Contenders."

My chronograph verified Garrett 45-70 ammo at over 1700 fps with a 415-grain hard silver-enriched bullet from a 22-inch barrel. The company recommends this ammo for big game including elk, moose, grizzly, buffalo, and even African lions. Another Garrett 45-70 load fired a 400-grain Barnes X bullet at 2020 fps for over 3600 foot pounds, but also with a warning. This one stated: "These are safe ONLY in Ruger

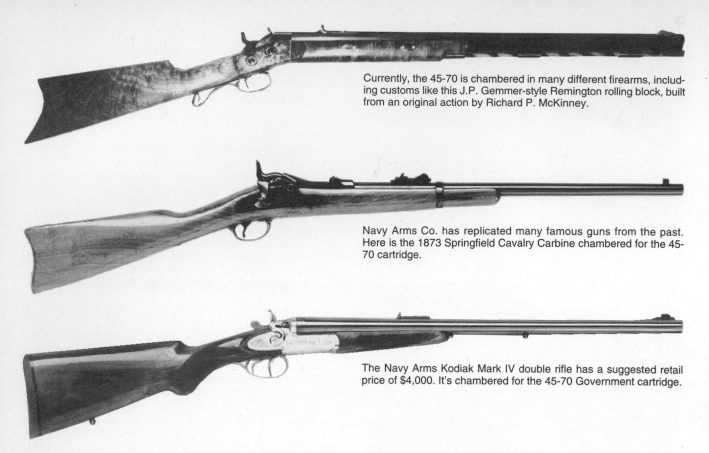

Currently, the 45-70 is chambered in many different firearms, including customs like this J.P. Gemmer-style Remington rolling block, built from an original action by Richard P. McKinney.

Navy Arms Co. has replicated many famous guns from the past. Here is the 1873 Springfield Cavalry Carbine chambered for the 45-70 cartridge.

The Navy Arms Kodiak Mark IV double rifle has a suggested retail price of $4,000. It's chambered for the 45-70 Government cartridge.

starting out at 1200 fps was sighted in dead-on at 100, 200, and 300 yards. In order to get it on target at 100 yards, the group had to strike almost 4 inches high at 50 yards. For a 200-yard sight-in, the group struck over 13 inches high at 100 yards. And to get the bullets to pattern at 300 yards, the group was 34 inches high at 150 yards. These are approximations only, because group size was part of the factor along with pure trajectory, but the figures give an idea of the rather looping trajectory of the 45-70 with its comparatively low muzzle velocity.

Accuracy

Obviously, the 45-70 is not your best choice for modern benchrest competition. On the other hand, the accuracy potential of the old round is high. Accuracy, after all, still is mostly a matter of precision-made bullets fired from precision-made barrels with proper powder and powder charges from good cases. I fired one single shot breechloader fitted with a scope that produced cloverleaf three-shot groups at 50 yards and under-an-inch center-to-center groups at 100 yards with 500-grain cast bullets. Accuracy varied widely with other test rifles and iron sights, not to mention different bullets and powder charges. The bullet is very important, of course. Ideally, the shooter interested in gaining the highest accuracy from his 45-70 should experiment with various loads, but not blindly.

Authors like Mike Venturino, Dave Scovill, Paul Matthews and Steve Garbe have already accomplished a ton of testing, and the work of these experts should be consulted. Two examples of good 45-70 accuracy work, just to get the reader started, are *Handloader's Bullet Making Annual,* Vol. I, 1990, from Wolfe Publishing Company of Prescott, Arizona, and *Handloader* Number 173 for February of 1995. The first contains many articles on making accurate bullets, including Paul A. Matthews' "Best Cast Bullets for the 45-70" and "Lubrication" by Dave Scovill. The second listing has an article by Steve Garbe entitled "45-70 Black Powder Target Loads." There are many other sources for 45-70 accuracy improvement.

Garbe points out that the 45-70 is to BPCR (Black Powder Cartridge Rifle) silhouette shooting what the 308 Winchester is to high-power target shooting. He also states that "Rifles chambered for the 45-70 have a well-deserved reputation for shooting decently with nearly any intelligently assembed handload." It's the "intelligently assembled" part of that statement that carries weight, and that means learning from research and personal testing. Garbe further points out that even the great barrelmaker Harry Pope considered the 45-70 an inaccurate cartridge, but that was in the old days before the accuracy potential of the round was discovered. Today, we know better.

Accuracy and the Bullet Base

Recovered bullets caught by various media downrange may show base deterioration caused by the hot gases from the blackpowder or Pyrodex charge. If a base is damaged, accuracy can suffer badly. Stuffing wads into a case may cause a safety problem, so the most certain way to protect a bullet's base when there is evidence of damage is with a bullet designed to take a gas check, which attaches to the base of the bullet and safeguards that base from hot gases.

Standard Deviation

We've discussed standard deviation as a pretty good measure of a load's potential for accuracy. A very low standard deviation from the average velocity indicates a good balance between bullet and powder. The 45-70 shows very low standard deviations with both Pyrodex and blackpowder. Initially, I suspected my own results, they were so good. I was getting standard deviations under 10 fps for all my loads. However, after looking into figures from other sources, I realized mine were accurate. Garbe, for example, had loads giving standard deviations as low as 4 fps, and in one series of tests, his poorest standard deviation was only 8 fps.

Loading the 45-70

Old Versus New Cases

The first thing to remember is that old-time cartridge cases were not constructed in the same way modern cases are. The balloon-head case, for example had less "meat" in the head section than modern brass. This is why it can be difficult to get a full 70 grains of blackpowder into a 45-70 case today, because the modern case, with more metal in

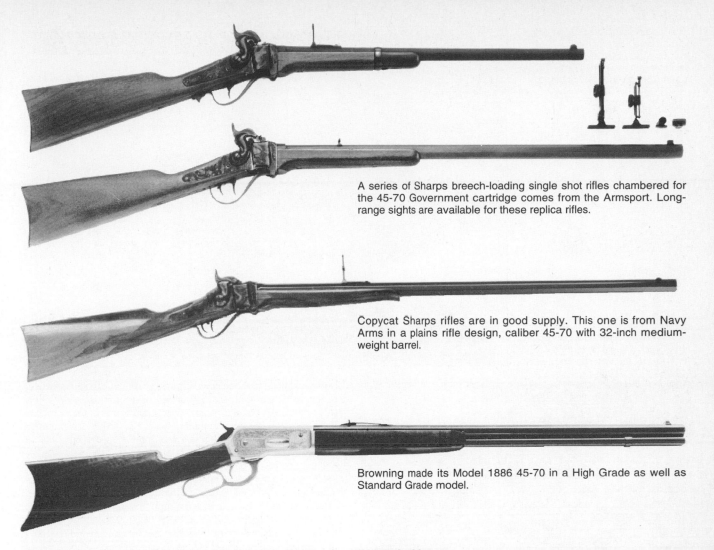

A series of Sharps breech-loading single shot rifles chambered for the 45-70 Government cartridge comes from the Armsport. Long-range sights are available for these replica rifles.

Copycat Sharps rifles are in good supply. This one is from Navy Arms in a plains rifle design, caliber 45-70 with 32-inch medium-weight barrel.

Browning made its Model 1886 45-70 in a High Grade as well as Standard Grade model.

the head area, cuts down on case capacity. This is not a problem to the shooter, but it is a point he needs to understand.

Load Density

In the name of best accuracy as well as safety, only 100 percent load density is broached in this work for any blackpowder cartridge. All this means is that the case is full of blackpowder or Pyrodex with no air space. Paul Matthews, one of the country's leading 45-70 cast-bullet shooters, uses the word "never" considering air space in the case. He feels that in the name of safety, the case should be filled with powder. This factor relates to our frequent discussion of the short-started load in the muzzleloader with consequent unwanted air space. The bullet determines how much blackpowder or Pyrodex can be installed in the 45-70 case, while still maintaining proper overall loaded length of the cartridge. For example, a full 70 grains by volume of blackpowder or Pyrodex can be installed with bullets in the 405-grain class, but that load may not work with heavier bullets. For example, in loading a 500-grain bullet to 100 percent load density in one particular case, the most powder I could install was 67 grains.

A drop tube can be used to evenly distribute powder into the 45-70 case, thereby allowing a full load.

The Best Bullet

Properly cast lead bullets shoot extremely well in the 45-70, and are also the rule for the BPCR match, which does not allow the use of jacketed bullets. There are literally dozens of different bullet styles and weights available from the various companies that offer 45-caliber moulds, such as NEI, Lyman, Hoch, Redding-SAECO, RCBS and others.

Proper Lubing

In all of the loading data presented by today's experts on the blackpowder cartridge, special attention is given to the type of lube used on the bullet. Dave Scovill pointed out in the *Handloader's Bullet Making Annual* noted above that lubrication "bears a close scrutiny. Without its protection, cast bullets are nearly useless. Even with it, accuracy and velocity can vary considerably, depending on the type or brand of lube used." This is not the forum for a full-blown discussion on lubes for the 45-70; however, it is important for the shooter to recognize how important lubes are and to study the subject fully for best accuracy in his own cast-bullet loads.

The Primer

Contrary to my original thinking and practice, today the magnum primer, not the standard rifle primer, is preferred for best overall results in the 45-70 loaded with blackpowder.

Crimping Bullets

Bullets for 45-70 ammo to be used in lever-action rifles must be crimped into the cartridge case or recoil will drive the bullet back into the case, since in the tubular magazine one round is loaded in line with another. Also, blunt-nosed bullets are used in tubular magazines to prevent the possibility of the point of one bullet detonating the primer of the round in front of it.

Correct Seating Depth

Bullets must be correctly seated for proper depth in order to create the right overall cartridge length. Some experts feel that with cast bullets in the 45-70, seating so that the bullet just barely engages the rifling

Shooting with crossed sticks helps to tame the recoil of the 45-70 cartridge, since the shooter's body can roll with the punch better than it can from the prone position.

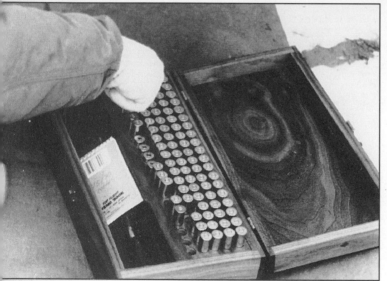

(Below) Today, shooters handload the 45-70 very seriously. This handsome box contains specially loaded 45-70 ammo for a silhouette shooting match.

is best if that option is available. Some rifles may have too much leade in the chamber to allow this.

Recoil

Having worked with big-bore muzzleloaders such as the Zephyr, which is allowed much greater powder charges than those of the 45-70, I've never considered the latter a problem in the recoil department.

However, silhouette shooters who fire 80 to 100 times in the course of a match, including sighters, warming up, and scoring, do consider the 45-70 a "kicker." Recoil can cause two big problems in such a match: fatigue and damage to concentration. Fatigue is simply tiring through thump after thump on the shoulder and into the cheekbone. Lack of concentration may follow, with possible flinching. The way I handle recoil may not be the best, but it works for me. I tell myself one thing: "You may as well maintain control over the rifle, squeezing the trigger carefully, because recoil is identical, whether you flinch or not."

Recoil in the 45-70 is no different from recoil in any other rifle or round. Powder charge weight and bullet weight account for some of the kick, while the weight of the rifle helps thwart rearward motion. I don't use sub-loads in the 45-70, as noted above, so cutting back on the blackpowder powder charge is out of the question in reducing recoil. A good bullet is a good bullet, so if it happens to weigh 500 grains I'll stick with it over a lighter missile. However, a heavier firearm can be selected over a lighter one where match rules are not violated. Furthermore, NRA Silhouette rules allow the use of a recoil pad on the blackpowder breechloader, as well as a shoulder pad for the shooter. These pads help reduce the effects of felt recoil. Finally, shooting from cross-sticks rather than from prone also helps reduce felt recoil, since cross-sticks allow the body to roll with the punch more than prone does.

This has been a one-chapter look at the most popular blackpowder big-bore cartridge of the day. It's chambered in dozens of different rifles, including customs. Components for reloading, as well as factory ammo, are easy to find all over the country. Add good accuracy and power, and it's no wonder that the 45-70 Government is still with us, and likely to remain so for a very long time to come. ●

chapter 46

Blackpowder Ivory Hunters

JUDGE THEM IN their own time frame, which was the 19th century. By today's standards, their actions would not always reflect brightly under the light of modern scrutiny. But in their world, they were known as explorers, adventurers, businessmen, and most of all, daring hunters. They shot pachyderms for the ivory, and they did it with what they had—soot-belching big-bore rifles, mostly muzzleloaders. Without the guns, their story wouldn't fit this work. But throw the firearms in and their tale is just as valid as the mountain man's, their guns just as interesting—maybe more so—for they were hunting the largest animal on four feet.

It's been said that the English in India and the Dutch in Africa turned to big-bore single shot blackpowder rifles in the early 1800s. That's too generic. However, it is true that muzzleloading 4-, 6-, and 8-bore guns were relied upon to stop the charge of the lion or bring down a fortune in ivory. Fortune? Few true fortunes were made by the time the smoke cleared, literally as well as figuratively. But most certainly the ivory hunter made his mark. Some say it was a blemish. Others realistically consider it within the context of history. Who were they? They did, indeed, come out of Europe, as their names imply. There was Major Shakespear and S.W. Baker, Harris, Oswell, Roualeyn Gordon-Cumming and G.P. Sanderson. And they were as brave a bunch, like them or not, as the fur trappers who left their safe homes "back East" to roam among the grizzlies and Blackfeet out West.

Our interest here is a brief encounter with three of these hardy souls: William Cotton Oswell, Sir Samuel White Baker, better known as S.W. Baker to his readers, and Frederick Courteney Selous, who signed his name F.C. Selous (pronounced Sell-oo). But before these three are introduced, a few words about Gordon-Cumming, Major Shakespear, and the lesser known hunter, George P. Sanderson, just to set the tone of the chapter. In his book, *Five Years Hunting Adventures in Africa*, covering 1843 to 1848, Gordon-Cumming wrote about his guns. We learn that he used a Dickson double-barrel 12-bore that eventually "burst from too much fouling," and that he also carried a Dutch-made 6-bore with six balls to the pound, meaning that each bullet weighed close to 1200 grains. The 6-bore was said to be loaded with ten to fifteen drams of powder. If so, that would be from roughly 270 to over 400 grains of blackpowder. Even in a rifle weighing 15 pounds, free recoil was tremendous.

Major Shakespear, from his book, *Wild Sports of India*, which dealt with 1834 to 1859, noted that:

> My own battery consists of two heavy double rifles and a double gun; the heaviest is a Westley Richards weighing twelve and a quarter

The early ivory hunters used big-bore blackpowder rifles, first muzzleloaders and later breechloaders, as shown in this illustration. They were hunting the world's largest four-footed animal; living dangerously was a way of life for them.

339

pounds, length of barrel, twenty-six inches, poly-grooved, carrying bullets ten to the pound [a 10-gauge or 10-bore]. It is a splendid weapon, bearing a large charge of powder without recoil; that is to say, its own bullet mould full of the strongest rifle powder.

Shakespear's notes are especially interesting because of his reference to a powder charge generated from his bullet mould. Even a 10-bore mould—conical as well as round ball—would not hold all that much powder. Perhaps that is why the hunter didn't consider recoil all that fierce. Shakespear complained about a high trajectory with his 10-bore, the bullet rising about five inches at 50 yards in order to strike dead on at about 100 yards. At the same time, the Westley Richards rifle was fitted with two rear sights, one for 150-yard shooting and the other for 250 yards. It's a bit puzzling—high trajectory, but long-range sights. Furthermore, the venerable elephant hunter considered the powder charge for his 10-bore "heavy." Perhaps something was lost in the translation concerning a powder charge that matched the bullet mould.

Shakespear spoke of a second rifle, another 10-bore, two-groove instead of poly-grooved, but this one, according to the author, shot "point blank from muzzle up to ninety yards." See below for remarks about the meaning of "point blank." The folding sights on this particular rifle, built by Wilkinson of Pall Mall, were set for 150, 250, and 400 yards. This 10.5-pound rifle with 30-inch barrel was a favorite of Shakespear's, but it leaves a modern shooter wondering about actual ballistics, which is important when we study any 19th century shooter's work. Certain things are a puzzle. Why Shakespear felt that his Wilkinson shot flatter than his Westley Richards is a mystery, but stranger is the famous hunter's remark about recoil being light. At ten balls to the pound, his 10-bore fired a 700-grain projectile, and if he used about eight drams of powder (over 200 grains) velocity had to be around the speed of sound or a bit faster, probably 1200 feet per second. Shakespear noted that his rifle fired "without recoil," but it probably developed around eighty foot pounds of free recoil even in a heavy rifle, while my own 30-06 sporter has around eighteen foot pounds of free recoil.

An explanation of "point blank" is in order, for it does not mean that Shakespear's rifle "shot flat" from muzzle to 90 yards. Point blank comes from the French. The bullseye was a white, or *blanc* mark, and the rifle shot "point blank" when its bullet did not stray either above or below that white target at a given range. The details of this practice are mainly lost to history, but it seems to have something to do with the parabola of the projectile. For our purpose, all we need to know is that the old-time shooter was not suggesting that his rifle "shot flat," but rather it could be counted on to keep its projectiles within a certain vertical limit at a specific distance.

A few words from George P. Sanderson, and we'll move on to our three major players. Sanderson wrote a book called, *Thirteen Years Among the Wild Beasts of India*. Comments on his first rifle, a 12-bore, are interesting. They go like this:

> I at first killed several elephants with a No. 12 spherical ball rifle, with hard bullets and six drams of powder, but I found it insufficient for many occasions. I then had a single-barreled, center-fire No. 4 bore rifle, weighing sixteen and one-half pounds and firing ten drams, made to order by Lang and Sons, Cockspur Street. A cartridge of this single barrel, however, missed fire on one occasion, and nearly brought me to grief, so I gave it up and had a center-fire No. 4 smoothbore, weighing nineteen and one-half pounds, built by W.W. Greener. This I have used ever since. I ordinarily fire twelve drams of powder with it. Without something of the cannon kind, game of the ponderous class cannot be brought to fighting quarters with even a moderate degree of safety or effect.

Several points of interest are generated from Sanderson's remarks, not the least of which is the W.W. Greener smoothbore, first because it was made by this famous gunsmith and writer, whose works are still read to this day, and second because it was a smoothbore, which offered certain advantages: namely ease of cleaning, with no rifling to hold fouling, and facility of loading for the same reason. Also, Sanderson rightly points out that you needed a big bore—a truly big bore—to

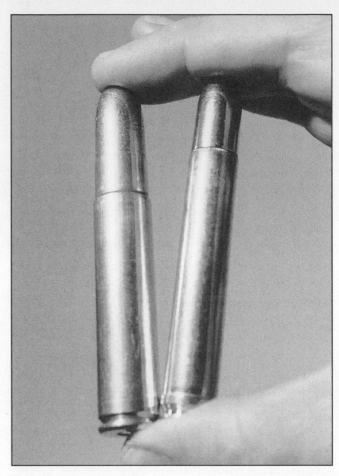

The concept of big bores for really big and/or dangerous game lives on to this day. While the 458 Winchester cartridge (right) is certainly powerful, it is no match for the 500 Van Horn Express round (left), which is actually 51-caliber and fires bullets weighing 700 grains.

feel at all safe in the work of elephant hunting during blackpowder days. Big bore was not then what it is now. A 12-gauge, for example, with its round ball under 500 grains weight, was simply not enough bullet for dangerous game. The Greener 4-bore fired a round ball weighing 1750 grains! Incidentally, the custom big bore of the 19th century could run from $750 to $1000, from what I understand. That's a heap of wampum considering the times, and could be the bulk of a year's wages for an ordinary workman.

William Cotton Oswell

Oswell's name is not recognized by the average modern hunter or shooter; however, in his own day he was quite well known. Sir Samuel W. Baker, whose own brief mini-bio appears here, said of Oswell, "His name will be remembered with tears of sorrow and profound respect." Oswell was one of the first professional ivory hunters in South Africa. His name is linked with David Livingstone and it appears that Oswell, along with a partner named Murray, financed a Livingstone expedition, which was recognized with a medal given by the French Geographical Society, for he, Oswell, was considered a part of the first mapping of Africa's lake systems.

Oswell hunted often on horseback, and was noted as a great rider. He took to the saddle, gave chase, then dismounted for the shot when he was within range, which was most often twenty or thirty paces. That's why Oswell did so well with a smoothbore. Supreme accuracy was not necessary close up. Oswell began his hunting career in Africa with a 12-bore double rifle built by Westley Richards, along with a single barrel 8-bore that fired belted round balls weighing around 875 grains or so, known in everyday parlance as a "two-ounce ball."

William Cotton Oswell was known in England as the "Pioneer of Civilisation." He was a friend and interpreter of Dr. Livingstone, who applauded Oswell in Livingstone's "Zambesi and its Tributaries." Oswell was born in 1818, died in 1893, and was remembered by S.W. Baker as a man "without a rival; and certainly without an enemy; the greatest hunter ever known in modern times."

Oswell's favorite firearm, however, was a Purdey smoothbore double that, according to S.W. Baker, weighed 10 pounds even. How did Baker know so much about Oswell's 10-bore? He borrowed it in 1861 for an expedition to the Nile. Oswell was retired at the time.

Baker at first heaped great praise on the 10-gauge smoothbore, but in later writings, put smoothbores down as too inaccurate to be counted on in the field. Oswell's favorite 10-bore was loaded with "six drachms of fine-grained powder," which we know meant six drams, or about 164 grains. It was, of course, a muzzleloader. There were no breechloaders when Oswell acquired the gun. He wrapped his 10-gauge round balls in fine leather or linen, tightly, cutting off excess material. Then he built a paper cartridge that contained both the powder charge and the patched ball. To reload, Oswell nipped the end of the paper cartridge off with his teeth and dropped the powder, with paper, downbore. Then he put the pre-patched ball on the muzzle and, with a loading rod noted as "powerful," pushed the bullet firmly into the breech on the powder charge.

Baker pointed out that Oswell's smoothbore "exhibited in an unmistakable degree the style of hunting which distinguished its determined owner. The hard walnut stock was completely eaten away for an inch of surface; the loss of wood suggested that rats had gnawed it, as there were minor traces of apparent teeth." Actually, the stock had been "chewed on" by the wait-a-bit thorn bushes common to Oswell's hunting grounds as he galloped his horse in hot pursuit of game. Baker reported that he returned Oswell's smoothbore in good condition, but minus the ramrod, which had been lost when one of Baker's native bearers was attacked by a group of marauders. The frightened man loaded the smoothbore just in time to save his own life, but he did not

have time to withdraw the ramrod, which he fired completely through the body of one of his assailants.

Oswell hunted hard. He wrote of some of his adventures in "African Game Rifles," as part of the *Badminton Library, Big Game Shooting,* published by Longmans Green & Company of London, 1902. He said, "I spent five years in Africa. I was never ill for a single day—laid up occasionally by accident, but that was all. I had the best of companions—Murray, Vardon, Livingstone—and capital servants, who stuck to me throughout. I never had occasion to raise a hand against a native, and my foot only once, when I found a long lazy fellow poking his paw into my sugar tin." He also noted that he "filled their stomachs," speaking of his native helpers. Baker was much more a commander than Oswell, and certainly not above dealing out punishment, since he saw himself in the superior role.

Sir Samuel White Baker

I warned that the blackpowder-shooting ivory hunters of the 19th century had to be viewed through the window of their own times. So let it be with Baker. He was an English gentleman, knighted by the Queen no less, and of course he considered the natives of foreign soil beneath him. He was known to thrash a servant who didn't do the "right thing." He also shot game at will, unlike Oswell, who spared female elephants, for example, and harvested only enough game to feed his followers. While he hunted Africa, Baker is better known for his adventures in Ceylon, and for the book that told about those times, *The Rifle and the Hound in Ceylon,* reprinted in modern times and still available through interlibrary loan, with some for sale at bookstores dealing in old titles. He was 24 years old when he arrived in Ceylon.

Sir Samuel White Baker, an English knight, explored the Nile from 1861 to 1865 and was a big man at 6 feet, 6 inches tall, and 250 pounds. He was born in London in 1821, married a Hungarian noblewoman when he was 41, and died in England in 1893. He authored *Ismalia* and *Eight Years' Wandering in Ceylon*.

Before he was finished, Baker also hunted the United States as well as Africa. He made two around-the-world hunting trips that included America, these taken between 1879 and 1888, which included hunting in Asia Minor and India.

Credited with being the first English-speaking person to travel the Nile, Baker's history is easier to locate than either Cotton Oswell's life story, or Selous' biography. Sometimes he's even called the "discoverer" of the Nile, just as Balboa is credited with "discovering" the Pacific Ocean. But our interest is in Baker's firearms and his shooting theories. His firearms were custom made, as money was no stumbling block for Baker. By his own claim, his first good rifle was also the first firearm to see action in Ceylon, in 1845. Perhaps. That good rifle was a 4-bore muzzleloader that weighed in at a trifling 21 pounds. The single-barrel rifle was two-grooved, made by Gibbs of Bristol. It was noted to shoot a "four ounce" ball, which makes sense as there are 16 ounces in pound, so four 4-gauge balls to the pound. In other words, each ball weighed 1750 grains. The cannon-like rifle was loaded with 16 drams of blackpowder, or over 430 grains.

Baker was a big man, and he apparently stood up to his 4-bore with impunity, for he was known to make hits with it, and other ponderous rifles, at very long range—meaning 300 yards and farther. His backup rifle was "only" an 8-gauge single barrel, built by Blisset, with poly-grooved bore and a 2-ounce ball, but also loaded with 16 drams of blackpowder. While in Ceylon, Baker put down a great number of elephants, along with considerable water buffalo and other game. His two single shot rifles were not enough for him, however, so he ordered four more rifles, all 10-gauge muzzleloaders weighing about fifteen pounds each. These rifles, made by Holland, became his elephant hunting battery. He carried them in Africa as well as Ceylon, but when the British army adopted the Snider breechloader, Baker had Holland build him a double-barrel 577 that

fired a 648-grain conical bullet at a reported 1650 fps. He carried this little toy in America for deer, bear, elk and bison.

It's difficult to say just how fast Baker's 4-bore round bullets were taking off at the muzzle, but with 16 drams of the best blackpowder of the day, I think it's safe to assume that velocity was over 1200 fps and could have been as high as 1600 fps or more. Let's call it 1400 fps just for the sake of conversation. That would be well over 7500 foot pounds of muzzle energy. If 1600 fps were achieved, the energy rating would be close to 10,000 foot pounds. The 577 breechloader with its 648-grain bullet at 1650 fps delivered 3918 foot pounds. So Baker's breechloader was a step down from his frontloaders in ballistic force, even though it fired a conical with better retention of downrange energy.

Baker did gravitate to smaller firearms, owning and shooting double-barrel 400 Holland Express rifles, which he used for deer-sized game in England and Scotland. However, when he wrote his book, *Wild Beasts and Their Ways* in 1891, two years before his death, he concluded that for the largest game he would have nothing smaller than an 8-bore rifle firing a 3-ounce projectile in front of 14 drams of blackpowder (about 380 grains weight). Also interesting is Baker's respect for the round ball. He actually considered the lead sphere more efficient against pachyderms than even the heaviest conical. One of his reasons, and certainly not a scientific one, was explained as the "conical making too neat a wound," sort of like a rapier sliding right through without imparting much of its energy in the target, whereas the round ball smacked hard, delivering its blow in the target instead of behind it. Some of today's round ball fans cite Baker's work when their beloved lead spheres are put down.

Selous

Here was a man among men. Suffice that he died a soldier, although he was not truly a professional military man. On January 4, 1917, during World War I, Selous was killed in action as he fought the Germans

Frederick Courteney Selous was a professional ivory hunter at 19 years of age in 1871. He was known in the heat of the chase to grab a handful of powder and pour it downbore to reload. Born in London in 1851, he died a hero in 1917 during World War I. Selous spanned the gap between blackpowder and smokeless, going from huge 4-bore muzzleloaders to a little 6.5mm rifle.

in Tanganyika as a volunteer. He was born in Regents Park, London, in 1851, making him 66 years old in 1917 when he was shot to death as he led his men against an enemy four times greater in strength. Frederick was educated mainly as a naturalist and was schooled in England, but finished his education in Switzerland and Germany. He was a good student and could have made a mark in society, but he read too much of Africa, and there he simply had to go. On September 4, 1871, he arrived on the Dark Continent.

He was 19 years old and he possessed only 400 English pounds to launch his career as an ivory hunter. He made his way into wild territory, often straying far afield from his wagons so that he could hunt on foot undisturbed. Tracking elephants with a Hottentot native known as Cigar, F.C. lived off the land. Cigar, by the way, became a friend and partner in the chase and was far more than a "hired man." Wandering afar with only a blanket against the night and his 4-bore muzzleloader over his shoulder, Selous was a genius in the art of what we would call woodsmanship. As for that 4-bore, Selous packed along a bag filled with powder and twenty-four 4-ounce "round bullets," to use his words. He admitted that sometimes he did not measure his charge, simply dropping a fistful of blackpowder downbore.

Anyone interested in more of the Selous story should read the man's book, *A Hunter's Wanderings in Africa*, a title still available through special book search agencies. Luckily, Selous's biography was set down by J.G. Millais, in a book entitled *The Life of Frederick Courtenay Selous, D.S.O.,* written in 1918. (Spelling of middle name seen today as Courteney, not Courtenay.) Naturalist, writer, settler, guide, explorer, and even soldier, Selous's life as an ivory hunter began most interestingly when his double 12-bore Reilly rifle was stolen from a wagon on his first trip out. He ended up taking seventy-eight elephants with a pair of Dutch-made Roer two-groove 4-bores. These were not the finest rifles of the era, but they were powerful. Selous loaded these with sixteen to eighteen drams of blackpowder, or 437 to almost 500

grains with the huge 1750-grain lead ball. Recoil was dreadful. Selous said:

> They kicked most frightfully and in my case the punishment I received from these guns has affected my nerves to such an extent as to have materially influenced my shooting ever since, and I am heartily sorry I ever had anything to do with them.

Selous went on to shoot much smaller firearms, and is remembered today as the elephant hunter who used a .256-caliber rifle for big game, even elephants. Although Selous complained that those big Dutch 4-bores harmed his shooting ability, in fact he was known as a superior marksman all his life, using the tiny 256, which is a 6.5mm rifle firing a 160-grain round-nose bullet at only 2300 fps MV, even for elephants. With a muzzle energy of only 1880 foot pounds, about like a 30-30, Selous usually downed his elephant with one perfectly placed shot. He also liked the ordinary 303 British round with a 215-grain bullet at 2000 fps MV for 1910 foot pounds of muzzle energy.

Selous loved to hunt. He hunted not only Africa as a professional, but also America as a sportsman. He hunted the Rocky Mountains, and also ended up in Alaska with Charles Sheldon. He also hunted Canada for moose and other game. F.C. Selous traveled the road from muzzleloader to breechloader, and from blackpowder to smokeless. He used them all, from the most hellish big bore blackpowder charcoal burning muzzleloader to the neat little Mannlicher-type smokeless powder smallbores.

The 19th-century ivory hunters relied on very heavy lead bullets fired with very heavy charges of blackpowder, mostly from muzzleloaders. Theirs are interesting stories, for they lived interesting lives. They also had much to say about the guns they used, firearms that are seldom exceeded in power to this day with the most modern bolt-action smokeless powder cartridge rifle. ●

chapter 47

American Sharpshooters Of the Past

UNFORTUNATELY, SOME OF the very best blackpowder shooting of the past was accomplished by men shooting at men, especially during the American Civil War, but even before that time. Snipers were on hand for the Seven Years War, the American Revolution, the War of 1812, and, of course, long-range shooting was accomplished when the white man and American Native crossed paths in anger. Some of the stories of super-long-range shooting are, I fear, just that—stories. Others are bold fact. The truth, however, is almost as unbelievable as the fiction in many instances. Long-range sniping by buffalo runners, for example, I find not at all difficult to believe. These men were well-practiced. They had good, accurate rifles. And their rifles often wore target-type telescopic sights of high magnification. Give them a clear target, even at several hundred yards, and the buffalo boys could "dope out" wind drift and drop, and lay a bullet on target, especially when the target was a human figure standing upright with plenty of vertical latitude. Shooters who preceded them could also hit the mark at incredible distances. I believe this fact is of interest to modern downwind shooters, therefore this chapter.

First True Snipers

Long-range shooting with Kentucky/Pennsylvania rifles did occur, and I have no doubt of it. I also doubt nothing about the far shooting accomplished with plains rifles. I've seen for myself both types of frontloaders at shooting matches today whacking gongs and similar targets out to 500 yards, 600 yards, and farther. But without a doubt the true sniper appeared during the Civil War. Here was a soldier, chosen because he was already a good shot, and then trained to be even better. Colonel Hiram Berdan has survived the longest in memory, although he was not the only soldier to organize long-range shooting on the battlefield. Berdan, however, was probably the first to convince military leadership to put a regiment of sharpshooters together. He was an amateur New York rifleman and target shooter, and in 1861 organized and assembled other good marksmen to join the North as snipers.

Berdan's Sharpshooters, as they were called, did a lot of damage. So did their fellow Union U.S. Sharpshooters under the command of Colonel Henry A.V. Post, also a New York resident. The two regi-

ments of long-range experts allowed only those who could pass a shooting test entry into this elite corps. The exact test may have been lost in time, and some stories are almost preposterous, if not entirely false. But a simple set of rules does seem plausible. It was ten shots into a 10-inch bullseye at 200 yards from any shooting position, which included prone, and apparently any rifle, target rifles accepted, using any sight, certainly to include the scope sight. Given a good heavy target rifle of the day with a scope, ten shots into a 10-inch bull at 200 yards seems reasonable enough.

The Sharpshooter could bring his own rifle, I understand, for which the government would reimburse him up to $60 for its use. It's easy

Hiram Berdan remains the best-known of the sharpshooter leaders, probably because he seems to be the first man to recognize how deadly a regiment of snipers would be.

344

enough to assume that some brought their best target-shooting guns, with long scope sights mounted on them. These, by the way, were true scopes, often of high magnification, although with very limited field of view. Malcolm and Vollmer scopes were quite well known by the time of the American Civil War. Looking like tube sights, metal tubes were actually precision instruments with glass lenses. And there is no doubt that they could turn a deadly long-range rifle into a far deadlier long-range rifle. Buffalo hunters later in the same century often mounted scopes on their breechloaders, these scopes very much like the ones used in the Civil War.

Overstated Claims—Maybe

An exhibition to show President Abraham Lincoln just how good the Sharpshooters really were was supposed to include a hundred men firing at a man-sized target a full 600 yards in the distance. This supposedly occurred in 1861. Berdan himself was said to be one of the marksmen. Out of 100 shooters, all placed bullets in the kill zone, Berdan firing a five-shot group that went about ten inches in spread. While it is noted that the rifles did wear scope sights, all 100 men hitting the target with all of their shots does seem hard to believe, although no one can say for certain that this did not occur. To top it off, someone asked Berdan to hit one of the targets, a figure of Jefferson Davis, in the eye. Berdan fired his rifle and a neat round hole appeared where once a pupil was represented. Lincoln was supposed to have remarked something about the shot being the luckiest he'd ever seen, or something like that. This story is repeated below in a somewhat different way from a *Harper's* magazine article.

Another possibly overstated claim concerns the one-mile hit credited to a Northern sharpshooter by the name of John H. Metcalf. Supposedly, the event took place in 1864 during the Red River Campaign, specifically the battle of Pleasant Hill in Louisiana. Metcalf took aim, the story goes, on a Confederate general named Lainhart, the distance being over one mile—and hit him. There are many possibilities. The first is that the Union sharpshooter fired and someone in the distance fell, at which point credit for a one-mile hit was awarded. Or, the distance could have been overjudged. Or, (stranger things have happened) the shot truly was executed. That the bullet could travel so far and remain deadly is not a problem. It could have. That anyone could judge a shot at a mile is on the incredible side. So if the shot was made, fluke would be a good name for it.

Documented Claims

In *The Battle of Gettysburg*, a scholarly work by Francis Marshal

printed in 1914 by the Neale Publishing Co. of New York, the death of Confederate Major-General John Sedgewick is related. Sedgewick was well known to the North as well as to the South, his reputation as a gentleman and soldier highly regarded. In fact, it was said after his death that Sedgewick had two mourners, "friends and his foe." His soldiers of the Sixth Corps considered Sedgewick their father more than their commander. But I'll let Professor Marshal tell the story in his own way:

> The numerical sacrifice of human life, however, terrible as it is, does not equal the loss to the Federal army of one life, which has issued from its ranks on its long furlough. Major-General John Sedgewick, one of its main bulwarks for years, the loved commander and father of the old reliable Sixth Corps, is among the dead. Smiling encouragement to some of his men new to battle, whom he saw dodging the bullets that whizzed past, he had just remarked, jokingly: 'Soldiers, don't dodge bullets. Why, they can't hit an elephant at this distance.' At that instant a veteran officer at his side heard the familiar thud of a bullet, and turned to remark it to Sedgewick, who at that moment gave him a smile and fell dead into his arms, shot through the head.

Whitworths and Sharps Rifles

Due to the timing of the Sedgewick episode, the best bet is that the rifle used to fell the great soldier was a Whitworth rifled musket. (See Chapter 33.) The Whitworth saw plenty of action, apparently, in the hands of sharpshooters on both sides of the fight. And why not? The rifle was accurate and it fired a long bullet capable of retaining its velocity/energy at great distance. But while the Whitworth was without doubt worthy of the sharpshooters, the Sharps rifle seems to have caught considerable favor with army personnel at the time. This would be the blackpowder breech-loading Sharps. Breech-loading, yes, but not a cartridge rifle. This Sharps, noted as the New Model 1859 Military Rifle, used paper or linen cartridges. Paper cartridges were nothing new at the time. The soldier nipped off the end and either poured the powder downbore followed by the bullet, or he could more or less ram the whole cartridge down after exposing the powder charge.

The breech-loading Sharps, however, worked quite differently. Its paper or linen cartridge was inserted into the chamber, bullet forward, of course, with the rearmost of the paper or linen cartridge sticking out just a bit beyond the chamber. When the rifle was put into the battery position, the breechblock, which had a very sharp end, cut off the back of the paper or linen cartridge, thus exposing the powder charge to the flash of the percussion cap. This rifle was about 52-caliber, but as I

General Patrick Cleburne can be thought of as the counterpart of Colonel Hiram Berdan. Berdan led sharpshooters of the Union Army, while Cleburne was responsible for putting a sniper corps together for the Confederate side.

General John Sedgewick was a well-loved leader in the Confederate Army; however, he had very little respect for the long-range shooting ability of the Yankees in the distance, and for this he paid the supreme price, as history clearly records.

The New Model 1859 Sharps rifle, shown here in replica form from Dixie Gun Works, is associated to this day with Berdan's Sharpshooters, who were furnished 2000 guns. Not a muzzleloader, the Model 1859 is also not a cartridge rifle, but a cross between the two, since it did use a cartridge, but one made of paper or linen, not metal.

(Right) Built with accuracy in mind, Joseph Whitworth's muzzleloader found its way into the hands of Southern snipers in the American Civil War. This Parker-Hale reproduction of the 45-caliber Whitworth is faithful to the original.

(Below) Although scope sights were certainly the best by far for sniping, some rifles did use sights like these during the Civil War.

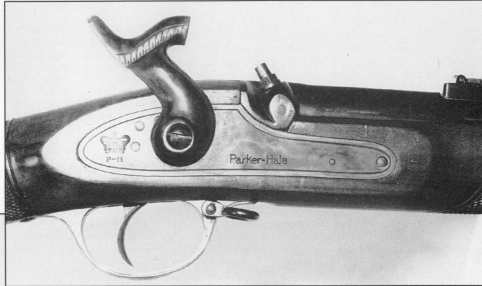

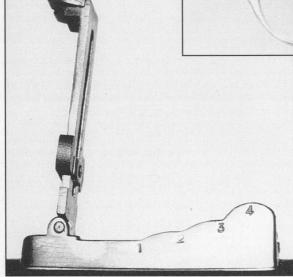

Every aspect of the Whitworth was, and is, geared toward accuracy, right down to a good bullet mould producing an accurate missile, plus accoutrements designed to promote close-shooting groups.

understand it, fired a 465-grain conical bullet that was 56-caliber in diameter. Barrel length was 30 inches, and the rifle weighed 9 pounds.

Today, the Sharps New Model 1859 Military Rifle is sold in replica form, the one I know of from Dixie Gun Works. Dixie's historical study shows that the rifle was first used by the First Connecticut Volunteers of Hartford, but it was mostly associated with the First United States (Berdan's) Sharpshooters. Two thousand were furnished to the Sharpshooters, with the U.S. Navy receiving 2780. The balance of a grand total of 6689 rifles built was spread among various army units. The Dixie model, by the way, has a 30-inch six-groove barrel, and is 54-caliber, not 52, with a 1:48-inch rate of twist. The 54-caliber part seems quite a bit more reasonable if indeed the Sharps bullet was 56-caliber. Asking a lead bullet to swage itself in the bore by two calibers is one thing, while four calibers is quite another. Sights, by the way, are ladder-style rear, adjustable for elevation, with a blade front. To adjust the rear sight for windage, it is moved in its dovetail notch.

The Confederate Sharpshooter

While today it is Berdan's Sharpshooters who are better known, the Confederate Army did have its own snipers. Everything suggests that they were at least as good as their Northern brothers. General Patrick Cleburne was in charge of them. The South apparently put its sharpshooters to work in 1862, but not officially until 1864 did these snipers get their Whitworths. In February of that year, Whitworth rifles were issued to the men, who heretofore were apparently armed with their own personal guns. Notes show a 530-grain, 45-caliber conical bullet for these Whitworths, with 2.5 drams of blackpowder. Recall that a dram is 27.34 grains weight, so the charge would have been about 68 grains of powder, a good target load.

Was Lincoln Fired Upon?

History says yes, Abraham Lincoln was fired upon in 1864 by Confederate sharpshooters during an attack on Washington. The Southern sharpshooters were apparently several hundred yards from the Union trenches, hidden in farm buildings, when a tall man in a black top hat was seen behind the Yankee lines. The Rebel sharpshooters lost no time sighting in on the figure, who was there to get a first-hand look at the battle. Apparently Honest Abe had forgotten what he saw his own sharpshooters accomplish, or he did not think there were any Confederate snipers on the scene. The Rebs fired a few rounds before Lincoln

Reminiscent of the long-range sharpshooter is today's blackpowder (cartridge) marksman, who can hit a silhouette at several hundred yards.

was dragged to cover by General Wright. The story goes that the shots were so close that wood near the President was struck, with splinters actually embedding in his clothing. But, no cigar for the Rebs. They missed! Some historians wonder if the course of the war may have been altered if Lincoln had been killed. We'll never know.

They Shot at Each Other, Too

There were apparently a number of duels between Yankee and Rebel sharpshooters who were sent forth only to discover the other and mow each other down. There can be no doubt that this happened many times. Some of the duels were recorded. One story includes a Private Ide, one of Berdan's sharpshooters, who engaged in a shootout with a Southern marksman in 1862. The bout became a show, as soldiers from both sides watched the two men shoot at each other from long range. The fight ended when the unnamed Confederate marksman put a bullet right on target and through the head of the unfortunate Yankee. Another duel had Rebel sharpshooters firing on Northern soldiers who were pinned down in a ditch, just their knapsacks showing above the trench. Bullets from the Rebs shot the knapsacks to pieces, and when one soldier exposed himself, he was toppled. There's also a colorful, if sad, account of a duel between one James Ragin, a Berdan sharpshooter, and a Southern sniper that ended when the two men fired simultaneously. The Reb's bullet clipped Ragin in the head, creating a furrow in his hair to the scalp. The Confederate marksman did not fare so well. Ragin's bullet ended the man's shooting for that day and for evermore.

Ned Roberts and His Uncle Alvaro

Ned Roberts, in his book *The Muzzle-Loading Cap Lock Rifle*, credits his uncle Alvaro (Alvaro F. Annis) with teaching him how to shoot. Annis, according to Roberts, was one of Berdan's Sharpshooters in the Civil War. Roberts was very proud of that, and he applauded the Sharpshooters, citing an account from *Harper's Weekly* magazine dated August 7th, 1861. The item read, in part:

We illustrate herewith the exploits of Colonel Berdan and his famous sharpshooting regiment, which will shortly be heard of at the war. On the 7th, the Colonel gave an exhibition of his skill at Weehawken, New Jersey, in the presence of a large crowd of spectators. The 'man target' christened Jeff Davis was set up at a distance of a little more than 200 yards. Colonel Berdan inaugurated the firing. . . .

Balancing his rifle for a moment, he fired at the head of the figure. When the smoke had cleared away, the hole made by the bullet was observed by the aid of a telescope, in the cheek, near the nose.

The *Harper's* story went on to say that Berdan hit the target several times, calling one shot in the eye, with the bullet striking "near enough to that organ to destroy its use had it been a real one." Here is also the story of the right eye hit mentioned earlier, incidentally, for someone calls out to hit it and Berdan does. The article continues that no man could enter the Sharpshooter regiment without proving that he could shoot, at 600 feet (feet, not yards) ten consecutive shots "at an average of five inches from the bull's eye."

But Was Uncle Alvaro Really a Berdan Sharpshooter?

Roberts mentioned his Uncle Alvaro so often that certain people decided to check the facts. On the one hand, what they learned was damaging to the story. On the other hand, there could be a reason for the problem. The problem is, the name Alvaro F. Annis does not appear on the roster of Berdan's Sharpshooters Regiment at all. Nor does the War Department in Washington show him. What does appear is an attempt to enlist by Alvaro, which proved negative due to failing the physical exam. A year later, Alvaro Annis entered the army all the same, the story goes, taking the place of a man who had passed the physical exam. After learning of his skilled marksmanship, the Union admitted Annis to the Sharpshooters under the name of the other soldier and Alvaro answered to that man's name. I venture a guess that we will never know if Uncle Alvaro made himself a Berdan Sharpshooter to impress his nephew, or if indeed he truly did take the place of another soldier, whose name has slipped away.

They Were Great Marksmen

Regardless of overstatements and historical flaws, the American sniper was quite a marksman with his blackpowder rifle. Of that there is no doubt. But recall that many Americans were great shooters in the 19th century, for shooting was nearly a daily way of life at the time. Interestingly, the shooting prowess of our forebears lives once again, if not always with the muzzleloader, certainly with the modern blackpowder cartridge and its shooting games. While a few of the feats credited to both Northern and Southern snipers may have been exaggerated, they were not that far fetched, as proved by the fact that good shooters today can repeat the marksmanship with rifle types from the past. ●

chapter 48

The Hawken Rifle

The Navy Arms Ithaca Hawken is a close copy of the real thing. It carries all of the major attributes of the original Hawken, including plain iron furniture, double keys, percussion lock, half-stock design, and so forth.

TODAY THE HAWKEN name is generic, like so many famous trademarks: Kleenex™, Jell-O™, Xerox™. Why did this rifle create so bold a pattern upon the fabric of shooting, not only in its day, but also in our own? Father Time's racing steeds have once again trampled the facts with relentless hooves. And so we have no absolutely clear story of the Hawken in all its glory. But we do have facts enough to tell an interesting and relatively accurate tale, because Hawken history was recorded in its own time, and details were later unearthed by men like James Serven, John Baird, John Barsotti, Charles Hanson, and other careful students of gun history.

Did Every Mountain Man Carry a Hawken?

The most exaggerated part of the tale is one of numbers. Charles E. Hanson, Jr., being the professional researcher that he is, came up with many rational conclusions about the Hawken rifle. His book *The Hawken Rifle: Its Place in History* is bravely told, not only with facts, but also with personal conclusions. In short, Hanson is not afraid to say what he believes after examining the evidence. "Estimates of enormous Hawken production have been put forth for years by implication," Hanson points out. "Many have assured us that every Mountain Man, including Asheley's crew, had one." Where are the examples? Hanson wants to know. Sure, Hawkens led a rough life, but guns survive comparatively well, even for a few hundred years, let alone a mere part of a century.

Why weren't Hawkens dug up all over the place in the early 1900s, for example, only three-quarters of a century after the fur trade era? Because they were never there to begin with, Hanson believes. Also, he points out that there are very few Hawken mentions in 19th century literature, which include diaries and daily logs directly out of the fur trade era. This does not bother me as much as it annoys Hanson. I've been reading 19th century hunting literature most of my life, and details about the guns were not, apparently, important to most authors. Tales of the chase in which great deeds were done make no mention of guns used. However, the Hawken shop records simply do not show a multitude of guns manufactured. Also, it's farfetched to believe that the first wave of trappers west had Hawkens with them. They came to the fur trade carrying what they owned, it seems to me, and it was unlike-

ly that their guns were marked "Hawken." Hanson concludes that "In summary, our research indicates that there were no J&S Hawken rifles before 1825 and does not conclusively document any rifle before 1831."

The Hawken Name on Guns

Also, some "Hawkens" may not have been. Hanson believes that many Hawken pistols weren't. He located several marked "J&S Hawken," but with proofmarks proving at least that parts of the guns were not even of American manufacture. "One that I examined," Hanson says, "had Belgian proofs under the barrel." Hanson believes that at least some of these pistols were simply marked "J&S Hawken" when that "brand" became important. There were also Hawken shotguns that may or may not have been the real thing. The Hawken shop did produce shotguns and pistols, to be sure, but not all guns so marked are true Hawkens. On the other hand, I've read that flintlocks marked "Hawken" must be phony, because everyone knows that the Hawken was a percussion firearm. This is not so. While rarer than snow on the Sahara, there were some early flintlock Hawkens.

The Hawken Rifle in General

None of this is to take away from the fact that the Hawken brand burned itself into the hearts and minds of 19th century shooters. The rifle deservedly achieved great fame, and that's why it lives on in name to this day. While "typical" may be a dangerous word to apply to any rifle that was made, essentially, one at a time, we can probably get away with noting a Hawken as a half-stock (usually) plains rifle. Hanson shows one on page 49 of his book as full-stock, the rifle appearing to be damaged in the forepart of the stock. The lock is marked "R. Ashmore & Son." The barrel is 31¼-inches long, but evidence of shortening is clear. The barrel is 1¹¹⁄₁₆-inch across the flats. It is a percussion rifle with a walnut stock, weighing, with shortened barrel, 8 pounds and 7 ounces.

Hawkens for Real

The Colorado State Museum in Denver has a Hawken on display. The rifle belonged to the noted trapper Mariano Modena of fur trade

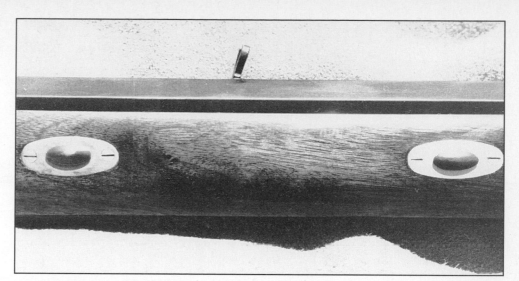

Two features noted on most Hawkens are the double keys (two barrel wedges) and the simple rear sight, both featured on this Ithaca Hawken replica.

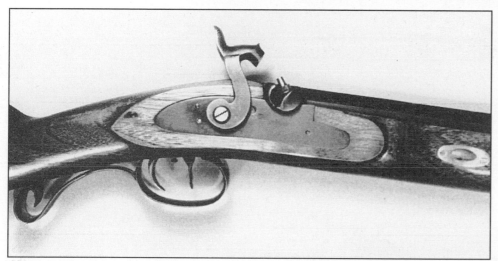

A feature attached to the plains rifle in general is the percussion cap system. Once percussion caps became reliable, it was inevitable that the caplock system would take over from the flintlock, which it did.

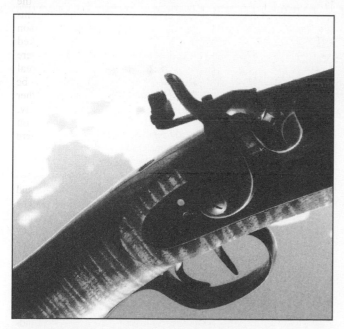

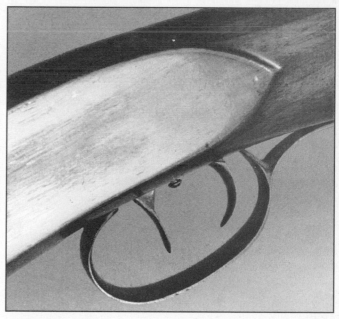

Reportedly, a few original Hawkens had peep sights, either installed at the Hawken shop or perhaps by their owners. Some researchers say that individual owners did modify the sights on their rifles as an ordinary thing to do. Also, the pistol grip, while rare, was found on some Hawkens. This custom shows both peep sight and the pistol grip features.

The double-set trigger on this Hawken replica allows a shooter to fine-tune his rifle.

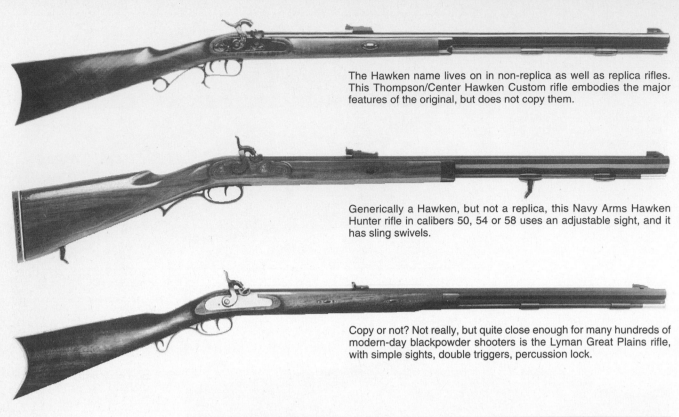

The Hawken name lives on in non-replica as well as replica rifles. This Thompson/Center Hawken Custom rifle embodies the major features of the original, but does not copy them.

Generically a Hawken, but not a replica, this Navy Arms Hawken Hunter rifle in calibers 50, 54 or 58 uses an adjustable sight, and it has sling swivels.

Copy or not? Not really, but quite close enough for many hundreds of modern-day blackpowder shooters is the Lyman Great Plains rifle, with simple sights, double triggers, percussion lock.

Hawken replicas also come from custom gunmakers. This Late S. Hawken is a copy made by Andy Fautheree from an original owned by the gunmaker himself.

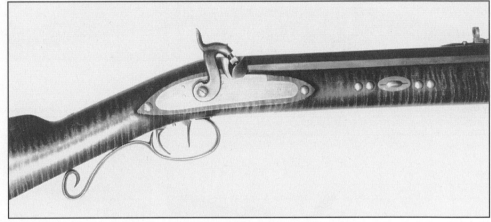

fame. Here is a real Hawken. So what is it like? The rifle is 58-caliber, weighs 12½ pounds, and has a barrel 34¾ inches long. It has double keys (two wedges holding stock and barrel together), a large patchbox and nine silver star inlays. The top barrel flat is stamped "S. Hawken, St. Louis." The lock is stamped "A. Meyer & Co., St. Louis." A silver plate is inlaid into the cheekpiece recording the purchase in 1833, St. Louis, and presented to General A.H. Jones in 1837.

Calibers ran quite a range in Hawken rifles, the largest on record apparently running about 66 to 68 or thereabouts. A retired Hawken employee said, "I made a rifle for his, Ashley's, special use. The barrel was 3 feet and 6 inches long and carried an ounce ball." If indeed this Ashley Hawken fired a 1-ounce round ball, it had to be around 66-caliber or so. A pure lead round ball .650-inch in diameter weighs 413 grains. A .660-inch round ball weighs 433 grains. There are 420 grains in 1 ounce. Considering windage, the bore would be about a caliber larger than the round ball.

The Kephart Hawken

The famous Kephart Hawken has caused as much trouble as good, in my estimation. Kephart, historian and shooter, bought his Hawken in 1896. It had been in storage for years. He loaded it with 204 grains

of blackpowder, a lot for a 53-caliber bore. That powder charge is bothersome for two reasons clearly pointed out in our study of ballistics. First, that much powder would be in the grip of the law of diminishing returns, making a lot of smoke and noise, but not gaining full benefit in velocity. Second, why 204 grains? The four grains would be a meaningless gesture, when there would be no discernible difference between 200 and 204 grains of powder. To be fair to Kephart, who wrote much on his Hawken into the middle 1920s and later, he also pointed out that best accuracy occurred with much less powder: "82 grains of FFg Deadshot." Some sources show FFFg, by the way. Kephart spoke of the "double charge," which was passed down to us in later times, some shooters believing that you could pour as much blackpowder as you pleased downbore for more and more power.

J.P. Gemmer is well known in any Hawken story, so he needs to be touched on briefly here. Gemmer worked for the Hawkens, along with a dozen or so other gunsmiths. Records show that Gemmer bought the Hawken operation in 1862.

The Hawken Brothers

This is the story of the rifle, not the men; however, the brothers were treated as one because they acted together in making the Hawken

While not an exact replica of a Hawken pistol, the Lyman Plains pistol embodies features of the original.

The plains rifle was short compared with the Pennsylvania/Kentucky long tom, but barrels were still generally over 30 inches in length.

stamp synonymous with plains rifle. They had impact on an important time in America, the era of westward expansion. (Oddly enough, they had impact on our era, too, although they could never know it.) The opening of the west did not begin with pioneer wagon trains squeaking over the Oregon Trail. It began with the beaver trapper, the mountain man, the hell-for-adventure boys who boldly headed into a territory unknown by the White Eyes. Their story has been told early in this book, and shall not be repeated here. But for the sake of understanding the rifle, recall that the Lewis and Clark Expedition of the early 1800s brought the first mountain men into the Far West.

Thomas Jefferson knew that the western half of America was up for grabs. The French were the first to see many parts of the territory west of the Mississippi River. Coming down from Canada to trap, they got an eyeful of a vast land under the control of no government at the time. The English were also working the region. But it was Jefferson's band, led by Lewis and Clark, that drove a stake marked "America" into the western earth. Some men from the expedition explored on their own, as well as with the group. I said the French and English were in the region, but the area was vast. Boots of the European trapper had made few tracks over the unexplored landscape of the Far West in the early 1800s.

The mountain man era was on, and the movement would bring fame to the brothers Hawken, for a new rifle was needed. The transition from the European Jaeger large-bore rifle, with big ball but smallish powder charge and low velocity, to the sleek Pennsylvania long rifle, with comparatively fast, but lightweight patched ball, would be continued with another kind of rifle—a big bore with large powder charge driving a heavy lead sphere with superior killing force. Wild animals were bigger out west. Not that bison didn't roam east of the Mississippi, but the real herds were out west. The buffalo, as it was

called, was the largest four-footed animal on the North American continent. The smallbore rifle was not right here. The black bear of the east was well represented out west, but so was his fierce cousin, the much larger grizzly. The rifle that was right back east was wrong out west. The Hawken brothers were not the first, nor hardly the only, gunsmiths to answer the call for a new rifle. However, they were the best at what they did, which was the manufacture of a working rifle that could be relied on under harsh conditions in the Rocky Mountains, a land explorers called the "marrow of the earth."

The brothers came by their gunmaking expertise naturally. Grandfather Henry Hawken was a smith of no small ability. His surname was Wee Hawken, but as was common with immigrants to America, the family name was changed. Wee Hawken became Hawkins. Finding that there were a number of Hawkins around, Grandpapa apparently altered the name back to Hawken. However, the Hawkins handle was not lost immediately. Years later, Hawken rifles would often be referred to as Hawkins. Serven notes that a court reference found in a Lancaster, Pennsylvania, paper makes reference to a Henry Hawkins. "In November of 1724 Henry Hawkins petitioned the Court—then in Chester County—for redress against John Burt to whom he had apprenticed himself for five years to learn the trade of a gunsmith. . . Henry Hawkins was probably an ancestor of the later Hawkins who made the famous Hawkins rifle in St. Louis, Missouri." Another source, Bruell's *Sir William Johnson*, noted that "Henry Hawkins was not only a great riflemaker himself but his sons and grandsons succeeded him in later years, establishing shops at Rochester, Louisville, Detroit and St. Louis, until during the last quarter of the eighteenth and first half of the nineteenth century, the 'Hawkins' rifle was famous all through the West."

Many modern hunters have enjoyed carrying replicas of original Hawken plains rifles. Here, Fadala packs an Ithaca 50-caliber Hawken.

Although the Hawken story is fairly complete as biographies go, it does have many gaps. However, the branches of the family tree probably take this form: Henry Hawkins, who would later be Henry Hawken, had a son named Henry Hawken who, with a wife named Julienne, had two boys, Jacob and Samuel. Jacob was born in 1783, Samuel in 1796. Both entered this world at Hagerstown, Maryland. The Hawken boys' grandfather had been a gunsmith of note, as already stated above. Their father, Henry, was also a gunsmith. Of their youth and growing up years, I could find nothing substantial. Father Time's horse churned these facts into the clay.

But we do know that Jacob, or "Jake," Hawken went to St. Louis in 1807. Hanson believes that Jacob worked at some mechanical enterprise until 1815 when he was able to turn his savings into a gunshop. Timing was right. The beginning of the fur trade era is considered to be 1815, with its heydey starting around 1822, per Ashley's advertise-

ment for trappers in that year. Meanwhile, Sam Hawken had a shop in Xenia, Ohio. When his first wife died, he closed that shop in 1822 and followed his brother Jake to St. Louis to become his full partner. Note the timing—1822, the very year of Ashley's advertisement. Sam was a great asset to Jake's gunmaking operation.

Another particle of history ties in perfectly with the development of the famous Hawken rifle—the percussion cap. Joshua Shaw, an Englishman residing in Philadelphia, had his version of a "copper detonating cap" in 1816, but it was not entirely reliable. If doused with water, it might not fire. There was no reason to switch from the tried and true flintlock system to an ignition mode that was not foolproof. Furthermore, the Shaw cap was highly corrosive. The resulting spark damaged both metal and wood. But there would be an important change in the cap. In 1824 in England, fulminate of mercury replaced the oxymuriate of potash that Shaw used in his percussion cap. The new cap was not terribly corrosive. Furthermore, it could be waterproofed (to a degree) by treatment with sealing wax and spirits of wine. By 1830, the new cap was readily available in America. Sam and Jake were enterprising. Their Rocky Mountain Rifle would be a caplock.

The fur trade was not the only event that brought work into the Hawken shop. Serven's details of a Sam Hawken interview for the newspaper contained these facts: "From Hagerstown I went to Xenia, Ohio, and kept a gun store there, and on June 3, 1822, I arrived in St. Louis. Our first shop was on the levee near Cherry Street. I didn't stop at the first place long; soon we had a new shop on the levee near Olive Street. . .when the California gold fever broke I had a bigger demand for guns than I could meet. But I did not raise my prices. From $22.50 to $25.00 was all I asked. Might just as well have got $50.00. Folks wanted me to raise my price, but I said no, those that bought would send back for more, and so they did...." Gold had its impact. But beavers had an even greater effect upon the Hawken gun business. "Every man going West wanted one," said Sam. "William Ashley's men were the next lot to go out; they started for the Rocky Mountains and were driven back by the Indians. The boys had a terrible hard time on that trip. . . . Fremont's company that went to California had my rifles." The Hawken Shop furnished all of the guns for the Missouri Fur Trade Company.

We know that every man going West may have wanted a Hawken, but far from every man had one. Be that as it may, the very idea that Hawken was *the* rifle to own made an impact. And the Hawken shop was a big success. The famous Hawken rifle was a reality, and the name survived into modern times, where any black-powder rifle that carried even similar traits was called a Hawken.

History, right or wrong, put Hawkens in famous hands, which may have furthered the mystique that lives on: Mariano Modena, Jim Baker, Jim Bridger, Kit Carson, and others. Bridger, rumors hold, owned several Hawken rifles.

The Hawken Story Ends

The story of Jake Hawken ends in 1849. Cholera was rampant, and the disease struck him down on May 8th. His body may have been cremated along with other victims of the tragedy. But Samuel Hawken lived on, a man of great character. Serven portrays him as civic-minded, one-time candidate for mayor of St. Louis and also instrumental in forming a volunteer fire company in the city, one of its first. Samuel had never seen the Rocky Mountains, the great territory in which the Hawken rifle was made famous. It was time to go to the marrow of the world. In 1859, ten years after Jacob's passing, Sam took the trail west at 67 years age.

His health was at low ebb, and he wanted to move, walking most of the distance from St. Louis to Denver. He was on the trail for several weeks, leaving St. Louis on April 20th and arriving in Denver on June 30th. Some report that the Hawken shop was left in the hands of trusted individuals. Probably, it was sold. Apparently, Samuel was interested in the Pike's Peak gold strike, but after looking around in that area, he returned to Denver to do what he did best—build guns. He worked out of a log cabin for a while, reporting, "Here I am once more at my old trade, putting guns and pistols in order 'how to shoot'." An advertisement in the *Rocky Mountain News*, January 25, 1860, read: "S. Hawken, for the last thirty-seven years engaged in the manufacture of the Rocky Mountain rifle in St. Louis, would respectfully say to the citizens of Denver, Auraria, and his old mountain

Although flintlocks were rare on Hawken rifles, they were not entirely unknown. This Navy Arms Ithaca Hawken is a flintlock, offered for those who want both the Hawken style and the spark-tosser ignition system.

Many hunters have enjoyed the Hawken rifle in modern times. Bill Fadala poses with a small mule deer taken with his 50-caliber Ithaca Hawken rifle.

friends, that he has established himself in the gun business on Ferry Street, between Fourth and Fifth, next door to Jones and Cartwright's, Auraria, and is now prepared to manufacture his style of rifles to order."

The change of residence proved to be what Sam Hawken needed. The *Rocky Mountain News* reported that "Our venerable friend, S. Hawken, whose rifle for years has had an unequaled celebrity among hunters, trappers and voyageurs of the plains and mountains, has raised a tall pole in front of his shop on Ferry Street, on the top of which a mammoth rifle is swung on a pivot. The big gun can be seen from all parts of the city—now pointing at the mountains, now away from them, as it is swayed by the breeze."

The newspaper story went on to say that "Mr. Hawkens (sic) is an old resident of St. Louis, having made guns there for thirty-seven years, and came to this country [Colorado] about a year ago for the benefit of his health, which he informs us has been completely restored."

Colorado was good to Sam Hawken, but home was St. Louis. Sam returned to that city in 1861. He seems to have have left his son, William S. Hawken, who had come to Denver, in charge of the new gunshop. The story goes that Sam sold the Hawken shop to J.P. Gemmer. But it appears more likely that Gemmer purchased the shop from William L. Watt and Joseph Eterle, who had bought it from Sam Hawken, but decided that they did not want to run it. A St. Louis directory lists an advertisement that says "William L Watt, Successor to W.S. Hawken, Rifle and Shotgun Manufacturer, 21 Washington Ave., Hawken Rifles always on hand." That advertisement, plus the remarks

of Gemmer, indicate that Gemmer did not buy the shop directly from Hawken.

What is important to the Hawken brothers story is the fact that the rifle they made famous did not die with Jacob in 1849, nor after the selling of the Hawken enterprise. Watt, by the way, no doubt sold some Hawken rifles. Baird believes that these rifles may have been marked simply "Hawken," and not "S. Hawken," as Sam would do. Perhaps Watt did not have the right to use the full "S. Hawken" stamp. Baird studied an authentic Hawken rifle of the Watt period which was marked "Hawken" and nothing more. Gemmer, on the other hand, was not restricted concerning the Hawken stamp.

Sam Hawken built rifles until he was 70 years old. He lived with his daughter Mrs. Fred Colburn after the passing of his second wife, Martha Richey Hawken. Retired, yes, but Samuel could not stay away from the gunshop. Gemmer reports that the old gentleman visited every day. He still believed in rifles the way he made them—the right way. Apparently, the cartridge gun never won him over completely. He passed up, for example, a chance to sell Colt revolvers. On May 9, 1884, Sam Hawken went to his final reward. Born on October 26, 1792, that made the great old gunmaker 92 years old.

The authors named above, as well as many others, have worked hard to uncover as many Hawken facts as possible. The reader interested in a deeper knowledge of the rifle and its makers is urged to locate and digest these writings. Sam and Jake Hawken made more than rifles. They made history that continues to this day in the sincerest form of flattery—imitation—not only in replicas, but non-replica muzzleloaders bearing the Hawken name. ●

Rendezvous Shooting Games

chapter **49**

REMEMBER THAT THE original rendezvous was a business enterprise of Yankee Trader ingenuity (see Chapter 2). The beaver hat was a hot item in the world of fashion, and before you could make such a hat to sell to London's fashion-conscious set, you simply had to have a beaver. Beavers lived over much of North America in the first half of the 19th century, but they were in especially good supply in the Far West, that mostly unmapped region west of the Mississippi River where the Indian was still landlord. The crux of the enterprise was rather simple and ingenious—advertise. You see, Americans already knew the value of advertising. William Ashley placed an advertisement in the Missouri Gazzette, a St. Louis newspaper, on February 13, 1822 (see Chapter 2).

Of course, these were the "enterprising young men," along with others, who would one day be known as the Mountain Men. Places would be named for them, like Jackson Hole, Wyoming; Bridger National Forest; Bridger, Montana; Fort Bridger; Henry's Fork of the Snake River; Bonneville Pass; Laramie, Wyoming (for the Laramie River, which was named in honor of slain mountain man, Jaques LaRamie).

And many more. As we know, the men trapped beaver during favorable times of the year, learning from the Indian how to survive the winter. Then their plews of beaver were traded at a pre-arranged summer meeting place that came to be known as the rendezvous. The furs were worth trade dollars. A trapper might have $1000 to $2000 in trade value for a season's labor at a time when a skilled workman was worth $1.50 a day. However, the beaver pelts were like tokens, to be traded in for goods only. Well, not only goods. One trapper was supposed to have given an Indian chief the sum value of $2000 for the hand of his daughter. She must have been a knockout.

Trade values for furs were preserved in *Journal of a Trapper*, a well-thought of compilation originally written by Osborne Russell, himself a trapper. Whisky purchased in St. Louis for 30 cents a gallon was cut with water and sold for $3 a pint in fur trade value. Coffee and sugar selling for a dime a pound in St. Louis brought $2 a pound at rendezvous. Lead for "running ball," which was the term for casting shooting projectiles, as well as gunpowder sold for $2 a pound, while the first was worth 6 cents a pound in St. Louis, and powder brought 7

Shooting games are the highlight of the modern day rendezvous. There's a great deal of comradeship and helpfulness during these shoots, with fellows competing to win, but also willing to share a tip with another shooter.

355

Lady shooters are welcome at rendezvous matches, and they're great competitors, too.

(Below) Shooters have to be prepared in many ways if they want to make a good showing at the rendezvous shoots. Selection of the right firearm is the first step. For offhand shooting, the longer barreled rifle does a good job, not only because of its extended sight radius, but also because of the up-front weight which helps to steady the gun.

Splitting the round ball on the blade of an axe to create two bullets out of one is no easy task. Here, one clay pigeon has been broken, but the other has survived.

cents a pound. But the mountain men did not necessarily feel cheated. It took great risk to run wagons of goods from the more civilized East to the wilds of the West. Such risk should carry profit, a lot of profit. And it did.

The first rendezvous was held in 1825 when General Ashley gathered his trappers on what is now known as Henry's Fork of the Green River. The actual site is two miles from the present townsite of Daniel, Wyoming, which in turn is not too far from the larger city of Pinedale, Wyoming (where every year there is a pageant held in honor of the mountain men). Although the original rendezvous, as noted, was a business venture, it turned out to be much more than that. Those meetings in the summer brought together the trappers and Indians in a way that nothing else had ever done before—although there were fights and disruptions, there was also comradeship. The rendezvous was a time for renewing old acquaintances and making new ones.

Along with the fur trading, there was story telling, knife and tomahawk throwing contests, swapping, new and old wares for sale, many games *and shooting matches*. The shooting matches are the interest

of this little chapter, for they continue to this day at the modern rendezvous. While some shooting matches are as exciting as watching cement dry, the firearm competitions set up by the buckskinners have real flare. They are ingenious and creative. Many of today's competitions are non-historically correct, by the way. So this is not a history of old rendezvous shooting games, just a look at a few of today's matches.

General Shooting Tips

Each contest listed below contains a tip or two, but here are some general notes that might help a shooter score higher no matter what the event is. Hopefully, the reader will study other chapters in the book geared to promote better shooting, too. Consider the following mere reminders, with the real meat in those previous chapters, such as Chapter 20 on loading techniques.

Cleaning During a Match

If cleaning at any time during a shoot presents a problem, remember that there are two ways to beat this. The first is the all-day lube, as

Matches are limited only by the imagination of the rendezvous leaders, and, of course, the range and effectiveness of the guns. These shooters are getting ready to fire at long range.

A number of matches at rendezvous are for offhand shooting only, with handgun as well as rifle. Ranges are as close as 25 yards, but can run to several hundred yards.

defined in Chapter 19. The second is the use of Pyrodex. See Chapter 15. Pyrodex does not demand cleaning between shots or even strings of shots until a lot of firing has taken place.

The Fouling Shot

A "fresh" bore may shoot slightly off the mark, even if it has been dried with a cleaning patch before and after loading. That first round ball or conical flying wide of the mark can be critical on many matches, obviously, opening up an otherwise tight group, missing the blade of the axe, or flying right by that charcoal briquette. So fire one fouling shot before the match begins. This holds for both blackpowder and Pyrodex.

Leftover Lube in the Bore

As related before, an oily, wet, or greasy bore may throw a projectile off course. Where the accuracy of the shot is absolutely vital, it's wise to sight in with a dry bore and maintain a dry bore during a match. This is easily accomplished by running a cleaning patch downbore after loading. Dangerous? No more so than running the patched ball or conical down on the powder charge, which was necessary to load the gun.

Altering Loads for a Match

By all means use the same components you sighted in with. Do not change bullet, powder type, or charge unless you know what the change will bring. Sometimes it is important, of course, to alter a load for a given match. No sense, for example, firing a heavy powder charge for the 25-yard offhand match. But do know what the lower powder charge will bring in bullet placement and trajectory.

Consistency

Try to maintain consistency in every regard for best results at a shooting match. While a pressure regulator is not necessary, do attempt to seat your bullets with about the same ramrod feel every time. Likewise with the powder measure. Establish a routine and stay with it.

Smoke is part of the fun. Most of the matches are designed to entertain the spectator as well as the shooter.

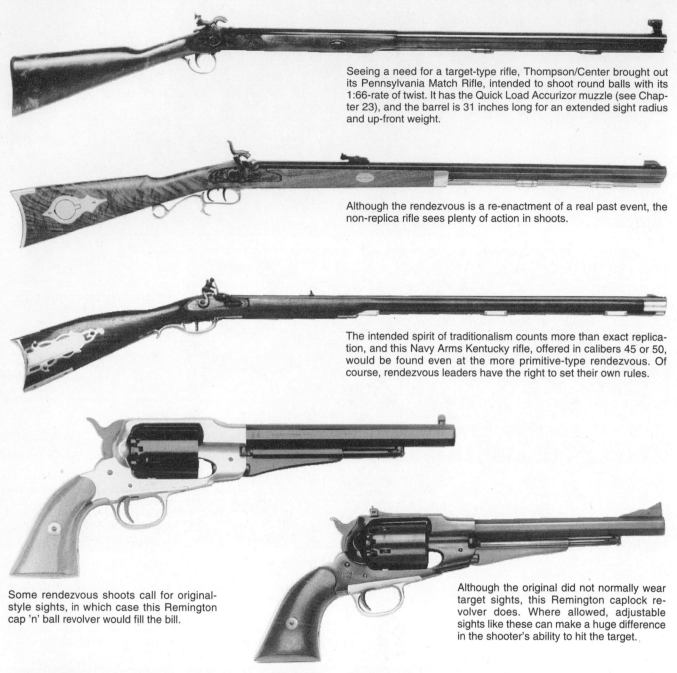

Seeing a need for a target-type rifle, Thompson/Center brought out its Pennsylvania Match Rifle, intended to shoot round balls with its 1:66-rate of twist. It has the Quick Load Accurizor muzzle (see Chapter 23), and the barrel is 31 inches long for an extended sight radius and up-front weight.

Although the rendezvous is a re-enactment of a real past event, the non-replica rifle sees plenty of action in shoots.

The intended spirit of traditionalism counts more than exact replication, and this Navy Arms Kentucky rifle, offered in calibers 45 or 50, would be found even at the more primitive-type rendezvous. Of course, rendezvous leaders have the right to set their own rules.

Some rendezvous shoots call for original-style sights, in which case this Remington cap 'n' ball revolver would fill the bill.

Although the original did not normally wear target sights, this Remington caplock revolver does. Where allowed, adjustable sights like these can make a huge difference in the shooter's ability to hit the target.

Lube that Revolver

If you don't pay attention to proper lube for your revolver, it may lock up right in the middle of a match. An all-day type of cream on top of the ball just before shooting can help keep the cylinder turning on your cap 'n' ball revolver.

Check Sights and/or Sight-In Before a Shoot

Before heading for rendezvous, ensure that the sights on your firearm are intact and unmoved. If in doubt, go to the range and re-sight the gun.

The Games

The Knock-Down Target

A steel target is secured on a hinge. When the target is struck, ping!—it literally folds over. **Tips:** Oftentimes, this is a round-ball-only event, and at ranges that can be very long, even 500 meters and more. Consider a large bore rifle over a smaller bore. It's going to be bullet weight that tips the target over. Also, consider a reasonable powder charge that will give the bullet a fairly good starting velocity.

The Silhouette

The silhouette game, although played more seriously these days with the single shot blackpowder cartridge rifle, can be a rendezvous event for muzzleloaders as well. **Tips:** The same as for hinged knock-down targets—go with a good-sized ball and a powder charge that gives good starting velocity.

Splitting the Ball on the Axe Blade

The old trick of splitting a round ball on an exposed axe blade is an excellent match. A double-bitted axe is buried into the center of a large piece of tree stump. The object is to strike the exposed blade of the axe so that the lead ball is sliced into two missiles. On either side of the blade are placed balloons, clay pigeons, or other breakable targets. If the shooter does his part, the halved ball breaks both targets. **Tips:** This is an offhand event, and as such, rifles of good weight with long sight radius help steadiness and ball placement. Also, consider a larger caliber for its size alone—if you come close with a 40-caliber ball, but miss the blade of the axe, you may have hit it with a 50-caliber ball. It also helps to sight-in for this event with horizontal grouping consid-

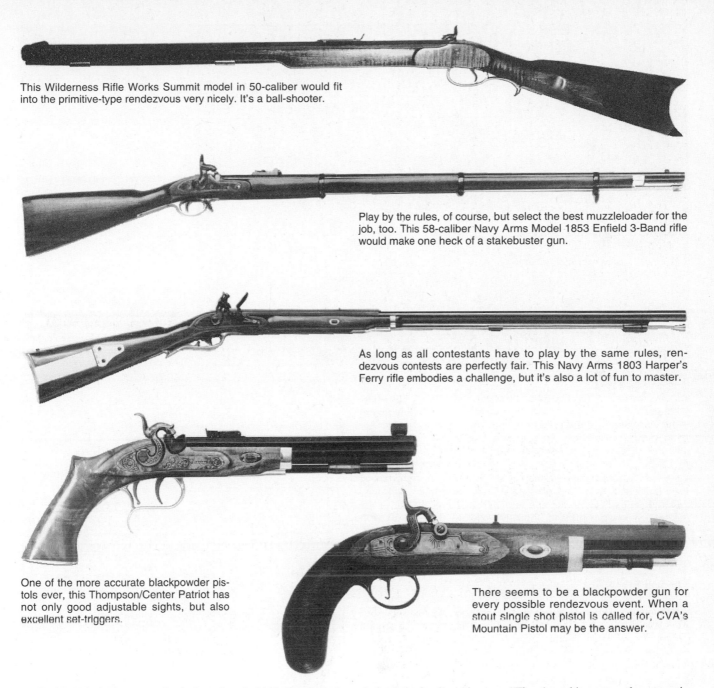

This Wilderness Rifle Works Summit model in 50-caliber would fit into the primitive-type rendezvous very nicely. It's a ball-shooter.

Play by the rules, of course, but select the best muzzleloader for the job, too. This 58-caliber Navy Arms Model 1853 Enfield 3-Band rifle would make one heck of a stakebuster gun.

As long as all contestants have to play by the same rules, rendezvous contests are perfectly fair. This Navy Arms 1803 Harper's Ferry rifle embodies a challenge, but it's also a lot of fun to master.

One of the more accurate blackpowder pistols ever, this Thompson/Center Patriot has not only good adjustable sights, but also excellent set-triggers.

There seems to be a blackpowder gun for every possible rendezvous event. When a stout single shot pistol is called for, CVA's Mountain Pistol may be the answer.

ered to be critical. There are a few inches of vertical latitude, but horizontal latitude is very narrow.

The Ricochet Target

There is a ricochet match in which a chunk of heavy metal, such as boiler plate, is placed so that a ball can be skipped off of it into a safety bank. Situated at the backstop is a target, a balloon, clay pigeon or other breakable object. The shooter must skillfully glance the round ball from the boiler plate so that it strikes the target. **Tips:** Just about any bullet size works here. The trick is more aim than gun-oriented.

The Distant Gong

A gong is another interesting target. One Montana rendezvous held a long-range gong shoot in which the participant had to aim his rifle so the lead sphere struck a metal gong 500 yards away. Considering the looping trajectory of a round ball, even from a large-bore rifle, it's no simple trick to guide one out to 500 yards with accuracy. **Tips:** Successful shooters say they hit the gong because they use a consistent

aiming point far above the gong. "There's a white spot on the mountain in the background that I aim at," one shooter admitted. Of course, the large and stable round ball is best for this match, with a reasonably strong powder charge behind it.

The Stakebuster Match

The stakebuster match incorporates a wooden plank, 6 inches wide (other sizes will also work). Two shooters generally team up for this one. The idea is to cut the plank in half with rifle fire. **Tips:** Larger calibers are best. Also, it doesn't hurt to be sighted in specifically for this close-range event. It takes skill to draw a line across the wood with successive bullet strikes in order to cut the stake in two, but with a smallbore, more shots must be fired than with a larger bore.

Swinging Charcoal Briquettes

Briquettes swinging on strings suspended from a horizontal wire make interesting targets, with a puff of smoke denoting a solid hit. **Tips:** A good-holding offhand rifle, preferably with a long barrel for up-front weight and long sight radius, help the shooter in this event.

The caplock revolver must be kept well lubricated during a shooting match to prevent lockup of the cylinder. One of the major reasons for placing lube on projectiles loaded into cylinder chambers is the prevention of lockup.

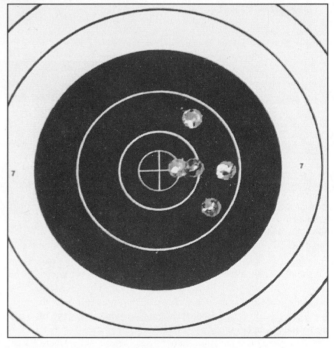

(Below) There is no shortage of regulation-type targets at the rendezvous. Most of the challenges are held offhand. When the range is 100 yards, and the rifle has iron sights, and offhand only is allowed, groups like this one are commendable.

The Seneca Run

The Seneca Run match is more involved. It reminds a little of the Olympic Biathlon. The shooter must successfully accomplish a series of tasks, running from station to station. He must run the course at good speeds, because he is up against the clock as well as the targets placed along the way. The marksman with the highest score in the least time is the winner. Some of the targets may be for rifle only; others are for sidearms. The tomahawk may be included as part of the run, as may the knife. Running from one target to the next requires reloading, either along the way, or in rapid fashion once the next target site is reached. Shooting is offhand. **Tips:** The obvious tip here is to use a firearm that is very familiar to the shooter, one he can reload readily without a hitch. Since shooting is once again offhand, that steady long rifle may be the best bet.

The Log Rest

This event simulates hunting conditions. It can be run in many different ways. One is to have a walking path, strictly regulated. The contestant strolls along the path, and at intervals various targets are set up. They can be metallic silhouettes, or paper targets with a dirt backstop, or cutouts of animals. The shooter takes the prone position with the forend of the rifle resting on the log. **Tips:** Although heavy-barrel rifles seem to do okay when the barrel itself is rested (take note of blackpowder cartridge rifle shooters), consistency is best maintained here by resting the forend rather than the barrel. Also, a heavy rifle, nine pounds and up, rests more solidly in place than a lighter rifle.

Regulation Targets

There are many regulation target matches at rendezvous. They are not, perhaps, as interesting as the novelty shoots, but they are valuable, and, in fact, show off the shooter and his equipment very well. Offhand shooting at regulation targets can be held at 25, 50, 100, 200 yards. **Tips:** Clearly, the more accurate the firearm the better, be it rifle, pistol, or revolver. And, as always, the stable firearm is at an advantage. With a rifle, that generally means the slightly heavier model with longer barrel that produces up front weight for steady holding, plus a long sight radius.

Shotgun Competition

The shotgun is an obvious tool of competition at any rendezvous. Events can range from trap to Skeet, and anything in between, the only limitation being the range of the shotgun and the imagination of the event designers.

Anything Safe Goes

As long as it's safe, rendezvous shooting matches can range from moving targets on a taut wire to multiple-skill competitions, where rifle, handgun, and even shotgun are put into play in a single match. Most events will have spectator interest, however. Something will go *clang* like a gong in the distance, or *poof*, when a charcoal briquette blows up, or *pop* as balloons do when the split-in-two round balls break them, or they may fall over when hit. This is good, because it promotes interest in the great sport of blackpowder shooting. "Hey, I'd like to get in on that!" you might hear from a spectator. Of course, the plain old paper target with a regulation-style bullseye tests the skill of the shooter, too, and should remain a strong event for a long time to come. ●

chapter 50

Blackpowder Cartridge Games

THE BLACKPOWDER CARTRIDGE has come of age—*again*. Hunters all over North America are enjoying success with old-time rounds, but more than hunting, it's blackpowder games that have escalated interest in this cartridge of old. Cowboy Action Shooting has six-shooter and lever-action fans beaming. There are many replica guns for those events, and even new handloading manuals. But the thrust of this chapter is not cowboy shooting, which is generally done with smokeless powder, but rather the single shot breech-loading blackpowder cartridge rifle of the late 19th century. One major arms company spokesman said, "We can't keep a Remington rolling block or a Sharps Rifle in the house." Why this great interest in the old-time rifle? It's the challenge and the fun that make the sport so popular.

Informal and Local Shoots

Shooting clubs all over the U.S. and Canada are fashioning their own special blackpowder cartridge games. Rules vary with each group. For example, I attended a "Buffalo Gong Shoot" where "Any single shot breech-loading cartridge rifle or replica in the spirit of the era—late 1800s to early 1900s," was acceptable. Only blackpowder or Pyrodex were allowed in this contest, with duplex loads (smokeless mixed with blackpowder) or smokeless powder loads forbidden. Only metallic sights were allowed—no scopes. Cast lead bullets are the only projectiles shot and without gas checks. The major competition was a metal gong at 200 yards, offhand shooting only, but as it turned out, the most watched and enjoyed aspect of this particular shoot was the buffalo (American bison) metallic cutout placed at 900 yards. Boom! Wait a while. Clang! Everybody loved it, shooter and spectator alike.

The buffalo metallic silhouette was not an off-hand event. All shooters used crossed sticks. The range was opened one day ahead of the match so that shooters could visit with each other, have some fun, and also practice a little before the real thing got underway. This local shoot-out was typical—plenty of rules, but nothing as stringent as national competition calls for, which we'll talk about next. The local shoot is always geared for a good time, even when prizes are offered. Oftentimes muzzle-loading events are included along with the blackpowder cartridge games. While gongs and metallic silhouettes are the usual targets, the local group is not bound to regulations, not to types or sizes of targets, or shooting distances.

The *Silueta* Game is Born

The roots of blackpowder cartridge silhoutte shooting reach down into the rich soil of a sport that began in Mexico. It's difficult to say just

That tiny black dot on the distant hill (arrow) is a metal cutout of a bison. It is located 900 yards from the shooter. Is that a challenge with iron sights and a 19th-century blackpowder cartridge rifle? You bet.

when, nor do I don't know who, exactly, got the original idea for the formal *silueta* match, but I do know that in the late 1950s and early 1960s, Victor Ruiz, a well-known marksman from Nogales, Sonora, Mexico, along with several of his shooting companions, was firing away at metal cutouts at long range from the offhand position only. Interest grew in the sport, partly because the Mexican marksmen invited north-of-the-border shooters to join them. The *siluta* match soon became a hands-across-the-border friendly competition, and from there an international game with literally hundreds of participants. It continues to thrive today.

Modern Rifle Silhouette Shoot

The original silhouette match was built around the centerfire sporting rifle. Rules called for a rifle that you or I might take into the hunting field for big game. As the shoot became more sophisticated, many rules were designed to keep the sport from deteriorating. A special scope-sighted silhouette rifle was born, most-often chambered for the 308 Winchester. Of course, the metallic cutouts became official in size, as well as the distances at which they were placed.

Blackpowder Cartridge Silhouette

The formal blackpowder cartridge metallic silhouette game is definitely an offshoot of the original smokeless powder cartridge match. It is NRA sponsored, and the first official shoot was held near Raton, New Mexico at the N.R.A. Whittington Center in 1985. The event has been run annually ever since, with ever increasing interest. Canada has joined in with its own national competition, as well, and Australians also shoot blackpowder cartridge silhouette.

The Rifles

Currently, the year 1892 has been established as the cut-off manufacturing date for legal blackpowder silhouette rifles for formal competition. Previously, 1895 and 1896 were noted as official dates. For now, rifles manufactured after 1892 are not allowed in the NRA Blackpowder Cartridge Rifle (BPCR) Silhouette match. The rule was slightly modified to include the Stevens 44$\frac{1}{2}$, with its falling block action and exposed hammer. Even though the 44$\frac{1}{2}$ dates from 1903, the rifle follows a design that matches the spirit, intention, and even the mechanics and design of the 19th century single shot breech-loading blackpowder cartridge rifle. The sport still has growing pains, and further changes in allowable rifles may be seen.

Browning, noting the great interest generated in the BPCR Sil-

A good look at a modern day blackpowder cartridge single shot breechloader fan, Sharps rifle and all. The Vernier tang peep sight, covered front sight, and crossed sticks are all part of the game.

houette sport, introduced its version of the 1885 single shot, calling it, appropriately, the Browning Model 1885 BPCR (Black Powder Cartridge Rifle). It even has a long range metallic sight suited to the game. A few original and rebuilt Sharps rifles, especially the model of 1874, are quite popular in silhouette shooting. But mainly it's the modern replica of this rifle that dominates. Currently, the Winchester Model 1885 High Wall is also popular. Add to these a number of Remington rolling blocks, most of them modern-made replicas, with a few originals as well as customs built on original or replica actions. There are also Hepburns and Ballards and even the Trapdoor Springfield.

Lined up at the firing line, these shooters get ready to fire at a bison target at 900 yards.

The Shiloh Model 1874 Sharps No. 3 Sporter shown here is a very popular rifle. It meets all of the requirements for the silhouette sport, including iron sights, a weight under 12 pounds, 2 ounces, American manufacture prior to 1892, and so forth.

In answer to a great demand for accurate long-range, single shot, blackpowder cartridge rifles that meet the requirements for silhouette competition, Browning brought out the Model 1885 BPCR (Black Powder Cartridge Rifle) with excellent sights and good trigger.

Another rifle born of the modern blackpowder silhouette game is Dixie Gun Works' 1874 Sharps Blackpowder Silhouette Rifle, which resembles a No. 1 Sporter model. It weighs 10 pounds, 3 ounces without target sights. It has double-set triggers, and calibers are 45-70 or 40-65.

The Cartridge

Since the legal rifle for the sport is a single shot breechloader of American design, military or sporting, with an exposed hammer, it only makes sense that the cartridge must follow. It's easy to see that the two rifles most carried by the buffalo runner of the latter part of the 19th century qualify immediately, these being the Remington rolling block and the Sharps, not to exclude the original Winchester single shot breechloader, or for that matter the Trapdoor mentioned in Chapter 45. What were the cartridges for these rifles? There were many, but the rules call for only original 19th century rounds, and they must be of American origin. For example, a great number of Remington rolling blocks were chambered for the 43 Egyptian, but that cartridge is not allowed in formal silhouette competition.

While the 45-70 Government cartridge continues as number one in the game, it is losing ground to the 40-65 Winchester. The reason is quite simple: many rounds are fired during a match. The 45-70 with a 400- or 500-grain bullet in front of a full package of fuel delivers a fairly strong blow to the shoulder. The 40-65 is milder, while still providing enough punch to knock a metallic silhouette over. Actually, there are dozens of rounds that qualify for this sport. The Sharps line alone includes a multitude of them. One shoot had the following cartridges: 40-70, 40-50, 45-100, 44-100, just to mention a few. The list also included, by the way, the huge 50-140 Sharps. While the accurate and mild 38-55 Winchester cartridge has shown up at some matches, it's just a touch shy of striking power to knock down the big ram target.

Sights

Only iron sights are allowed. Open sights are approved, but they would create an insurmountable handicap. The Vernier tang sight is the rule in this game, because it mounts on the tang, close enough to the eye to provide a long sight radius, and it is, after all, a peep sight. Furthermore, it can be adjusted for great distances, with quick and accurate sighting changes. It is not, however, the only sight used. There are other allowable designs. But no scopes.

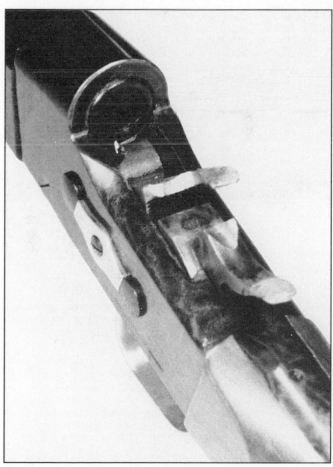

While the Sharps is number one on the blackpowder silhouette range, it is not the only viable rifle for the sport. The Remington rolling block also meets the requirements.

This Vernier tang rear sight has a disc with multiple apertures for various sightings.

(Below) Sights must be iron—no scopes allowed. However, they need not be primitive. These fine sights are typical, with a Shiloh tang sight (lower left), a custom Vernier sporting tang sight (upper right), and a hooded globe front sight.

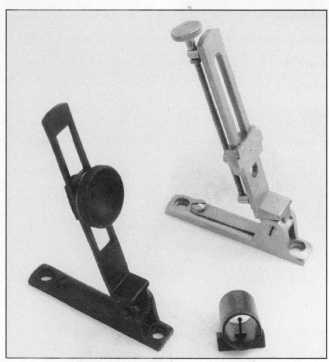

Lead bullets only are the general rule. Here, the 45-70 is shown on the right with a 500-grain lead projectile. The 30-06 Springfield on the left is for comparison.

The Targets, Distances, and Rules

The targets are the same as used in the modern rifle silhouette match, being the chicken, pig, turkey, and sheep. The chicken is placed at 200 meters, the pig at 300, the turkey at 385, and the sheep at a full 500 meters. Cross-sticks (also called crossed sticks) are allowed for the pig, turkey, and sheep, but the chicken is shot offhand and is considered the toughest target to hit, even though it's the closest. The sport is amazingly difficult. Every target looks like a fly speck on a window "away out there," and remember that only iron sights are allowed—no magnification! Even the sheep, the largest target, is a mere 13 inches high and 32 inches long. The turkey is taller—23 inches—but that includes head and neck. The body itself is only 11 by 13 inches in size; picture that at 385 meters.

Shooters simply have to be highly skilled in order to have a chance at competing in this competition. While heavy rifles are allowed, and heavy rifles do sit well from offhand and rest steady on the cross-sticks, maximum legal weight is 12 pounds, 2 ounces, which is a far cry from a blackpowder bench gun, for example, that can weigh much more. No one has ever shot a perfect score in the blackpowder cartridge silhouette game. At least not yet. It is that challenging.

Allowable Powder and Ballistics

Ballistics play a strong role in the difficulty of this fascinating shooting sport. Velocities range from around 1100 fps to about 1300 fps. With only Pyrodex and blackpowder allowed, hopes for higher velocities are eliminated. This means delivered energies at long range are relatively low. A hit target counts for nothing. It must fall over. That's why the larger calibers are imperative. The sheep weighs about fifty pounds, and it's a full 500 meters from the muzzle. Bullets much under 40-caliber simply don't have the remaining punch to *always* knock the sheep off its feet, so to speak. It's difficult trying to hit the target at 500 meters with iron sights and a high trajectory firearm. But hitting it and getting no score, is very discouraging. The 45-caliber bullet, weighing around 400 or 500 grains, has enough punch left, even at long range, to knock over the target when the bullet hits.

Drop Tube Powder Loading

Consider that you have a hole in the ground and you want to fill it with a bunch of rocks. You can haphazardly dump a whole batch of rocks into the hole, or you can toss them into the hole a few at a time. Tossing all the rocks in at once will not allow them to settle evenly, which means there will be air gaps between the rocks. Settling them in a few at a time reduces space between the rocks, creating greater uniformity. The same is true of powder. If you dump a charge of blackpowder or Pyrodex into the case all at once, it will not settle nearly as well as trickling in a few grains at a time. The shape of the kernal, being irregular, has something to do with this. Now lengthen the fall of the kernals and you further improve upon a firmly packed charge.

That's where the drop tube comes in. It allows powder to enter the case from as high as thirty inches or so, but it also introduces the powder a few granules at a time. It can take several seconds to drop a powder charge into a case, in fact. For competition, and even for good hunting loads, the drop tube method comes highly recommended. There is nothing new about it. Drop tubes have been around for a very long time. It's simply an old trick that works very well in producing an evenly packed powder charge.

The Primer

After packing the powder into the case via the drop tube method, the usual condition is 100 percent load density, or very close to it. While many gun writers, including this one, recommended mild

Blackpowder cartridge games are so popular that shooters will come out in any weather—even wintertime!

(Above) This shooter uses the crossed-sticks method to take aim at a far-off target. Notice the box of preloaded ammo.

The single shot is the name of the game. The blackpowder breechloader takes rounds one at a time, directly into the chamber, as shown here.

The spotter can be a great asset to a shooter. He watches through a spotting scope to see where the bullet strikes, and then he passes the information on to the shooter for a sight change, if necessary.

In order to keep blackpowder fouling softer, a neoprene tube is inserted into the breech and the shooter blows through the tube, deposting moisture in the bore.

primers in the past, due to the fact that blackpowder truly does ignite readily, we have had to reverse our position these days, strongly recommending hot primers. In brief, match winners use magnum primers, such as Federal's 215.

The Bullet

The typical bullet for the 19th century blackpowder cartridge did not have a rocket-like profile. Therefore, it did not enjoy a high ballistic coefficient. Velocities are already quite low in the blackpowder cartridge, which means a looping trajectory. Adding blunt-nosed bullets promotes even greater drop. Of course, the tubular magazine of the blackpowder lever-action rifle demanded, and still demands, the blunt-nosed bullet. But the single shot breechloader does not. That's why serious silhouette shooters have gravitated to projectiles with much higher ballistic coefficient. True spitzers may not prevail, but a look at some of the popular missiles shows a tendency toward streamlining. Two examples are Redding/SAECO's 40-caliber cast lead bullets in 370 and 410 grains weight. While it may be argued that a blunt bullet is better at delivering a knock-down blow to the metal silhouette, there's much to be said for a bit flatter trajectory and higher striking energy that a the streamlined bullet affords.

The Wind and the Sun

Both the wind and the sun play prominent rolls in blackpowder cartridge competitive shooting. The wind can drift bullets completely off target. Part of the reason is time of flight, which is very long. With bullets leaving the muzzle at 1100 to 1300 fps, and ranges up to 500 meters and more, it takes quite a while for a bullet to go from muzzle to target. The wind has all that time to play on the projectile. It's not uncommon for a breeze to blow a big blackpowder projectile way off the mark. A mere 10 mph zephyr can drift a bullet off course by 4 feet from muzzle to sheep silhouette at 500 meters. The sun also plays a role. Light striking the sights can greatly alter bullet placement. That's why covered front sights are prominent.

The Spotter

Because of severe bullet drop, the use of iron sights, wind and light problems, and a host of other gremlins, the spotter becomes a tremendous asset. He's allowed to sit behind the shooter watching through a telescope to see where the bullets hit. If he's good, he'll be able to tell the marksman where his missile landed, which prompts a sight or hold adjustment for the next shot.

Minimum Accuracy

Mike Venturino, an avid blackpowder cartridge fan for years, concludes that a 4-inch group at 200 yards is about minimum accuracy for competition in the silhouette game.

Triggers

Double-set triggers are common. While the rule of hold and squeeze always pertains, being able to touch off a shot at just the right moment is invaluable. The set-trigger, with a very light breaking point, allows just that. It's not a trigger jerk, really, but a clean, controlled squeeze made possible by that set trigger.

Growing Sport

The blackpowder cartridge and single shot rifle are back in full force. Blackpowder silhouette shooting on a formal level will probably undergo a few more changes in the future, but that's healthy. Hopefully, the usual trend to upgrade that prevails in any sport will be carefully regulated. So far, so good. The people steering the sport have recognized that if they allow the rules to grow lax, the game will lose a great deal of its appeal. Of course, it's terribly difficult to hit targets at very long range with old-time ballistics, and certainly things could be made easier. But should they be? Not if making the sport easier waters it down. The highest single-day score I know of, supplied by Mike Venturino, was 33 hits in a 40-shot match. Ideally, scores will improve as shooters work with the old-time rifle and its cartridges, as well as their own skills, but not because of high-tech paraphernalia. ●

Catalog of Blackpowder Firearms

Blackpowder Handguns 368
Blackpowder Muskets & Rifles 374
Blackpowder Shotguns 389

ARMY 1851 PERCUSSION REVOLVER
Caliber: 44, 6-shot.
Barrel: 7¹/₂″.
Weight: 45 oz. **Length:** 13″ overall.
Stocks: Walnut finish.
Sights: Fixed.
Features: 44-caliber version of the 1851 Navy. Imported by The Armoury, Armsport.

ARMY 1860 PERCUSSION REVOLVER
Caliber: 44, 6-shot.
Barrel: 8″.
Weight: 40 oz. **Length:** 13⁵/₈″ overall.
Stocks: Walnut.
Sights: Fixed.
Features: Engraved Navy scene on cylinder; brass trigger guard; case-hardened frame, loading lever and hammer. Some importers supply pistol cut for detachable shoulder stock, have accessory stock available. Also available in Hartford model with steel frame, German silver trim, cartouches (E.M.F.) or as single or double cased set (Navy Arms). Imported by American Arms, Cabela's (1860 Lawman), E.M.F., Navy Arms, The Armoury, Cimarron, Dixie Gun Works (half-fluted cylinder, not roll engraved), Euroarms of America (brass or steel model), Armsport, Traditions (brass or steel), Uberti U.S.A. Inc.

American Arms 1860 Army

Colt 1847 Walker

COLT 1849 POCKET DRAGOON REVOLVER
Caliber: 31.
Barrel: 4″.
Weight: 24 oz. **Length:** 9¹/₂″ overall.
Stocks: One-piece walnut.
Sights: Fixed. Brass pin front, hammer notch rear.
Features: Color case-hardened frame. No loading lever. Unfluted cylinder with engraved scene. Exact reproduction of original. From Colt Blackpowder Arms Co.

Colt 1851 Navy

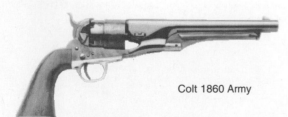

Colt 1860 Army

COLT 1860 CAVALRY MODEL REVOLVER
Caliber: 44.
Barrel: 8″, 7-groove, left-hand twist.
Weight: 42 oz.
Stocks: One-piece walnut.
Sight: German silver front, hammer notch rear.
Features: Fluted cylinder; color case-hardened frame, hammer, loading lever and plunger; blued barrel, backstrap and cylinder; brass trigger guard. Has four-screw frame cut for optional shoulder stock. From Colt Blackpowder Arms Co.

BABY DRAGOON 1848, 1849 POCKET, WELLS FARGO
Caliber: 31.
Barrel: 3″, 4″, 5″, 6″; seven-groove, RH twist.
Weight: About 21 oz.
Stocks: Varnished walnut.
Sights: Brass pin front, hammer notch rear.
Features: No loading lever on Baby Dragoon or Wells Fargo models. Unfluted cylinder with stagecoach holdup scene; cupped cylinder pin; no grease grooves; one safety pin on cylinder and slot in hammer face; straight (flat) mainspring. From Armsport, Dixie Gun Works, Uberti USA Inc., Cabela's.

CABELA'S PATERSON REVOLVER
Caliber: 36, 5-shot cylinder.
Barrel: 7¹/₂″.
Weight: 24 oz. **Length:** 11¹/₂″ overall.
Stocks: One-piece walnut.
Sights: Fixed.
Features: Recreation of the 1836 gun. Color case-hardened frame, steel backstrap; roll-engraved cylinder scene. Imported by Cabela's.

COLT 1847 WALKER PERCUSSION REVOLVER
Caliber: 44.
Barrel: 9″, 7 groove, right-hand twist.
Weight: 73 oz.
Stocks: One-piece walnut.
Sights: German silver front sight, hammer notch rear.
Features: Made in U.S. Faithful reproduction of the original gun, including markings. Color case-hardened frame, hammer, loading lever and plunger. Blue steel backstrap, brass square-back trigger guard. Blue barrel, cylinder, trigger and wedge. From Colt Blackpowder Arms Co.

COLT 1851 NAVY PERCUSSION REVOLVER
Caliber: 36.
Barrel: 7¹/₂″, octagonal, 7 groove left-hand twist.
Weight: 40¹/₂ oz.
Stocks: One-piece oiled American walnut.
Sights: Brass pin front, hammer notch rear.
Features: Faithful reproduction of the original gun. Color case-hardened frame, loading lever, plunger, hammer and latch. Blue cylinder, trigger, barrel, screws, wedge. Silver-plated brass backstrap and square-back trigger guard. From Colt Blackpowder Arms Co.

COLT 1860 ARMY PERCUSSION REVOLVER
Caliber: 44.
Barrel: 8″, 7 groove, left-hand twist.
Weight: 42 oz.
Stocks: One-piece walnut.
Sights: German silver front sight, hammer notch rear.
Features: Steel backstrap cut for shoulder stock; brass trigger guard. Cylinder has Navy scene. Color case-hardened frame, hammer, loading lever. Reproduction of original gun with all original markings. From Colt Blackpowder Arms Co.

COLT 1861 NAVY PERCUSSION REVOLVER
Caliber: 36.
Barrel: 7¹/₂″.
Weight: 42 oz. **Length:** 13¹/₈″ overall.
Stocks: One-piece walnut.
Sights: Blade front, hammer notch rear.
Features: Color case-hardened frame, loading lever, plunger; blued barrel, backstrap, trigger guard; roll-engraved cylinder and barrel. From Colt Blackpowder Arms Co.

COLT 1862 POCKET POLICE "TRAPPER MODEL" REVOLVER

Caliber: 36.
Barrel: 3¹/₂".
Weight: 20 oz. **Length:** 8¹/₂" overall.
Stocks: One-piece walnut.
Sights: Blade front, hammer notch rear.
Features: Has separate 4⁵/₈" brass ramrod. Color case-hardened frame and hammer; silver-plated backstrap and trigger guard; blued semi-fluted cylinder, blued barrel. From Colt Blackpowder Arms Co.

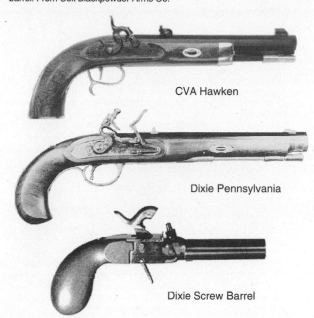

CVA Hawken

Dixie Pennsylvania

Dixie Screw Barrel

FRENCH–STYLE DUELING PISTOL

Caliber: 44.
Barrel: 10".
Weight: 35 oz. **Length:** 15³/₄" overall.
Stock: Carved walnut.
Sights: Fixed.
Features: Comes with velvet-lined case and accessories. Imported by Mandall Shooting Supplies.

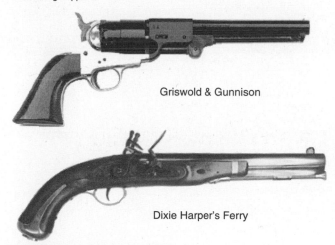

Griswold & Gunnison

Dixie Harper's Ferry

HARPER'S FERRY 1806 PISTOL

Caliber: 58 (.570" round ball).
Barrel: 10".
Weight: 40 oz. **Length:** 16" overall.
Stock: Walnut.
Sights: Fixed.
Features: Case-hardened lock, brass-mounted browned barrel. Replica of the first U.S. Gov't.-made flintlock pistol. Also available from Dixie in kit form and from Navy Arms as cased set. Imported by Navy Arms, Dixie Gun Works.

COLT THIRD MODEL DRAGOON

Caliber: 44.
Barrel: 7¹/₂".
Weight: 66 oz. **Length:** 13³/₄" overall.
Stocks: One-piece walnut.
Sights: Blade front, hammer notch rear.
Features: Color case-hardened frame, hammer, lever and plunger; round trigger guard; flat mainspring; hammer roller; rectangular bolt cuts; Three or four-screw frame; Four-screw frame with blued steel grip straps, shoulder stock cuts, dove-tailed folding leaf rear sight. From Colt Blackpowder Arms Co.

CVA HAWKEN PISTOL

Caliber: 50.
Barrel: 9³/₄"; ¹⁵/₁₆" flats.
Weight: 50 oz. **Length:** 16¹/₂" overall.
Stock: Select hardwood or laminate.
Sights: Beaded blade front, fully adjustable open rear.
Features: Color case-hardened lock, polished brass wedge plate, nose cap, ramrod thimble, trigger guard, grip cap. Available in kit form. Imported by CVA.

DIXIE PENNSYLVANIA PISTOL

Caliber: 44 (.430" round ball).
Barrel: 10" (⁷/₈" octagon).
Weight: 2¹/₂ lbs.
Stock: Walnut-stained hardwood.
Sights: Blade front, open rear drift-adjustable for windage; brass.
Features: Available in flint only. Brass trigger guard, thimbles, nosecap, wedge-plates; high-luster blue barrel. Available in kit form. Imported from Italy by Dixie Gun Works.

DIXIE SCREW BARREL PISTOL

Caliber: .445".
Barrel: 2¹/₂".
Weight: 8 oz. **Length:** 6¹/₂" overall.
Stock: Walnut.
Features: Trigger folds down when hammer is cocked. Close copy of the originals once made in Belgium. Uses No. 11 percussion caps. Available in kit form. From Dixie Gun Works.

DIXIE WYATT EARP REVOLVER

Caliber: 44.
Barrel: 12" octagon.
Weight: 46 oz. **Length:** 18" overall.
Stocks: Two-piece walnut.
Sights: Fixed.
Features: Highly polished brass frame, backstrap and trigger guard; blued barrel and cylinder; case-hardened hammer, trigger and loading lever. Navy-size shoulder stock ($45) will fit with minor fitting. From Dixie Gun Works.

GRISWOLD & GUNNISON PERCUSSION REVOLVER

Caliber: 36 or 44, 6-shot.
Barrel: 7¹/₂".
Weight: 44 oz. (36-cal.). **Length:** 13" overall.
Stocks: Walnut.
Sights: Fixed.
Features: Replica of famous Confederate pistol. Brass frame, backstrap and trigger guard; case-hardened loading lever; rebated cylinder (44-cal. only). Rounded Dragoon-type barrel. Available in kit form or single or double-cased sets. Imported by Navy Arms as Reb Model 1860.

KENTUCKY FLINTLOCK PISTOL

Caliber: 44, 45.
Barrel: 10¹/₈".
Weight: 32 oz. **Length:** 15¹/₂" overall.
Stock: Walnut.
Sights: Fixed.
Features: Specifications, including caliber, weight and length may vary with importer. Case-hardened lock, blued barrel. Also available in kit form, as single (Navy Arms) or double cased set. Imported by Navy Arms (44 only), The Armoury.

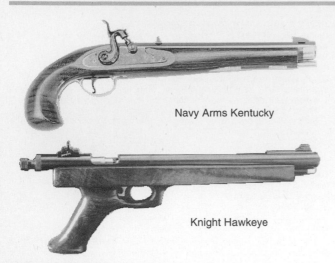

Navy Arms Kentucky

Knight Hawkeye

KENTUCKY PERCUSSION PISTOL
Caliber: 44, 45.
Barrel: 10$^1/_8$".
Weight: 32 oz. **Length:** 15$^1/_2$" overall.
Stock: Walnut.
Sights: Fixed.
Features: Similar to flint version but percussion lock. Imported by The Armoury, Navy Arms, CVA (50-cal.).

KNIGHT HAWKEYE PISTOL
Caliber: 50.
Barrel: 12", 1:20" twist.
Weight: 3$^1/_4$ lbs. **Length:** 20" overall.
Stock: Black composite, autumn brown or shadow black laminate.
Sights: Bead front on ramp, open fully adjustable rear.
Features: In-line ignition design; patented double safety system; removable breech plug; fully adjustable trigger; receiver drilled and tapped for scope mounting. Blue or steel finish. Made in U.S. by Modern Muzzle Loading, Inc.

LE MAT REVOLVER
Caliber: 44/65.
Barrel: 6$^3/_4$" (revolver); 4$^7/_8$" (single shot).
Weight: 3 lbs., 7 oz.
Stocks: Hand-checkered walnut.
Sights: Post front, hammer notch rear.
Features: Exact reproduction with all-steel construction; 44-cal. 9-shot cylinder, 65-cal. single barrel; color case-hardened hammer with selector; spur trigger guard; ring at butt; lever-type barrel release. Available in Cavalry model (lanyard ring, spur trigger guard); Army model (round trigger guard, pin-type barrel release; Naval-style (thumb selector on hammer); Engraved 18th Georgia cased set; Engraved Beauregard cased set. From Navy Arms.

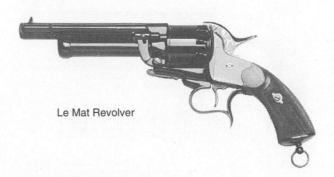

Le Mat Revolver

LE PAGE PERCUSSION DUELING PISTOL
Caliber: 44.
Barrel: 10", rifled.
Weight: 40 oz. **Length:** 16" overall.
Stock: Walnut, fluted butt.
Sights: Blade front, notch rear.
Features: Double-set triggers. Blued barrel; trigger guard and buttcap are polished silver. Imported by Dixie Gun Works.

LYMAN PLAINS PISTOL
Caliber: 50 or 54.
Barrel: 8", 1:30" twist, both calibers.
Weight: 50 oz. **Length:** 15" overall.
Stock: Walnut half-stock.
Sights: Blade front, square notch rear adjustable for windage.
Features: Polished brass trigger guard and ramrod tip, color case-hardened coil spring lock, spring-loaded trigger, stainless steel nipple, blackened iron furniture. Hooked patent breech, detachable belt hook. Also available in kit form. Introduced 1981. From Lyman Products.

NAVY ARMS DELUXE 1858 REMINGTON-STYLE REVOLVER
Caliber: 44.
Barrel: 8".
Weight: 2 lbs., 13 oz.
Stocks: Smooth walnut.
Sights: Dovetailed blade front.
Features: First exact reproduction—correct in size and weight to the original, with progressive rifling; highly polished with blue finish, silver-plated trigger guard. From Navy Arms.

NAVY ARMS LE PAGE DUELING PISTOL
Caliber: 44.
Barrel: 9", octagon, rifled.
Weight: 34 oz. **Length:** 15" overall.
Stock: European walnut.
Sights: Adjustable rear.
Features: Single-set trigger; percussion or flintlock; rifled or smoothbore barrel (flintlock); single or double cased sets. Polished metal finish. From Navy Arms.

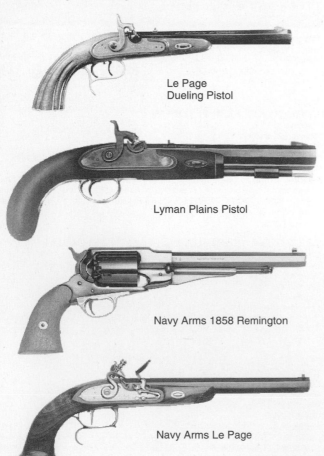

Le Page
Dueling Pistol

Lyman Plains Pistol

Navy Arms 1858 Remington

Navy Arms Le Page

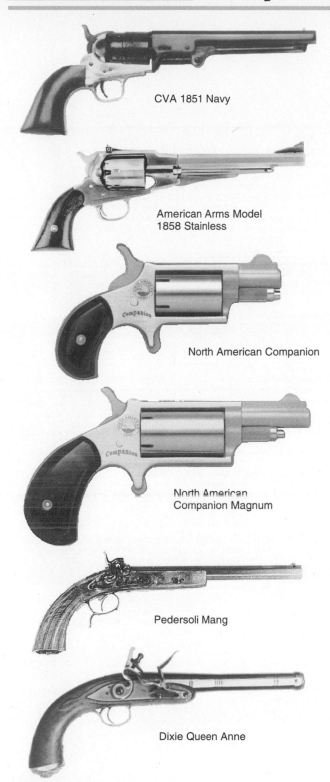

CVA 1851 Navy

American Arms Model
1858 Stainless

North American Companion

North American
Companion Magnum

Pedersoli Mang

Dixie Queen Anne

QUEEN ANNE FLINTLOCK PISTOL
Caliber: 50 (.490″ round ball).
Barrel: 7½″, smoothbore.
Stock: Walnut.
Sights: None.
Features: Browned steel barrel, fluted brass trigger guard, brass mask on butt. Lockplate left in the white. Made by Pedersoli in Italy. Also available in kit form. Introduced 1983. Imported by Dixie Gun Works.

NAVY MODEL 1851 PERCUSSION REVOLVER
Caliber: 36, 44, 6-shot.
Barrel: 7½″.
Weight: 44 oz. **Length:** 13″ overall.
Stocks: Walnut finish.
Sights: Post front, hammer notch rear.
Features: Brass backstrap and trigger guard; some have 1st Model squareback trigger guard, engraved cylinder with navy battle scene; case-hardened frame, hammer, loading lever. Available as single or double cased sets, Confederate Navy model, Hartford model or in kit form. Imported by American Arms, The Armoury, Cabela's, Navy Arms, E.M.F., Dixie Gun Works, Euroarms of America, Armsport, CVA (36-cal. only), Traditions (44 only), Uberti USA Inc., Stone Mountain Arms.

NEW MODEL 1858 ARMY PERCUSSION REVOLVER
Caliber: 36 or 44, 6-shot.
Barrel: 6½″ or 8″.
Weight: 38 oz. **Length:** 13½″ overall.
Stocks: Walnut.
Sights: Blade front, groove-in-frame rear.
Features: Replica of Remington Model 1858. Also available from some importers as Army Model Belt Revolver in 36-cal., a shortened and lightened version of the 44. Target Model (Uberti USA Inc., Navy Arms) has fully adjustable target rear sight, target front, 36 or 44. Imported by American Arms, Cabela's, Cimarron, CVA (as 1858 Army, steel or brass frame, 44 only), Dixie Gun Works, Navy Arms, The Armoury, E.M.F., Euroarms of America (engraved, stainless and plain), Armsport, Traditions (44 only), Uberti USA Inc. Stone Mountain Arms.

NORTH AMERICAN COMPANION PERCUSSION REVOLVER
Caliber: 22.
Barrel: 1⅛″.
Weight: 5.1 oz. **Length:** 4⁵/₁₀″ overall.
Stocks: Laminated wood.
Sights: Blade front, notch fixed rear.
Features: All stainless steel construction. Uses standard #11 percussion caps. Comes with bullets, powder measure, bullet seater, leather clip holster, gun rag. Long Rifle or Magnum frame size. Introduced 1996. Made in U.S. by North American Arms.

NORTH AMERICAN MAGNUM COMPANION PERCUSSION REVOLVER
Caliber: 22.
Barrel: 1⅛″.
Weight: 7.2 oz. **Length:** 5⁷/₁₆″ overall.
Stocks: Laminated wood.
Sights: Blade front, notch fixed rear.
Features: Similar to the Companion except has larger frame. Weighs 7.2 oz., has 1⅝″ barrel, measures 5⁷/₁₆″ overall. Comes with bullets, powder measure, bullet seater, leather clip holster, gun rug. Introduced 1996. Made in U.S. by North American Arms.

PEDERSOLI MANG TARGET PISTOL
Caliber: 38.
Barrel: 10.5″, octagonal; 1:15″ twist,
Weight: 2.5 lbs. **Length:** 17.25″ overall.
Stock: Walnut with fluted grip.
Sights: Blade front, open rear adjustable for windage.
Features: Browned barrel, polished breech plug, rest color case-hardened. Imported from Italy by Dixie Gun Works.

POCKET POLICE 1862 PERCUSSION REVOLVER
Caliber: 36, 5-shot.
Barrel: 4½″, 5½″, 6½″, 7½″.
Weight: 26 oz. **Length:** 12″ overall (6½″ bbl.).
Stocks: Walnut.
Sights: Fixed.
Features: Round tapered barrel; half-fluted and rebated cylinder; case-hardened frame, loading lever and hammer; silver or brass trigger guard and backstrap. Imported by CVA (7½″ only), Navy Arms (5½″ only or as single-cased set with accessories), Uberti USA Inc. (5½″, 6½″ only); E.M.F.

ROGERS & SPENCER PERCUSSION REVOLVER
Caliber: 44.
Barrel: 7¹/₂″.
Weight: 47 oz. **Length:** 13³/₄″ overall.
Stocks: Walnut.
Sights: Cone front, integral groove in frame for rear.
Features: Accurate reproduction of a Civil War design. Solid frame; extra large nipple cut-out on rear of cylinder; loading lever and cylinder easily removed for cleaning. Also available in kit form and target version. From Euroarms of America (standard blue, engraved, burnished, target models), Navy Arms, Stone Mountain Arms.

Euroarms Rogers & Spencer

Ruger Old Army

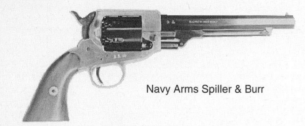

Navy Arms Spiller & Burr

STONE MOUNTAIN ARMS SHERIFF'S MODEL
Caliber: 44
Barrel: 5¹/₂″.
Weight: 40¹/₂ oz. **Length:** 10¹/₂″
Stocks: One-piece oiled American walnut.
Sights: Fixed.
Features: Blued steel frame; semi-fluted cylinder. From Stone Mountain Arms.

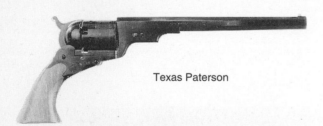

Texas Paterson

THOMPSON/CENTER SCOUT PISTOL
Caliber: 45, 50 and 54.
Barrel: 12″, interchangeable.
Weight: 4 lbs., 6 oz. **Length:** NA.
Stocks: American black walnut stocks and forend.
Sights: Blade on ramp front, fully adjustable Patridge rear.
Features: Patented in-line ignition system with special vented breech plug. Patented trigger mechanism consists of only two moving parts. Interchangeable barrels. Wide grooved hammer. Brass trigger guard assembly. Introduced 1990. From Thompson/Center.

TRADITIONS BUCKSKINNER PISTOL
Caliber: 50.
Barrel: 10″ octagonal, ⁷/₈″ flats, 1:20″ twist.
Weight: 40 oz. **Length:** 15″ overall.
Stocks: Stained beech or laminated wood.
Sights: Blade front, fixed rear.
Features: Percussion ignition. Blackened furniture. Imported by Traditions.

RUGER OLD ARMY PERCUSSION REVOLVER
Caliber: 45, 6-shot. Uses .457″ dia. lead bullets.
Barrel: 7¹/₂″ (6-groove, 16″ twist).
Weight: 46 oz. **Length:** 13³/₄″ overall.
Stocks: Smooth walnut.
Sights: Ramp front, rear adjustable for windage and elevation; or fixed (groove).
Features: Stainless steel or blue finish; standard size nipples, chrome-moly steel cylinder and frame, same lockwork as in original Super Blackhawk. Also available in stainless steel. Made in USA. From Sturm, Ruger & Co.

SHERIFF MODEL 1851 PERCUSSION REVOLVER
Caliber: 36, 44, 6-shot.
Barrel: 5″.
Weight: 40 oz. **Length:** 10¹/₂″ overall.
Stocks: Walnut.
Sights: Fixed.
Features: Brass backstrap and trigger guard; engraved navy scene; case-hardened frame, hammer, loading lever. Also available with steel frame. Imported by E.M.F., Stone Mountain Arms (5¹/₂″ barrel).

SPILLER & BURR REVOLVER
Caliber: 36 (.375″ round ball).
Barrel: 7″, octagon.
Weight: 2¹/₂ lbs. **Length:** 12¹/₂″ overall.
Stocks: Two-piece walnut.
Sights: Fixed.
Features: Reproduction of the C.S.A. revolver. Brass frame and trigger guard. Also available as a kit and in single or double cased sets. From Cabela's, Dixie Gun Works, Navy Arms.

TEXAS PATERSON 1836 REVOLVER
Caliber: 36 (.375″ round ball).
Barrel: 7¹/₂″.
Weight: 42 oz.
Stocks: One-piece walnut.
Sights: Fixed.
Features: Copy of Sam Colt's first commercially-made revolving pistol. Has no loading lever but comes with loading tool. From Dixie Gun Works (engraved), Navy Arms, Uberti USA Inc. (with loading lever).

Thompson/Center Scout

Traditions Buckhunter

TRADITIONS BUCKHUNTER PRO IN-LINE PISTOL
Caliber: 50, 54.
Barrel: 10″ round.
Weight: 48 oz. **Length:** 14″ overall.
Stock: Smooth walnut or black epoxy coated grip and forend.
Sights: Beaded blade front, folding adjustable rear.
Features: Thumb safety; removable stainless steel breech plug; adjustable trigger, barrel drilled and tapped for scope mounting. From Traditions.

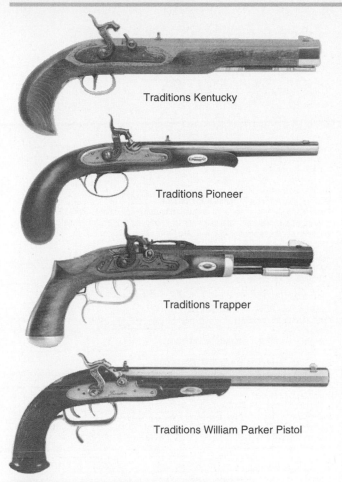

Traditions Kentucky

Traditions Pioneer

Traditions Trapper

Traditions William Parker Pistol

UBERTI 1861 NAVY PERCUSSION REVOLVER
Caliber: 36.
Barrel: 7 1/2".
Weight: 42 oz. **Length:** 13 1/8" overall.
Stocks: One-piece walnut.
Sights: Blade front, hammer notch rear.
Features: Blued barrel, backstrap, trigger guard; color case-hardened loading lever and plunger; roll-engraved cylinder. From Uberti USA Inc.

UBERTI 1862 POCKET NAVY PERCUSSION REVOLVER
Caliber: 36, 5-shot.
Barrel: 5 1/2", 6 1/2", octagonal, 7-groove, LH twist.
Weight: 27 oz. (5 1/2" barrel). **Length:** 10 1/2" overall (5 1/2" bbl.).
Stocks: One-piece varnished walnut.
Sights: Brass pin front, hammer notch rear.
Features: Rebated cylinder, hinged loading lever, brass or silver-plated backstrap and trigger guard, color-cased frame, hammer, loading lever, plunger and latch, rest blued. Has original-type markings. From Uberti USA Inc.

UBERTI 2ND MODEL DRAGOON REVOLVER
Caliber: 44.
Barrel: 7 1/2", part round, part octagon.
Weight: 64 oz.
Stocks: One-piece walnut.
Sights: German silver blade front, hammer notch rear.
Features: Similar to the 1st Model; distinguished by rectangular bolt cuts in the cylinder.

UBERTI 3RD MODEL DRAGOON REVOLVER
Caliber: 44.
Barrel: 7 1/2", part round, part octagon.
Weight: 64 oz.
Stocks: One-piece walnut.
Sights: German silver blade front, hammer notch rear.
Features: Oval trigger guard, long trigger, modifications to the loading lever and latch. Imported by Uberti USA Inc.

TRADITIONS KENTUCKY PISTOL
Caliber: 50.
Barrel: 10"; octagon with 7/8" flats; 1:20" twist.
Weight: 40 oz. **Length:** 15" overall.
Stock: Stained beech.
Sights: Blade front, fixed rear.
Features: Birds-head grip; brass thimbles; color case-hardened lock. Percussion only. Also available in kit form. Introduced 1995. From Traditions.

TRADITIONS PIONEER PISTOL
Caliber: 45.
Barrel: 9 5/8", 13/16" flats, 1:16" twist.
Weight: 31 oz. **Length:** 15" overall.
Stock: Beech.
Sights: Blade front, fixed rear.
Features: V-type mainspring. Single trigger. German silver furniture, blackened hardware. Also available in kit form. From Traditions.

TRADITIONS TRAPPER PISTOL
Caliber: 50.
Barrel: 9 3/4", 7/8" flats, 1:20" twist.
Weight: 2 3/4 lbs. **Length:** 16" overall.
Stock: Beech.
Sights: Blade front, adjustable rear.
Features: Double-set triggers; percussion or flintlock; brass buttcap, trigger guard, wedge plate, forend tip, thimble. Also available in kit form. From Traditions.

TRADITIONS WILLIAM PARKER PISTOL
Caliber: 50.
Barrel: 10 3/8", 15/16" flats; polished steel.
Weight: 37 oz. **Length:** 17 1/2" overall.
Stock: Walnut with checkered grip.
Sights: Brass blade front, fixed rear.
Features: Replica dueling pistol with 1:20" twist, hooked breech. Brass wedge plate, trigger guard, cap guard; separate ramrod. Double-set triggers. Polished steel barrel, lock. Imported by Traditions.

UBERTI 1ST MODEL DRAGOON
Caliber: 44.
Barrel: 7 1/2", part round, part octagon.
Weight: 64 oz.
Stocks: One-piece walnut.
Sights: German silver blade front, hammer notch rear.
Features: First model has oval bolt cuts in cylinder, square-back flared trigger guard, V-type mainspring, short trigger. Ranger and Indian scene roll-engraved on cylinder. Color case-hardened frame, loading lever, plunger and hammer; blue barrel, cylinder, trigger and wedge. Available with old-time charcoal blue or standard blue-black finish. Polished brass backstrap and trigger guard. From Uberti USA Inc.

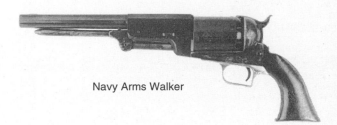

Navy Arms Walker

WALKER 1847 PERCUSSION REVOLVER
Caliber: 44, 6-shot.
Barrel: 9".
Weight: 84 oz. **Length:** 15 1/2" overall.
Stocks: Walnut.
Sights: Fixed.
Features: Case-hardened frame, loading lever and hammer; iron backstrap; brass trigger guard; engraved cylinder. Imported by Cabela's, CVA, Navy Arms (deluxe model with fitted case or single cased set), Dixie Gun Works, Uberti USA Inc., E.M.F. (Hartford model), Cimarron, Traditions.

Armoury R140 Hawken

ARMSPORT 1863 SHARPS RIFLE, CARBINE
Caliber: 45, 54.
Barrel: 22″, 28″, round.
Weight: 8.4 lbs. **Length:** 46″ overall.
Stock: Walnut.
Sights: Blade front, folding adjustable rear. Tang sight set optionally available.
Features: Replica of the 1863 Sharps. Color case-hardened frame, rest blued. Imported by Armsport.

CABELA'S BLUE RIDGE RIFLE
Caliber: 32, 36, 45, 50, 54.
Barrel: 28″ (percussion carbine), 39″, octagonal.
Weight: About 7¾ lbs. **Length:** 55″ overall.
Stock: American black walnut.
Sights: Blade front, rear drift adjustable for windage.
Features: Color case-hardened lockplate and cock/hammer, brass trigger guard and buttplate, double set, double-phased triggers. Percussion or flintlock. From Cabela's.

CABELA'S ROLLING BLOCK MUZZLELOADER
Caliber: 50, 54.
Barrel: 26½″ octagonal; 1:32″ (50), 1:48″ (54) twist.
Weight: About 9¼ lbs. **Length:** 43½″ overall.
Stock: American walnut, rubber butt pad.
Sights: Blade front, adjustable buckhorn rear.
Features: Uses in-line ignition system, Brass trigger guard, color case-hardened hammer, block and buttplate; black-finished, engraved receiver; easily removable screw-in breech plug; black ramrod and thimble. From Cabela's.

CABELA'S SHARPS SPORTING RIFLE
Caliber: 45, 54.
Barrel: 31″, octagonal.
Weight: About 10 lbs. **Length:** 49″ overall.
Stock: American walnut with checkered grip and forend.
Sights: Blade front, ladder-type adjustable rear.
Features: Color case-hardened lock and buttplate. Adjustable double set, double-phased triggers. From Cabela's.

CABELA'S TRADITIONAL HAWKEN
Caliber: 45, 50, 54, 58.
Barrel: 29″.
Weight: About 9 lbs.
Stock: Walnut.
Sights: Blade front, open adjustable rear.
Features: Flintlock or percussion. Adjustable double-set triggers. Polished brass furniture, color case-hardened lock. Available in right- or left-hand percussion, right-hand only in flintlock. Imported by Cabela's.

Cook & Brother

ARMOURY R140 HAWKEN RIFLE
Caliber: 45, 50 or 54.
Barrel: 29″.
Weight: 8¾ to 9 lbs. **Length:** 45¾″ overall.
Stock: Walnut, with cheekpiece.
Sights: Dovetail front, fully adjustable rear.
Features: Octagon barrel, removable breech plug; double set triggers; blued barrel, brass stock fittings, color case-hardened percussion lock. From Armsport, The Armoury.

BOSTONIAN PERCUSSION RIFLE
Caliber: 45.
Barrel: 30″, octagonal
Weight: 7¼ lbs. **Length:** 46″ overall.
Stock: Walnut.
Sights: Blade front, fixed notch rear.
Features: Color case-hardened lock, brass trigger guard, buttplate, patchbox. Imported from Italy by E.M.F.

CABELA'S RED RIVER RIFLE
Caliber: 45, 50, 54, 58.
Barrel: NA.
Weight: About 7 lbs. **Length:** 45″ overall.
Stock: Walnut-stained hardwood.
Sights: Blade front, adjustable buckhorn rear.
Features: Brass trigger guard, forend cap, thimbles; color case-hardened lock and hammer; rubber recoil pad. Introduced 1995. Imported by Cabela's.

CABELA'S ROLLING BLOCK MUZZLELOADER CARBINE
Caliber: 50, 54.
Barrel: 22¼″.
Weight: 8¼ lbs.
Stock: American walnut.
Sights: Bead on ramp front, modern fully adjustable rear.
Features: Uses in-line ignition system. Brass trigger guard, color case-hardened hammer, block and buttplate; black-finished, engraved receiver; easily removable screw-in breech plug; black ramrod and thimble. From Cabela's.

CABELA'S SPORTERIZED HAWKEN HUNTER
Caliber: 45, 50, 54, 58.
Barrel: 29″.
Weight: About 9 lbs.
Stock: Walnut. Modern style with rubber recoil pad.
Sights: Blade front, open adjustable rear.
Features: Percussion only. Blued furniture; Swing swivels. Available as carbine or rifle, right-hand only. Imported by Cabela's.

COLT MODEL 1861 MUSKET
Caliber: 58.
Barrel: 40″.
Weight: 9 lbs., 3 oz. **Length:** 56″ overall.
Stock: Oil-finished walnut.
Sights: Blade front, adjustable folding leaf rear.
Features: Made to original specifications and has authentic Civil War Colt markings. Bright-finished metal, blued nipple and rear sight. Bayonet and accessories available. From Colt Blackpowder Arms Co.

COOK & BROTHER CONFEDERATE CARBINE, RIFLE
Caliber: 58.
Barrel: 24″ (carbine), 33″ (rifle).
Weight: 7½ lbs. **Length:** 40½″ overall.
Stock: Select walnut.
Features: Recreation of the 1861 New Orleans-made artillery carbine. Color case-hardened lock, browned barrel. Buttplate, trigger guard, barrel bands, sling swivels and nose cap of polished brass. From Euroarms of America.

Cumberland Mountain

CUMBERLAND MOUNTAIN BLACKPOWDER RIFLE

Caliber: 50.
Barrel: 26″, round.
Weight: 9½ lbs. **Length:** 43″ overall.
Stock: American walnut.
Sights: Bead front, open rear adjustable for windage.
Features: Falling block action fires with shotshell primer. Blued receiver and barrel. Introduced 1993. Made in U.S. by Cumberland Mountain Arms, Inc.

CVA APOLLO BROWN BEAR

Caliber: 50, 54.
Barrel: 24″; round with octagon integral receiver; 1:32″ twist.
Weight: 7½ lbs. **Length:** 42″ overall.
Stock: Hardwood with oil finish; raised comb and cheekpiece.
Sights: Bead front, Williams Hunter rear; drilled and tapped for scope mounting.
Features: Blued barrel and action. In-line ignition; modern-style trigger with automatic safety; oversize trigger guard. From CVA.

CVA APOLLO COMET RIFLE

Caliber: 50, 54.
Barrel: 24″; round with octagon integral receiver; 1:32″ twist.
Weight: 7-7½ lbs. **Length:** 42″ overall.
Stock: Synthetic with black finish.
Sights: Blade on ramp front, fully adjustable rear; drilled and tapped for scope mounting.
Features: Stainless steel barrel and action. In-line ignition; modern-style trigger with automatic safety; oversize trigger guard. From CVA.

CVA APOLLO DOMINATOR

Caliber: 50, 54.
Barrel: 24″; round with octagon integral receiver; 1:32″ twist.
Weight: 7-7½ lbs. **Length:** 42″ overall.
Stock: Bell & Carlson synthetic thumbhole.

Sights: Bead on blade front, Williams micro-adjustable rear; drilled and tapped for scope mounting.
Features: Stainless steel barrel and action. In-line ignition; modern-style trigger with automatic safety; oversize trigger guard. From CVA.

CVA Apollo Classic

CVA APOLLO SHADOW, CLASSIC RIFLES

Caliber: 50, 54.
Barrel: 24″; round with octagon integral receiver; 1:32″ twist.
Weight: 7-7½ lbs. **Length:** 42″ overall.
Stock: Synthetic Dura-Grip (Shadow); brown laminate with swivel studs (Classic); pistol grip, solid rubber buttpad.
Sights: Blade on ramp front, fully adjustable rear; drilled and tapped for scope mounting.
Features: In-line ignition; modern-style trigger with automatic safety; oversize trigger guard; synthetic ramrod. From CVA.

CVA BOBCAT HUNTER

Caliber: 50 and 54.
Barrel: 26″; 1:48″ twist.
Weight: 6½ lbs. **Length:** 40″ overall.
Stock: Black synthetic with checkered waist and forend.
Sights: Blade front, sporter adjustable rear; drilled and tapped for scope mounting.
Features: Oversize trigger guard; wood ramrod; engraved, blued lockplate and offset hammer. From CVA.

CVA BOBCAT RIFLE

Caliber: 50 and 54.
Barrel: 26″; 1:48″ twist.
Weight: 6½ lbs. **Length:** 40″ overall.
Stock: Dura-Grip synthetic.
Sights: Blade front, open rear.
Features: Oversize trigger guard; wood ramrod; matte black finish. Introduced 1995. From CVA.

CVA BUCKMASTER RIFLE

Caliber: 50 and 54.
Barrel: 24″; round with octagon integral receiver; 1:32″ twist.
Weight: 7-7½ lbs. **Length:** 42″ overall.
Stock: Synthetic Dura-Grip with Advantage camouflage pattern.
Sights: Blade on ramp front, fully adjustable rear; drilled and tapped for scope mounting.
Features: Blued barrel and action; in-line ignition; modern-style trigger with automatic safety; oversize trigger guard. From CVA.

CVA FRONTIER HUNTER CARBINE

Caliber: 50, 54.
Barrel: 24″, 15/16″ flats; 1:32″ twist.
Weight: 6¾ lbs. **Length:** 40″ overall.
Stock: Laminated hardwood.
Sights: Bead front, Patridge-style click-adjustable rear.
Features: Offset hammer; black-chromed furniture; solid buttpad; barrel drilled and tapped for scope mounting. From CVA.

CVA Express Rifle

CVA EXPRESS RIFLE

Caliber: 50.
Barrel: 28″, round; 1:48″ twist.
Weight: 10 lbs.
Stock: Select hardwood; ventilated rubber recoil pad.
Sights: Bead and blade front, adjustable rear.
Features: Double rifle with twin percussion locks and triggers, adjustable barrels. Button breech. Introduced 1989. From CVA.

CVA St. Louis Hawken

CVA KENTUCKY RIFLE

Caliber: 50.
Barrel: 33 1/2", rifled, octagon; 7/8" flats.
Weight: 7 1/2 lbs. **Length:** 48" overall.
Stock: Select hardwood.
Sights: Brass Kentucky blade-type front, fixed open rear.
Features: Available in percussion only. Color case-hardened lockplate. Stainless steel nipple included. Available finished or in kit form. From CVA.

CVA HAWKEN RIFLE

Caliber: 50, 54.
Barrel: 28", octagon; 15/16" across flats; 1:48" twist.
Weight: 8 lbs. **Length:** 44" overall.
Stock: Select hardwood.
Sights: Beaded blade front, fully adjustable open rear.
Features: Fully adjustable double-set triggers; synthetic ramrod (kits have wood); brass patch box, wedge plates, nosecap, thimbles, trigger guard and buttplate; blued barrel; color case-hardened, engraved lockplate. V-type mainspring. Button breech. Introduced 1981. Available as a combo kit (both 50- and 54-caliber barrels), left-hand percussion or flintlock (50-cal. only), right-hand flintlock (50-cal. only), percussion kit (50-cal., blued, wood ramrod), or as St. Louis Hawken Classic with laminated stock. From CVA.

CVA Lynx

CVA PLAINSHUNTER RIFLE

Caliber: 50.
Barrel: 26", octagonal; 15/16" flats; 1:48" twist.
Weight: About 6 1/2 lbs. **Length:** 40" overall.
Stock: Select hardwood.
Sights: Brass blade front, semi-buckhorn rear.
Features: Brass nosecap, thimbles, wedge plates; wood ramrod. Introduced 1995. From CVA.

CVA LYNX RIFLE

Caliber: 50 and 54.
Barrel: 26", octagonal; 15/16" flats; 1:48" twist.
Weight: About 6 1/2 lbs. **Length:** 40" overall.
Stock: Dura-Grip synthetic.
Sights: Beaded blade front, rear adjustable for windage.
Features: Oversize trigger guard; color case-hardened lock, blued barrel, Realtree All Purpose® camo stock. Drilled and tapped for scope mounting. Synthetic ramrod. Introduced 1995. From CVA.

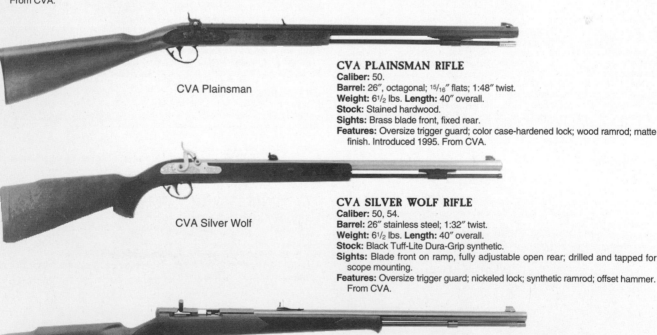

CVA Plainsman

CVA PLAINSMAN RIFLE

Caliber: 50.
Barrel: 26", octagonal; 15/16" flats; 1:48" twist.
Weight: 6 1/2 lbs. **Length:** 40" overall.
Stock: Stained hardwood.
Sights: Brass blade front, fixed rear.
Features: Oversize trigger guard; color case-hardened lock; wood ramrod; matte finish. Introduced 1995. From CVA.

CVA Silver Wolf

CVA SILVER WOLF RIFLE

Caliber: 50, 54.
Barrel: 26" stainless steel; 1:32" twist.
Weight: 6 1/2 lbs. **Length:** 40" overall.
Stock: Black Tuff-Lite Dura-Grip synthetic.
Sights: Blade front on ramp, fully adjustable open rear; drilled and tapped for scope mounting.
Features: Oversize trigger guard; nickeled lock; synthetic ramrod; offset hammer. From CVA.

CVA Staghorn

CVA STAGHORN RIFLE

Caliber: 50, 54.
Barrel: 24"; round with octagon integral receiver; 1:32" twist.
Weight: 7-7 1/2 lbs. **Length:** 42" overall.
Stock: Synthetic with black finish.
Sights: Blade on ramp front, fully adjustable rear; drilled and tapped for scope mounting or receiver sight.
Features: Blued barrel and action. In-line ignition; modern-style trigger with automatic safety; oversize trigger guard. From CVA.

CVA Timber Wolf

CVA VARMINT RIFLE
Caliber: 32.
Barrel: 24″ octagonal; 7/8″ flats; 1:48″ rifling.
Weight: 6 3/4 lbs. **Length:** 40″ overall.
Stock: Select hardwood.
Sights: Blade front, Partridge-style click adjustable rear.
Features: Brass trigger guard, nose cap, wedge plate, thimble and buttplate. Drilled and tapped for scope mounting. Color case-hardened lock. Single trigger. Aluminum ramrod. Imported by CVA.

DIXIE 1863 SPRINGFIELD MUSKET
Caliber: 58 (.570″ patched ball or .575″ Minie).
Barrel: 50″, rifled.
Stocks: Walnut stained.
Sights: Blade front, adjustable ladder-type rear.
Features: Bright-finish lock, barrel, furniture. Reproduction of the last of the regulation muzzleloaders. Available finished or in kit form. Imported from Japan by Dixie Gun Works.

CVA WOLF SERIES RIFLES
Caliber: 50, 54.
Barrel: 26″ octagonal; 1:32″ twist; 15/16″ flats; blue finish.
Weight: 6 1/2 lbs. **Length:** 40″ overall.
Stock: Tuff-Lite polymer—gray finish, solid buttplate (Grey Wolf); Realtree All Purpose® camo finish, solid buttplate (Timber Wolf); checkered grip.
Sights: Blade front on ramp, fully adjustable open rear; drilled and tapped for scope mounting.
Features: Oversize trigger guard; synthetic ramrod; offset hammer. Grey Wolf in 50 or 54; Timber Wolf in 50-cal. only. From CVA.

DIXIE DELUX CUB RIFLE
Caliber: 40, 50 (Deerslayer).
Barrel: 28″.
Weight: 6 1/2 lbs.
Stock: Walnut.
Sights: Fixed.
Features: Short rifle for small game and beginning shooters. Brass patchbox and furniture. Flint or percussion. Available finished and in kit form. From Dixie Gun Works.

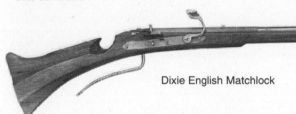

Dixie English Matchlock

DIXIE ENGLISH MATCHLOCK MUSKET
Caliber: 72.
Barrel: 44″.
Weight: 8 lbs. **Length:** 57.75″ overall.
Stock: Walnut with satin oil finish.
Sights: Blade front, open rear adjustable for windage.
Features: Replica of circa 1600-1680 English matchlock. Getz barrel with 11″ octagonal area at rear, rest is round with cannon-type muzzle. All steel finished in the white. Imported by Dixie Gun Works.

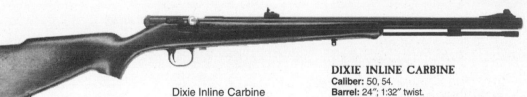

Dixie Inline Carbine

DIXIE INLINE CARBINE
Caliber: 50, 54.
Barrel: 24″; 1:32″ twist.
Weight: 6.5 lbs. **Length:** 41″ overall.
Stock: Walnut-finished hardwood with Monte Carlo comb.
Sights: Ramp front with red insert, open fully adjustable rear.
Features: Sliding "bolt" fully encloses cap and nipple. Fully adjustable trigger, automatic safety. Aluminum ramrod. Imported from Italy by Dixie Gun Works.

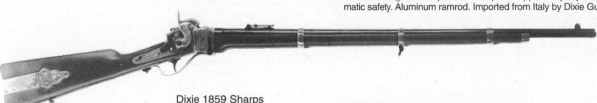

Dixie 1859 Sharps

DIXIE SHARPS NEW MODEL 1859 MILITARY RIFLE
Caliber: 54.
Barrel: 30″, 6-groove; 1:48″ twist.
Weight: 9 lbs. **Length:** 45 1/2″ overall.
Stock: Oiled walnut.
Sights: Blade front, ladder-style rear.
Features: Blued barrel, color case-hardened barrel bands, receiver, hammer, nose cap, lever, patchbox cover and buttplate. Introduced 1995. Imported from Italy by Dixie Gun Works.

DIXIE TENNESSEE MOUNTAIN RIFLE
Caliber: 32 or 50.
Barrel: 41 1/2″, 6-groove rifling, brown finish. **Length:** 56″ overall.
Stock: Walnut, oil finish; Kentucky-style.
Sights: Silver blade front, open buckhorn rear.
Features: Recreation of the original mountain rifles. Early Schultz lock, interchangeable flint or percussion with vent plug or drum and nipple. Tumbler has fly. Double-set triggers. All metal parts browned. Available as right- or left-handed, flint or percussion, finished or in kit form; Squirrel Rifle, 32-cal., flint or percussion, finished or as kit. From Dixie Gun Works.

Dixie Model 1816

DIXIE U.S. MODEL 1861 SPRINGFIELD

Caliber: 58.
Barrel: 40".
Weight: About 8 lbs. **Length:** 55¹³/₁₆" overall.
Stock: Oil-finished walnut.
Sights: Blade front, step adjustable rear.
Features: Exact recreation of original rifle. Sling swivels attached to trigger guard bow and middle barrel band. Lockplate marked "1861" with eagle motif and "U.S. Springfield" in front of hammer; "U.S." stamped on top of buttplate. Available finished or in kit form. From Dixie Gun Works, Stone Mountain Arms.

DIXIE U.S. MODEL 1816 FLINTLOCK MUSKET

Caliber: 69.
Barrel: 42", smoothbore.
Weight: 9.75 lbs. **Length:** 56.5" overall.
Stock: Walnut with oil finish.
Sights: Blade front.
Features: All metal finished "National Armory Bright"; three barrel bands with springs; steel ramrod with button-shaped head. Imported by Dixie Gun Works.

E.M.F. 1863 SHARPS MILITARY CARBINE

Caliber: 54.
Barrel: 22", round.
Weight: 8 lbs. **Length:** 39" overall.
Stock: Oiled walnut.
Sights: Blade front, military ladder-type rear.
Features: Color case-hardened lock, rest blued. Imported by E.M.F.

Euroarms 1861

EUROARMS BUFFALO CARBINE

Caliber: 58.
Barrel: 26", round.
Weight: 7¾ lbs. **Length:** 42" overall.
Stock: Walnut.
Sights: Blade front, open adjustable rear.
Features: Shoots .575" round ball. Color case-hardened lock, blue hammer, barrel, trigger; brass furniture. Brass patchbox. Imported by Euroarms of America.

EUROARMS 1861 SPRINGFIELD RIFLE

Caliber: 58.
Barrel: 40".
Weight: About 10 lbs. **Length:** 55.5" overall.
Stock: European walnut.
Sights: Blade front, three-leaf military rear.
Features: Reproduction of the original three-band rifle. Lockplate marked "1861" with eagle and "U.S. Springfield." Metal left in the white. Imported by Euroarms of America.

Euroarms Volunteer

EUROARMS VOLUNTEER TARGET RIFLE

Caliber: .451.
Barrel: 33" (Two-band), 36" (Three-band).
Weight: 11 lbs. (Two-band). **Length:** 48.75" overall (two-band).
Stock: European walnut with checkered wrist and forend.
Sights: Hooded bead front, adjustable rear with interchangeable leaves.
Features: Alexander Henry-type rifling with 1:20" twist. Color case-hardened hammer and lockplate, brass trigger guard and nose cap, rest blued. Available as Two-band or Three-band model. Imported by Euroarms of America.

Gonic GA-87

GONIC GA-87 M/L RIFLE

Caliber: 45, 50.
Barrel: 26".
Weight: 6 to 6½ lbs. **Length:** 43" overall (Carbine).
Stock: American walnut with checkered grip and forend, or laminated stock.
Sights: Optional bead front, open or peep rear adjustable for windage and elevation; drilled and tapped for scope bases (included).
Features: Closed-breech action with straight-line ignition. Modern trigger mechanism with ambidextrous safety. Satin blue finish on metal, satin stock finish. Introduced 1989. Available with or without sights. From Gonic Arms, Inc.

GONIC GA-93 MAGNUM M/L RIFLE

Caliber: 50.
Barrel: 22".
Weight: 6-6½ lbs. **Length:** 39" overall.
Stock: Black wrinkle-finish wood, or gray or brown, standard or thumbhole laminate.
Sights: Optional bead front, open or peep fully adjustable rear; drilled and tapped for included scope bases.
Features: Open-bolt mechanism; single safety; in-line ignition; modern trigger with ambidextrous safety. Many metal finish, sight and stock options available. From Gonic Arms, Inc.

HATFIELD MOUNTAIN RIFLE

Caliber: 50, 54.
Barrel: 32".
Weight: 8 lbs. **Length:** 49" overall.
Stock: Select American fancy maple. Half-stock with nose cap.
Sights: Silver blade front on brass base, fixed buckhorn rear.
Features: Traditional leaf spring and fly lock with extra-wide tumbler of 4140 steel. Slow rust brown metal finish. Double-set triggers. From Hatfield Gun Co.

Blackpowder Muskets & Rifles

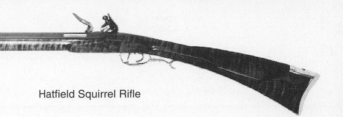

HATFIELD SQUIRREL RIFLE
Caliber: 36, 45, 50.
Barrel: 39¹⁄₂", octagon, 32" on half-stock.
Weight: 7¹⁄₂ lbs. (32-cal.).
Stock: American fancy maple.
Sights: Silver blade front, buckhorn rear.
Features: Recreation of the traditional squirrel rifle. Available in flint or percussion with brass trigger guard and buttplate. Available with full lstock, Grade II or III, percussion or flintlock. From Hatfield Rifle Works. Introduced 1983.

Hatfield Squirrel Rifle

Navy Arms 1803
Harper's Ferry

HARPER'S FERRY 1803 FLINTLOCK RIFLE
Caliber: 54 or 58.
Barrel: 35".
Weight: 9 lbs. **Length:** 59¹⁄₂" overall.
Stock: Walnut with cheekpiece.
Sights: Brass blade front, fixed steel rear.
Features: Brass trigger guard, sideplate, buttplate; steel patch box. Imported by Euroarms of America, Navy Arms (54-cal. only), Cabela's, Stone Mountain Arms.

HAWKEN RIFLE
Caliber: 45, 50, 54 or 58.
Barrel: 28", blued, 6-groove rifling.
Weight: 8³⁄₄ lbs. **Length:** 44" overall.
Stock: Walnut with cheekpiece.
Sights: Blade front, fully adjustable rear.
Features: Coil mainspring, double-set triggers, polished brass furniture. From Armsport, Navy Arms, E.M.F.

Ithaca-Navy Hawken

ITHACA-NAVY HAWKEN RIFLE
Caliber: 50.
Barrel: 32" octagonal, 1" dia.
Weight: About 9 lbs.
Stocks: Walnut.
Sights: Blade front, rear adjustable for windage.
Features: Hooked breech, 1⁷⁄₈" throw percussion lock. Attached twin thimbles and under-rib. German silver barrel key inlays. Hawken-style toe and buttplates, lock bolt inlays, barrel wedges, entry thimble, trigger guard, ramrod and cleaning jag, nipple and nipple wrench. Introduced 1977. From Navy Arms.

KENTUCKIAN RIFLE & CARBINE
Caliber: 44.
Barrel: 35" (Rifle), 27¹⁄₂" (Carbine).
Weight: 7 lbs. (Rifle), 5¹⁄₂ lbs. (Carbine). **Length:** 51" overall (Rifle), 43" (Carbine).
Stock: Walnut stain.
Sights: Brass blade front, steel V-ramp rear.
Features: Octagon barrel, case-hardened and engraved lockplates. Brass furniture. Available as percussion or flintlock. Imported by Dixie Gun Works.

KENTUCKY PERCUSSION RIFLE
Caliber: 45, 50.
Barrel: 35".
Weight: 7 lbs. **Length:** 50" overall.
Stock: Walnut stained hardwood; brass fittings.
Sights: Fixed.
Features: Percussion lock. Finish and features vary with importer. Imported by Navy Arms, The Armoury, CVA.

KENTUCKY FLINTLOCK RIFLE
Caliber: 44, 45, or 50.
Barrel: 35".
Weight: 7 lbs. **Length:** 50" overall.
Stock: Walnut stained; brass fittings.
Sights: Fixed.
Features: Available in carbine model also, 28" bbl. Some variations in detail, finish. Kits also available from some importers. Imported by Navy Arms, The Armoury.

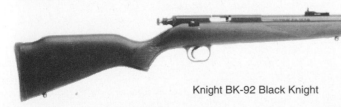

Knight BK-92 Black Knight

KNIGHT BK-92 BLACK KNIGHT RIFLE
Caliber: 50, 54.
Barrel: 24", blued.
Weight: 6¹⁄₂ lbs.
Stock: Black composition.
Sights: Blade front on ramp, open adjustable rear.
Features: Patented double safety system; removable breech plug for cleaning; adjustable Accu-Lite trigger; Green Mountain barrel; receiver drilled and tapped for scope bases. Made in U.S. by Modern Muzzleloading, Inc.

Blackpowder Muskets & Rifles

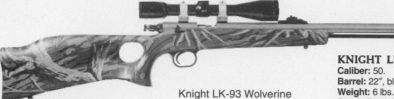

Knight LK-93 Wolverine

KNIGHT MK-85 RIFLE
Caliber: 50, 54.
Barrel: 24″.
Weight: 6³/₄ lbs.
Stock: Walnut, laminated or composition.
Sights: Hooded blade front on ramp, open adjustable rear.
Features: Patented double safety; Sure-Fire in-line percussion ignition; Timney Featherweight adjustable trigger; aluminum ramrod; receiver drilled and tapped for scope bases. Available as Hunter with walnut stock; Stalker with laminated or composition stock; Predator with stainless steel, laminated or composition stock; Knight Hawk, blued, or stainless with composition thumbhole stock. Made in U.S. by Modern Muzzleloading, Inc.

KNIGHT LK-93 WOLVERINE RIFLE
Caliber: 50.
Barrel: 22″, blued.
Weight: 6 lbs.
Stock: Black Fiber-Lite synthetic.
Sights: Blade front on ramp, open adjustable rear.
Features: Patented double safety system; removable breech plug; Sure-Fire in-line percussion ignition system. Also available as LK-93 Stainless and LK-93 Thumbhole. Made in U.S. by Modern Muzzleloading, Inc.

KNIGHT MK-95 MAGNUM ELITE RIFLE
Caliber: 50, 54.
Barrel: 24″, stainless.
Weight: 6³/₄ lbs.
Stock: Composition; black or Realtree All-Purpose camouflage.
Sights: Hooded blade front on ramp, open adjustable rear.
Features: Enclosed Posi-Fire ignition system uses large rifle primers; Timney Featherweight adjustable trigger; Green Mountain barrel; receiver drilled and tapped for scope bases. Made in U.S. by Modern Muzzleloading, Inc.

Kodiak Mk. III Double Rifle

KODIAK MK. III DOUBLE RIFLE
Caliber: 54x54, 58x58, 50x50.
Barrel: 28″, 5-groove, 1:48″ twist.
Weight: 9¹/₂ lbs. **Length:** 43¹/₄″ overall.
Stock: Czechoslovakian walnut, hand-checkered.
Sights: Adjustable bead front, adjustable open rear.
Features: Hooked breech allows interchangeability of barrels. Comes with sling, swivels, bullet mould and bullet starter. Engraved lockplates, top tang and trigger guard. Locks and top tang polished, rest browned. Introduced 1976. Imported from Italy by Navy Arms.

London Armory 1861

LONDON ARMORY 2-BAND 1858 ENFIELD
Caliber: .577″ Minie, .575″ round ball.
Barrel: 33″.
Weight: 10 lbs. **Length:** 49″ overall.
Stock: Walnut.
Sights: Folding leaf rear adjustable for elevation.
Features: Blued barrel, color case-hardened lock and hammer, polished brass buttplate, trigger guard, nosecap. From Navy Arms, Euroarms of America, Dixie Gun Works.

LYMAN COUGAR IN-LINE RIFLE
Caliber: 50 or 54.
Barrel: 22″; 1:24″ twist.
Weight: NA. **Length:** NA.
Stock: Smooth walnut; swivel studs.
Sights: Bead on ramp front, folding adjustable rear. Drilled and tapped for Lyman 57WTR receiver sight and Weaver scope bases.
Features: Blued barrel and receiver. Has bolt safety notch and trigger safety. Rubber recoil pad. Delrin ramrod. Introduced 1996. From Lyman.

LYMAN DEERSTALKER CUSTOM CARBINE
Caliber: 50.
Barrel: 21″ stepped octagon; 1:24″ twist.
Weight: 6³/₄ lbs. **Length:** 38¹/₂″ overall.
Stock: Walnut with black rubber buttpad.
Sights: Lyman 37MA front, Lyman 16A folding rear.
Features: All metal parts are blackened; color case-hardened lock; single trigger. Comes with Delrin ramrod, modern sling and swivels. From Lyman.

LONDON ARMORY 1861 ENFIELD MUSKETOON
Caliber: 58, Minie ball.
Barrel: 24″, round.
Weight: 7-7¹/₂ lbs. **Length:** 40¹/₂″ overall.
Stock: Walnut, with sling swivels.
Sights: Blade front, graduated military-leaf rear.
Features: Brass trigger guard, nose cap, buttplate; blued barrel, bands, lockplate, swivels. Finished or kit. Imported by Euroarms of America, Navy Arms.

LONDON ARMORY 3-BAND 1853 ENFIELD
Caliber: 58 (.577″ Minie, .575″ round ball, .580″ maxi ball).
Barrel: 39″.
Weight: 9¹/₂ lbs. **Length:** 54″ overall.
Stock: European walnut.
Sights: Inverted "V" front, traditional Enfield folding ladder rear.
Features: Recreation of the famed London Armory Company Pattern 1853 Enfield Musket. One-piece walnut stock, brass buttplate, trigger guard and nose cap. Lockplate marked "London Armoury Co." and with a British crown. Blued Baddeley barrel bands. Available finished or as assembled kit. From Dixie Gun Works, Euroarms of America, Navy Arms.

LYMAN DEERSTALKER RIFLE
Caliber: 50, 54.
Barrel: 24″, octagonal; 1:48″ rifling.
Weight: 7¹/₂ lbs.
Stock: Walnut with black rubber buttpad.
Sights: Lyman #37MA beaded front, fully adjustable fold-down Lyman #16A rear.
Features: Stock has less drop for quick sighting. All metal parts are blackened, with color case-hardened lock; single trigger. Comes with sling and swivels. Available as percussion right- or left-hand; flintlock left-hand in 50-cal. only. Introduced 1990. From Lyman.

Lyman Great Plains

LYMAN TRADE RIFLE
Caliber: 50, 54.
Barrel: 28″ octagon, 1:48″ twist.
Weight: 8¾ lbs. **Length:** 45″ overall.
Stock: European walnut.
Sights: Blade front, open rear adjustable for windage or optional fixed sights.
Features: Fast twist rifling for conical bullets. Polished brass furniture with blue steel parts, stainless steel nipple. Hook breech, single trigger, coil spring percussion lock. Steel barrel rib and ramrod ferrules. Introduced 1980. Percussion or flintlock. From Lyman.

LYMAN GREAT PLAINS RIFLE
Caliber: 50- or 54-cal.
Barrel: 32″, 1:66″ twist.
Weight: 9 lbs.
Stock: Walnut.
Sights: Steel blade front, buckhorn rear adjustable for windage and elevation and fixed notch primitive sight included.
Features: Blued steel furniture. Stainless steel nipple. Coil spring lock, Hawken-style trigger guard and double-set triggers. Round thimbles recessed and sweated into rib. Steel wedge plates and toe plate. Available as percussion or flintlock, finished or kit, right- or left-hand. Introduced 1979. From Lyman.

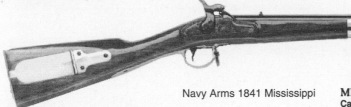

Navy Arms 1841 Mississippi

MOWREY 1 N 30 CONICAL RIFLE
Caliber: 45, 50, 54.
Barrel: 28″; 1:24″ twist.
Weight: About 7½ lbs. **Length:** 43″ overall.
Stock: Curly maple.
Sights: German silver blade front, semi-buckhorn rear.
Features: Steel frame; overall browned finish; cut-rifled barrel; adjustable sear and trigger pull; also available in kit form. Made in U.S. by Mowrey Gan Works.

MOWREY PLAINS RIFLE
Caliber: 50, 54.
Barrel: 28″ (Rocky Mountain Hunter), 32″.
Weight: About 8 lbs. **Length:** 47″ overall.
Stock: Curly maple.
Sights: German silver front, semi-buckhorn rear.
Features: Brass or steel boxlock action; cut-rifled barrel; steel rifle has overall browned finish, brass has browned barrel; adjustable sear and trigger pull. Also available in kit form. Made in U.S. by Mowrey Gun Works.

MISSISSIPPI 1841 PERCUSSION RIFLE
Caliber: 58.
Barrel: 32½″.
Weight: About 9½ lbs. **Length:** 48½″ overall.
Stock: Walnut-finished hardwood; brass patchbox and buttplate.
Sights: Fixed front, rear adjustable for elevation.
Features: Patterned after the U.S. Model 1841 musket. Imported by Dixie Gun Works, Euroarms of America, Navy Arms, Stone Mountain Arms.

MOWREY SILHOUETTE RIFLE
Caliber: 40.
Barrel: 32″; 13/16″ flats; 1:66″ twist.
Weight: About 7½ lbs. **Length:** 47″ overall.
Stock: Curly maple; crescent buttplate.
Sights: German silver blade front, semi-buckhorn rear.
Features: Brass or steel boxlock action; cut-rifled barrel, steel rifle has browned finish; brass has browned barrel. Adjustable sear and trigger pull. Also available in kit form. Made in U.S. by Mowrey Gun Works.

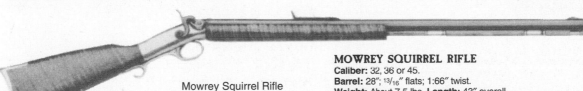

Mowrey Squirrel Rifle

MOWREY SQUIRREL RIFLE
Caliber: 32, 36 or 45.
Barrel: 28″; 13/16″ flats; 1:66″ twist.
Weight: About 7.5 lbs. **Length:** 43″ overall.
Stock: Curly maple; crescent buttplate.
Sights: German silver blade front, semi-buckhorn rear.
Features: Brass or steel boxlock action; cut-rifled barrel. Steel rifles have browned finish, brass have browned barrel. Adjustable sear and trigger pull. Also available in kit form. Made in U.S. by Mowrey Gun Works.

Navy Arms J.P. Murray

J.P. MURRAY 1862–1864 CAVALRY CARBINE
Caliber: 58 (.577″ Minie).
Barrel: 23″.
Weight: 7 lbs., 9 oz. **Length:** 39″ overall.
Stock: Walnut.
Sights: Blade front, rear drift adjustable for windage.
Features: Browned barrel, color case-hardened lock, blued swivel and band springs, polished brass buttplate, trigger guard, barrel bands. From Navy Arms, Euroarms of America.

Blackpowder Muskets & Rifles

NAVY ARMS 1777 CHARLEVILLE MUSKET
Caliber: 69.
Barrel: 44⅝″.
Weight: 10 lbs., 4 oz. **Length:** 59¾″ overall.
Stock: Walnut.
Sights: Brass blade front.
Features: Exact copy of the musket used in the French Revolution. All steel is polished, in the white. Brass flashpan. Also available as the 1816 M.T. Wickham musket. Introduced 1991. Imported by Navy Arms.

NAVY ARMS 1861 SPRINGFIELD RIFLE
Caliber: 58.
Barrel: 40″
Weight: 10 lbs., 4 oz. **Length:** 56″ overall.

Navy Arms 1863

NAVY ARMS 1863 C.S. RICHMOND RIFLE
Caliber: 58.
Barrel: 40″.
Weight: 10 lbs. **Length:** NA.
Stock: Walnut.
Sights: Blade front, adjustable rear.
Features: Copy of the three-band rifle musket made at Richmond Armory for the Confederacy. All steel polished bright. Imported by Navy Arms.

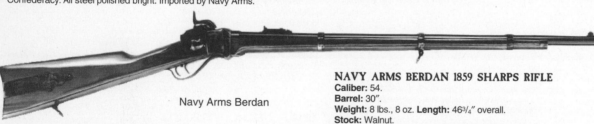

Navy Arms Berdan

Navy Arms Country Boy

NAVY ARMS HAWKEN HUNTER RIFLE/CARBINE
Caliber: 50, 54, 58.
Barrel: 22½″ or 28″; 1:48″ twist.
Weight: 6 lbs., 12 oz. **Length:** 39″ overall.
Stock: Walnut with cheekpiece.
Sights: Blade front, fully adjustable rear.
Features: Double-set triggers; all metal has matte black finish; rubber recoil pad; detachable sling swivels. Imported by Navy Arms.

Navy Arms Mortimer Match

NAVY ARMS 1859 SHARPS CAVALRY CARBINE
Caliber: 54.
Barrel: 22″.
Weight: 7¾ lbs. **Length:** 39″ overall.
Stock: Walnut.
Sights: Blade front, military ladder-type rear.
Features: Color case-hardened action, blued barrel. Has saddle ring. Introduced 1991. Imported from Navy Arms.

Stock: Walnut.
Sights: Blade front, military leaf rear.
Features: Steel barrel, lock and all furniture have polished bright finish. Has 1855-style hammer. Imported by Navy Arms.

NAVY ARMS 1863 SPRINGFIELD
Caliber: 58, uses .575″ Minie.
Barrel: 40″, rifled.
Weight: 9½ lbs. **Length:** 56″ overall.
Stock: Walnut.
Sights: Open rear adjustable for elevation.
Features: Full-size three-band musket. Polished bright metal, including lock. From Navy Arms.

NAVY ARMS BERDAN 1859 SHARPS RIFLE
Caliber: 54.
Barrel: 30″.
Weight: 8 lbs., 8 oz. **Length:** 46¾″ overall.
Stock: Walnut.
Sights: Blade front, folding military ladder-type rear.
Features: Replica of the Union sniper rifle used by Berdan's 1st and 2nd Sharpshooter regiments. Color case-hardened receiver, patch box, furniture. Double-set triggers. Also available as three-band Infantry model. Imported by Navy Arms.

NAVY ARMS COUNTRY BOY IN-LINE RIFLE
Caliber: 50.
Barrel: 24″.
Weight: 8 lbs. **Length:** 41″ overall.
Stock: Black composition.
Sights: Bead front, fully adjustable open rear.
Features: Chrome-lined barrel; receiver drilled and tapped for scope mount; buttstock has trap containing takedown tool for nipple and breech plug removal. Introduced 1996. From Navy Arms.

NAVY ARMS MORTIMER FLINTLOCK RIFLE
Caliber: 54.
Barrel: 36″.
Weight: 9 lbs. **Length:** 52¼″ overall.
Stock: Checkered walnut.
Sights: Bead front, rear adjustable for windage.
Features: Waterproof pan, roller frizzen; sling swivels; browned barrel; external safety. Also available as Match Rifle with hooded globe front sight, fully adjustable target aperature rear, color case-hardened lock. Introduced 1991. Imported by Navy Arms.

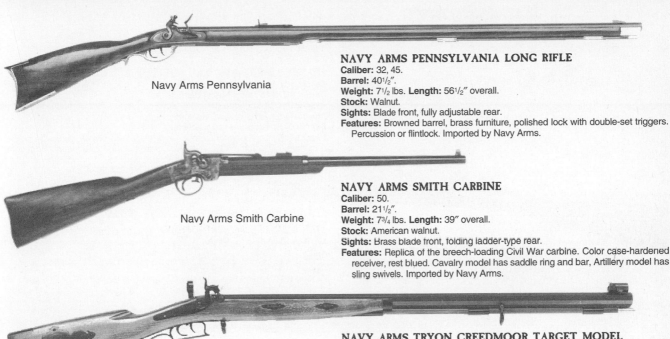

Navy Arms Pennsylvania

NAVY ARMS PENNSYLVANIA LONG RIFLE
Caliber: 32, 45.
Barrel: 40¹/₂".
Weight: 7¹/₂ lbs. **Length:** 56¹/₂" overall.
Stock: Walnut.
Sights: Blade front, fully adjustable rear.
Features: Browned barrel, brass furniture, polished lock with double-set triggers. Percussion or flintlock. Imported by Navy Arms.

Navy Arms Smith Carbine

NAVY ARMS SMITH CARBINE
Caliber: 50.
Barrel: 21¹/₂".
Weight: 7³/₄ lbs. **Length:** 39" overall.
Stock: American walnut.
Sights: Brass blade front, folding ladder-type rear.
Features: Replica of the breech-loading Civil War carbine. Color case-hardened receiver, rest blued. Cavalry model has saddle ring and bar, Artillery model has sling swivels. Imported by Navy Arms.

Navy Arms Tryon Creedmore

NAVY ARMS TRYON CREEDMOOR TARGET MODEL
Caliber: 45.
Barrel: 33", octagonal.
Weight: About 9 lbs. **Length:** 51" overall.
Stock: European walnut.
Sights: Globe front, fully adjustable match rear.
Features: Double-set triggers; back action lock; hooked breech with long tang; sling swivels. Imported by Navy Arms.

NAVY ARMS VOLUNTEER RIFLE
Caliber: .451".
Barrel: 32".
Weight: 9¹/₂ lbs. **Length:** 49" overall.
Stock: Walnut, checkered wrist and forend.
Sights: Globe front, adjustable ladder-type rear.
Features: Recreation of the type of gun issued to volunteer regiments during the 1860s. Rigby-pattern rifling, patent breech, detented lock. Stock is glass bedded for accuracy. Imported by Navy Arms.

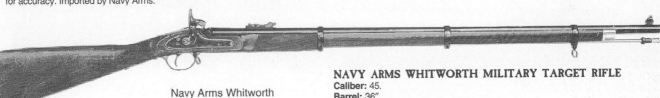

Navy Arms Whitworth

NAVY ARMS WHITWORTH MILITARY TARGET RIFLE
Caliber: 45.
Barrel: 36".
Weight: 9¹/₄ lbs. **Length:** 52¹/₂" overall.
Stock: Walnut. Checkered at wrist and forend.
Sights: Hooded post front, open step-adjustable rear.
Features: Faithful reproduction of the Whitworth rifle, only bored for 45-cal. Trigger has a detented lock, capable of being adjusted very finely without risk of the sear nose catching on the half-cock bent and damaging both parts. Introduced 1978. Imported by Navy Arms.

Peifer Model TS-93

PEIFER MODEL TS-93 RIFLE
Caliber: 45, 50.
Barrel: 24" Douglas premium; 1:20" twist in 45, 1:28" in 50.
Weight: 7 lbs. **Length:** 43¹/₄" overall.
Stock: Bell & Carlson solid composite, with recoil pad, swivel studs.
Sights: Williams bead front on ramp, fully adjustable open rear. Drilled and tapped for Weaver scope mounts with dovetail for rear peep.
Features: In-line ignition uses #209 shotshell primer; extremely fast lock time; fully enclosed breech; adjustable trigger; automatic safety; removal primer holder. Blue or stainless, black, composite or camouflage composite stock. Made in U.S. by Peifer Rifle Co. Introduced 1996.

PENNSYLVANIA FULL-STOCK RIFLE
Caliber: 45 or 50.
Barrel: 32" rifled, ¹⁵/₁₆" dia.
Weight: 8¹/₂ lbs.
Stock: Walnut.
Sights: Fixed.
Features: Available in flint or percussion. Blued lock and barrel, brass furniture. Offered complete or in kit form. Percussion or flintlock. From The Armoury.

Prairie River Bullpup

PRAIRIE RIVER ARMS PRA BULLPUP RIFLE
Caliber: 50, 54.
Barrel: 28″; 1:28″ twist; blue or stainless.
Weight: 7¹/₂ lbs. **Length:** 31¹/₂″ overall.
Stock: Hardwood or black all-weather.
Sights: Blade front, open adjustable rear.
Features: Bullpup design thumbhole stock. Patented internal percussion ignition system. Left-hand model available. Dovetailed for scope mount. Introduced 1995. Made in U.S. by Prairie River Arms.

Prairie River Classic

PRAIRIE RIVER ARMS PRA CLASSIC RIFLE
Caliber: 50, 54.
Barrel: 26″; 1:28″ twist; blue or stainless steel.
Weight: 7¹/₂ lbs. **Length:** 40¹/₂″ overall.
Stock: Hardwood or black all-weather.
Sights: Blade front, open adjustable rear.
Features: Patented internal percussion ignition system. Drilled and tapped for scope mount. Introduced 1995. Made in U.S. by Prairie River Arms, Ltd.

Remington 700 ML

REMINGTON 700 ML, MLS RIFLE
Caliber: 50, 54.
Barrel: 24″; 1:28″ twist.
Weight: 7³/₄ lbs. **Length:** 44¹/₂″ overall.
Stock: Black fiberglass-reinforced synthetic with checkered grip and forend; magnum-style buttpad.
Sights: Ramped bead front, open fully adjustable rear. Drilled and tapped for scope mounts.
Features: Uses the Remington 700 bolt action, stock design, safety and trigger mechanisms; removable stainelss steel breech plug, No. 11 nipple; solid aluminum ramrod. Comes with cleaning tools and accessories. Model 700 ML is blued, 700 MLS is stainless steel.

C.S. Richmond

STONE MOUNTAIN 1853 ENFIELD MUSKET
Caliber: 58.
Barrel: 39″.
Weight: About 9 lbs. **Length:** 54″ overall.
Stock: Walnut.
Sights: Inverted V front, rear step adjustable for elevation.
Features: Three-band musket. Barrel, tang, breech plug are blued, color case-hardened lock, brass nose cap, trigger guard and buttplate. From Stone Mountain Arms.

C.S. RICHMOND 1863 MUSKET
Caliber: 58.
Barrel: 40″.
Weight: 11 lbs. **Length:** 56¹/₄″ overall.
Stock: European walnut with oil finish.
Sights: Blade front, adjustable folding leaf rear.
Features: Reproduction of the three-band Civil War musket. Sling swivels attached to trigger guard and middle barrel band. Lock plate marked "1863" and "C.S. Richmond." All metal left in the white. Brass buttplate and forend cap. Imported by Euroarms of America.

Navy Arms Brown Bess

SECOND MODEL BROWN BESS MUSKET
Caliber: 75, uses .735″ round ball.
Barrel: 42″, smoothbore.
Weight: 9¹/₂ lbs. **Length:** 59″ overall.
Stock: Walnut (Navy); walnut-stained hardwood (Dixie).
Sights: Fixed.
Features: Polished barrel and lock with brass trigger guard and buttplate. Bayonet and scabbard available. Carbine model available from Navy Arms. Finished or in kit form. From Navy Arms, Dixie Gun Works, Cabela's.

STONE MOUNTAIN SILVER EAGLE RIFLE
Caliber: 50.
Barrel: 26″, octagonal; ¹⁵/₁₆″ flats; 1:48″ twist.
Weight: About 6¹/₂ lbs. **Length:** 40″ overall.
Stock: Dura-Grip synthetic; checkered grip and forend.
Sights: Blade front, fixed rear.
Features: Weatherguard nickel finish on metal; oversize trigger guard. Also available as Silver Eagle Hunter with adjustable sight, drilled and tapped for scope mounting, swivel studs, synthetic ramrod. Introduced 1995. From Stone Mountain Arms.

Thompson/Center Fire Hawk

THOMPSON/CENTER BIG BOAR RIFLE

Caliber: 58.
Barrel: 26″ octagon; 1:48″ twist.
Weight: 7³/₄ lbs. **Length:** 42¹/₂″ overall.
Stock: American black walnut; rubber buttpad; swivels.
Sights: Bead front, fullt adjustable open rear.
Features: Percussion lock; single trigger with wide bow trigger guard. Comes with soft leather sling. Introduced 1991. From Thompson/Center.

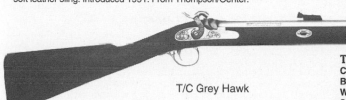

T/C Grey Hawk

T/C Hawken

THOMPSON/CENTER HAWKEN SILVER ELITE RIFLE

Caliber: 50 only.
Barrel: 28″ octagon, hooked breech.
Stock: Semi-fancy American walnut; no patchbox.
Sights: Blade front, fully adjustable open rear.
Features: Percussion only. Solid brass furniture, double-set triggers, all metal is satin-finished stainless steel; button-rifled barrel; coil-type mainspring. Introduced 1996. From Thompson/Center.

THOMPSON/CENTER PENNSYLVANIA HUNTER CARBINE

Caliber: 50.
Barrel: 21″, 1:66″ twist.
Weight: 6¹/₂ lbs. **Length:** 38″ overall.
Stock: Black walnut.
Sights: Open, adjustable.
Features: Designed for shooting patched round balls. Available in percussion or flight. Introduced 1992. From Thompson/Center Arms.

THOMPSON/CENTER PENNSYLVANIA HUNTER RIFLE

Caliber: 50.
Barrel: 31″, half-octagon, half-round.
Weight: About 7¹/₂ lbs. **Length:** 48″ overall.
Stock: Black walnut.
Sights: Open, adjustable.
Features: Rifled 1:66″ for round ball shooting. Available in flintlock or percussion. From Thompson/Center.

THOMPSON/CENTER SCOUT CARBINE

Caliber: 50 and 54.
Barrel: 21″, interchangeable, 1:38″ twist.
Weight: 7 lbs., 4 oz. **Length:** 38⁵/₈″ overall.
Stocks: American black walnut stock and forend, or black Rynite synthetic.
Sights: Bead front, adjustable semi-buckhorn rear.
Features: Patented in-line ignition system with special vented breech plug. Patented trigger mechanism consists of only two moving parts. Interchangeable barrels. Wide grooved hammer. Brass trigger guard assembly, brass barrel band and buttplate. Ramrod has blued hardware. Comes with quick detachable swivels and suede leather carrying sling. Drilled and tapped for standard scope mounts. Introduced 1990. From Thompson/Center.

THOMPSON/CENTER FIRE HAWK RIFLE

Caliber: 32, 50, 54, 58.
Barrel: 24″; 1:38″ twist.
Weight: 7 lbs. **Length:** 41³/₄″ overall.
Stock: American black walnut or black Rynite; Rynite thumbhole style; all with cheekpiece and swivel studs.
Sights: Ramp front with bead, adjustable leaf-style rear.
Features: In-line ignition with sliding thumb safety; free-floated barrel; exposed nipple; adjustable trigger. Available in blue or stainless, and as Bantam model. Comes with Weaver-style scope mount bases. Introduced 1995. Made in U.S. by Thompson/Center Arms.

THOMPSON/CENTER GREY HAWK PERCUSSION RIFLE

Caliber: 50, 54.
Barrel: 24″; 1:48″ twist.
Weight: 7 lbs. **Length:** 41″ overall.
Stock: Black Rynite with rubber recoil pad.
Sights: Bead front, fully adjustable open hunting rear.
Features: Stainless steel barrel, lock, hammer, trigger guard, thimbles; blued sights. Percussion only. Introduced 1993. From Thompson/Center Arms.

THOMPSON/CENTER HAWKEN RIFLE

Caliber: 45, 50 or 54.
Barrel: 28″ octagon, hooked breech.
Stock: American walnut.
Sights: Blade front, rear adjustable for windage and elevation.
Features: Solid brass furniture, double-set triggers, button rifled barrel, coil-type mainspring. Percussion or flintlock. Also available in kit form. From Thompson/Center.

THOMPSON/CENTER NEW ENGLANDER RIFLE

Caliber: 50, 54.
Barrel: 20″, round.
Weight: 7 lbs., 15 oz.
Stock: American walnut or Rynite.
Sights: Open, adjustable.
Features: Color case-hardened percussion lock with engraving, rest blued. Also accepts 12-ga. shotgun barrel. Right- or left-hand models. Introduced 1987. From Thompson/Center.

THOMPSON/CENTER PENNSYLVANIA MATCH RIFLE

Caliber: 50.
Barrel: 31″; half-octagon, half-round; 1:66″ twist.
Weight: 7¹/₂ lbs. **Length:** 48″ overall.
Stock: Black walnut.
Sights: Globe front with seven interchangeable inserts, tang peep rear.
Features: Designed for round ball shooting. Percussion or flintlock. Introduced 1996. From Thompson/Center Arms.

THOMPSON/CENTER RENEGADE RIFLE

Caliber: 50 and 54.
Barrel: 26″, 1″ across the flats.
Weight: 8 lbs.
Stock: American walnut.
Sights: Open hunting (Patridge) style, fully adjustable for windage and elevation.
Features: Coil spring lock, double-set triggers, blued steel trim. Percussion (finished or kit, right- or left-hand) or flintlock (50-cal. only). From Thompson/Center.

THOMPSON/CENTER RENEGADE HUNTER

Caliber: 50 and 54.
Barrel: 26″, 1″ flats.
Weight: 8 lbs.
Stock: American walnut.
Sights: Open hunting (Patridge) style, fully adjustable for windage and elevation.
Features: Coil spring lock; single trigger in a large-bow shotgun-style trigger guard; color case-hardened lock, rest blued. Introduced 1987. From Thompson/Center.

Blackpowder Muskets & Rifles

T/C Scout Rifle

THOMPSON/CENTER SCOUT RIFLE
Caliber: 50 and 54.
Barrel: 24″ part round, part octagon; 1:38″ twist.
Weight: 7½ lbs. **Length:** 41⅝″ overall.
Stock: American black walnut stock and forend, or black Rynite.
Sights: Bead front, adjustable semi-buckhorn rear.
Features: Patented in-line ignition system with vented breech plug; interchangeable barrels; wide grooved hammer; drilled and tapped for scope mounts. Introduced 1995. From Thompson/Center.

T/C Thunderhawk

THOMPSON/CENTER THUNDERHAWK CARBINE
Caliber: 50, 54.
Barrel: 21″, 24″; 1:38″ twist.
Weight: 6.75 lbs. **Length:** 38.75″ overall.
Stock: American walnut or black Rynite with rubber recoil pad.
Sights: Bead on ramp front, adjustable leaf rear.
Features: Uses modern in-line ignition system, adjustable trigger. Knurled striker handle indicators for Safe and Fire. Black wood ramrod, Drilled and tapped for T/C scope mounts. Available blued with walnut or Rynite stock, or stainless with Rynite stock. Introduced 1993. From Thompson/Center Arms.

THOMPSON/CENTER THUNDERHAWK SHADOW
Caliber: 50 and 54.
Barrel: 24″ only.
Weight: 6¾ lbs. **Length:** 38¾″ overall.
Stock: Composite.
Sights: Bead on ramp front, polycarbonate adjustable rear.
Features: Uses modern in-line ignition system; adjustable trigger; knurled striker handle indicators for Safe and Fire; drilled and tapped for T/C scope mounts. Introduced 1996. From Thompson/Center Arms.

Traditions Model 1853

TRADITIONS 1853 THREE-BAND ENFIELD
Caliber: 58.
Barrel: 39″; 1:48″ twist.
Weight: 10 lbs. **Length:** 55″ overall.
Stock: Walnut.
Sights: Military front, adjustable ladder-type rear.
Features: Color case-hardened lock; brass buttplate, trigger guard, nose cap. Has V-type mainspring; steel ramrod; sling swivels. Introduced 1995. From Traditions.

TRADITIONS 1861 U.S. SPRINGFIELD RIFLE
Caliber: 58.
Barrel: 40″; 1:66″ twist.
Weight: 10 lbs. **Length:** 56″ overall.
Stock: Walnut.
Sights: Military front, adjustable ladder-type rear.
Features: Full-length stock with white steel barrel, buttplate, ramrod, trigger guard, barrel bands, swivels, lockplate. Introduced 1995. From Traditions.

Traditions In-Line Buckhunter

TRADITIONS BUCKHUNTER IN-LINE SCOUT
Caliber: 50 only.
Barrel: 22″; 1:32″ twist.
Weight: 7 lbs. **Length:** 39″ overall.
Stock: Black epoxied beech with 13″ pull length.
Sights: Beaded blade front, click adjustable rear; drilled and tapped for scope mounting.
Features: Removable breech plug; PVC ramrod; sling swivels; C-Nickel barrel. Introduced 1996. Omported by Traditions.

Traditions Buckhunter Pro In-Line

TRADITIONS BUCKHUNTER PRO IN-LINE RIFLES
Caliber: 50 (1:32″ twist), 54 (1:48″ twist).
Barrel: 24″ tapered round.
Weight: 7½ lbs. **Length:** 42″ overall.
Stock: Beech, composite or laminated; thumbhole available in black Mossy Oak Treestand or Realtree® Advantage camouflage.
Sights: Beaded blade front, fully adjustable open rear. Drilled and tapped for scope mounting.
Features: In-line percussion ignition system; adjustable trigger; manual thumb safety; removable stainless steel breech plug. Seventeen models available. Introduced 1996. From Traditions.

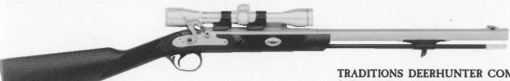

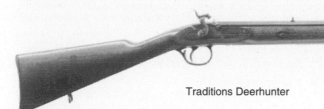

Traditions Deerhunter Composite

TRADITIONS DEERHUNTER COMPOSITE RIFLE
Caliber: 50 (flintlock), 50, 54 (percussion).
Barrel: 24″ octagonal; $^{15}/_{16}$″ flats; 1:48″ or 1:66″ twist.
Weight: 6 lbs. **Length:** 40″ overall.
Stock: Black composite; checkered grip and forend.
Sights: Blade front, fixed rear; drilled and tapped for scope mounting.
Features: Flint or percussion; color case-hardened lock; blue or C-Nickel barrel. Introduced 1996. Imported by Traditions.

TRADITIONS BUCKSKINNER CARBINE
Caliber: 50.
Barrel: 21″, $^{15}/_{16}$″ flats, half octagon, half round; 1:20″ or 1:66″ twist.
Weight: 6 lbs. **Length:** 37″ overall.
Stock: Beech or black laminated.
Sights: Beaded blade front, hunting-style open rear click adjustable for windage and elevation.
Features: Uses V-type mainspring, single trigger. Non-glare hardware. Percussion or flintlock. From Traditions.

Traditions Deerhunter

TRADITIONS DEERHUNTER SCOUT RIFLE
Caliber: 32.
Barrel: 22″ octagonal; 1:48″ twist.
Weight: 5 lbs., 10 oz. **Length:** 36 $^{1}/_{2}$″ overall.
Stock: Beech.
Sights: Blade front, fixed rear; drilled and tapped for scope mounting.
Features: Hooked breech; percussion only; PVC ramrod; color case-hardened lock; over size trigger guard; blackened furniture. Introduced 1996. Imported by Traditions.

TRADITIONS HAWKEN WOODSMAN RIFLE
Caliber: 50 and 54.
Barrel: 28″; $^{15}/_{16}$″ flats.
Weight: 7 lbs., 11 oz. **Length:** 44$^{1}/_{2}$″ overall.
Stock: Walnut-stained hardwood.
Sights: Beaded blade front, hunting-style open rear adjustable for windage and elevation.
Features: Percussion only. Brass patchbox and furniture. Double triggers. Left-hand in 50-cal. only. From Traditions.

TRADITIONS IN-LINE BUCKHUNTER SERIES RIFLES
Caliber: 50, 54.
Barrel: 24″, round; 1:32″ (50), 1:48″ (54) twist.
Weight: 7 lbs., 6 oz. to 8 lbs. **Length:** 41″ overall.
Stock: Beech, epoxy coated beech, laminated or fiberglass thumbhole; rubber recoil pad.
Sights: Beaded blade front, click adjustable rear. Drilled and tapped for scope mounting.
Features: Removable breech plug; PVC ramrod; sling swivels. Fifteen models available with blackened furniture, blued, C-nickel barrels, thumbhole stock. Introduced 1995. From Traditions.

TRADITIONS DEERHUNTER RIFLE SERIES
Caliber: 32, 50 or 54.
Barrel: 24″, octagonal, $^{15}/_{16}$″ flats; 1:48″ or 1:66″ twist.
Weight: 6 lbs. **Length:** 40″ overall.
Stock: Stained beech with rubber buttpad, sling swivels.
Sights: Blade front, fixed rear.
Features: Flint or percussion with color case-hardened lock. Hooked breech, over-sized trigger guard, blackened furniture, PVC ramrod. All-Weather has epoxied beech stock and C-Nickel barrel. Drilled and tapped for scope mounting. Percussion available in kit form. Imported by Traditions, Inc.

TRADITIONS KENTUCKY RIFLE
Caliber: 50.
Barrel: 33$^{1}/_{2}$″; $^{7}/_{8}$″ flats; 1:66″ twist.
Weight: 7 lbs. **Length:** 49″ overall.
Stock: Beech, inletted toe plate.
Sights: Blade front, fixed rear.
Features: Full length, two-piece stock; brass furniture; color case-hardened lock. Finished or in kit form. Introduced 1995. From Traditions.

TRADITIONS PENNSYLVANIA RIFLE
Caliber: 50.
Barrel: 40$^{1}/_{4}$″, $^{7}/_{8}$″ flats; 1:66″ twist, octagon.
Weight: 9 lbs. **Length:** 57$^{1}/_{2}$″ overall.
Stock: Walnut.
Sights: Blade front, adjustable rear.
Features: Flint or percussion. Brass patchbox and ornamentation. Double-set triggers. From Traditions.

TRADITIONS PIONEER RIFLE
Caliber: 50, 54.
Barrel: 28″, $^{15}/_{16}$″ flats.
Weight: 7 lbs. **Length:** 44″ overall.
Stock: Beech with pistol grip, recoil pad.
Sights: German silver blade front, buckhorn rear with elevation ramp.
Features: V-type mainspring, adjustable single trigger; blackened furniture; color case-hardened lock; large trigger guard. Percussion only. From Traditions.

Traditions Shenandoah

TRADITIONS SHENANDOAH RIFLE
Caliber: 50.
Barrel: 33$^{1}/_{2}$″ octagon, 1:66″ twist.
Weight: 7 lbs., 3 oz. **Length:** 49$^{1}/_{2}$″ overall.
Stock: Walnut.
Sights: Blade front, buckhorn rear.
Features: V-type mainspring; double-set trigger; solid brass buttplate, patchbox, nose cap, thimbles, trigger guard. Flint or percussion. Introduced 1996. From Traditions.

Traditions Tennessee

TRADITIONS TENNESSEE RIFLE
Caliber: 50.
Barrel: 24", octagon with 15/16" flats; 1:32" twist.
Weight: 6 lbs. **Length:** 40 1/2" overall.
Stock: Stained beech.
Sights: Blade front, fixed rear.
Features: One-piece stock has inletted brass furniture, cheekpiece; double-set trigger; V-type mainspring. Flint or percussion. Introduced 1995. From Traditions.

TRYON TRAILBLAZER RIFLE
Caliber: 50, 54.
Barrel: 28", 30".
Weight: 9 lbs. **Length:** 48" overall.
Stock: European walnut with cheekpiece.
Sights: Blade front, semi-buckhorn rear.
Features: Reproduction of a rifle made by George Tryon about 1820. Double-set triggers, back action lock, hooked breech with long tang. From Armsport.

UFA GRAND TETON RIFLE
Caliber: 45 and 50.
Barrel: 30" tapered octagon.
Weight: 9 lbs. **Length:** 46" overall.
Stock: Brown or black laminated stock and forend.
Sights: Marble's bead front, Marbles's fully adjustable rear.
Features: Available in blue or stainless steel with brushed or matte finish. Removable, interchangeable barrel; removable one-piece breech plug/nittle, hammer/trigger assembly; hammer blowback block; glass-bedded stock and forend. Introduced 1994. Made in U.S. by UFA, Inc.

UFA TETON BLACKSTONE RIFLE
Caliber: 50 only.
Barrel: 26"; shallow-groove 1:26" rifling.
Weight: 7 1/2 lbs. **Length:** 42" overall.
Stock: Hardwood with black epoxy coating; 1" recoil pad.
Sights: Marble's bead front, Marble's fully adjustable rear.
Features: Stainless steel with matte finish. Removable, interchangeable barrel; removable one-piece breech plug/nittle, hammer/trigger assembly; hammer blowback block; glass-bedded stock and forend. Introduced 1994. Made in U.S. by UFA, Inc.

UFA TETON RIFLE
Caliber: 45, 50, 12-bore (rifled, 72-cal.), 12-gauge.
Barrel: 26".
Weight: 8 lbs. **Length:** 42" overall.
Stock: Black or brown laminated wood; 1" recoil pad. Premium walnut or maple available.
Sights: Marble's bead front, Marble's fully adjustable rear.
Features: Removable, interchangeable barrel; removable one-piece breech plug/nipple, hammer/trigger assembly; hammer blowback block; glass-bedded stock and forend. Stainless or blue. Introduced 1994. Made in U.S. by UFA, Inc.

Ultra Light Model 90

ULTRA LIGHT ARMS MODEL 90 MUZZLELOADER
Caliber: 45, 50.
Barrel: 28", button rifled; 1:48" twist.
Weight: 6 lbs.
Stock: Kevlar/graphite, colors optional.
Sights: Hooded blade front on ramp, Williams aperture rear adjustable for windage and elevation.
Features: In-line ignition system with top loading port. Timney trigger; integral side safety. Comes with recoil pad, sling swivels and hard case. Introduced 1990. Made in U.S. by Ultra Light Arms.

WHITE SHOOTING SYSTEMS BISON BLACKPOWDER RIFLE
Caliber: 54.
Barrel: 22" bull; 1:28" twist.
Weight: 7 1/4 lbs. **Length:** 39 1/2" overall.
Stock: Black-finished hardwood.
Sights: Bead front on ramp, fully adjustable open rear; drilled and tapped for scope mounting.
Features: Uses Insta-Fire in-line ignition system; double safety; matte blue finish; Delrin ramrod; adjustable trigger; action and trigger safeties. Introduced 1993. Made in U.S. by White Shooting Systems, Inc.

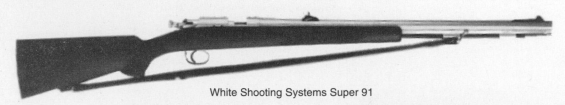

White Shooting Systems Super 91

WHITE SHOOTING SYSTEMS SUPER SAFARI RIFLE
Caliber: 41, 45 or 50.
Barrel: 26".
Weight: 7 1/2 lbs. **Length:** 43 1/2" overall.
Stock: Black composite, Mannlicher style.
Sights: Bead front on ramp, fully adjustable open rear.
Features: Insta-Fire straight-line ignition system; all stainless steel construction; side-swing safety; fully adjustable trigger; full barrel under-rib with two ramrod thimbles. Introduced 1993. Made in U.S. by White Shooting Systems, Inc.

WHITE SHOOTING SYSTEMS SUPER 91 BLACKPOWDER RIFLE
Caliber: 41, 45 or 50.
Barrel: 26".
Weight: 7 1/2 lbs. **Length:** 43.5" overall.
Stock: Black laminate or black composite; recoil pad, swivel studs.
Sights: Bead front on ramp, fully adjustable open rear.
Features: Insta-Fire straight-line ignition system; all stainless steel construction; side-swing safety; fully adjustable trigger; full barrel under-rib with two ramrod thimbles. Introduced 1991. Made in U.S. by White Shooting Systems, Inc.

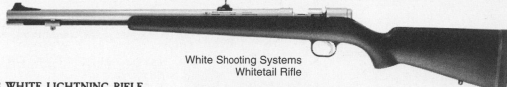

White Shooting Systems
Whitetail Rifle

WHITE SHOOTING SYSTEMS WHITE LIGHTNING RIFLE

Caliber: 50.
Barrel: 22".
Weight: 6¼ lbs. **Length:** 40" overall.
Stock: Black hardwood.
Sights: Bead on ramp front, fully adjustable open rear.
Features: Small action with cocking lever and secondary safety on right side, primary safety on left. Stainless steel construction. Introduced 1996. Made in U.S. by White Shooting Systems, Inc.

ZOUAVE PERCUSSION RIFLE

Caliber: 58, 59.
Barrel: 32½".
Weight: 9½ lbs. **Length:** 48½" overall.

WHITE SHOOTING SYSTEMS WHITETAIL RIFLE

Caliber: 41, 45 or 50.
Barrel: 22".
Weight: 6.5 lbs. **Length:** 39.5" overall.
Stock: Black composite or laminate; classic style; recoil pad, swivel studs.
Sights: Bead front on ramp, fully adjustable open rear.
Features: Insta-Fire straight-line ignition; action and trigger safeties; adjustable trigger; blue or stainless steel. Introduced 1992. Made in U.S. by White Shooting Systems, Inc.

Stock: Walnut finish, brass patchbox and buttplate.
Sights: Fixed front, rear adjustable for elevation.
Features: Color case-hardened lockplate, blued barrel. From Navy Arms, Dixie Gun Works, Euroarms of America (kit, 58-cal., M1863), E.M.F., Cabela's.

Blackpowder Shotguns

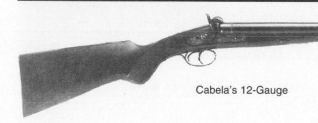

Cabela's 12-Gauge

CVA CLASSIC TURKEY DOUBLE SHOTGUN

Gauge: 12.
Barrel: 28".
Weight: 9 lbs. **Length:** 45" overall.
Stock: European walnut; classic English style with checkered straight grip, wraparound forend with bottom screw attachment.
Sights: Bead front.
Features: Hinged double triggers; color case-hardened and engraved lockplates, trigger guard and tang. Polymer-coated fiberglass ramrod. Rubber recoil pad. Not suitable for steel shot. Introduced 1990. Imported by CVA.

CABELA'S BLACKPOWDER SHOTGUNS

Gauge: 10, 12, 20.
Barrel: 28½" (10-, 12-ga.), Imp. Cyl., Mod., Full choke tubes; 27½" (20-ga.), Imp. Cyl., Mod. choke tubes.
Weight: 6½ to 7 lbs. **Length:** 45" overall (28½" barrel).
Stock: American walnut with checkered grip; 12- and 20-gauge have straight stock, 10-gauge has pistol grip.
Features: Blued barrels, engraved, color case-hardened locks and hammers, brass ramrod tip. From Cabela's.

CVA TRAPPER PERCUSSION

Gauge: 12.
Barrel: 28".
Weight: 6 lbs. **Length:** 46" overall.
Stock: English-style checkered straight grip of walnut-finished hardwood.
Sights: Brass bead front.
Features: Single blued barrel; color case-hardened lockplate and hammer; screw adjustable sear engagements, V-type mainspring; brass wedge plates; color case-hardened and engraved trigger guard and tang. From CVA.

Dixie Magnum

DIXIE MAGNUM PERCUSSION SHOTGUN

Gauge: 10, 12, 20.
Barrel: 30" (Imp. Cyl. & Mod.) in 10-gauge; 28" in 12-gauge.
Weight: 6¼ lbs. **Length:** 45" overall.
Stock: Hand-checkered walnut, 14" pull.
Features: Double triggers; light hand engraving; case-hardened locks in 12-gauge, polished steel in 10-gauge; sling swivels. Available in kit form in 10- and 12-gauge. From Dixie Gun Works.

Mowrey Shotgun

MOWREY SHOTGUN

Gauge: 12, 28.
Barrel: 28" (28-gauge, Cyl.); 32" (12-gauge, Cyl.); octagonal.
Weight: About 8 lbs. **Length:** 48" overall (32" barrel).
Stock: Curly maple.
Sights: Bead front.
Features: Brass or steel frame; shotgun butt. Available finished or in kit form. Made in U.S. by Mowrey Gun Works.

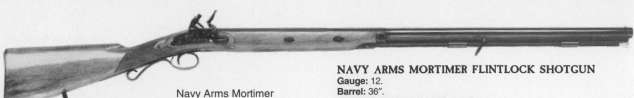

Navy Arms Mortimer

NAVY ARMS MORTIMER FLINTLOCK SHOTGUN
Gauge: 12.
Barrel: 36″.
Weight: 7 lbs. **Length:** 53″ overall.
Stock: Walnut, with cheekpiece.
Features: Waterproof pan, roller frizzen, external safety. Color case-hardened lock, rest blued. Imported by Navy Arms.

NAVY ARMS STEEL SHOT MAGNUM SHOTGUN
Gauge: 10.
Barrel: 28″ (Cyl. & Cyl.).
Weight: 7 lbs., 9 oz. **Length:** 45½″ overall.
Stock: Walnut, with cheekpiece.
Features: Designed specifically for steel shot. Engraved, polished locks; sling swivels; blued barrels. Imported by Navy Arms.

NAVY ARMS FOWLER SHOTGUN
Gauge: 10, 12.
Barrel: 28″.
Weight: 7 lbs., 12 oz. **Length:** 45″ overall.
Stock: Walnut-stained hardwood.
Features: Color case-hardened lockplates and hammers; checkered stock. Imported by Navy Arms.

Navy Arms T&T

NAVY ARMS T&T SHOTGUN
Gauge: 12.
Barrel: 28″ (Full & Full).
Weight: 7½ lbs.
Stock: Walnut.
Sights: Bead front.
Features: Color case-hardened locks, double triggers, blued steel furniture. From Navy Arms.

T/C New Englander

THOMPSON/CENTER NEW ENGLANDER SHOTGUN
Gauge: 12.
Barrel: 28″ (Imp. Cyl.), round.
Weight: 5 lbs., 2 oz.
Stock: Select American black walnut with straight grip.
Features: Percussion lock is color case-hardened, rest blued. Also accepts 26″ round 50- and 54-cal. rifle barrel. Introduced 1986. From Thompson/Center.

Traditions Buckhunter Pro

TRADITIONS BUCKHUNTER PRO SHOTGUN
Gauge: 12.
Barrel: 24″; choke tube.
Weight: 6 lbs., 4oz. **Length:** 43″ overall.
Stock: Composite matte black, Mossy Oak Treestand or Advantage camouflage.
Features: In-line action with removable stainless steel breech plug; thumb safety; adjustable trigger; rubber buttpad. Introduced 1996. From Traditions.

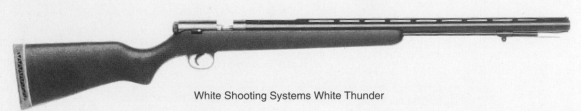

White Shooting Systems White Thunder

WHITE SHOOTING SYSTEMS "TOMINATOR" SHOTGUN
Gauge: 12.
Barrel: 26″ (Imp. Cyl., Mod., Full and Super Full Turkey choke tubes).
Weight: About 5¾ lbs.
Stock: Black laminate.
Features: Insta-Fire in-line ignition system; double safeties; match-grade trigger; Delon ramrod. Introduced 1995. Made in U.S. by White shooting Systems, Inc.

WHITE SHOOTING SYSTEMS WHITE THUNDER SHOTGUN
Gauge: 12.
Barrel: 26″ (Imp. Cyl., Mod., Full choke tubes); ventilated rib.
Weight: About 5¾ lbs.
Stock: Black hardwood.
Features: InstaFire in-line ignition; double safeties; match-grade trigger; Delron ramrod. Introduced 1995. From White Shooting Systems, Inc.

Appendicies

Appendix A: Manufacturers' Directory of Blackpowder Guns
and Accessories 392
Appendix B: Frontloader's Library 394
Appendix C: Blackpowder Periodical Publications 397
Appendix D: Blackpowder Arms Associations 397
Appendix E: Glossary of Blackpowder Terms 398

Appendix A
Manufacturers' Directory
Of Blackpowder Guns And Accessories

Accuracy Unlimited, 7479 S. DePew St., Littleton, CO 80123

Adkins, Luther, 1292 E. McKay Rd., Shelbyville, IN 46176-9353/317-392-3795

Allen Mfg., 6449 Hodgson Rd., Circle Pines, MN 55014/612-429-8231

American Arms, Inc., 715 Armour Rd., N. Kansas City, MO 64116/816-474-3161; FAX: 816-474-1225

Anderson Manufacturing Co., Inc., 22602 53rd Ave. SE, Bothell, WA 98021/206-481-1858; FAX: 206-481-7839

Antique Arms, P.O. Box 10794, Costa Mesa, CA 92627

Armi San Paolo, via Europa 172-A, I-25062 Concesio, 030-2751725 (BS) ITALY

Armsport, Inc., 3950 NW 49th St., Miami, FL 33142/305-635-7850; FAX: 305-633-2877

Bagmaster Mfg., Inc., 2731 Sutton Ave., St. Louis, MO 63143/314-781-8002; FAX: 314-781-3363

Ballistic Products, Inc., 20015 75th Ave. North, Hamel, MN 55340-9456/612-494-9237; FAX: 612-494-9236

Barnes Bullets, Inc., P.O. Box 215, American Fork, UT 84003/801-756-4222, 800-574-9200; FAX: 801-756-2465; WEB: http://www.itsnet.com/home/bbullets

Bauska Barrels, 105 9th Ave. W., Kalispell, MT 59901/406-752-7706

Beauchamp & Son, Inc., 160 Rossiter Rd., P.O. Box 181, Richmond, MA 01254/413-698-3822; FAX: 413-698-3866

Beaver Lodge (See Fellowes, Ted)

Bentley, John, 128-D Watson Dr., Turtle Creek, PA 15145

Big Bore Express, Ltd., 7154 W. State St., #200, Boise, ID 83703

Birchwood Casey, 7900 Fuller Rd., Eden Prairie, MN 55344/800-328-6156, 612-937-7933; FAX: 612-937-7979

Birdsong & Assoc., W.E., 1435 Monterey Rd., Florence, MS 39073-9748/601-366-8270

Blackhawk West, Box 285, Hiawatha, KS 66434

Blue and Gray Products, Inc. (See Ox-Yoke Originals, Inc.)

Bracklyn Products, 4400 Stillman Blvd., Suite C, Tuscaloosa, AL 35401

Bradley Company, 4330 N. State Road 110, Oshkosh, WI 54904

Bridgers Best, P.O. Box 1410, Berthoud, CO 80513

Browning Arms Co. (Gen. Offices), One Browning Place, Morgan, UT 84050/801-876-2711; FAX: 801-876-3331

Buckskin Machine Works, A. Hunkeler, 3235 S. 358th St., Auburn, WA 98001/206-927-5412

Buffalo Arms, 123 S. Third, Suite 6, Sandpoint, ID 83864/208-263-6953; FAX: 208-265-2096

Buffalo Bullet Co., Inc., 12637 Los Nietos Rd., Unit A, Santa Fe Springs, CA 90670/310-944-0322; FAX: 310-944-5054

Burgess & Son Gunsmiths, R.W., P.O. Box 3364, Warner Robins, GA 31099/912-328-7487

Butler Creek Corporation, 290 Arden Dr., Belgrade, MT 59714/800-423-8327, 406-388-1356; FAX: 406-388-7204

Cache La Poudre Rifleworks, 140 N. College, Ft. Collins, CO 80524/303-482-6913

California Sights (See Fautheree, Andy)

Cape Outfitters, 599 County Rd. 206, Cape Girardeau, MO 63701/314-335-4103; FAX: 314-335-1555

Cash Mfg. Co., Inc., P.O. Box 130, 201 S. Klein Dr., Waunakee, WI 53597-0130/608-849-5664; FAX: 608-849-5664

CenterMark, P.O. Box 4066, Parnassus Station, New Kensington, PA 15068/412-335-1319

Chambers Flintlocks Ltd., Jim, Rt. 1, Box 513-A, Candler, NC 28715/704-667-8361

Chopie Mfg., Inc., 700 Copeland Ave., LaCrosse, WI 54603/608-784-0926

Cimarron Arms, P.O. Box 906, Fredericksburg, TX 78624-0906/210-997-9090; FAX: 210-997-0802

Cogar's Gunsmithing, P.O. Box 755, Houghton Lake, MI 48629/517-422-4591

Colonial Repair, P.O. Box 372, Hyde Park, MA 02136-9998/617-469-4951

Colt Blackpowder Arms Co., 5 Centre Market Place, New York, NY 10013/212-925-2159; FAX: 212-966-4986

Cousin Bob's Mountain Products, 7119 Ohio River Blvd., Ben Avon, PA 15202/412-766-5114; FAX: 412-766-5114

Cumberland Arms, 514 Shafer Road, Manchester, TN 37355/800-797-8414

Cumberland Knife & Gun Works, 5661 Bragg Blvd., Fayetteville, NC 28303/919-867-0009

Curly Maple Stock Blanks (See Tiger-Hunt)

CVA, 5988 Peachtree Corners East, Norcross, GA 30071/800-251-9412; FAX: 404-242-8546

Dangler, Homer L., Box 254, Addison, MI 49220/517-547-6745

Davis Co., R.E., 3450 Pleasantville NE, Pleasantville, OH 43148/614-654-9990

Day & Sons, Inc., Leonard, P.O. Box 122, Flagg Hill Rd., Heath, MA 01346/413-337-8369

Dayton Traister, 4778 N. Monkey Hill Rd., P.O. Box 593, Oak Harbor, WA 98277/206-679-4657; FAX:206-675-1114

Deer Creek Products, P.O. Box 246, Waldron, IN 46182

deHaas Barrels, RR 3, Box 77, Ridgeway, MO 64481/816-872-6308

Delhi Gun House, 1374 Kashmere Gate, Delhi, INDIA 110 006/(011)237375 239116; FAX: 91-11-2917344

Desert Industries, Inc., P.O. Box 93443, Las Vegas, NV 89193-3443/702-597-1066; FAX: 702-871-9452

Dewey Mfg. Co., Inc., J., P.O. Box 2014, Southbury, CT 06488/203-264-3064; FAX: 203-262-6907

DGS Gunsmithing, Inc., Dale A. Storey, 1117 E. 12th, Casper, WY 82601/307-237-2414

Dixie Gun Works, Inc., Hwy. 51 South, Union City, TN 38261/901-885-0561, order 800-238-6785; FAX: 901-885-0440

Dyson & Son Ltd., Peter, 29-31 Church St., Honley Huddersfield, W. Yorkshire HD7 2AH, ENGLAND/44-1484-661062; FAX: 44-1484-663709

EMF Co., Inc., 1900 E. Warner Ave. Suite 1-D, Santa Ana, CA 92705/714-261-6611; FAX: 714-756-0133

Euroarms of America, Inc., P.O. Box 3277, Winchester, VA 22604/540-662-1863; FAX: 540-662-4464

Eutaw Co., Inc., The, P.O. Box 608, U.S. Hwy. 176 West, Holly Hill, SC 29059/803-496-3341

Fautheree, Andy, P.O. Box 4607, Pagosa Springs, CO 81157/303-731-5003

Feken, Dennis, Rt. 2 Box 124, Perry, OK 73077/405-336-5611

Fellowes, Ted, Beaver Lodge, 9245 16th Ave. SW, Seattle, WA 98106/206-763-1698

Fire'n Five, P.O. Box 11 Granite Rt., Sumpter, OR 97877

Flintlocks, Etc. (See Beauchamp & Son, Inc.)

Forgett Jr., Valmore J., 689 Bergen Blvd., Ridgefield, NJ 07657/201-945-2500; FAX: 201-945-6859

Forster Products, 82 E. Lanark Ave., Lanark, IL 61046/815-493-6360; FAX: 815-493-2371

Fort Hill Gunstocks, 12807 Fort Hill Rd., Hillsboro, OH 45133/513-466-2763

Frontier, 2910 San Bernardo, Laredo, TX 78040/210-723-5409; FAX: 210-723-1774

Getz Barrel Co., P.O. Box 88, Beavertown, PA 17813/717-658-7263

Ghost Scent/DeWitt USA, 1604 Stockton Street, Jacksonville, FL 32204

GOEX, Inc., 1002 Springbrook Ave., Moosic, PA 18507/717-457-6724; FAX: 717-457-1130

Golden Age Arms Co., 115 E. High St., Ashley, OH 43003/614-747-2488

Gonic Arms, Inc., 134 Flagg Rd., Gonic, NH 03839/603-332-8456, 603-332-8457

Green Mountain Rifle Barrel Co., Inc., P.O. Box 2670, 153 West Main St., Conway, NH 03818/603-447-1095; FAX: 603-447-1099

Hastings Barrels, 320 Court St., Clay Center, KS 67432/913-632-3169; FAX: 913-632-6554

Hatfield Gun Co., Inc., 224 N. 4th St., St. Joseph, MO 64501/816-279-8688; FAX: 816-279-2716

Hawken Shop, The (See Dayton Traister)

Hege Jagd-u. Sporthandels, GmbH, P.O. Box 101461, W-7770 Ueberlingen a. Bodensee, GERMANY

Hodgdon Powder Co., Inc., P.O. Box 2932, 6231 Robinson, Shawnee Mission, KS 66202/913-362-9455; FAX: 913-362-1307; WEB: http://www.uni-com.net/hpc

Hopkins & Allen Arms, P.O. Box 246, Waldron, IN 46182

Hoppe's Div., Penguin Industries, Inc., Airport Industrial Mall, Coatesville, PA 19320/610-384-6000

Hornady Mfg. Co., P.O. Box 1848, Grand Island, NE 68802/800-338-3220, 308-382-1390; FAX: 308-382-5761

House of Muskets, Inc., The, P.O. Box 4640, Pagosa Springs, CO 81157/303-731-2295

Hubbard's Outdoor Products, P.O. Box 338, Helena, AL 35080

Hunkeler, A. (See Buckskin Machine Works)

Jamison's Forge Works, 4527 Rd. 6.5 NE, Moses Lake, WA 98837/509-762-2659

Jonad Corp., 2091 Lakeland Ave., Lakewood, OH 44107/216-226-3161

Jones Co., Dale, 680 Hoffman Draw, Kila, MT 59920/406-755-4684

K&M Industries, Inc., Box 66, 510 S. Main, Troy, ID 83871/208-835-2281; FAX: 208-835-5211

Kennedy Firearms, 10 N. Market St., Muncy, PA 17756/717-546-6695

Knight Rifles (See Modern MuzzleLoading, Inc.)

Kruse Knives, P.O. Box 487, Reseda, CA 91335

Kwik-Site Co., 5555 Treadwell, Wayne, MI 48184/313-326-1500; FAX: 313-326-4120

L&R Lock Co., 1137 Pocalla Rd., Sumter, SC 29150/803-775-6127

Lee Precision, Inc., 4275 Hwy. U, Hartford, WI 53027/414-673-3075

Legend Products Corp., 1555 E. Flamingo Rd., Suite 404, Las Vegas, NV 89119/702-228-1808, 702-796-5778; FAX: 702-228-7484

Log Cabin Sport Shop, 8010 Lafayette Rd., Lodi, OH 44254/216-948-1082

Lothar Walther Precision Tool, Inc., 2190 Coffee Rd., Lithonia, GA 30058/770-482-4253; Fax: 770-482-9344

Lutz Engraving, Ron, E. 1998 Smokey Valley Rd., Scandinavia, WI 54977/715-467-2674

Lyman Products Corporation, 475 Smith Street, Middletown, CT 06457-1541/860-632-2020, 800-22-LYMAN; FAX: 860-632-1699

McCann's Muzzle-Gun Works, 14 Walton Dr., New Hope, PA 18938/215-862-2728

Michaels of Oregon Co., P.O. Box 13010, Portland, OR 97213/503-255-6890; FAX: 503-255-0746

Midway Arms, Inc., 5875 W. Van Horn Tavern Rd., Columbia, MO 65203/800-243-3220, 314-445-6363; FAX: 314-446-1018

MMP, Rt. 6, Box 384, Harrison, AR 72601/501-741-5019; FAX: 501-741-3104

Modern MuzzleLoading, Inc., 234 Airport Rd., P.O. Box 130, Centerville, IA 52544/515-856-2626; FAX: 515-856-2628

Montana Armory, Inc., 100 Centennial Dr., Big Timber, MT 59011/406-932-4353

Montana Precision Swaging, P.O. Box 4746, Butte, MT 59702/406-782-7502

Mountain State Muzzleloading Supplies, Box 154-1, Rt. 2, Williamstown, WV 26187/304-375-7842; FAX: 304-375-3737

Mowrey Gun Works, P.O. Box 246, Waldron, IN 46182/317-525-6181; FAX: 317-525-9595

MSC Industrial Supply Co., 151 Sunnyside Blvd., Plainview, NY 11803-9915/516-349-0330

Mt. Alto Outdoor Products, Rt. 735, Howardsville, VA 24562

Mushroom Express Bullet Co., 601 W. 6th St., Greenfield, IN 46140-1728/317-462-6332

Muzzleloaders Etcetera, Inc., 9901 Lyndale Ave. S., Bloomington, MN 55420/612-884-1161

Narragansett Armes, Ltd., 3025 North Meridian St., #801, Indianapolis, IN 46208

Navy Arms Co., 689 Bergen Blvd., Ridgefield, NJ 07657/201-945-2500; FAX: 201-945-6859

North American Arms, Inc., 2150 South 950 East, Provo, UT 84606-6285/800-821-5783, 801-374-9990; FAX: 801-374-9998

North Star West, P.O. Box 488, Glencoe, CA 95232/209-293-7010

Northeast Industrial, Inc., P.O. Box 249, Canyon City, OR 97820

October Country, P.O. Box 969, Dept. GD, Hayden, ID 83835/208-772-2068; FAX: 208-772-9230

Oklahoma Leather Products, Inc., 500 26th NW, Miami, OK 74354/918-542-6651; FAX: 918-542-6653

Olson, Myron, 989 W. Kemp, Watertown, SD 57201/605-886-9787

Omark Industries, Div. of Blount, Inc., 2299 Snake River Ave., P.O. Box 856, Lewiston, ID 83501/800-627-3640, 208-746-2351

Orion Rifle Barrel Co., RR2, 137 Cobler Village, Kalispell, MT 59901/406-257-5649

Ox-Yoke Originals, Inc., 34 Main St., Milo, ME 04463/800-231-8313, 207-943-7351; FAX: 207-943-2416

Pacific Rifle Co., 1040-D Industrial Parkway, Newberg, OR 97132/503-538-7437

Pack Idaho, HC 61, Box 40, Salmon, ID 83467

Parker Reproductions, 124 River Rd., Middlesex, NJ 08846/908-469-0100; FAX: 908-469-9692

Pedersoli Davide & C., Via Artigiani 57, Gardone V.T., Brescia, ITALY 25063/030-8912402; FAX: 030-8911019 (U.S. importers—Beauchamp & Son, Inc.; Cabela's; Cape Outfitters; Dixie Gun Works; EMF Co., Inc.; Navy Arms Co.)

Penguin Industries, Inc., Airport Industrial Mall, Coatesville, PA 19320/610-384-6000; FAX: 610-857-5980

Petro-Explo, Inc., 7650 U.S. Hwy. 287, Suite 100, Arlington, TX 76017/817-478-8888

Pioneer Arms Co., 355 Lawrence Rd., Broomall, PA 19008/215-356-5203

Powder Horn Trading Post, 138 S. Kimball, Casper, WY 82601

Prairie River Arms, 1220 N. Sixth St., Princeton, IL 61356/815-875-1616, 800-445-1541; FAX: 815-875-1402

R.K. Lodges, Box 567, Hector, MN 55342

Radical Concepts, P.O. Box 1473, Lake Grove, OR 97035/503-538-7437

Rapine Bullet Mould Mfg. Co., 9503 Landis Lane, East Greenville, PA 18041/215-679-5413; FAX: 215-679-9795

Remington Arms Co., Inc., P.O. Box 700, 870 Remington Drive, Madison, NC 27025-0700/800-243-9700

Rusty Duck Premium Gun Care Products, 7785 Foundation Dr., Suite 6, Florence, KY 41042/606-342-5553; FAX: 606-342-5556

R.V.I. (See Fire'n Five)

S&B Industries, 11238 McKinley Rd., Montrose, MI 48457/810-639-5491

S&S Firearms, 74-11 Myrtle Ave., Glendale, NY 11385/718-497-1100; FAX: 718-497-1105

Selsi Co., Inc., P.O. Box 10, Midland Park, NJ 07432-0010/201-935-0388; FAX: 201-935-5851

Sharps Arms Co., Inc., C. (See Montana Armory, Inc.)

Shooter's Choice, 16770 Hilltop Park Place, Chagrin Falls, OH 44023/216-543-8808; FAX: 216-543-8811

Sile Distributors, Inc., 7 Centre Market Pl., New York, NY 10013/212-925-4111; FAX: 212-925-3149

Single Shot, Inc. (See Montana Armory, Inc.)

Sklany, Steve, 566 Birch Grove Dr., Kalispell, MT 59901/406-755-4257

Slings 'N Things, Inc., 8909 Bedford Circle, Suite 11, Omaha, NE 68134/402-571-6954; FAX: 402-571-7082

Smokey Valley Rifles (See Lutz Engraving, Ron E.)

South Bend Replicas, Inc., 61650 Oak Rd., South Bend, IN 46614/219-289-4500

Southern Bloomer Mfg. Co., P.O. Box 1621, Bristol, TN 37620/615-878-6660; FAX: 615-878-8761

Special Projects, Div. of Cold Steel, 2128-D Knoll Dr., Ventura, CA 93003

Starr Trading Co., Jedediah, P.O. Box 2007, Farmington Hills, MI 48333/810-683-4343; FAX: 810-683-3282

Stone Mountain Arms, 5988 Peachtree Corners E., Norcross, GA 30071/800-251-9412

Storey, Dale A. (See DGS, Inc.)

Sturm, Ruger & Co., Inc., Lacey Place, Southport, CT 06490/203-259-4537; FAX: 203-259-2167

Taylor's & Co., Inc., 304 Lenoir Dr., Winchester, VA 22603/540-722-2017; FAX: 540-722-2018

Tennessee Valley Mfg., P.O. Box 1175, Corinth, MS 38834/601-286-5014

Thompson Bullet Lube Co., P.O. Box 472343, Garland, TX 75047-2343/214-271-8063; FAX: 214-840-6743

Thompson/Center Arms, P.O. Box 5002, Rochester, NH 03866/603-332-2394; FAX: 603-332-5133

Thunder Mountain Arms, P.O. Box 593, Oak Harbor, WA 98277/206-679-4657; FAX: 206-675-1114

Tiger-Hunt, Box 379, Beaverdale, PA 15921/814-472-5161

Track of the Wolf, Inc., P.O. Box 6, Osseo, MN 55369-0006/612-424-2500; FAX: 612-424-9860

Traditions, Inc., P.O. Box 776, 1375 Boston Post Rd., Old Saybrook, CT 06475/860-388-4656; FAX: 860-388-4657

Trail Guns Armory, 1422 E. Main, League City, TX 77573

Treso, Inc., P.O. Box 4640, Pagosa Springs, CO 81157/303-731-2295

Uberti, Aldo, Casella Postale 43, I-25063 Gardone V.T., ITALY (U.S. importers—American Arms, Inc.; Cimarron Arms; Dixie Gun Works; EMF Co., Inc.; Forgett Jr., Valmore J.; Navy Arms Co; Taylor's & Co., Inc.; Uberti USA, Inc.)

UFA, Inc., 6927 E. Grandview Dr., Scottsdale, AZ 85254/800-616-2776

Uberti USA, Inc., P.O. Box 469, Lakeville, CT 06039/860-435-8068; FAX: 860-435-8146

Uncle Mike's (See Michaels of Oregon Co.)

Upper Missouri Trading Co., 304 Harold St., Crofton, NE 68730/402-388-4844

Venco Industries, Inc. (See Shooter's Choice)

Walters, John, 500 N. Avery Dr., Moore, OK 73160/405-799-0376

Warren Muzzleloading Co., Inc., Hwy. 21 North, P.O. Box 100, Ozone, AR 72854/501-292-3268

Wescombe, P.O. Box 488, Glencoe, CA 95232/209-293-7010

West Penn Lapidary, 278 Hazel Drive, Pittsburgh, PA 15228

White Owl Enterprises, 2583 Flag Rd., Abilene, KS 67410/913-263-2613; FAX: 913-263-2613

White Shooting Systems, Inc., 25 E. Hwy. 40, Box 330-12, Roosevelt, UT 84066/801-722-3085, 800-213-1315; FAX: 801-722-3054

Williams Gun Sight Co., 7389 Lapeer Rd., Box 329, Davison, MI 48423/810-653-2131, 800-530-9028; FAX: 810-658-2140

Woodworker's Supply, 1108 North Glenn Rd., Casper, WY 82601/307-237-5354

Young Country Arms, P.O. Box 3615, Simi Valley, CA 93093

Appendix B
Blackpowder Library

Advanced Muzzleloader's Guide, by Toby Bridges, Stoeger Publishing Co., So. Hackensack, NJ, 1985. 256 pp., illus. Paper covers. $14.95.

The complete guide to muzzle-loading rifles, pistols and shotguns—flintlock and percussion.

The African Adventures: A Return to the Silent Places, by Peter Hathaway Capstick, St. Martin's Press, New York, NY, 1992. 220 pp., illus. $22.95.

This book brings to life four turn-of-the-century adventurers and the savage frontier they braved. Frederick Selous, Constatine "Iodine" Ionides, Johnny Boyes and Jim Sutherland.

African Hunting and Adventure, by William Charles Baldwin, Books of Zimbabwe, Bulawayo, 1981. 451 pp., illus. $75.00.

Facsimile reprint of the scarce 1863 London edition. African hunting and adventure from Natal to the Zambezi.

After Big Game in Central Africa, by Edouard Foa, St. Martin's Press, New York, NY, 1989. 400 pp., illus. $16.95.

Reprint of the scarce 1899 edition. This sportsman covered 7200 miles, mostly on foot—from Zambezi delta on the east coast to the mouth of the Congo on the west.

America's Great Gunmakers, by Wayne van Zwoll, Stoeger Publishing Co., So. Hackensack, NJ, 1992. 288 pp., illus. Paper covers. $16.95.

This book traces in great detail the evolution of guns and ammunition in America and the men who formed the companies that produced them.

American Military Shoulder Arms: Volume 1, Colonial and Revolutionary War Arms, by George D. Moller, University Press of Colorado, Niwot, CO, 1993. 538 pp., illus. $75.00.

A superb in-depth study of the shoulder arms of the United States. This volume covers the pre-colonial period to the end of the American Revolution.

American Military Shoulder Arms: Volume 2, From the 1790's to the End of the Flintlock Period, by George D. Moller, University Press of Colorado, Niwot, CO, 1994. 496 pp., illus. $75.00.

Describes the rifles, muskets, carbines and other shoulder arms used by the armed forces of the United States from the 1790s to the end of the flintlock period in the 1840s.

Antique Guns, the Collector's Guide, 2nd Edition, edited by John Traister, Stoeger Publishing Co., S. Hackensack, NJ, 1994. 320 pp., illus. Paper covers. $19.95.

Covers a vast spectrum of pre-1900 firearms: those manufactured by U.S. gunmakers as well as Canadian, French, German, Belgian, Spanish and other foreign firms.

Arms Makers of Maryland, by Daniel D. Hartzler, George Shumway, York, PA, 1975. 200 pp., illus. $50.00.

A thorough study of the gunsmiths of Maryland who worked during the late 18th and early 19th centuries.

Australian Military Rifles & Bayonets, 200 Years of, by Ian Skennerton, I.D.S.A. Books, Piqua, OH, 1988. 124 pp., 198 illus. Paper covers. $19.50.

Australian Service Machineguns, 100 Years of, by Ian Skennerton, I.D.S.A. Books, Piqua, OH, 1989. 122 pp., 150 illus. Paper covers. $19.50.

Baker's Remarks on the Rifle, by Ezekiel Baker, Standard Publications, Inc.

Reproduction of his 1835 work, dealing his views on blackpowder shooting.

Black Powder Cartridge Rifle Magazine, edited by John D. Baird, Spider Hill Press, 1995.

Articles excerpted from *Black Powder Cartridge Rifles* magazine from 1980 to 1983.

Black Powder Guide, 2nd Edition, by George C. Nonte, Jr., Stoeger Publishing Co., So. Hackensack, NJ, 1991. 288 pp., illus. Paper covers. $14.95.

How-to instructions for selection, repair and maintenance of muzzleloaders, making your own bullets, restoring and refinishing, shooting techniques.

Blackpowder Hobby Gunsmithing, by Sam Fadala and Dale Storey, DBI Books, Inc., Northbrook, IL., 1994. 256 pp., illus. Paper covers. $18.95.

A how-to-guide for gunsmithing blackpowder pistols, rifles and shotguns from two men at the top of their respective fields.

Blackpowder Loading Manual, 3rd Edition, edited by Sam Fadala, DBI Books, Inc., Northbrook, IL, 1995. 368 pp., illus. Paper covers. $19.95.

Revised and expanded edition of this landmark blackpowder loading book. Covers hundreds of loads for most of the popular blackpowder rifles, handguns and shotguns.

The Blackpowder Notebook, by Sam Fadala, Wolfe Publishing Co., Prescott, AZ, 1994. 212 pp., illus. $22.50.

For anyone interested in shooting muzzleloaders, this book will help improve scores and obtain accuracy and reliability.

The Blunderbuss 1500-1900, by James D. Forman, Museum Restoration Service, Bloomfield, Ont., Canada, 1995. 40 pp., illus. Paper covers. $4.95.

The guns that had no peer as an anti-personal weapon throughout the flintlock era.

Boarders Away, Volume II: Firearms of the Age of Fighting Sail, by William Gilkerson, Andrew Mowbray, Inc. Publishers, Lincoln, RI, 1993. 331 pp., illus. $65.00.

Covers the pistols, muskets, combustibles and small cannon used aboard American and European fighting ships, 1626-1826.

Boss & Co. Builders of Best Guns Only, by Donald Dallas, Safari Press, Huntington Beach, CA, 1996. 336 pp., illus. $75.00

The famous London gunmaker Boss & Company is chronicled for its founding by Thomas Boss (1790 - 1857) to the present day.

Breech-Loading Carbines of the United States Civil War Period, by Brig. Gen. John Pitman, Armory Publications, Tacoma, WA, 1987. 94 pp., illus. $29.95.

The first in a series of previously unpublished manuscripts originated by the late Brigadier General John Putnam. Exploded drawings showing parts actual size follow each sectioned illustration.

The Breech-Loading Single-Shot Rifle, by Major Ned H. Roberts and Kenneth L. Waters, Wolfe Publishing Co., Prescott, AZ, 1995. 333 pp., illus. $28.50.

A comprehensive and complete history of the evolution of the Schutzen and single-shot rifle.

British Military Firearms 1650-1850, by Howard L. Blackmore, Stackpole Books, Mechanicsburg, PA, 1994. 224 pp., illus. $50.00.

The definitive work on British military firearms.

The British Shotgun, Volume 1, 1850-1870, by I.M. Crudington and D.J. Baker, Barrie & Jenkins, London, England, 1979. 256 pp., illus. $59.95.

An attempt to trace, as accurately as is now possible, the evolution of the shotgun during its formative years in Great Britain.

The British Shotgun, Volume 2, 1871-1890, by I.M. Crudginton and D.J. Baker, Ashford Press, Southampton, England, 1989. 250 pp., illus. $59.95.

The second volume of a definitive work on the evolution and manufacture of the British shotgun.

The British Soldier's Firearms from Smoothbore to Rifled Arms, 1850-1864, by Dr. C.H. Roads, R&R Books, Livonia, NY, 1994. 332 pp., illus. $49.00.

A reprint of the classic text covering the development of British military hand and shoulder firearms in the crucial years between 1850 and 1864.

Carbines of the Civil War, by John D. McAulay, Pioneer Press, Union City, TN, 1981. 123 pp., illus. Paper covers. $7.95.

A guide for the student and collector of the colorful arms used by the Federal cavalry.

Cartridges of the World, 8th Edition, by Frank Barnes, edited by M. L. McPherson, DBI Books, Inc., Northbrook, IL, 1996. 480 pp., illus. Paper covers. $24.95.

Completely revised edition of the general purpose reference work for which collectors, police, scientists and laymen reach first for answers to cartridge identification questions. Available October, 1996.

Civil War Breech Loading Rifles, by John D. McAulay, Andrew Mowbray, Inc., Lincoln, RI, 1991. 144 pp., illus. Paper covers. $15.00.

All the major breech-loading rifles of the Civil War and most, if not all, of the obscure types are detailed, illustrated and set in their historical context.

Civil War Carbines Volume 2: The Early Years, by John D. McAulay, Andrew Mowbray, Inc., Lincoln, RI, 1991. 144 pp., illus. Paper covers. $15.00.

Covers the carbines made during the exciting years leading up to the outbreak of war and used by the North and South in the conflict.

Civil War Pistols, by John D. McAulay, Andrew Mowbray Inc., Lincoln, RI, 1992. 166 pp., illus. $38.50.

A survey of the handguns used during the American Civil War.

Collector's Illustrated Encyclopedia of the American Revolution, by George C. Neumann and Frank J. Kravic, Rebel Publishing Co., Inc., Texarkana, TX, 1989. 286 pp., illus. $29.95.

A showcase of more than 2,300 artifacts made, worn, and used by those who fought in the War for Independence.

Colonial Frontier Guns, by T.M. Hamilton, Pioneer Press, Union City, TN, 1988. 176 pp., illus. Paper covers. $13.95.

A complete study of early flint muskets of this country.

The Colt Armory, by Ellsworth Grant, Man-at-Arms Bookshelf, Lincoln, RI, 1996. 232 pp., illus. $35.00.

A history of Colt's Manufacturing Company.

Colt Heritage, by R.L. Wilson, Simon & Schuster, 1979. 358 pp., illus. $75.00.

The official history of Colt firearms 1836 to the present.

Colt Peacemaker British Model, by Keith Cochran, Cochran Publishing Co., Rapid City, SD, 1989. 160 pp., illus. $35.00.

Covers those revolvers Colt squeezed in while completing a large order of revolvers for the U.S. Cavalry in early 1874, to those magnificent cased target revolvers used in the pistol competitions at Bisley Commons in the 1890s.

Colt Peacemaker Encyclopedia, by Keith Cochran, Keith Cochran, Rapid City, SD, 1986. 434 pp., illus. $65.00.

A must book for the Peacemaker collector.

Colt Peacemaker Encyclopedia, Volume 2, by Keith Cochran, Cochran Publishing Co., SD, 1992. 416 pp., illus. $60.00.

Included in this volume are extensive notes on engraved, inscribed, historical and noted revolvers, as well as those revolvers used by outlaws, lawmen, movie and television stars.

Colt Percussion Accoutrements 1834-1873, by Robin Rapley, Robin Rapley, Newport Beach, CA, 1994. 432 pp., illus. Paper covers. $39.95.

The complete collector's guide to the identification of Colt percussion accoutrements; including Colt conversions and their values.

Colt Revolvers and the U.S. Navy 1865-1889, by C. Kenneth Moore, Dorrance and Co., Bryn Mawr, PA, 1987. 140 pp., illus. $29.95.

The Navy's use of all Colt handguns and other revolvers during this era of change.

Colt Rifles and Muskets from 1847-1870, by Herbert Houze, Krause Publications, Iola, WI, 1996. 192 pp., illus. $34.95.

Discover previously unknown Colt models along with an extensive list of production figures for all models.

Colt's Dates of Manufacture 1837-1978, by R.L. Wilson, published by Maurie Albert, Coburg, Australia; N.A. distributor I.D.S.A. Books, Hamilton, OH, 1983. 61 pp. $10.00.

An invaluable pocket guide to the dates of manufacture of Colt firearms up to 1978.

Colt's 100th Anniversary Firearms Manual 1836-1936: A Century of Achievement, Wolfe Publishing Co., Prescott, AZ, 1992. 100 pp., illus. Paper covers. $12.95.

Originally published by the Colt Patent Firearms Co., this booklet covers the history, manufacturing procedures and the guns of the first 100 years of the genius of Samuel Colt.

The Colt Whitneyville-Walker Pistol, by Lt. Col. Robert D. Whittington, Brownlee Books, Hooks, TX, 1984. 96 pp., illus. Limited edition. $20.00.

A study of the pistol and associated characters 1846-1851.

The Complete Blackpowder Handbook, 3rd Edition, by Sam Fadala, DBI Books, Inc., Northbrook, IL, 1996. 416 pp., illus. Paper covers. $21.95.

Expanded and refreshed edition of the definitive book on the subject of blackpowder. (Available September, 1996)

The Complete Guide to Game Care and Cookery, 3rd Edition, by Sam Fadala, DBI Books, Inc., Northbrook, IL, 1994. 320 pp., illus. Paper covers. $18.95.

Over 500 photos illustrating the care of wild game in the field and at home with a separate recipe section providing over 400 tested recipes.

Confederate Revolvers, by William A. Gary, Taylor Publishing Co., Dallas, TX, 1987. 174 pp., illus. $49.95.

Comprehensive work on the rarest of Confederate weapons.

Cowboy Action Shooting, by Charly Gullett, Wolfe Publishing Co., Prescott, AZ, 1995. 400 pp., illus. Paper covers. $24.50.

The fast growing of the shooting sports is comprehensively covered in this text—the guns, loads, tactics and the fun and flavor of this Old West era competition.

Development of the Henry Cartridge and Self-Contained Cartridges for the Toggle-Link Winchesters, by R. Bruce McDowell, A.M.B., Metuchen, NJ, 1984. 69 pp., illus. Paper covers. $10.00.

From powder and ball to the self-contained metallic cartridge.

Early American Waterfowling, 1700's-1930, by Stephen Miller, Winchester Press, Piscataway, NJ, 1986. 256 pp., illus. $27.95.

Two centuries of literature and art devoted to the nation's favorite hunting sport—water-fowling.

Early Indian Trade Guns: 1625-1775, by T.M. Hamilton, Museum of the Great Plains, Lawton, OK, 1968. 34 pp., illus. Paper covers. $12.95.

Detailed descriptions of subject arms, compiled from early records and from the study of remnants found in Indian country.

East Africa and its Big Game, by Captain Sir John C. Willowghby, Wolfe Publishing Co., Prescott, AZ, 1990. 312 pp., illus. $52.00.

A deluxe limited edition reprint of the very scarce 1889 edition of a narrative of a sporting trip from Zanzibar to the borders of the Masai.

English Pistols: The Armories of H.M. Tower of London Collection, by Howard L. Blackmore, Arms and Armour Press, London, England, 1985. 64 pp., illus. Soft covers. $14.95.

All the pistols described and pictured are from this famed collection.

European Firearms in Swedish Castles, by Kaa Wennberg, Bohuslaningens Boktryckeri AB, Uddevalla, Sweden, 1986. 156 pp., illus. $50.00.

The famous collection of Count Keller, the Ettersburg Castle collection, and others. English text.

Fifteen Years in the Hawken Lode, by John D. Baird, The Gun Room Press, Highland Park, NJ, 1976. 120 pp., illus. $24.95.

A collection of thoughts and observations gained from many years of intensive study of the guns from the shop of the Hawken brothers.

'51 Colt Navies, by Nathan L. Swayze, The Gun Room Press, Highland Park, NJ, 1993. 243 pp., illus. $59.95.

The Model 1851 Colt Navy, its variations and markings.

Flayderman's Guide to Antique American Firearms...and Their Values, 6th Edition, by Norm Flayderman, DBI Books, Inc., Northbrook, IL, 1994. 624 pp., illus. Paper covers. $29.95.

Updated edition of this bible of the antique gun field.

Frank and George Freund and the Sharps Rifle, by Gerald O. Kelver, Gerald O. Kelver, Brighton, CO, 1986. 60 pp., illus. Paper covers. $12.00.

Pioneer gunmakers of Wyoming Territory and Colorado.

French Military Weapons, 1717-1938, Major James E. Hicks, N. Flayderman & Co., Publishers, New Milford, CT, 1973. 281 pp., illus. $35.00.

Firearms, swords, bayonets, ammunition, artillery, ordnance equipment of the French army.

The Frontier Rifleman, by H.B. LaCrosse Jr., Pioneer Press, Union City, TN, 1989. 183 pp., illus. Soft covers. $14.95.

The Frontier rifleman's clothing and equipment during the era of the American Revolution, 1760-1800.

Game Guns & Rifles: Percussion to Hammerless Ejector in Britain, by Richard Akehurst, Trafalgar Square, N. Pomfret, VT, 1993. 192 pp., illus. $34.95.

Long considered a classic this important reprint covers the period of British gunmaking between 1830-1900.

George Schreyer, Sr. and Jr., Gunmakers of Hanover, Pennsylvania, by George Shumway, George Shumway Publishers, York, PA, 1990. 160pp., illus. $50.00.

This monograph is a detailed photographic study of almost all known surviving long rifles and smoothbore guns made by highly regarded gunsmiths George Schreyer, Sr. and Jr.

The Golden Age of Remington, by Robert W.D. Ball, Krause publications, Iola, WI, 1995. 208 pp., illus. $29.95.

For Remington collectors or firearms historians, this book provides a pictorial history of Remington through World War I. Includes value guide.

Grand Old Shotguns, by Don Zutz, Shotgun Sports Magazine, Auburn, CA, 1995. 136 pp., illus. Paper covers. $19.95

A study of the great smoothbores, their history and how and why they were discontinued. Find out the most sought-after and which were the best shooters.

Great British Gunmakers: The Mantons 1782-1878, by D.H.L. Back, Historical Firearms, Norwich, England, 1994. 218 pp., illus. Limited edition of 500 copies. $175.00.

Contains detailed descriptions of all the firearms made by members of this famous family.

Great Irish Gunmakers: Messrs. Rigby 1760-1869, by D.H.L. Back, Historical Firearms, Norwich, England, 1993. 196 pp., illus. $150.00.

The history of this famous firm of Irish gunmakers illustrated with a wide selection of Rigby arms.

Great Shooters of the World, by Sam Fadala, Stoeger Publishing Co., So. Hackensack, NJ, 1991. 288 pp., illus. Paper covers. $18.95.

This book offers gun enthusiasts an overview of the men and women who have forged the history of firearms over the past 150 years.

A Guide to the Maynard Breechloader, by George J. Layman, George J. Layman, Ayer, MA, 1993. 125 pp., illus. Paper covers. $17.95.

The first book dedicated entirely to the Maynard family of breech-loading firearms. Coverage of the arms is given from the 1850s through the 1880s.

Gun and Camera in Southern Africa, by H. Anderson Bryden, Wolfe Publishing Co., Prescott, AZ, 1989. 201 pp., illus. $37.00.

A limited edition reprint. The year was 1893 and author Bryden wandered for a year in Bechuanaland and the Kalahari Desert hunting the white rhino, lechwe, eland, and more.

The Gun and Its Development, by W.W. Greener, Sampson Low, 1910.

Greener's classic on firearms through the ages.

Gun Collecting, by Geoffrey Boothroyd, Sportsman's Press, London, 1989. 208 pp., illus. $29.95.

The most comprehensive list of 19th century British gunmakers and gunsmiths ever published.

Gun Collector's Digest, 5th Edition, edited by Joseph J. Schroeder, DBI Books, Inc., Northbrook, IL, 1989. 224 pp., illus. Paper covers. $17.95.

The latest edition of this sought-after series.

Gun Digest, 1997, 51st Edition, edited by Ken Warner, DBI Books, Northbrook, IL, 1996. 544 pp., illus. Paper covers. $23.95.

All-new edition of the world's biggest selling gun book.

Gun Digest Treasury, 7th Edition, edited by Harold A. Murtz, DBI Books, Inc., Northbrook, IL, 1994, 320 pp., illus. Paper covers. $17.95.

A collection of some of the most interesting articles which have appeared in Gun Digest over its first 45 years.

Gun Tools, Their History and Identification by James B. Shaffer, Lee A. Rutledge and R. Stephen Dorsey, Collector's Library, Eugene, OR, 1992. 375 pp., illus. $32.00.

Written history of foreign and domestic gun tools from the flintlock period to WWII.

Gunmakers of London 1350-1850, by Howard L. Blackmore, George Shumway Publisher, York, PA, 1986. 222 pp., illus. $35.00.

A listing of all the known workmen of gun making in the first 500 years, plus a history of the guilds, cutlers, armourers, founders, blacksmiths, etc. 260 gunmarks are illustrated.

Guns and Gunmaking Tools of Southern Appalachia, by John Rice Irwin, Schiffer Publishing Ltd., 1983. 118 pp., illus. Paper covers. $9.95.

The story of the Kentucky rifle.

Guns of the Wild West, by George Markham, Sterling Publishing Co., New York, NY, 1993. 160 pp., illus. Paper covers. $19.95.

Firearms of the American Frontier, 1849-1917.

Gunsmiths of Illinois, by Curtis L. Johnson, George Shumway Publishers, York, PA, 1995. 160 pp., illus. $50.00.

Genealogical information is provided for nearly one thousand gunsmiths. Contains hundreds of illustrations of rifles and other guns, of handmade origin, from Illinois.

The Gunsmiths of Manhattan, 1625-1900: A Checklist of Tradesmen, by Michael H. Lewis, Museum Restoration Service, Bloomfield, Ont., Canada, 1991. 40 pp., illus. Paper covers. $4.95.

This listing of more than 700 men in the arms trade in New York City prior to about the end of the 19th century will provide a guide for identification and further research.

The Handgun, by Geoffrey Boothroyd, David and Charles, North Pomfret, VT, 1989. 566 pp., illus. $60.00.

Every chapter deals with an important period in handgun history from the 14th century to the present.

The Hawken Rifle: Its Place in History, by Charles E. Hanson, Jr., The Fur Press, Chadron, NE, 1979. 104 pp., illus. Paper covers. $15.00.

A definitive work on this famous rifle.

Hawken Rifles, The Mountain Man's Choice, by John D. Baird, The Gun Room Press, Highland Park, NJ, 1976. 95 pp., illus. $29.95.

Covers the rifles developed for the Western fur trade. Numerous specimens are described and shown in photographs.

Historic Pistols: The American Martial Flintlock 1760-1845, by Samuel E. Smith and Edwin W. Bitter, The Gun Room Press, Highland Park, NJ, 1986. 353 pp., illus. $45.00.

Covers over 70 makers and 163 models of American martial arms.

Historical Hartford Hardware, by William W. Dalrymple, Colt Collector Press, Rapid City, SD, 1976. 42 pp., illus. Paper covers. $10.00.

Historically associated Colt revolvers.

The History and Development of Small Arms Ammunition, Volume 1, by George A. Hoyem, Armory Publications, Oceanside, CA, 1991. 230 pp., illus. $60.00.

Military musket, rifle, carbine and primitive machine gun cartridges of the 18th and 19th centuries, together with the firearms that chambered them.

The History and Development of Small Arms Ammunition, Volume 2, by George A. Hoyem, Armory Publications, Oceanside, CA, 1991. 303 pp., illus. $60.00.

Covers the blackpowder military centerfire rifle, carbine, machine gun and volley gun ammunition used in 28 nations and dominions, together with the firearms that chambered them.

The History of Winchester Firearms 1866-1992, sixth edition, updated, expanded, and revised by Thomas Henshaw, New Win Publishing, Clinton, NJ, 1993. 280 pp., illus. $24.95.

This classic is the standard reference for all collectors and others seeking the facts about any Winchester firearm, old or new.

History of Winchester Repeating Arms Company, by Herbert G. Houze, Krause Publications, Iola, WI, 1994. 800 pp., illus. $50.00.

The complete Winchester history from 1856-1981.

Hodgdon Data Manual No. 26, Hodgdon Powder Co., Shawnee Mission, KS, 1993. 797 pp. $22.95.

Includes Hercules, Winchester and Dupont powders; data on cartridge cases, loads; silhouette; shotshell; pyrodex and blackpowder; conversion factors; weight equivalents, etc.

Home Gunsmithing the Colt Single Action Revolvers, by Loren W. Smith, Ray Riling Arms Books, Co., Phila., PA, 1995. 119 pp., illus. $24.95.

Affords the Colt Single Action owner detailed, pertinent information on the operating and servicing of this famous and historic handgun.

How-To's for the Black Powder Cartridge Rifle Shooter, by Paul A. Matthews, Wolfe Publishing Co., Prescott, AZ, 1995. 45 pp. Paper covers. $22.50.

Covers lube recipes, good bore cleaners and over-powder wads. Tips include compressing powder charges, combating wind resistance, improving ignition and much more.

Hunting in Many Lands, by Theodore Roosevelt and George Bird Grinnel, The Boone and Crockett Club, Dumfries, VA, 1987. 447 pp., illus. $40.00.

Limited edition reprint of this 1895 classic work on hunting in Africa, India, Mongolia, etc.

Illustrations of United States Military Arms 1776-1903 and Their Inspector's Marks, compiled by Turner Kirkland, Pioneer Press, Union City, TN, 1988. 37 pp., illus. Paper covers. $4.95.

Reprinted from the 1949 Bannerman catalog. Valuable information for both the advanced and beginning collector.

Indian Hunts and Indian Hunters of the Old West, by Dr. Frank C. Hibben, Safari Press, Long Beach, CA, 1989. 228 pp., illus. $24.95.

Tales of some of the most famous American Indian hunters of the Old West as told to the author by an old Navajo hunter.

Indian War Cartridge Pouches, Boxes and Carbine Boots, by R. Stephen Dorsey, Collector's Library, Eugene, OR, 1993. 156 pp., illus. Paper Covers. $25.00.

The key reference work to the cartridge pouches, boxes, carbine sockets and boots of the Indian War period 1865-1890.

An Introduction to the Civil War Small Arms, by Earl J. Coates and Dean S. Thomas, Thomas Publishing Co., Gettysburg, PA, 1990. 96 pp., illus. Paper covers. $10.00.

The small arms carried by the individual soldier during the Civil War.

Jaeger Rifles, by George Shumway, George Shumway Publisher, York, PA, 1994. 108 pp., illus. Paper covers. $25.00.

Thirty-six articles previously published in *Muzzle Blasts* are reproduced here. They deal with late-17th, and 18th century rifles from Vienna, Carlsbad, Bavaria, Saxony, Brandenburg, Suhl, North-Central Germany, and the Rhine Valley.

The Kentucky Rifle, by Captain John G.W. Dillin, George Shumway Publisher, York, PA, 1993. 221 pp., illus. $50.00.

This well-known book was the first attempt to tell the story of the American longrifle. This edition retains the original text and illustrations with supplemental footnotes provided by Dr. George Shumway.

Loading the Black Powder Rifle Cartridge, by Paul A Matthews, Wolfe Publishing Co., Prescott, AZ, 1993. 121 pp., illus. Paper covers. $22.50.

Author Matthews brings the blackpowder cartridge shooter valuable information on the basics, including cartridge care, lubes and moulds, powder charges and developing and testing loads in his usual authoritative style.

Loading the Peacemaker—Colt's Model P, by Dave Scovill, Wolfe Publishing Co., Prescott, AZ, 1995. $24.95.

A comprehensive work about the most famous revolver ever made, including the most extensive load data ever published.

Longrifles of North Carolina, by John Bivens, George Shumway Publisher, York, PA, 1988. 256 pp., illus. $50.00.

Covers art and evolution of the rifle, immigration and trade movements. Committee of Safety gunsmiths, characteristics of the North Carolina rifle.

Longrifles of Pennsylvania, Volume 1, Jefferson, Clarion & Elk Counties, by Russel H. Harringer, George Shumway Publisher, York, PA, 1984. 200 pp., illus. $50.00.

First in series that will treat in great detail the longrifles and gunsmiths of Pennsylvania.

Lyman Cast Bullet Handbook, 3rd Edition, edited by C. Kenneth Ramage, Lyman Publications, Middlefield, CT, 1980. 416 pp., illus. Paper covers. $19.95.

Information on more than 5000 tested cast bullet loads and 19 pages of trajectory and wind drift tables for cast bullets.

Lyman Black Powder Handbook, ed. by C. Kenneth Ramage, Lyman Products for Shooters, Middlefield, CT, 1975. 239 pp., illus. Paper covers. $14.95.

Comprehensive load information for the modern blackpowder shooter.

The Manufacture of Gunflints, by Sydney B.J. Skertchly, facsimile reprint with new introduction by Seymour de Lotbiniere, Museum Restoration Service, Ontario, Canada, 1984. 90 pp., illus. $24.50.

Limited edition reprinting of the very scarce London edition of 1879.

Massachusetts Military Shoulder Arms 1784-1877, by George D. Moller, Andrew Mowbray Publisher, Lincoln, RI, 1989. 250 pp., illus. $24.00.

A scholarly and heavily researched study of the military shoulder arms used by Massachusetts during the 90-year period following the Revolutionary War.

Military Bolt Action Rifles, 1841-1918, by Donald B. Webster, Museum Restoration Service, Alexander Bay, NY, 1993. 150 pp., illus. $34.50.

A photographic survey of the principal rifles and carbines of the European and Asiatic powers of the last half of the 19th century and the first years of the 20th century.

Military Handguns of France 1858-1958, by Eugene Medlin and Jean Huon, Excalibur Publications, Latham, NY, 1994. 124 pp., illus. Paper covers. $24.95.

The first book written in English that provides students of arms with a thorough history of French military handguns.

The More Complete Cannoneer, by M.C. Switlik, Museum & Collectors Specialties Co., Monroe, MI, 1990. 199 pp., illus. $19.95.

Compiled agreeably to the regulations for the U.S. War Department, 1861, and containing current observations on the use of antique cannon.

More Single Shot Rifles, by James C. Grant, The Gun Room Press, Highland Park, NJ, 1976. 324 pp., illus. $29.95.

Details the guns made by Frank Wesson, Milt Farrow, Holden, Borchardt, Stevens, Remington, Winchester, Ballard and Peabody-Martini.

Mortimer, the Gunmakers, 1753-1923, by H. Lee Munson, Andrew Mowbray Inc., Lincoln, RI, 1992. 320 pp., illus. $65.00.

Seen through a single, dominant, English gunmaking dynasty this fascinating study provides a window into the classical era of firearms artistry.

The Muzzle-Loading Cap Lock Rifle, by Ned H. Roberts, reprinted by Wolfe Publishing Co., Prescott, AZ, 1991. 432 pp., illus. $30.00.

Originally published in 1940, this fascinating study of the muzzle-loading cap lock rifle covers rifles on the frontier to hunting rifles, including the famous Hawken.

The Muzzle-Loading Rifle...Then and Now, by Walter M. Cline, National Muzzle Loading Rifle Association, Friendship, IN, 1991. 161 pp., illus. $32.00.

This extensive compilation of the muzzleloading rifle exhibits accumulative preserved data concerning the development of the "hallowed old arms of the Southern highlands."

Naval Percussion Locks and Primers, by Lt. J. A. Dahlgren, Museum Restoration Service, Bloomfield, Canada, 1996. 140 pp., illus. $35.00.

First published as an Ordnance Memoranda in 1853, this is the finest existing study of percussion locks and primers origin and development.

Ned H. Roberts and the Schuetzen Rifle, edited by Gerald O. Kelver, Brighton, CO, 1982. 99 pp., illus. $15.00.

A compilation of the writings of Major Ned H. Roberts which appeared in various gun magazines.

The Paper Jacket, by Paul Matthews, Wolfe Publishing Co., Prescott, AZ, 1991. Paper covers. $13.50.

Up-to-date and accurate information about paper-patched bullets.

The Paper Patched Bullet, by Randolph S. Wright, C. Sharps Arms Publishers, 1985.

A 19-page booklet concerning paper patches, bullets and swaging.

Patents for Inventions, Class 119 (Small Arms), 1855-1930. British Patent Office, Armory Publications, Oceanside, CA, 1993. 7 volume set. $350.00.

Contains 7980 abridged patent descriptions and their sectioned line drawings, plus a 37-page alphabetical index of the patentees.

Paterson Colt Pistol Variations, by R.L. Wilson and R. Phillips, Jackson Arms Co., Dallas, TX, 1979. 250 pp., illus. $35.00.

A book about the different models and barrel lengths in the Paterson Colt story.

Pennsylvania Longrifles of Note, by George Shumway, George Shumway, Publisher, York, PA, 1977. 63 pp., illus. Paper covers. $15.00.

Illustrates and describes rifles from a number of Pennsylvania rifle-making schools.

The Pennsylvania Rifle, by Samuel E. Dyke, Sutter House, Lititz, PA, 1975. 61 pp., illus. Paper covers. $5.00.

History and development, from the hunting rifle of the Germans who settled the area. Contains a full listing of all known Lancaster, PA, gunsmiths from 1729 through 1815.

The Pennsylvania-Kentucky Rifle, by Henry J. Kaufman, Stackpole Books, 1950.

The development and use of the rifle in Pennsylvania in early America.

The Pitman Notes on U.S. Martial Small Arms and Ammunition, 1776-1933, Volume 2, Revolvers and Automatic Pistols, by Brig. Gen. John Pitman, Thomas Publications, Gettysburg, PA, 1990. 192 pp., illus. $29.95.

A most important primary source of information on United States military small arms and ammunition.

The Plains Rifle, by Charles Hanson, Gun Room Press, Highland Park, NJ, 1989. 169 pp., illus. $29.95.

All rifles that were made with the plainsman in mind, including pistols.

The Powder Flask Book, by Ray Riling, R&R Books, Livonia, NY, 1993. 514 pp., illus. $70.00.

The complete book on flasks of the 19th century. Exactly scaled pictures of 1,600 flasks are illustrated.

Purdey's, the Guns and the Family, by Richard Beaumont, David and Charles, Pomfert, VT, 1984. 248 pp., illus. $39.95.

Records the history of the Purdey family from 1814 to today, how the guns were and are built and daily functioning of the factory.

The Rare and Valuable Antique Arms, by James E. Serven, Pioneer Press, Union City, TN, 1976. 106 pp., illus. Paper covers. $4.95.

A guide to the collector in deciding which direction his collecting should go, investment value, historic interest, mechanical ingenuity, high art or personal preference.

The Recollections of an Elephant Hunter 1864-1875, by William Finaughty, Books of Zimbabwe, Bulawayo, Zimbabwe, 1980. 244 pp., illus. $85.00.

Reprint of the scarce 1916 privately published edition. The early game hunting exploits of William Finaughty in Matabeleland and Nashonaland.

Recreating the American Longrifle, by William Buchele, et al., George Shumway, Publisher, York, PA, 1983. 175 pp., illus. $30.00.

Includes full-scale plans for building a Kentucky rifle.

Revolvers of the British Services 1854-1954, by W.H.J. Chamberlain and A.W.F. Taylerson, Museum Restoration Service, Ottawa, Canada, 1989. 80 pp., illus. $27.50.

Covers the types issued among many of the United Kingdom's naval, land or air services.

The Revolving Rifles, by Edsall James, Pioneer Press, Union City, TN, 1975. 23 pp., illus. Paper covers. $2.50.

Valuable information on revolving cylinder rifles, from the earliest matchlock forms to the latest models of Colt and Remington.

Rhode Island Arms Makers & Gunsmiths, by William O. Archibald, Andrew Mowbray, Inc., Lincoln, RI, 1990. 108 pp., illus. $16.50.

A serious and informative study of an important area of American arms making.

The Rifle and the Hound in Ceylon, by S.W. Baker, Arno Press, 1967.

Reprint of 19th-century work, a classic by an ivory hunter.

Sam Colt's Own Record 1847, by John Parsons, Wolfe Publishing Co., Prescott, AZ, 1992. 167 pp., illus. $24.50.

Chronologically presented, the correspondence published here completes the account of the manufacture, in 1847, of the Walker Model Colt revolver.

Schuetzen Rifles, History and Loading, by Gerald O. Kelver, Gerald O. Kelver, Publisher, Brighton, CO, 1972. Illus. $15.00.

Reference work on these rifles, their bullets, loading, telescopic sights, accuracy, etc. A limited, numbered ed.

Scottish Firearms, by Claude Blair and Robert Woosnam-Savage, Museum Restoration Service, Bloomfield, Ont., Canada, 1995. 52 pp., illus. Paper covers. $4.95.

This revision of the first book devoted entirely to Scottish firearms is supplemented by a register of surviving Scottish long guns.

Sharps Firearms, by Frank Seller, Frank M. Seller, Denver, CO, 1982. 358 pp., illus. $50.00.

Traces the development of Sharps firearms with full range of guns made including all martial variations.

Sharps Rifle—The Gun That Shaped American Destiny, by Martin Rywell, Pioneer Press, 1979.

The history, use and functioning of the Sharps rifle.

Shooting the Blackpowder Cartridge Rifle, by Paul A. Matthews, Wolfe Publishing Co., Prescott, AZ, 1994. 129 pp., illus. Paper covers. $22.50.

A general discourse on shooting the blackpowder cartridge rifle and the procedure required to make a particular rifle perform.

The Shotgun: History and Development, by Geoffrey Boothroyd, Safari Press, Huntington Beach, CA, 1995. 240 pp., illus. $35.00.

The first volume in a series that traces the development of the British shotgun from the 17th century onward.

Sidelocks & Boxlocks, by Geoffrey Boothroyd, Sand Lake Press, Amity, OR, 1991. 271 pp., illus. $35.00.

The story of the classic British shotgun.

Simeon North: First Official Pistol Maker of the United States, by S. North and R. North, The Gun Room Press, Highland Park, NJ, 1972. 207 pp., illus. $15.95.

Reprint of the rare first edition.

Sixgun Cartridges and Loads, by Elmer Keith, The Gun Room Press, Highland Park, NJ, 1986. 151 pp., illus. $24.95.

A manual covering the selection, uses and loading of the most suitable and popular revolver cartridges. Originally published in 1936. Reprint.

Spencer Firearms, by Roy Marcot, R&R Books, Livonia, NY, 1995. 237 pp., illus. $60.00.

The definitive work on one of the most famous Civil War firearms.

The Sporting Rifle and Its Projectiles, by James Forsyth, Buckskin Press, 1978.

This reprint of an 1863 text is a study in early ballistic data.

Springfield Shoulder Arms 1795-1865, by Claud E. Fuller, S. & S. Firearms, Glendale, NY, 1986. 76 pp., illus. Paper covers. $17.95.

Exact reprint of the scarce 1930 edition of one of the most definitive works on Springfield flintlock and percussion muskets ever published.

Standard Catalog of Firearms, 6th Edition, compiled by Ned Schwing and Herbert Houze, Krause Publications, Iola, WI, 1996. 1,116 pp., illus. Paper covers. $29.95.

1996 pricing guide in six grades with more than 2,300 photos and over 1,100 manufacturers.

Steel Canvas: The Art of American Arms, by R. L. Wilson, Random House, NY, 1995, 384 pp., illus. $65.00.

Presented here for the first time is the breathtaking panorama of America's extraordinary engravers and embellishers of arms, from the 1700s to modern times.

The Sumptuous Flaske, by Herbert G. Houze, Andrew Mowbray, Inc., Lincoln, RI, 1989. 158 pp., illus. Soft covers. $35.00.

Catalog of a recent show at the Buffalo Bill Historical Center bringing together some of the finest European and American powder flasks of the 16th to 19th centuries.

Tales of the Big Game Hunters, selected and introduced by Kenneth Kemp, The Sportsman's Press, London, 1986. 209 pp., illus. $15.00.

Writings by some of the best known hunters and explorers, among them: Frederick Courteney Selous, R.G. Gordon Cumming, Sir Samuel Baker, and elephant hunters Neumann and Sutherland.

United States Martial Flintlocks, by Robert M. Reilly, Andrew Mowbray, Inc., Lincoln, RI, 1986. 263 pp., illus. $39.50.

A comprehensive illustrated history of the flintlock in America from the Revolution to the demise of the system.

U.S. Breech-Loading Rifles and Carbines, Cal. 45, by Gen. John Pitman, Thomas Publications, Gettysburg, PA, 1992. 192 pp., illus. $29.95.

The third volume in the Pitman Notes on U.S. Martial Small Arms and Ammunition, 1776-1933. This book centers on the "Trapdoor Springfield" models.

U.S. Military Arms Dates of Manufacture from 1795, by George Madis, David Madis, Dallas, TX, 1989. 64 pp. Soft covers. $5.00.

Lists all U.S. military arms of collector interest alphabetically, covering about 250 models.

U.S. Military Small Arms 1816-1865, by Robert M. Reilly, The Gun Room Press, Highland Park, NJ, 1983. 270 pp., illus. $39.95.

Covers every known type of primary and secondary martial firearms used by Federal forces.

Weapons of the Highland Regiments 1740-1780, by Anthony D. Darling, Museum Restoration Service, Bloomfield, Canada, 1996. 28 pp., illus. Paper covers. $5.95.

This study deals with the formation and arming of the famous Highland regiments.

The Whitney Firearms, by Claud Fuller, Standard Publications, Huntington, WV, 1946, 334 pp., many plates and drawings, $50.00.

An authoritative history of all Whitney arms and their maker. Highly recommended. An exclusive with Ray Riling Arms Books Co.

Winchester: An American Legend, by R.L. Wilson, Random House, New York, NY, 1991. 403 pp., illus. $65.00.

The official history of Winchester firearms from 1849 to the present.

The Winchester Book, by George Madis, David Madis Gun Book Distributor, Dallas, TX, 1986. 650 pp., illus. $47.00.

A new, revised 25th anniversary edition of this classic book on Winchester firearms. Complete serial ranges have been added.

Winchester Dates of Manufacture 1849-1984, by George Madis, Art & Reference House, Brownsboro, TX, 1984. 59 pp. $5.95.

A most useful work, compiled from records of the Winchester factory.

The Winchester Era, by David Madis, Art & Reference House, Brownsville, TX, 1984. 100 pp., illus. $14.95.

Story of the Winchester company, management, employees, etc.

The Winchester Handbook, by George Madis, Art & Reference House, Lancaster, TX, 1982. 287 pp., illus. $19.95.

The complete line of Winchester guns, with dates of manufacture, serial numbers, etc.

Winchester Lever Action Repeating Firearms, Vol. 1, The Models of 1866, 1873 and 1876, by Arthur Pirkie, North Cape Publications, Tustin, CA, 1995. 112 pp., illus. Paper covers. $19.95.

Complete, part-by-part description, including dimensions, finishes, markings and variations throughout the production run of these fine, collectible guns.

The Winchester Single-Shot, by John Cambell, Andrew Mowbray, Inc., Lincoln RI, 1995. 272 pp., illus. $55.00.

Covers every important aspect of this highly-collectible firearm.

Appendix C
Blackpowder Periodical Publications

American Hunter (M)
National Rifle Assn., 11250 Waples Mill Rd., Fairfax, VA 22030 (Same address for both.) Publications Div. $35.00 yr. Wide scope of hunting articles.

American Rifleman (M)
National Rifle Assn., 11250 Waples Mill Rd., Fairfax, VA 22030 (Same address for both). Publications Div. $35.00 yr. Firearms articles of all kinds.

The Backwoodsman Magazine
P.O. Box 627, Westcliffe, CO 81252. $16.00 for 6 issues per yr.; $30.00 for 2 yrs.; sample copy $2.75. Subjects include muzzle-loading, woodslore, primitive survival, trapping, homesteading, blackpowder cartridge guns, 19th century how-to.

Black Powder Cartridge News (Q)
SPG, Inc., P.O. Box 761, Livingston, MT 59047. $17 yr. (4 issues). For the blackpowder cartridge enthusiast.

Black Powder Times
P.O. Box 234, Lake Stevens, WA 98258. $20.00 yr.; add $5 per year for Canada, $10 per year other foreign. Tabloid newspaper for blackpowder activities; test reports.

Blackpowder Hunting (Q)
P.O. Box 1900, Glenrock, WY 82637. $20 yr. (4 issues). For those who enjoy blackpowder hunting.

The Cast Bullet*(M)
Official journal of The Cast Bullet Assn. Director of Membership, 4103 Foxcraft Dr., Traverse City, MI 49684. Annual membership dues $14, includes 6 issues.

Fur-Fish-Game
A.R. Harding Pub. Co., 2878 E. Main St., Columbus, OH 43209. $15.95 yr. "Gun Rack" column by Don Zutz.

Gun List†
700 E. State St., Iola, WI 54990. $29.95 yr. (26 issues); $54.95 2 yrs.

(52 issues). Indexed market publication for firearms collectors and active shooters; guns, supplies and services.

Gun World
Gallant/Charger Publications, Inc., 34249 Camino Capistrano, Capistrano Beach, CA 92624. $22.50 yr. For the hunting, reloading and shooting enthusiast.

Guns
Guns Magazine, P.O. Box 85201, San Diego, CA 92138. $19.95 yr.; $34.95 2 yrs.; $46.95 3 yrs. In-depth articles on a wide range of guns, shooting equipment and related accessories for gun collectors, hunters and shooters.

Muzzle Blasts (M)
National Muzzle Loading Rifle Assn., P.O. Box 67, Friendship, IN 47021. $30.00 yr. annual membership. For the blackpowder shooter.

Muzzleloader Magazine*
Scurlock Publishing Co., Inc., Dept. Gun, Route 5, Box 347-M, Texarkana, TX 75501. $18.00 U.S.; $22.50 U.S. for foreign subscribers a yr. The publication for blackpowder shooters.

Outdoor Life
Times Mirror Magazines, Two Park Ave., New York, NY 10016. Special 1-yr. subscription, $11.97. Extensive coverage of hunting and shooting. Shooting column by Jim Carmichel.

Rifle*
Wolfe Publishing Co., 6471 Airpark Dr., Prescott, AZ 86301. $19.00 yr. The sporting firearms journal.

The Shotgun News‡
Snell Publishing Co., Box 669, Hastings, NE 68902/800-345-6923. $29.00 yr.; foreign subscription call for rates. Sample copy $4.00. Gun ads of all kinds.

Sports Afield
The Hearst Corp., 250 W. 55th St., New York, NY 10019. $13.97 yr. Tom Gresham on firearms, ammunition; Grits Gresham on shooting and Thomas McIntyre on hunting.

---

*Published bi-monthly †Published weekly ‡Published three times per month. All others are published monthly.
M=Membership requirements; write for details. Q=Published Quarterly.

Appendix D
Blackpowder Arms Associations

American Custom Gunmakers Guild
Jan Billeb, Exec. Director, P.O. Box 812, Burlington, IA 52601-0812/319-752-6114 (Phone or Fax)

American Single Shot Rifle Assn.
Gary Staup, Secy., 709 Carolyn Dr., Delphos, OH 45833/419-692-3866

American Society of Arms Collectors
George E. Weatherly, P.O. Box 2567, Waxahachie, TX 75165

The Cast Bullet Assn., Inc.
Ralland J. Fortier, Membership Director, 4103 Foxcraft Dr., Traverse City, MI 49684

Colt Collectors Assn.
25000 Highland Way, Los Gatos, CA 95030

Hopkins & Allen Arms & Memorabilia Society (HAAMS)
1309 Pamela Circle, Delphos, OH 45833

International Blackpowder Hunting Assn.
P.O. Box 1180, Glenrock, WY 82637/307-436-9817

National Muzzle Loading Rifle Assn.
Box 67, Friendship, IN 47021

National Rifle Assn. of America
11250 Waples Mill Rd., Fairfax, VA 22030

North-South Skirmish Assn., Inc.
Stevan F. Meserve, Exec. Secretary, 507 N. Brighton Court, Sterling, VA 20164-3919

Remington Society of America
Leon W. Wier Jr., President, 8268 Lone Feather Ln., Las Vegas, NV 89123

Single-Action Shooting Society
1938 North Batavia St., Suite C, Orange, CA 92665/714-998-0209;Fax:714-998-1992

Southern California Schuetzen Society
Dean Lillard, 34657 Ave. E., Yucaipa, CA 92399

U.S. Revolver Assn.
Brian J. Barer, 40 Larchmont Ave., Taunton, MA 02780/508-824-4836

Winchester Arms Collectors Assn.
Richard Berg, Executive Secy., P.O. Box 6754, Great Falls, MT 59406

Appendix E
Glossary of Blackpowder Terms

Accoutrements: Also accouterments. The equipment and trappings normally associated with soldiering, but not to include firearms or clothing. More broadly, here, the accessories used in blackpowder shooting.

Ampco Nipple: This nipple is no longer popular, but the term does come up in blackpowder literature. The Ampco nipple was made of a special alloy called beryllium. It looked like brass, but was much harder, which thwarted burnout. Also, the base pinhole was smaller than standard, which concentrated the flame, it was said, for better ignition.

Aperture Sight: Also peep sight. A disc with a small hole in its center. The eye naturally focuses in the center of the hole, because that's where the most light is concentrated. The shooter simply looks *through*, not at the hole, without consciously trying to center anything. The front sight is held on target and the trigger is squeezed. Not a new idea. There are examples of crossbows with aperture sights dating back several centuries.

Ball: In modern terms this may be military full-metal-jacket ammunition, but for muzzle-loading the word refers to a bullet in a general sense: a round ball is a spherical bullet, while a conical ball is an elongated bullet.

Ballistic Coefficient: Noted as "C," this is the ratio of a bullet's weight to the product of the square of its diameter combined with its form factor, and expressed as a number. C tells us how the shape, weight, length, diameter, and design of a bullet will affect its flight through the atmosphere in terms of retained velocity, hence extreme range and trajectory. The higher C is, the more efficient the projectile in velocity retention.

Barleycorn Sight: A simple front sight of early design usually mounted low on the barrel. Its rounded appearance from the side gives it a barleycorn shape.

Bolster: A lump of metal integral with the breech, or welded or screwed onto the breech area of a percussion muzzleloader, containing a nipple seat into which a nipple is screwed. Noted for strength.

Boot: Originally a leather covering for a flintlock, often treated with a waterproofing agent and fitted over the lock in bad weather. The term may be used for a similar device to cover a percussion lock.

Breech Plug: The threaded metal plug that screws into the breech to form the closed part of the barrel that becomes the chamber of the gun.

Bullet: A missile, round or conical, that is fired from a gun. Not exclusive to conicals.

Chamber: Not truly applicable to a muzzleloader, but sometimes used to denote the breech section. On blackpowder cartridge rifles, the chamber holds the cartridge.

Charger: A non-adjustable powder or shot measure that holds a specific volume. Can be a metal tube, piece of horn, etc.

Cone: Upper part of the nipple.

Conical: An elongated bullet, as opposed to a round bullet or round ball, its shape referred to as "cylindro-conoidal." Root word: cone.

CUP: Normally all capital letters or all lower case, as in "cup," means Copper Units of Pressure, used in conjunction with LUP, Lead Units of Pressure. The two are arrived at in the same manner and both refer to a pressure *value*. CUP and LUP are used on a continuum. The terms lie in direct comparison to one another. LUP becomes CUP at a specific pressure level. LUP is generally used for figures at or below the 10,000 unit mark, while CUP is used above the 10,000 unit mark. Both LUP and CUP are generated with a pressure barrel and a piston

crusher. Gas is bled off from the pressure barrel, and this gas activates a piston. A pellet, either of lead (hence LUP) or copper (hence CUP) is crushed. The *degree* of flattening is measured and translated to a figure. As an example, a mildly crushed lead pellet might translate into 5000 LUP, while more pronounced flattening of the lead pellet by the crusher/piston could result in 9000 LUP. A copper pellet replaces the lead pellet for higher pressures; however, the resulting CUP data are used in exactly the same context as LUP data. LUP and CUP are not the same as psi (pounds per square inch pressure) which is generated from a transducer (a quartz crystal device).

Detent: See Fly (in tumbler).

Drachm: In blackpowder shooting, drachm is interchangeable with dram, meaning exactly the same thing. A single drachm or a single dram both equate to 27.34 grains weight. There is an apothecary drachm which is *not* related to shooting. The apothecary dram is 60 grains weight. An old text may point out that a rifle used a load of 4 drachms of Fg blackpowder. This means the charge weighed approximately 109 grains.

Dram: A dram is equal to 27.34 grains weight. The terms remains in use as "dram equivalent" for modern shotshells. Although not exact, a 3-dram equivalent shotshell would theoretically equal in velocity to the old blackpowder shotgun shell loaded with 3 drams of blackpowder, although the new shell would contain a much lighter charge of smokeless powder.

Drivell: A word for ramrod. No longer used, but sometimes found in old books.

Duplex Load: Simply two different powders in one load. Sometimes shooters used a small dose of finer granulation powder to "kick off" the main charge in the breech. This practice later led to a tiny bit of smokeless powder mixed into the blackpowder charge. *Do not use smokeless powder/blackpowder duplex loads*. They can be dangerous. Furthermore, tests conducted under safe conditions revealed *no velocity improvement* with 10 grains volume smokeless Bulk shotgun powder added to 90 volume FFg in a 50-caliber barrel. Also, there was no major difference in bore fouling with the duplex load—it was just as dirty. The only duplex load now accepted is a small charge of blackpowder downbore in the flintlock prior to loading the gun with Pyrodex. Typically, 5 grains volume FFFg is dropped downbore first, and is included as part of the overall charge. If the charge is 100 volume RS, for example, then 5 grains volume FFFg is used with 95 volume RS.

Entry Thimble: The metal tube in the forestock through which the ramrod enters its station between the barrel and the stock.

Escutcheon: A metal inlay with a hole in its center, normally oval or rectangular in shape. Escutcheons are mounted on both sides of the stock. Barrel keys slip through these escutcheons to hold stock and barrel together. The key passes through the hole in one escutcheon, through a barrel tennon, and out the opposite side of the stock through another escutcheon.

False Muzzle: A short, separate section of barrel with protruding pins that fit into corresponding holes in the muzzle of a rifle. The false muzzle essentially becomes a temporary part of the muzzle of the rifle. It promotes aligned entry of the bullet in an undamaged condition.

Flash-in-the-pan: The condition whereby sparks from the frizzen set off the FFFFg in the pan of the flintlock, but the gun does not go off. The powder merely "flashes in the pan."

Fly: Also called a "detent," this tiny metal protrusion fits on the tumbler. It allows the sear to override the half-cock in the tumbler for smooth operation. The fly does not prevent the gun from being put on half-cock for safety.

Fouling Ring: Literally fouling that may become a corroded ring in the bore. Fouling builds up at the forepart of the load in the bore. Sometimes it can be felt with a patch on a jag. Judicious cleaning with a bristle brush helps to prevent a fouling ring in the bore.

Freshening: Also called "freshing out," it is the act of reboring a shot-out barrel. A 50-caliber, for example, may become a 52- or 53-caliber after freshing out.

Frizzen: Long ago called a hammer, and also known as a frizzle, battery or a steel, this hard piece of metal is struck by the flint in the flintlock, thereby giving off sparks to ignite the pan powder.

Frizzen Spring: The external spring on the flintlock that controls the frizzen's movement.

Furniture: Metal trim, often in iron, brass, or German silver, that is mounted on the muzzleloader's gunstock. An example is the patch box.

Gang Mould: A bullet mould that has two or more cavities so that more than one bullet can be cast at a time.

German Silver: Also known as nickel silver as well as electrum, German silver is an alloy of sixty parts copper, twenty parts zinc, and twenty parts nickel. Popular for sights and furniture on blackpowder guns, as well as magazine cappers, tins, tinder boxes, and other small accoutrements.

Globe Sight: A front sight normally hooded to prevent light reflection, often with easily interchanged inserts of various configurations. Also a spherical-shaped front sight resting on a thin stem.

Grain: Not to be confused with kernel, the grain is not a particle of powder, but rather a weight measurement. One grain equals $1/7000$ of 1 pound. To reverse this: 1 pound equals 7000 grains weight. An ounce is 437.5 grains.

Hammer Cup: The indented portion of the hammer nose. On a percussion gun the hammer cup should be deep enough to aid in deflecting cap particles downward rather than toward the shooter.

Hammer Nose: The forward portion of the hammer that protrudes downward to strike the cap on the nipple. The flintlock's cock is also known as a hammer. It holds the flint in jaws. The hammer is also found on blackpowder cartridge rifles, in which case it normally strikes a firing pin that in turn indents a primer.

Hang-fire: Also hangfire. Delayed ignition. Unlike a misfire, a gun suffering a hangfire does go off, but not right away. Should a firearm fail to fire, the muzzle should be pointed in a safe direction for at least one full minute before any attempt is made to fire the gun again with a new cap or new priming powder. If it is a hangfire, the gun will go off! If it is a misfire, it will not (see Misfire).

Hygroscopic: Attractor of moisture. Blackpowder is hygroscopic. Correct spelling is with a "g" and not a "d."

Jag: The jagged-edge device attached to the end of a loading rod, cleaning rod, or ramrod to hang onto a cleaning patch via many ridges.

Key: A flat, wedge-shaped piece of metal also known as a barrel wedge. It holds barrel and stock together, fitting through escutcheons (see Escutcheons).

Lockplate: The metal base, usually oval-shaped, that serves as a platform for lock parts. Normally held in place with bolts entering from offside of the gun.

LUP: See CUP.

Misfire: Unlike the hangfire, with a misfire the gun does not go off at all. It is not delayed ignition because there is no ignition.

Obturation: The same as internal bullet upset (in the bore) whereby the projectile actually grows slightly in its breadth. Inertia is the cause. Inertia "wants" a body at rest to remain at rest, or one in motion to keep on going. Obturation can affect an improved gas seal for the bullet because the projectile widens out in the bore, better holding gases behind it.

Patent Breech: Breech plug and nipple seat are made as a single unit.

Patina: A yellowing or golden color associated with the oxidation of metal, usually brass. Found on original old-time guns, where it should be left as is.

Pin: Usually, but not always, on full-stock rifles, this is the small metal rod that holds stock and barrel together. Several pins are normally required. They must be removed very carefully, and sometimes from one direction only, or the stock will be split.

Pipes: Short metal tubes usually attached to a barrel lug through which a ramrod passes. May also be called thimbles.

Pitch: The angle taken by the butt of the stock in relation to the line of sight, determined by resting buttstock of rifle on a flat surface to see if barrel pitches forward, no pitch, or backward. Pitch can be altered by a gunsmith.

Powder Measure, Adjustable: Much the same as a charger (see Charger), but with some means of altering the volume of powder held in the body. Sometimes used for shot as well.

Rod: A rod equals 5.5 yards in length. The "40 rod rifle" was used for grouping at 220 yards. A "20 rod rifle" was employed at 110 yards.

Rouge: A fine, abrasive paste or powder used with a buffing wheel to put a highly polished finish on metal. A leather rubbing pad may be used instead of a buffing wheel.

Sabot: Today, a plastic bore-sized cup that holds an undersized bullet, usually a pistol (handgun) projectile. Once meant "wooden work shoe."

Screw: Resembling a wood screw without a head, the shank of this screw is threaded so that it can be fitted to the end of a ramrod, cleaning rod, or loading rod for the purpose of withdrawing a stuck ball. The screw is centered in the bore and literally turned into the lead ball to engage it and then the ball is pulled free.

Sectional Density: The ratio of a bullet's weight, measured in pounds, to the square of the same bullet's diameter measured in inches, and expressed as a decimal number. The greater a bullet's sectional density, the less velocity loss from drag, and the deeper it penetrates the target (see Ballistic Coefficient).

Sizing: A coating or glaze, sometimes a type of starch, soaked into material so the material can be more easily woven, as well as having more sales appeal in the store. Sizing must be removed from material intended for shooting patches.

Slug Gun: A heavy bench target rifle with relatively fast rate of twist designed to stabilize conical projectiles with great accuracy. One slug gun was noted as caliber 69, firing a 1750-grain bullet in front of 280 grains of blackpowder. The rifle weighed 67 pounds.

Sprue: The projection or blemish left on the bullet from the sprue cutter plate during the casting process.

Tennessee Rifle: Also called the Poor Boy rifle, a plain gun, often with full-stock design.

Tenon: A metal loop or piece of flatwork fitted to the bottom of the barrel. The tenon has a hole through it. The key goes through this hole to secure the barrel to the stock.

Tenon pin: Another term for pin (see both Pin and Tenon).

Thimble: Sometimes called a ferrule or pipe, the thimble is a small metal loop through which the ramrod passes (see Entry Thimble).

Tinder Box: In blackpowder shooting, a handsome small tin used to carry tinder for starting a fire. Brass or German silver usually, but can be of other metal.

Tinder Box: Same term as above, but with a different meaning. This is the small box fitted to a flintlock gun for fire starting.

Triplex Load: The same as a duplex load, but with three different propellants in one load instead of two. Unwarranted. *Never mix smokeless powder with blackpowder or Pyrodex.*

Tube: Tube is an old-time word for the nipple on a percussion firearm.

Windage: The space that exists between a bare round ball and the walls of the bore. Windage is taken up by a patch.

Windage: In sighting, windage is horizontal movement, as opposed to elevation, which is vertical movement.

Worm: A corkscrew-like attachment that fits the end of a ramrod, loading rod, or cleaning rod. It is used to withdraw a stuck patch (not bullet).

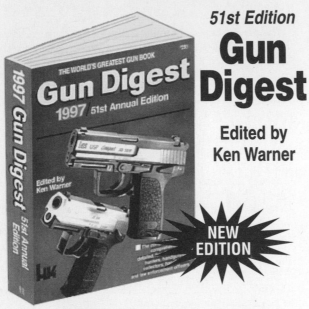